ENVIRONMENTAL BIOTECHNOLOGY

A New Approach

The Editors

Dr. Rajan Kumar Gupta (b. 25-10-1963) has worked on Ecophysiology of Antarctic Cyanobacteria for his Ph.D. degree with Late Prof. A.K. Kashyap of Centre of Advanced Study in Botany, Banaras Hindu University, Varanasi. For the past twenty years he has been working on various aspects of Antarctic microflora. Dr. Gupta was deputed by Govt. of India for his participation as Biological Scientist in Antarctica twice. He has participated in XI and XIV Indian Scientific Expeditions to Antarctica during 1991-92 and 1994-95. He has visited several countries like USA, Europe Belgium, Mauritius, Japan, Nepal, Thailand, South Africa, Singapore and Srilanka etc. for presentation of his work on different aspects of algae. Dr. Gupta has worked on various aspects of cyanobacteria, *i.e.*, morphology, ecology and nitrogen fixation, biotechnological applications and published more than 100 technical papers in various national and overseas journals and 7 reference (research) books entitled "Glimpses of Cyanobacteria", "Advances in Applied Phycology", "Soil Microflora", "Microbial Biotechnology and Ecology Vol-1 Vol-2", "Diversity: An Overview", "Diversity of Lower Plants" and "Biotechnology: An Overview". Nine students have been awarded the D.Phil degree and Six are working under his supervision for their D.Phil degree of Various Universities of Uttarakhand. Dr. Gupta is a recipient of Research Award from University Grants Commission, New Delhi. Dr. Gupta is member of number of organizations in India and abroad. He is the Fellow of the Society for Environment and Ecoplanning and International Botanical Society and chaired various sessions in the conferences in India and abroad. He is in the editorial and advisory board of various journals. Presently, Dr. Gupta is teaching Microbiology and Biotechnology in Department of Botany, Dr. P.D.B.H. Govt. P.G. College, Kotdwar (Pauri), Uttarakhand.

Dr. Satya Shila Singh (b. 28.10.1967) B.Sc. (Gorakhpur University), M.Sc. (Purvanchal University) and Ph.D. (Botany, Banaras Hindu University). She did her Ph.D. on the research problem related to the *Physiological and Biochemical Studies of Azolla- Anabaena under salt stress*. Presently, Dr. Singh is serving as an Assistant Professor in the Department of Botany, Guru Ghasidas Vishwavidyalaya (Central University), Bilaspur, Chhattisgarh. She has 3 years of teaching and 16 years of research experience with specialization in Stress Physiology and Molecular Genetics of Cyanobacteria. She has published number of papers in the journals of International and National reputes. She is recipients of Junior & Senior Research Fellowship (University of Grants Commision), Research Associateship (Council of Scientific and Industrial Research, New Delhi) and DST Young Scientist award. Her research activities mainly focuses on systematics of diazotrophic microorganisms *viz.* cyanobacterial strains collected from different parts of India and *Frankia* strains isolated from root nodules of *Hippophae salicifolia* D. Don located at high altitude of Eastern Himalayas, North Sikkim, India and both of them comprise the largest global suppliers of biologically fixed nitrogen in the environment through the enzyme nitrogenase complex encoded by three structural genes *i.e.* nif-HDK, which has been highly conserved throughout the evolution. She has also detected molecular diversity and phylogenetic relationships among the *Frankia* strains and heterocystous cyanobacteria based on amplification and sequence analyses of highly conserved 16S rRNA molecule, nitrogen fixing genes (*nif*H and *nif*D), short tandem repetitive repeats and highly iterated palindromic sequences. Siderophore production and their interaction with specific metal ions are also one of the prime targets of his research programme. She has also investigated on various factors regulating acid/alkaline phosphatase activity in cyanobacteria and *Azolla-Anabaena* system and also their potential application as a suitable biofertilizer. Interest in salt tolerance of *Anabaena azollae* (*Trichormus azollae*) forming symbiotic association with *Azolla* and *Frankia* strains with Seabuckthorn (*Hippöphae salicifolia* D. Don) arises for its possible contribution in nitrogen economy especially under saline environment. Recently, her laboratory also focusing on the molecular approaches to study the community diversity of paddy fields especially on diazotrophic cyanobacteria of Chhattisgarh.

ENVIRONMENTAL BIOTECHNOLOGY

A New Approach

— *Editors* —

Dr. Rajan Kumar Gupta

Dr. Satya Shila Singh

2016

Daya Publishing House®

A Division of

Astral International Pvt. Ltd.

New Delhi – 110 002

Cataloging in Publication Data--DK
Courtesy: D.K. Agencies (P) Ltd. <docinfo@dkagencies.com>

Environmental biotechnology : a new approach / editors, Dr. Rajan Kumar Gupta, Dr. Satya Shila Singh.
pages cm
Contributed articles.

ISBN 978-93-5130-962-8 (International Edition)

1. Bioremediation.   I. Gupta, Rajan Kumar, 1963- editor.
II. Singh, Satya Shila, 1967- editor.

TD192.5.E58 2016   DDC 628.5   23

Published by	:	**Daya Publishing House®**
		A Division of
		Astral International Pvt. Ltd.
		– ISO 9001:2008 Certified Company –
		4760-61/23, Ansari Road, Daryaganj
		New Delhi-110 002
		Ph. 011-43549197, 23278134
		E-mail: info@astralint.com
		Website: www.astralint.com
Laser Typesetting	:	**Classic Computer Services,** Delhi - 110 035
Printed at	:	**Thomson Press India Limited**

Acknowledgements

The editors are thankful to all the academicians and scientists whose contributions have enriched this volume. We also express our deep sense of gratitude to our parents whose blessings have always prompted us to pursue academic activities deeply. It is quite possible that in a work of this nature, some mistakes might have crept in text inadvertently and for these we owe undiluted responsibility.

We are grateful to all authors for their contribution to present book. We are also thankful to Professor G. S. Paliwal (former Heads), Department of Botany, HNB Garhwal University,Srinagar (Uttarakhand) and Professor D.N. Tiwari, Department of Botany, Banaras Hindu University for their cooperation and valuable suggestions for the preparation of this volume more informative and useful. We thank them from the core of our heart.

The author (RKG) is thankful to University Grants Commission, New Delhi for the financial support and to Uttarakhand Council for Science & Technology for providing financial assistance in the form of project. RKG wish to place on record his special thanks to his wife Mrs Alka and two little daughters Akriti and Ayushi and his research team for their cooperation in all his academic and scientific endeavours.

One of the author (SSS) is grateful to honourable Vice-Chancellor, Guru Ghasidas Vishwavidyalaya, Bilaspur, Chhattisgarh. Financial assistance from University Grant Commission, New Delhi and Council of Scientific and Industrial Research (CSIR), New Delhi are gratefully acknowledged. SSS express heartiest thanks to her husband Prof. A. K. Mishra for his incessant encouragement and constant inspiration during entire period of this task. SSS wish to express her thanks to her research scholars, parents and all family members for their constant wishes and blessings.

Finally, we will always remain debtor to all our well wishers for their blessings without which this book would not have come to light.

Dr. Rajan Kumar Gupta
Dr. Satya Shila Singh

Dr. P.K. Singh, *FNAAS, FNA*
INSA Senior Scientist
Department of Botany,B.H.U., Varanasi
[Formerly Vice-Chancellor, CSAU&T, Kanpur;
Joint Director (Research), Project Director,
Director, IARI, New Delhi; Joint Director,
Director, CRRI, Cuttack.]

House No. B3/22, Shivala. Post Shivala
Varanasi-221001, U.P.
Ph. No.: 0542-2277705;
Mobile: 09415714584,
Fax No.: 0542-6701124,
Email: pksalgae@yahoo.co.in

Foreword

For last few decades, biotechnology has been proved to be immensely used in different regimes of science and benefited humankind in various ways. Continued advances in biotechnology have driven forward the harnessing of plants and micro-organisms to protect our fragile environment from the harmful impacts of biotic and abiotic factors. Biotechnology has expanded its horizon and merged with the new and diverse sciences with the responsibilities to provide economic, social and environmental benefits to the society. In the late 20th and early 21st century, scientific community witnessed the emergence of biotechnological research from traditional fermentation technology to modern techniques like genomics, recombinant gene technologies, applied immunology, development of pharmaceutical therapies, diagnostic tests and environmental biotechnology.

Environmental biotechnology is science that is applied to and used to study the natural environment. The International Society for Environmental Biotechnology defines environmental biotechnology as "the development, use and regulation of biological systems for remediation of contaminated environments (land, air, water), and for environment-friendly processes (green manufacturing technologies and sustainable development)". Environmental Biotechnology deals with preservation and restoration of the natural resources and will be one of the most valuable tools for improving the quality of life for future generations.

I pen this message with great pleasure that the publication of the "Environmental Biotechnology: A New Approach", a book is indeed a glorious endeavour. This book consists of 22 chapters which mainly focus on Nanobiotechnology, Calcium ion regulation, cyanobacterial application in biotechnology, detoxification of herbicides through mycorrhizal fungi, bioprospecting of Himalayan lichens for antimicrobial activity, heavy metal stress and phytoremediation. In nutshell this book will serve as a storehouse of knowledge for the scientists, teachers and students and in a single jacket which offers innovative study materials in compact and lucid way.

I am happy to know that Dr. Rajan Kumar Gupta, Department of Botany, Govt. P.G. College, Kotdwar and Dr. Satya Shila Singh, Department of Botany, Guru Ghasidas Vishwavidyalaya (A Central University), Bilaspur, Chhattisgarh are bringing out the volume of "Environmental Biotechnology: A New Approach" to cater the long felt need of scientists, research scholars and students. I congratulate the editors for their maiden effort and wish them all success in their endeavor.

The devotion and dedication of the editors has enabled the Book to occupy a prominent position in the academic arena. I am sure that it will attain much greater heights in the days to come.

Dr. Gupta and Dr. Singh deserve for my sincere blessings on this occasion. I wish them all success.

Dr. P.K. Singh

Preface

Much before the beginning of civilization our ancestors were fermenting grains and fruits to create alcoholic beverages. This applied aspect of microorganisms, like that of animals, was completely trivial based on taste, smell, and vision, and not on any knowledge of genetic mechanisms for selective breeding. Biological inventions towards the entire twenty century have changed the sustainability in the present century. The discovery of DNA structure in 1953, genetically modified tomato in 1995, cloning of Dolly the sheep in 1996, and the sequencing of the entire human genome at the end of millennium testify awareness that the biotech miracle is inescapable today.

Biotechnology is a diverse field dealing with the application of biological discoveries to transform industry, agriculture, and medicine in a new dimension. This biotechnological application is much hyped created by overzealous promotion of biotechnology companies gravitated by press and media to cure diseases, develop new drugs, and feed the world's hunger through genetically modified organisms, resulting in confusion over what is real and what is fanciful speculation. Biotechnology is variably portrayed as either the next dot-com ride for those with excess capital to invest or as simply not worth following as an investment vehicle. There is no debate that biotech is a global business phenomenon. Worldwide successes and failures of biotech companies are tracked even for a weak signal that could boom like dot-com boom of 1990s.

Being multidisciplinary nature, biotechnology does not necessarily offer the single, best route; rather it can effectively be used as one of the tools or integrated into other processes. The strict rules on selection of programs, organisms, methods, technologies and biosafety as well as new legislation on restricted use of genome modifications of vertebrates, higher plants, genetically modified food, patenting of transgenic animals or sequenced parts of genomes, biotechnology has emerged with a very high standard high-tech and safe technology. Definitely the application of biotechnology invariably has reduced either operating costs or capital costs or both. This has led to a more sustainable process with lowered ecological disturbances by reducing some or many energy use, water use, waste-water or greenhouse gas production.

The ethical and social problems arising in agriculture and medicine are still controversial. The acceptance of "Biotechnology" in medicine, agriculture, food and pharma production has become a political matter.

The aim of this volume on Environmental Biotechnology: A New Approach is to keep the reader informed on the latest progress made in environmental biotechnology, nanotechnology, eutrophication, phytoremediation and extent of environmental management. The volume is consisting of 22 different chapters contributed by the author(s) having vast experience in teaching and research in the wide arena of biotechnology.

The aim of this volume on Commercial Biotechnology is to keep the reader abreast with the latest developments and progress taken place in the area of environmental biotechnology, nanotechnology, environmental management and phytoremediation technology to the extent of applying at industrial and production technology. A brief introduction of Nanotechnology, optical fiber biosensors and nanoparticles has been presented in chapter 1,2 and 3. In chapter 4,5 and 6 gives diversity and application of cyanobacteria in biotechnology and eutrophication. Chapter 7 describes how Azolla and bga affects the mycorrhization, chapter 8,9,12 and 22 explains application of biotechnology and biocomposting, organic farming and environmental management. Chapter 11 gives an idea about bioprospecting himalayam lichens. Biofuels have been discussed in chapter 15 again phytoremediation has been presented in chapter 17. In chapter 19, phytotechnology and its techniques are presented in chapter 19. Chapter 20 gives idea about the environmental stresses.

In modern age, environment is affected by anthropogenic activities at each and every second to maximize the benefits for the fulfillment of human need. At every moment, newer technologies, as an emerging field, have been continuously introduced to protect environment from pollution and contamination so that sustainable environment would be ensured. Environmental science is the study of environmental disharmonies created by the interactive effect of humans and the natural world and the solutions related to environmental problems by using newer technologies i.e. Environmental biotechnology which is most historic and eminently modern technical discipline.

This book deals with the technologies used for the improvement of the quality of the environment for the human welfare by using microbes and plants. It also includes constantly new technologies to introduce contemporary problems such as detoxification of hazardous chemicals. This book also targets at providing ideas for the generation or exploration of valuable resources from plants for human society.

This book consists of 22 chapters which mainly focus on Nanobiotechnology, Calcium ion regulation, cyanobacterial application in biotechnology, detoxification of herbicide through mycorrhizal fungi, Bioprospecting Himalayan lichens for antimicrobial activity, Heavy metal stress and phytoremediation. Modern and recent innovative techniques have been discussed and mentioned in this book through the contribution of the specialized research findings of the eminent scientists and researchers for rendering multidimensional approach to be used in the field of environmental biotechnology. Recent developments and ideas given by the experts in the various aspects of environmental bio-technology have been discussed, to make this book more suitable for the students of B.Sc and M.Sc as well as reference book for Ph.D. students.

We express our deep felt indebtedness to Prof. A. K. Mishra, Department of Botany, Banaras Hindu University for his sustained guidance, invaluable prolonged discussion, criticism and all that went to contribute substantially in the completion of this task.

We sincerely thank Dr. D.N. Tiwari and Dr. P.K.Singh for their constant and continued help and encouragement at all levels.

We extend our thanks to the Head, Department of Botany, Guru Ghasidas Vishwavidyalay, Blaspur, Chhattisgarh. The editor (RKG) wish to thank Principal, Govt. P.G. College, Kotdwar (Uttarakhand).

Dr. Rajan Kumar Gupta
Dr. Satya Shila Singh

Contents

2016, Environmental Biotechnology: A New Approach   *Pages 1–15*
Editors: Dr. Rajan Kumar Gupta and Dr. Satya Shila Singh
Published by: DAYA PUBLISHING HOUSE, NEW DELHI

Chapter 1

Nanobiotechnology: An Introduction

Ramna Tripathi[1]* and Akhilesh Kumar[2]

*[1]Department of Physics, THDC-Institute of Hydropower Engineering
and Technology, Tehri, Uttarakhand*
*[2]Department of Physics, Govt. Girls P.G. College,
Rajajipuram, Lucknow, U.P.*

ABSTRACT

Nanobiotechnology is the application of nanotechnology in biological fields. Nanotechnology is a multidisciplinary field that currently recruits approach, technology and facility available in conventional as well as advanced avenues of engineering, physics, chemistry and biology. Nanobiotechnology has multitude of potentials for advancing medical science thereby improving health care practices around the world. Many novel nanoparticles and nanodevices are expected to be used, with an enormous positive impact on human health. While true clinical applications of nanotechnology are still practically inexistent, a significant number of promising medical projects are in an advanced experimental stage. Implementation of nanotechnology in medicine and physiology means that mechanisms and devices are so technically designed that they can interact with sub-cellular (i.e. molecular) levels of the body with a high degree of specificity. Thus therapeutic efficacy can be achieved to maximum with minimal side effects by means of the targeted cell or tissue-specific clinical intervention. More detailed research and careful clinical trials are still required to introduce diverse components of nanobiotechnology in random clinical applications with success. Ethical and moral concerns also need to be addressed in parallel with the new developments.

Keywords: Nanobiotechnology, Bionanotechnology, Applications, Medical.

Introduction

Nanobiotechnology, bionanotechnology and nanobiology are terms that refer to the intersection of nanotechnology and biology [1]. Given that the subject is one that has only emerged very recently, bionanotechnology and nanobiotechnology serve as blanket terms for various related technologies.

Nanotechnology is a multidisciplinary field that currently recruits approach, technology and facility available in conventional as well as advanced avenues of engineering, physics, chemistry and biology. Nanotechnology involves research and development on materials and species with dimensions of roughly 1 to 100 nm, where properties of matter differ fundamentally from those of individual atoms or molecules or bulk molecules. The term *nano* is derived from the Greek word meaning "dwarf". In dimensional scaling, nano refers to 10^{-9}, *i.e.*, one billionth of a unit. Therefore, nanotechnology involves techniques and methods for studying, designing, fabricating and manipulating things at the nanoscale to understand and create materials, devices and systems to exploit these phenomena for novel applications [2].

The terms are often used interchangeably. When a distinction is intended, though, it is based on whether the focus is on applying biological ideas or on studying biology with nanotechnology.

Nanobiotechnology

Nanobiotechnology, refers to the ways that nanotechnology is used to create devices to study biological systems [3]. In other words, it is essentially miniaturized biotechnology.

Nanobiotechnology is considered to be the unique fusion of biotechnology (uses the knowledge and techniques of biology to manipulate molecular, genetic and cellular processes to develop products and services and is used in diverse fields from medicine to agriculture) and nanotechnology (novel scientific approach that involves materials and equipments capable of manipulating physical as well as chemical properties of a substance at molecular levels) by which classical micro-technology can be merged to a molecular biological approach in real. Through this methodology, atomic or molecular grade machines can be made by mimicking or incorporating biological systems, or by building tiny tools to study or modulate diverse properties of a biological system on molecular basis. Nanobiotechnology may, therefore, ease many avenues of life sciences by integrating cutting-edge applications of information technology and nanotechnology into contemporary biological issues. This technology has potential to remove obvious boundaries between biology, physics and chemistry to some extent, and shape up our current ideas and understanding.

Nanobiotechnology takes most of its fundamentals from nanotechnology. Most of the devices designed for nanobiotechnological use are directly based on other existing nanotechnologies. Nanobiotechnology is often used to describe the overlapping multidisciplinary activities associated with biosensors, particularly where photonics, chemistry, biology, biophysics, nano-medicine, and engineering converge.

Nanobiotechnology is the application of nanotechnology in biological fields. Nanobiotechnology (sometimes referred to as nanobiology) is best described as helping modern medicine progress from treating symptoms to generating cures and regenerating biological tissues. Three American patients have received whole cultured bladders with the help of doctors who use nanobiology techniques in their practice. Also, it has been demonstrated in animal studies that a uterus can be grown outside the body and then placed in the body in order to produce a baby. Stem cell treatments have been used to fix diseases that are found in the human heart and are in clinical trials in the United States. There is also funding for research into allowing people to have new limbs without having to resort to prosthesis. Artificial proteins might also become available to manufacture without the need for harsh chemicals and expensive machines. It has even been surmised that by the year 2055, computers may be made out of biochemicals and organic salts [4].

Another example of current nanobiotechnological research involves nanospheres coated with fluorescent polymers. Researchers are seeking to design polymers whose fluorescence is quenched when they encounter specific molecules. Different polymers would detect different metabolites. The polymer-coated spheres could become part of new biological assays, and the technology might someday lead to particles which could be introduced into the human body to track down metabolites associated with tumors and other health problems. Another example, from a different perspective, would be evaluation and therapy at the nanoscopic level, *i.e.* the treatment of Nanobacteria (25-200 nm sized) as is done by NanoBiotech Pharma.

Bionanotechnology

Bionanotechnology generally refers to the study of how the goals of nanotechnology can be guided by studying how biological "machines" work and adapting these biological motifs into improving existing nanotechnologies or creating new ones [5, 6].

Bionanotechnology is a specific application of nanotechnology. For example, DNA nanotechnology or cellular engineering would be classified as bionanotechnology because they involve working with biomolecules on the nanoscale. Conversely, many new medical technologies involving nanoparticles as delivery systems or as sensors would be examples of nanobiotechnology since they involve using nanotechnology to advance the goals of biology.

Most of the scientific concepts in bionanotechnology are derived from other fields. Biochemical principles that are used to understand the material properties of biological systems are central in bionanotechnology because those same principles are to be used to create new technologies. Material properties and applications studied in bionanoscience include mechanical properties(*e.g.* deformation, adhesion, failure), electrical/electronic (*e.g.* electromechanical stimulation, capacitors, energy storage/ batteries), optical (*e.g.* absorption, luminescence, photochemistry), thermal (*e.g.* thermomutability, thermal management), biological (*e.g.* how cells interact with nanomaterials, molecular flaws/defects, biosensing, biological mechanisms s.a. mechanosensing), nanoscience of disease (*e.g.* genetic disease, cancer, organ/tissue failure), as well as computing (*e.g.* DNA computing). DNA nanotechnology is one important example of bionanotechnology [7]. The utilization of the inherent properties of nucleic acids like DNA to create useful materials is a promising area of modern research. Another important area of research involves taking advantage of membrane properties to generate synthetic membranes. Protein folding studies provide a third important avenue of research, but one that has been largely inhibited by our inability to predict protein folding with a sufficiently high degree of accuracy. Given the myriad uses that biological systems have for proteins, though, research into understanding protein folding is of high importance and could prove fruitful for bionanotechnology in the future.

Lipid nanotechnology is another major area of research in bionanotechnology, where physico-chemical properties of lipids such as their antifouling and self-assembly is exploited to build nanodevices with applications in medicine and engineering [8].

The impact of bionanoscience, achieved through structural and mechanistic analyses of biological processes at nanoscale, is their translation into synthetic and technological applications through nanotechnology.

Nanobiology

While nanobiology is in its infancy, there are a lot of promising methods that will rely on nanobiology in the future. Biological systems are inherently nano in scale; nanoscience must merge with biology in order to deliver biomacromolecules and molecular machines that are similar to nature. Controlling and mimicking the devices and processes that are constructed from molecules is a tremendous challenge to face the converging disciplines of nanotechnology [9]. All living things, including humans, can be considered to be nanofoundries. Natural evolution has optimized the "natural" form of nanobiology over millions of years. In the 21st century, humans have developed the technology to artificially tap into nanobiology. This process is the best described as "organic merging with synthetic." Colonies of live neurons can live together on a biochip device; according to research from Dr. Gunther Gross at the University of North Texas. Self-assembling nanotubes have the ability to be used as a structural system. They would be composed together with rhodopsins; which would facilitate the optical computing process and help with the storage of biological materials. DNA (as the software for all living things) can be used as a structural proteomic system - a logical component for molecular computing. Ned Seeman - a researcher at New York University - along with other researchers are currently researching concepts that are similar to each other [10].

Concepts that are enhanced through nanobiology include: nanodevices, nanoparticles, and nanoscale phenomena that occurs within the discipline of nanotechnology. This technical approach to biology allows scientists to imagine and create systems that can be used for biological research. Biologically inspired nanotechnology uses biological systems as the inspirations for technologies not yet created [2].

The most important objectives that are frequently found in nanobiology involve applying nanotools to relevant medical/biological problems and refining these applications. Developing new tools, such as peptoid nanosheets, for medical and biological purposes is another primary objective in nanotechnology. New nanotools are often made by refining the applications of the nanotools that are already being used. The imaging of native biomolecules, biological membranes, and tissues is also a major topic for the nanobiology researchers. Other topics concerning nanobiology include the use of cantilever array sensors and the application of nanophotonics for manipulating molecular processes in living cells [3].

Advantage of Nanobiotechnology

Applications of bionanotechnology are extremely widespread. Insofar as the distinction holds, nanobiotechnology is much more commonplace in that it simply provides more tools for the study of biology. Bionanotechnology, on the other hand, promises to recreate biological mechanisms and pathways in a form that is useful in other ways.

The pathophysiological conditions and anatomical changes of diseased or inflamed tissues can potentially trigger a great deal of scopes for the development of various targeted nanotechnological products. This development is like to be advantageous in the following ways:

1. Drug targeting can be achieved by taking advantage of the distinct pathophysiological features of diseased tissues [11].
2. Various nanoproducts can be accumulated at higher concentrations than normal drugs [12].
3. Increased vascular permeability coupled with an impaired lymphatic drainage in tumors improve the effect of the nanosystems in the tumors or inflamed tissues through better transmission and retention [13, 14].
4. Nanosystems have capacity of selective localization in inflammed tissues [15].
5. Nanoparticles can be effectively used to deliver/transport relevant drugs to the brain overcoming the presence of blood–brain barrier (meninges) [16, 17].
6. Drug loading onto nanoparticles modifies cell and tissue distribution and leads to a more selective delivery of biologically active compounds to enhance drug efficacy and reduces drug toxicity [18, 19].

Application of Nanobiotechnology in Medical and Clinical Field

The interface of nanotechnology and biotechnology (nanobiotechnology) has produced tremendous applications in the domain of human health. Examples include the use of non-bleaching fluorescent nanocrystals (quantum dots) in place of dyes pulled glass capillaries with nanoscale tips utilized for injection of membrane-impermeable molecules (proteins, DNA). Another example is the use of carbon nanotubes (CNTs) for the delivery of biomolecular components to cells and smart biohybrid materials that enhance technologies by providing new avenues for regulating the activity of protein and DNA components.

A number of clinical applications of nanobiotechnology, such as disease diagnosis, target-specific drug delivery, and molecular imaging are being laboriously investigated at present. Some new promising products are also undergoing clinical trials [20, 21]. Such advanced applications of this approach to biological systems will undoubtedly transform the foundations of diagnosis, treatment, and prevention of disease in future. Some of these applications are discussed below.

(a) Diagnostic Applications

Current diagnostic methods for most diseases depend on the manifestation of visible symptoms before medical professionals can recognize that the patient suffers from a specific illness. But by the time those symptoms have appeared, treatment may have a decreased chance of being effective. Therefore the earlier a disease can be detected, the better the chance for a cure is. Optimally, diseases should be diagnosed and cured before symptoms even manifest themselves. Nucleic acid diagnostics will play a crucial role in that process, as they allow the detection of pathogens and diseases/diseased cells at such an early symptomless stage of disease progression that effective treatment is more feasible. Current technology, such as- polymerase chain reaction (PCR) leads toward such tests and devices, but nanotechnology is expanding the options currently available, which will result in greater sensitivity and far better efficiency and economy.

1. Detection

Many currently used/conventional clinical tests reveal the presence of a molecule or a disease causing organism by detecting the binding of a specific antibody to the disease-related target. Traditionally, such tests are performed by conjugating the antibodies with inorganic/organic dyes and visualizing the signals within the samples through fluorescence microscopy or electronic microscopy. However, dyes often limit the specificity and practicality of the detection methods. Nanobiotechnology offers a solution by using semiconductor nanocrystals (also referred to as "quantum dots"). These minuscule probes can withstand significantly more cycles of excitations and light emissions than typical organic molecules, which more readily decompose [22].

2. Individual Target Probes

Despite the advantages of magnetic detections, optical and colorimetric detections will continue to be chosen by the medical community. Nanosphere (Northbrook, Illinois) is one of the companies that developed techniques that allow/allowing doctors to optically detect the genetic compositions of biological specimens. Nano gold particles studded with short segments of DNA form the basis of the easy-to-read test for the presence of any given genetic sequence. If the sequence of interest in the samples, it binds to complementary DNA tentacles on multiple nanospheres and forms a dense web of visible gold balls. This technology allows/facilitates the detection of pathogenic organisms and has shown promising results in the detection of anthrax, giving much higher sensitivity than tests that are currently being used [23].

3. Protein Chips

Proteins play the central role in establishing the biological phenotype of organisms in healthy and diseased states and are more indicative of functionality. Hence, proteomics is important in disease diagnostics and pharmaceutics, where drugs can be developed to alter signaling pathways. Protein chips can be treated with chemical groups, or small modular protein components, that can specifically bind to proteins containing a certain structural or biochemical motif [16]. Two companies currently operating in this application space are Agilent, Inc. and NanoInk, Inc. Agilent uses a non-contact ink-jet technology to produce microarrays by printing oligos and whole cDNAs onto glass slides at the nanoscale. NanoInk uses dip-pen nanolithography (DPN) technology to assemble structure on a nanoscale of measurement [24].

4. Sparse Cell Detection

Sparse cells are both rare and physiologically distinct from their surrounding cells in normal physiological conditions (*e.g.* cancer cells, lymphocytes, fetal cells and HIV-infected T cells). They are significant in the detection and diagnosis of various genetic defects. However, it is a challenge to identify and subsequently isolate these sparse cells. Nanobiotechnology presents new opportunities for advancement in this area. Scientists developed nanosystems capable of effectively sorting sparse cells from blood and other tissues. This technology takes advantage of/exploits the unique properties of sparse cells manifested in differences in deformation, surface charges and affinity for specific receptors and/or ligands. For example, by inserting electrodes into microchannels, cells can be precisely sorted based on surface charge. They can also be sorted by using biocompatible surfaces with precise nanopores. The nanobiotechnology center at Cornell University (NBTC) is currently using these technologies to develop powerful diagnostic tools for the isolation and diagnosis of various diseases [25].

5. Nanotechnology as a Tool in Imaging

Intracellular imaging can be made possible through labelling of target molecules with quantum dots (QDs) or synthetic chomophores, such as fluorescent proteins that will facilitate direct investigation of intracellular signalling complex by optical techniques, *i.e.* confocal fluorescence microscopy or correlation imaging [26,27].

(b) Therapeutic Applications

Nanotechnology can provide new formulations of drugs with less side effects and routes for drug delivery.

1. Drug Delivery

Nanoparticles as therapeutics can be delivered to targeted sites, including locations that cannot be easily reached by standard drugs. For instance, if a therapeutic can be chemically attached to a nanoparticle, it can then be guided to the site of the disease or infection by radio or magnetic signals. These nanodrugs can also be designed to "release" only at times when specific molecules are present or when external triggers (such as infrared heat) are provided. At the same time, harmful side effects from potent medications can be avoided by reducing the effective dosage needed to treat the patient. By encapsulating drugs in nanosized materials (such as organic dendrimers, hollow polymer capsules, and nanoshells), release can be controlled much more precisely than ever before. Drugs are designed to carry a therapeutic payload (radiation, chemotherapy or gene therapy) as well as for imaging applications [28]. Many agents, which cannot be administered orally due to their poor bioavailability, will now have scope of use in therapy with the help of nanotechnology [29, 30]. Nano-formulations offer protection for agents vulnerable to degradation or denaturation when exposed to extreme pH, and also prolong half-life of a drug by expanding retention of the formulation through bioadhesion [31, 32]. Another broad application of nanotechnology is the delivery of antigens for vaccination [33, 34]. Recent advances in encapsulation and development of suitable animal models have demonstrated that microparticles and nanoparticles are capable of enhancing immunization [35].

2. Gene Delivery

Current gene therapy systems suffer from the inherent difficulties of effective pharmaceutical processing and development, and the chance of reversion of an engineered mutant to the wild type. Potential immunogenicity of viral vectors involved in gene delivery is also problematic [36, 37]. To address this issue, nanotechnological tools in human gene therapy have been tested and nanoparticle-based nonviral vectors (usully 50-500 nm in size) in transportation of plasmid DNA described. Therefore, successful introduction of less immunogenic nanosize gene carriers as a substitution of the disputed viral vectors seems beneficial in repairing or replacing impaired genes in human [38].

3. Liposomes

A liposome being composed of a lipid bilayer can be used in gene therapy due to its ability to pass through lipid bilayers and cell membranes of the target. Recent use of several groups of liposomes in a local delivery has been found to be convincingly effective [39, 40]. Liposomes can also help achieve targeted therapy. Zhang *et al.,* demonstrated widespread reporter expression in the brains of rhesus monkeys by linking nanoparticle (such as polyethylene glycol) treated liposomes to a monoclonal antibody for human insulin reporter [41]. These successful trials reflect the future of targeted therapy and the importance of nanometer-sized constructs for the advancement of molecular medicine.

4. Surfaces

In nature, there are a multitude of examples of the complicated interactions between molecules and surfaces. For example, the interactions between blood cells and the brain or between fungal pathogens and infection sites rely on complex interplays between cells and surface characteristics. Nanofabrication unravels the complexity of these interactions by modifying surface characteristics with nanoscale resolutions, which can lead to hybrid biological systems. This hybrid material can be used to screen drugs, as sensors, or as medical devices and implants. Nanosystems, owned by the Irish drug company Elan, developed a polymer coating capable of changing the surface of drugs that have poor water solubility [42].

5. Biomolecular Engineering

The expense and time involved in traditional biomolecule designing limit the availability of bioactive molecules. Nanoscale assembly and synthesis techniques provide an alternative to traditional methods. Improvements can be achieved due to the ability to carry out chemical and biological reactions on solid substrates, rather than through the traditional solution based processes. The use of solid substrate usually means less waste and the ability to manipulate the biomolecule far more precisely. EngeneOS (Waltham, Massachusetts) pioneered the field of biomolecular engineering. The company developed the engineered genomic operating systems that create programmable biomolecular machines employing natural and artificial building blocks. These biomolecule machines have broad range of commercial applications-as biosensors, in chemical synthesis and processing, as bioelectronic devices and materials, in nanotechnology, in functional genomics and in drug discovery.

6. Biopharmaceuticals

Nanobiotechnology can develop drugs for diseases that conventional pharmaceuticals cannot target. The pharmaceutical industry traditionally focuses on developing drugs to treat a defined universe of about five hundred confirmed disease targets. But approximately 70 to 80 percent of the new candidates for drug development fail, and these failures are often discovered late in the development process, with the loss of millions of dollars in R&D investments. Nanoscale techniques for drug development will be a boon to small companies, which cannot employ hundreds of organic chemists to synthesize and test thousands of compounds. Nanobiotechnology brings the ability to physically manipulate targets, molecules and atoms on solid substrates by tethering them to biomembranes and controlling where and when chemical reactions take place, in a fast process that requires few materials (reagents and solutions). This advancement will reduce drug discovery costs, will provide a large diversity of compounds, and will facilitate the development of highly specific drugs. Potentia Pharmaceuticals (Louisville, Kentucky) is an early-stage company that is attempting to streamline the drug development process with the use of nanotechnologies (Harvard Business School 2001).

7. Cardiac Therapy

Nanotechnology is currently offering promising tools for applications in modern cardiovascular science to explore existing frontiers at the cellular level and treat challenging cardiovascular diseases more effectively. These tools can be applied in diagnosis, imaging and tissue engineering [43]. Miniaturized nanoscale sensors like quantum dots (QDs), nanocrystals, and nanobarcodes are capable of sensing and monitoring complex immune signals in response to cardiac or inflammatory events [20]. Nanotechnology can also help detect and describe clinically-significant specific mechanisms implicated in cardiac disorders. In addition, it is useful in designing atomic-scale machines that can be incorporated into biological systems at the molecular level. Introduction of these newly designed nanomachines may positively change many ideas and hypotheses in the treatment of critical cardiovascular diseases.

Nanotechnology could also have great impact in tackling issues like unstable plaques and clarification of valves. Thus, this approach could be a real milestone of success in achieving localized and sustained arterial and cardiac drug therapy for the management of cardiovascular diseases [44].

8. Dental Care

Nanotechnology will have future medical applications in the field of dentistry. The role of nanodentistry by means of the use of nanomaterials [45, 46], biotechnology [47, 48], and nanorobotics will ensure better oral health. Millions of people currently receiving poor dental care will benefit from such remarkable breakthrough in the science of dental health [49, 50]. Moreover, nanodental techniques in major tooth repair may also evolve. Reconstructive dental nanorobots could be used in selective and precise occlusion of specific tubules within minutes, and this will facilitate quick and permanent recovery. The advantage of nanodentistry in natural tooth maintenance could also be significant [51]. Covalently-bonded artificial materials like sapphire may replace upper enamel layer to boost the appearance and durability of teeth [52].

9. Orthopedic Applications

Nanomaterials sized between 1 and 100 nm have role to play as new and functional constituents of bones being also made up of nanosized organic and mineral phases [53, 54]. Nanomaterials, nanopolymers, carbon nanofibers, nanotubes, and ceramic nanocomposites may help with more efficient deposition of calcium-containing minerals on implants. Based on these evidences and observations, nanostructure materials represent a unique realm of research and development that may improve the attachment of an implant to the surrounding bone matters by enhancing bone cell interactions and this will indeed aid in improving orthopedic implant efficacy while drastically minimizing patient-compliance problems.

Future Prospects of Nanobiotechnology

There is much debate on the future implications of nanobiotechnology. It could create and suggest implementation of a choice of various new materials and devices potentially useful in the field of medicine, electronics, biomaterials and energy production. Nevertheless, this approach raises many of the same issues as any new technology, including problems with toxicity and environmental impact of nanomaterials [55] and their potential effects on global economics, as well as speculation about various doomsday scenarios. These concerns have accounted for a debate among advocacy groups and governments on whether special statutory regulation of nanobiotechnology is warranted.

Despite the existence of some disputes, this technology renders immense hope for the future. It may lead to innovations by playing a prominent role in various biomedical applications ranging from drug delivery and gene therapy to molecular imaging, biomarkers and biosensors. One of these applications being the prime research objective of the present time would be target-specific drug therapy and methods for early diagnosis and treatment of diseases [56]. Two types of medical applications are already emerging, both in clinical diagnosis and in R&D. Imaging applications, such as quantum dot technology are already being licensed and applications for monitoring cellular activities in tissue are coming soon. The second major type of application involves the development of highly specific and sensitive means of detecting nucleic acids and proteins [57]. By 2015 to 2020, we will see that products being tested in academic and government laboratories will be creeping into commercialization. Sparse cell isolation and molecular filtration applications should, by then, make it to market. Some of the drug delivery systems should be commercialized or in advanced clinical trials. For example - drug delivery systems have been developed by Nano Systems or by American Pharmaceutical Partners, which is testing the encapsulation of Taxol, a cancer drug in a nanopolymer called paclitaxel. Most medical devices and therapeutics are a decade or more away from market. Therefore, drug target manipulation as well as device implantation requires a complex technical infrastructure like nanotechnology as well as complex regulatory management [58].

Continuous advancements in nanomedicine have opened up its opportunities for application in a variety of medical disciplines. Its future application as diagnostic and regenerative medicine is currently being investigated. In diagnosis, detection of diseased cells would be faster, possibly at the point of a single sick cell, while allowing diseased cells to be cured at once before they spread into and affect other parts of the body. Also, individuals suffering from major traumatic injuries or impaired organ functions could benefit from the use of nanomedicine.

Challenges for Nanobiotechnology

No single person can provide the answers to challenges that nanotechnology brings, nor can any single group or intellectual discipline. The five main challenges are to develop instruments to assess exposure to engineered nano-materials in the air and water. It is fairly understood that exposure of humans and animals to the environment potentially contaminated with nano-materials may need to be monitored for any adverse consequence. The challenge becomes increasingly difficult in more complex matrices like food. The second challenge would be to develop applicable methods to detect and determine the toxicity of engineered nano-materials within next 5 to 15 years. Then again, proposing models for predicting effects of these nano-materials on human health and the environment would be an inevitable issue. The next challenge would be to develop reverse systems to evaluate precise impact of engineered nano-materials on health and the environment over the entire life span that speaks to the life cycle issue. The fifth being more of a grand challenge would be to develop the tools to properly assess risk to human health and to the environment. Commercialization challenges of nanobiotechnology include uncertainty of effectiveness of innovation, scalability, funding, scarce resources, patience etc. A broad majority of company recognizes a great potential in nanotechnology for the development of new products and the improvement of existing products. A new potentially disruptive technology like nanotechnology raises fundamental questions about the need for new regulations. Authorities around the world should evaluate possible risks and an appropriate regulatory response to the extensive use of this advanced technology.

References

1. Ehud Gazit (2007). Plenty of room for biology at the bottom. *An introduction to bionanotechnology*. Imperial College Press.

2. Emerich DF, Thanos CG (2003). Nanotechnology and medicine. *Expert Opin Biol Ther*. (3): 655–663.

3. http://www.wordiq.com/definition/Bionanotechnology.

4. Nolting B (2005). "Biophysical Nanotechnology". In: "Methods in Modern Biophysics". Springer.

5. "Nanobiology". Nanotech-Now.com.

6. "Nanobiology". Swiss Nanoscience Institute.

7. "The future of nanobiology". ZD Net.

8. "Nanobiology: from physics and engineering to biology". IOP Science.

9. Sivakumar K; Liu X; Madhaiyan M; Ji L; Yang L; Tang C; Song H; Kjelleberg S; Cao B (2013). "Influence of outer membrane c-type cytochromes on particle size and activity of extracellular nanoparticles produced by Shewanella oneidensis.". *Biotechnology and Bioengineering*. (7): 1831–7.

10. "The Nanobiology Imperative". HistorianoftheFuture.com.

11. Vasir JK, Labhasetwar V (2005). Targeted drug delivery in cancer therapy. *Technol Cancer Res Treat*. (4): 363–374.

12. Vasir JK, Reddy MK, Labhasetwar V (2005). Nanosystems in drug targeting: opportunities and challenges. *Curr Nanosci*. (1): 47–64.

13. Maeda H, Wu J, Sawa T, Matsumura Y, Hori K. *et al*. (2000). Tumor vascular permeability and the EPR effect in macromolecular therapeutics: a review. *J Control Release*. (65): 271–284.

14. Matsumura Y, Maeda H (1986). A new concept for macromolecular therapeutics in cancer chemotherapy: mechanism of tumoritropic accumulation of proteins and the antitumor agent smancs. *Cancer Res*. (46): 6387–6392.

15. Allen TM, Cullis PR (2004). Drug delivery systems: entering the mainstream. *Science*. (303): 1818–1822.

16. Alyautdin RN, Tezikov EB, Ramge P, Kharkevich DA, Begley DJ, Kreuter J. *et al*. (1998). Significant entry of tubocurarine into the brain of rats by adsorption to polysorbate 80-coated polybutylcyanoacrylate nanoparticles: An *in situ* brain perfusion study. *J Microencapsul*. (15): 67–74.

17. Garcia-Garcia E, Gil S, Andrieux K, Desmaële D, Nicolas V, Taran F, Georgin D, Andreux JP, Roux F, Couvreur P (2005). A relevant *in vitro* rat model for the evaluation of blood–brain barrier translocation of nanoparticles. *Cell Mol Life Sci*. (12): 1400–1408.

18. Feng SS, Mu L, Win KY. *et al*. (2004). Nanoparticles of biodegradable polymers for clinical administration of paclitaxel. *Curr Med Chem*. 2004. (11): 413–424.

19. de Kozak Y, Andrieux K, Villarroya H, Klein C, Thillaye-Goldenberg B, Naud MC. *et al.* (2004). Intraocular injection of tamoxifen-loaded nanoparticles: a new treatment of experimental autoimmune uveoretinitis. *Eur. J. Immunol.* (34): 3702–3712.

20. Shaffer C. Nanomedicine transforms drug delivery (2005). *Drug Discov Today.* (10): 1581–1582.

21. Moghimi SM, Hunter AC, Murray JC (2005). Nanomedicine: current status and future prospects. *FASEB J.* (19): 311–330.

22. Drexler EK (1992). *Nanosytems: Molecular Machinery, Manufacturing and Computation.* John Wiley and Sons, New York.

23. Hu L, Tang X, Cui F (2004). Solid lipid nanoparticles (SLNs) to improve oral bioavailability of poorly soluble drugs. *J Pharm Pharmacol.* (56): 1527–1535.

24. Arangoa MA, Campanero MA, Renedo MJ, Ponchel G, Irache JM (2001). Gliadin nanoparticles as carriers for the oral administration of lipophilic drugs. Relationships between bioadhesion and pharmacokinetics. *Pharm Res.* (18): 1521–1527.

25. Arbos P, Campanero MA, Arangoa MA, Irache JM (2004). Nanoparticles with specific bioadhesive properties to circumvent the pre-systemic degradation of fluorinated pyrimidines. *J Control Release.* (96): 55–65.

26. Diwan M, Elamanchili P, Lane H, Gainer A, Samuel J (2003). Biodegradable nanoparticle mediated antigen delivery to human cord blood derived dendritic cells for induction of primary T cell responses. *J Drug Target.* (11): 495–507.

27. Koping-Hoggard M, Sanchez A, Alonso MJ (2005). Nanoparticles as carriers for nasal vaccine delivery. *Expert Rev Vaccines.* (4): 185–196.

28. Lutsiak ME, Robinson DR, Coester C, Kwon GS, Samuel J (2002). Analysis of poly (d, l-lactic-co-glycolic acid) nanosphere uptake by human dendritic cells and macrophages *in vitro. Pharm Res.* (19): 1480–1487.

29. Yotsuyanagi T, Hazemoto N. Cationic liposomes in gene delivery (1998). *Nippon Rinsho.* (56): 705–712.

30. Young LS, Searle PF, Onion D, Mautner V (2006). Viral gene therapy strategies: from basic science to clinical application. *J Pathol.* (208): 299–318.

31. Davis SS (1997). Biomedical applications of nanotechnology–implications for drug targeting and gene therapy. *Trends Biotechnol.* (15): 217–224.

32. Hart SL (2005). Lipid carriers for gene therapy. Curr Drug Deliv. (2): 423–428.

33. Ewert K, Evans HM, Ahmad A, Slack NL, Lin AJ, Martin-Herranz A. *et al.* (2005). Lipoplex structures and their distinct cellular pathways. *Adv Genet.* (53): 119–155.

34. Zhang Y, Schlachetzki F, Li JY, Boado RJ, Pardridge WM. *et al.* (2003). Organ specific gene expression in the rhesus monkey eye following intravenous nonviral gene transfer. *Mol Vis.* (9): 465–472.

35. Elan Corporation, PLC. (2004). Available at http://www.elan.com.

36. Wickline SA, Neubauer AM, Winter P, Caruthers S, Lanza G (2006). Applications of nanotechnology to atherosclerosis, thrombosis, and vascular biology. *Arterioscler Thromb Vasc Biol.* (26): 435–441

37. Panyam J, Labhasetwar V (2003). Biodegradable nanoparticles for drug and gene delivery to cells and tissue. *Adv Drug Deliv Rev.* (55): 329–347.

38. West JL, Halas NJ (2000). Applications of nanotechnology to biotechnology commentary. *Curr Opin Biotechnol.* (11): 215–217.

39. Shi H, Tsai WB, Garrison MD, Ferrari S, Ratner BD (1999). Templateimprinted nanostructured surfaces for protein recognition. *Nature.* (398): 593–597.

40. Sims MR (1999). Brackets, epitopes and flash memory cards: a futuristic view of clinical orthodontics. *Aust Orthod J.* (15): 260–268.

41. Slavkin HC (1999). Entering the era of molecular dentistry. *J Am Dent Assoc.* (130): 413–417.

42. Ure D, Harris J (2003). Nanotechnology in dentistry: reduction to practice. *Dent Update.* (30): 10–15.

43. Fartash B, Tangerud T, Silness J, Arvidson K (1996). Rehabilitation of mandibular edentulism by single crystal sapphire implants and overdentures: 3-12

44. Year results in 86 patients. A dual center international study. *Clin Oral Implants Res.* (7): 220–229.

45. Shellhart WC, Oesterle LJ (1999). Uprighting molars without extrusion. *J Am Dent Assoc.* (130): 381–385.

46. Webster TJ, Waid MC, McKenzie JL, Price RL, Ejiofor JU (2004). Nanobiotechnology: carbon nanofibres as improved neural and orthopedic implants. *Nanotechnology.* (15): 48–54.

47. Price RL, Waid MC, Haberstroh KM, Webster TJ (2003). Selective bone cell adhesion on formulations containing carbon nanofibers. *Biomaterials.* (24): 1877–1887.

48. Buzea C, Pacheco I, Robbie K (2007). Nanomaterials and nanoparticles: Sources and Toxicity. *Biointerphases.* (2): MR17.

49. Milunovich S, Roy J (2001). The next small thing: *An Introduction to nanotechnology.* Merrill Lynch Report.

50. Hamad-Schifferli K, Schwartz J, Santos AT, Zhang S, Jacobson J (2002). Remote electronic control of DNA hybridization through inductive coupling to an attached metal nanocrystal antenna. *Nature.* (415): 152–155.

51. Li Z, Hulderman T, Salmen R, Chapman R, Leonard SS, Young SH. *et al.* (2007). Cardiovascular effects of pulmonary exposure to single-wall carbon nanotubes. *Environ Health Perspect.* (115): 377–382.

52. Nijhara R, Balakrishnan K (2006). Bringing nanomedicines to market: regulatory challenges, opportunities, and uncertainties. *Nanomedicine.* (2): 127–136.

53. Lam CW, James JT, McCluskey R, Hunter RL (2004). Pulmonary toxicity of single-wall carbon nanotubes in mice 7 and 90

54. days after intratracheal instillation. *Toxicol Sci.* (77): 126–134.

55. Williams D (2004). The risks of nanotechnology. *Med Device Technol.* (15): 9–10.

56. Oberdorster E (2004). Manufactured nanomaterials (fullerenes, C60) induce oxidative stress in the brain of juvenile largemouth bass. *Environ Health Perspect.* (112): 1058–1062.

57. Elder A, Gelein R, Silva V, Feikert T, Opanashuk L, Carter J. *et al.* (2006) Translocation of inhaled ultrafine manganese oxide particles to the central nervous system. *Environ Health Perspect.* (114): 1172–1178.

58. Radomski A, Jurasz P, Onso-Escolano D, Drews M, Morandi M, Malinski T. *et al.* (2005). Nanoparticle-induced platelet aggregation and vascular thrombosis. *Br J Pharmacol.* (146): 882–893.

59. Medina C, Santos-Martinez MJ, Radomski A, Corrigan OI, Radomski MW (2007). Nanoparticles: pharmacological and toxicological significance. *Br J. Pharmacol.* (150): 552–558.

60. Chen Z, Meng H, Xing G, Chen C, Zhao Y, Jia G. *et al.* (2006). Acute toxicological effects of copper nanoparticles *in vivo*. *Toxicol Lett.* (163): 109–120.

2016, Environmental Biotechnology: A New Approach   *Pages 17–31*
Editors: Dr. Rajan Kumar Gupta and Dr. Satya Shila Singh
Published by: DAYA PUBLISHING HOUSE, NEW DELHI

Chapter 2

Optical Fiber Biosensors: An Overview

Vijendra Lingwal*

Pt. L.M.S. Government P.G. College (An Autonomous College),
Rishikesh, Dehradun

ABSTRACT

Remarkable developments can be seen in the field of optical fibre biosensors in the last decade. More sensors for specific analytes have been reported, novel sensing chemistries or transduction principles have been introduced, and applications in various analytical fields have been realised. Among different biosensors enzyme optical fiber-based biosensors continue to be the most commonly reported types of optical fiber-based biosensor in the literature. The abundance of enzyme optical fiber based biosensors reported in the literature is likely due to the inherent selectivity that enzymes have for a specific analyte, and the ease with which they are employed in sensing. Introduction to biosensors, classification and explaination of optical fiber based biosensors in particular has been given in this review.

Keywords: Biosensor, Enzymes, Optical fiber, Refractive index, Transducer.

Introduction

Biological recognition elements have attracted extraordinary interest in recent years, because of the key role they play in the development of highly sensitive and selective chemical analysis. According to IUPAC (International Union of Pure and

---

* *Corresponding author.* E-mail: lingwalv@yahoo.co.in
# This article is not a research work of author, it is inspire by the article of refer [2] and [3[.

Applied Chemistry) definition [1], "A biosensor is a self-contained integrated device which is capable of providing specific quantitative or semi-quantitative analytical information using a biological recognition element (biochemical receptor) which is in direct spatial contact with a transducer element. A biosensor should be clearly distinguished from a bioanalytical system, which requires additional processing steps, such as reagent addition. Furthermore, a biosensor should be distinguished from a bioprobe which is either disposable after one measurement, *i.e.* single use, or unable to continuously monitor the analyte concentration". Biosensors that include transducers based on integrated circuit microchips are known as biochips.

Specificity, precision, speed and sensitivity should be the main properties of any proposed biosensor. The first depends entirely on the inherent binding capabilities of the bioreceptor molecule whereas other parameters will depend on both the nature of the biological element and the type of transducer used to detect this reaction. Because of these improved properties biosensors can be regarded as an interesting alternative to conventional techniques for these applications. Biosensor development is driven by the continuous need for simple, rapid, and continuous *in situ* monitoring techniques in a broad range of areas, *e.g.* medical, pharmaceutical, environmental, defense, bioprocessing, or food technology. Several reviews and books published in last decade [2-12] summarize the main achievements of biosensor research in these areas.

Biosensors can be classified according to either the nature of the bioreceptor element or the principle of operation of the transducer. The main types of transducer used in the development of biosensors can be divided into four groups [13]: (i) optical, (ii) electrochemical, (iii) mass-sensitive, and (iv) thermometric. Each group can be further subdivided into different categories, because of the broad spectrum of methods used to monitor analyte-receptor interactions. The bioreceptor component can be classified into five groups [14]:

1. *Enzymes*: Proteins that catalyze specific chemical reactions. These can be used in a purified form or be present in a microorganism or in a slice of intact tissue. The mechanisms of operation of these bioreceptors can involve: (i) conversion of the analyte into a sensor-detectable product, (ii) detection of an analyte that acts as enzyme inhibitor or activator, or (iii) evaluation of the modification of enzyme properties upon interaction with the analyte.

2. *Antibodies and antigens*: An antigen is a molecule that triggers the immune response of an organism to produce an antibody, a glycoprotein produced by lymphocyte B cells which will specifically recognize the antigen that stimulated its production [15].

3. *Nucleic acids*: The recognition process is based on the complementarity of base pairs (adenine and thymine or cytosine and guanine) of adjacent strands in the double helix of DNA. These sensors are usually known as genosensors. Alternatively, interaction of small pollutants with DNA can generate the recognition signal [16].

4. *Cellular structures or whole cells*: The whole microorganism or a specific cellular component, for example a non-catalytic receptor protein, is used as the biorecognition element.

5. *Biomimetic receptors*: Recognition is achieved by use of receptors, for instance, genetically engineered molecules [17], artificial membranes [18], or molecularly imprinted polymers (MIP), that mimic a bioreceptor.

Optical biosensor that exploit light absorption, fluorescence, luminescence, reflectance, Raman scattering and refractive index are powerful alternatives to conventional analytical techniques. The development of optical-fiber sensors during recent years is related to two of the most important scientific advances: the laser and modern low-cost optical fibers. Their use as a probe or as a sensing element is increasing in clinical, pharmaceutical, industrial and military applications. Excellent light delivery, long interaction length, low cost and ability not only to excite the target molecules but also to capture the emitted light from the targets are the main points in favour of the use of optical fibers in biosensors [19-26] and also provide rapid, highly sensitive, real time, and high-frequency monitoring without any time-consuming sample concentration and/or prior sample pre-treatment steps. It is, in fact, usually said that application of optical fibers enables the scientist "to bring the spectrometer to the sample".

The great success of optical-fiber sensors is because they can be used to tackle difficult measurement situations where use of conventional sensors is not appropriate. The sensors are usually compact and lightweight, minimally invasive, and can be multiplexed effectively on a single fiber network. They are immune to electromagnetic interference, because there are no electrical currents flowing at the sensing point. OFS can survive in difficult environments, *e.g.* in the presence of high doses of radiation, and so might have potential in the nuclear industry [27].

Fundamentals of Fiber Optics

Optical fibers transmit light on the basis of the principle of total internal reflection (TIR). When this phenomenon occurs the light rays are guided through the core of the fiber with very little loss to the surroundings. The optical fiber is formed by a core with a refractive index n_1 and a cladding with a refractive index n_2. For light propagation by TIR the refractive index of the core (n_1) must be larger than that of the cladding (n_2), *i.e.* $n_1 > n_2$. When a ray of light strikes the boundary interface between these transparent media of different refractive index and the angle of incidence is larger than the critical angle, defined by the Snell's law [$q_c = \sin^{-1}(n_2/n_1)$], it will be totally internally reflected and propagated through the fiber. When the incident light is totally internally reflected, its intensity does not abruptly decay to zero at the interface. A small portion of light penetrates the reflecting medium by a fraction of wavelength, far enough for recognition of the different refractive index. This electromagnetic field, called the evanescent wave, has an intensity that decays exponentially with distance, starting at the interface and extending into the medium of lower refractive index [2]. The penetration depth (d_p), defined as the distance required for the electric field amplitude to fall to $1/e$ (0.37) of its value at the interface, increases with closer index matching and it is also a function of the wavelength of the light and the angle of incidence. The evanescent wave can interact with molecules within the penetration depth, thereby producing a net flow of energy across the reflecting surface in the surrounding medium (*i.e.* that with refractive index n_2) to maintain the evanescent field. This transfer of energy will lead to attenuation in reflectance which can be used to develop absorption sensors based on evanescent waves (attenuated total reflection (ATR) sensors). When the evanescent light selectively excites a fluorophore, the fluorescence emitted can be directed back into the fiber and guided to the detector. This principle is called total internal reflection of fluorescence (TIRF) and has been widely applied in the design of immunosensors. Optical fiber biosensors can be used in combination with different types of spectroscopic technique, *e.g.* absorption, fluorescence, phosphorescence, Raman, surface Plasmon resonance (SPR), etc. When absorbance is measured the biological receptor can be immobilized close to the optical fiber or directly on its surface. On interaction with the analyte, variation of the absorbance properties of the sensitive layer (Beer's law) will occur and will be related to the concentration of the species analyzed. One or several fibers can be used to guide the light from the source to the sensitive tip, and the emerging radiation back to the detector [2].

Although fluorescence measurements can be used whenever a naturally fluorescent analyte is detected, this is not common in biosensor development and the technique is usually applied in combination with artificially labelled compounds, for instance in competitive immunosensors, or with fluorescence quenching measurements. This type of transduction is usually applied in enzymatic biosensors in which the analyte is biocatalytically converted to a product, or reacts with a compound with optical properties, or that induces an optical signal. This is so, for instance, with oxidase-type enzymes, for which enzymatic consumption of oxygen, as a result of the transformation of the analyte, is evaluated by measuring the decrease in the luminescent signal of a fluorescent dye formed by this molecule. Bioor chemiluminescent sensors, in which the analyte induces emission of light on interaction with a bioreceptor, can also be used [2].

The measurement scheme for an extrinsic fluorescence sensor will be similar to that used in absorbance measurements. In this instance the same, or a different, fiber will collect the light emitted by the sensing element containing the fluorescent indicator; this light will then be filtered and delivered to the detector. Alternatively, molecular recognition can occur at the surface of the fiber core, accompanied by binding or release of a chromophore or luminophore that can be excited by the evanescent wave. In all instances light intensity, decay time, polarization, or phase of the emitted radiation can be selected as the analytical property used to evaluate the concentration of the analyte. Surface plasmon resonance (SPR) is an optical phenomenon caused by charge density oscillation at the interface of two media with dielectric constants of opposite sign, for example a metal and a dielectric [28]. When light of an appropriate wavelength interacts with the dielectric– metal interface at a defined angle, called the resonance angle, there is a match of resonance between the energy of the light photons and the electrons at the metal surface. In this situation the photon energy is transferred to the surface of the metal as packets of electrons called plasmons and the reflection of light from the metal film will usually be attenuated. This resonance is experimentally observed as a sharp minimum of light reflectance when the angle of incidence is varied. Alternatively, SPR can also be generated by use of a fixed angle, white light, and spectral detection [29].

The resonance angle will depend on wavelength of the incident light, the metal, and the nature of the media in contact with the surface [30-31]. Any change in the refractive index of the adsorbed layers at the metal surface will affect the SPR coupling conditions, and produce a shift in the resonance angle. The most common configurations used to couple light rays into a surface plasmon mode that exists on the surface of a thin metal solid film are prisms and diffraction gratings, but SPR sensors can also be based on optical fibers or integrated optical wave guides [30, 32].

Enzyme-Based Optical Biosensors

Enzyme-based optical biosensors have bee extensively studied in the last decades due to vital practical needs of industry, medicine and environmental control and monitoring [33-35] and most commonly used biological components of fiber-optic biosensors. The main reasons are: (i) the large number of reactions they catalyze (ii) the possibility of detecting a broad range of analytes (substrates, products, inhibitors, and modulators of the catalytic activity) and (iii) the different transduction principles that can be used to detect the analyte of interest. Because enzymes are natural proteins which transform a specific substrate molecule into a product without being consumed in the reaction, they can easily be used for continuous biosensing of a specific compound. Among other advantages [36] enzymes are highly selective and sensitive compared with chemical reactions, they are fairly fast-acting in comparison with other biological receptors, and they can be used in combination with different transduction mechanisms. The stability of the enzyme, on the other hand, determines the lifetime of the biosensor.

Enzyme optical fiber-based biosensors commonly operate in indirect or direct detection mode, based on the optical properties of the products or intermediates of the enzymatic reaction. When enzymatic reactions have products or intermediates that exhibit an optical property (*e.g.* absorbance, reflectance, fluorescence, etc.), biosensors directly monitor optical changes. When enzymatic reactions have products or intermediates that possess no intrinsic optical properties (*e.g.* O_2, H_2O_2, NH_3, CO_2, etc. commonly called transducers), biosensors use an indicator or reactant to indirectly generate optical changes. These two methods of detection may be further grouped by whether the enhancement or inhibition of the signal is monitored. In enhancement mode, the signal is proportional to the increase in the analyte concentration. In inhibition mode, the signal is proportional to the decrease of enzymatic activity relative to analyte concentration [3].

The optical measurement method is critical to the sensitivity and detection limit of the sensor. Typical methods include absorbance or reflectance, fluorescence, chemiluminescence (CL), or electrogenerated chemiluminescence. The enzyme optical fiber-based biosensors discussed in this review will be categorized by these different methods of interrogation.

Absorbance-Based and Reflectivity-Based Methods

Absorbance and reflectivity measurements are commonly used with enzyme optical fiber-based biosensors, as many reaction products or intermediates are colored. When they are not colored, a broad range of indicators can be used. The determination of oxalate in food samples, as described by Sotomayor *et al.* [37], is an example of the use of an indicator to interact with a colorless product to produce a change in color. Their optical sensor system was based on the enzymatic reaction of oxalic acid and oxalate oxidase; the oxalate concentration was determined indirectly via the pH change resulting from the production of carbon dioxide:

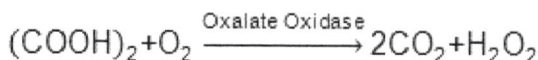

$$(COOH)_2 + O_2 \xrightarrow{\text{Oxalate Oxidase}} 2CO_2 + H_2O_2$$

Catalase was used in conjunction with the oxidase to consume peroxide in order to drive the reaction to the right and thereby increase the production of carbon dioxide:

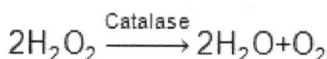

$$2H_2O_2 \xrightarrow{\text{Catalase}} 2H_2O + O_2$$

The fiber bundle was covered with a CO_2-permeable membrane and placed facing an enzymatic reactor containing oxalate oxidase and catalase, which were immobilized with glutaraldehyde on ground barley seeds (to provide a natural enzyme environment and extend its useful lifetime). The pH changes were monitored using bromothymol blue immobilized via a bifunctional silane on the surface of the optical fiber bundle. A pH change resulted when the CO_2 produced by the enzymatic reaction diffused across the membrane and produced carbonic acid. A linear response was demonstrated between 8.0 and 100 mM for experiments carried out in 0.050 M succinate buffer at pH 4 and 25°C. Potential interferences due to ascorbic, citric, acetic, and salicylic acid were analyzed, and citric acid was found to be the only significant interferent. The application of this method for determining oxalate in food samples was demonstrated by analyzing spinach leaves. The lifetime of the sensor was shown to be around 120 h. Enzyme optical fiber-based biosensors have recently been examined for sensing pesticides. Andreou and Clonis [38] developed a layer-based biosensor for the organophosphorous pesticides carbaryl and dichlorvos using acetylcholinesterase. In this biosensor, the substrate acetylcholine was added to the sample, and enzyme activity was measured by monitoring the pH change (resulting from the hydrolysis of acetylcholine to acetic acid) using the pH absorbance indicator bromocresol purple encapsulated in a sol–gel. Carbaryl and dichlorvos inhibited the formation of acetate by acetylcholinesterase, thereby decreasing the pH relative to the uninhibited reaction. The calibration curve for carbaryl was useful for concentrations between 0.11 – 8.0 mg L^{-1}, and the limit of detection was 108 µg L^{-1}. The calibration curve for dichlorvos was useful between 5.0– 30 µg L^{-1}, and the limit of detection was 5.2 µg L^{-1}. This method was successfully applied to the analysis of real water samples with no sample preparation. Sensor lifetime was demonstrated to be approximately 3 weeks, with a 30 per cent activity decrease. Acetylcholinesterase immobilized in a Langmuir– Blodgett film was similarly used for the detection of aqueous Paraoxon (an organophosphorous pesticide) by Choi *et al.* [39]. The fiber optic system operated in flow cell mode and monitored the enzymatic formation of yellow *o*-nitrophenol from o-nitrophenyl acetate. The inhibition of acetylcholinesterase activity due to Paraoxon exposure was monitored as a decrease in absorbance proportional to the concentration of the organophosphorous compound. The sensor was useful for concentrations 0.0 and 2.0 ppm with a response time of 10 min. An optical fiber-based biosensor for the detection of the chlorinated herbicide atrazine was described by Andreou and Clonis [40]. The sensing material comprised multiple layers, including sol–gel-immobilized bromocresol green and a poly (vinylidene fluoride) membrane coated with glutathione S-transferase. Glutathione S-transferase catalyzed the nucleophilic attack of glutathione on atrazine, liberating HCl and resulting in a localized pH change. Detection was performed by monitoring the absorbance change of the bromocresol green due to the change in pH. The sensor displayed a linear range of 2.52–125 µM with a limit of detection of 0.84 µM. The sensor was additionally employed for the detection of atrazine in water samples with no sample preparation. The results agreed with officially accepted methods. The sensor retained 75 per cent of its original activity after 1 month of use, although the layer design facilitates rapid changing of the sensing chemistry for complete regeneration.

Glutathione S-transferase was also used to detect the fungicide captan. A flow cell fiber optic system containing a gel film of immobilized glutathione S-transferase was used by Choi *et al.* [41] to detect captan in aqueous media by monitoring the decrease in enzyme activity. The reaction of glutathione S-transferase with the substrates 1-chloro-2,4-dinitrobenzene and glutathione produced S-(2,4-dinitrobenzene) glutathione, a yellow-colored compound. Inhibition of this process by captan decreased the absorbance rapidly; a steady signal was achieved in 15 min. The method was found to be useful for detecting captan at up to 2 ppm, and sufficient enzyme activity was maintained for 30 days. Kuswandi [42] developed an innovative, enzyme optical fiber-based biosensor for the detection of heavy metals. Dutta *et al.* [43] developed a novel enzyme optical fiber-based biosensor for measuring enzyme activity.

Fluorescence-Based Methods

Fluorescence measurements are not used as often as absorbance and reflectance with enzyme optical fiberbased biosensors, as it is not common for enzyme reactions to produce fluorescent products or intermediates. Most fluorescence techniques employ a fluorescent dye to indirectly monitor formation or consumption of a transducer.

A fluorescence optical fiber-based biosensor was developed by Doong and Tsai [44] for the determination of acetylcholine. This approach employed acetylcholinesterase immobilized in sol–gel on the distal end of an optical fiber. The enzymatic reaction of acetylcholinesterase and acetylcholine formed acetic acid, resulting in a change in fluorescence intensity of the pH-sensitive fluorescent dye fluorescein isothiocyanate-dextran. After optimization of the sol–gel composition, buffer solution, and mixing protocol, this reversible system showed linearity between 0.5 and 20 mM, required roughly 5 min for analysis, and took 30 min to recover to baseline levels. The potential use of this system for detecting organophosphorous pesticides was demonstrated by analyzing a sample containing Paraoxon, which inhibited acetylcholinesterase and thus increased fluorescence.

Issberner *et al.* [45] designed an enzyme optical fiberbased biosensor for L-glutamate to study its localized release and re-uptake from the foregut of the tobacco worm Manduca sexta. The distal end of an optical fiber was silanized with (3-aminopropyl) triethoxysilane, treated with poly (acrylamide-co-N-acryloxysuccinimide), and coated with a gel containing L-glutamate oxidase and the fluorescent dye seminaphthofluorescein (SNAFL). L-Glutamate was quantified by the decrease in fluorescence intensity of the SNAFL dye caused by liberated NH_3, a product of the L-glutamate and L-glutamate oxidase reaction. This method demonstrated an effective quantification range of 1–100 mM glutamate, with a limit of detection of 10–100 µM. Sensor specificity was examined with other physiologically active amino acids present, and no significant interferences were observed.

An optical fiber-based biosensor employing cytoplasmic membranes as the source of enzyme was described by Ignatov *et al.* [46]. The cytoplasmic membranes used in this study were derived from Escherichia coli that had been cultured in the presence of sodium lactate to induce formation of the enzyme L-lactate oxidase. This enzyme catalyzed the conversion of lactate to pyruvate, consuming oxygen and liberating peroxide in the process. The fluorescence of ruthenium tris (diphenylphenanthroline) immobilized in polysiloxane on the surface of an optical fiber was used to measure oxygen consumption. Oxygen quenches the fluorescence of ruthenium complexes, so fluorescence increased with decreasing oxygen concentration. This sensor was used to determine lactate concentrations between 0 and 5 mM, which bracketed the physiological concentration of interest, 2 mM. Interference was not observed, even in the presence of 50 mM concentrations of glucose, fructose, and glutamic acid. The sensor demonstrated good selectivity, reversibility, and response time. Durrieu and Tran-Minh [47] developed an enzyme optical fiber-based biosensor for the detection of heavy metals that employed alkaline phosphatase present on the outside membrane of live *Chlorella vulgaris* microalgae.

Chemiluminescence and Electrogenerated Chemilumine-scence-Based Methods

Chemiluminescence (CL) and electrogenerated chemiluminescence (ECL) methods are commonly used for detecting hydrogen peroxide formed by enzymatic reactions. These processes are characterized by simplified optical instrumentation; no excitation light source is required, and therefore no additional apparatus is necessary to remove scattered excitation light. The lack of an excitation source is advantageous, and these methods have distinct advantages in selectivity and sensitivity over traditional absorbance, fluorescence, and reflectance optical techniques, as the background is greatly reduced [48].

An enzyme optical fiber-based biosensor for p-iodophenol, 2-naphthol, and p-coumaric acid using horseradish peroxidase (HRP) immobilized in sol–gel on the end of an optical fiber was developed by Ramos *et al.* [49]. The sensor monitored the enhancement in CL of the luminol/H_2O_2/HRP system due to the addition of p-iodophenol, p-coumaric acid, and 2-naphthol (enhancers accelerate the formation of luminol radicals, thus increasing light emission [50]). The linear ranges were 2.67 – 9.33 µM, 40 – 200 nM, 33 – 266 nM, and the detection limits were 0.83 µM, 48 nM, and 15 nM for p-iodophenol, 2-naphthol, and p-coumaric acid, respectively, using immobilized HRP. The CL enhancement effect of the phenolic compounds on H_2O_2

detection was examined as well. The enhancer p-coumaric acid was found to give the highest signal increase, yielding a detection limit of 16.6 µM H_2O_2 for immobilized HRP, an improvement over the detection limit of 0.667 mM H_2O_2 performed in the absence of enhancer [51]. The use of solution-based HRP resulted in lower detection limits than immobilized HRP; however, immobilized HRP demonstrated an increased linear range relative to free HRP. Thus, this system enhanced the detection limit for enzyme-catalyzed reactions that produced H_2O_2.

Sensing via ECL requires the incorporation of an electrode into the design of the sensing system to electrochemically trigger CL. This triggering is usually accomplished by placing a conventional electrode opposite the distal end of an optical fiber. The optical fiber is used to carry the ECL signal to a detector. Zhu *et al.* [52] used this approach to create a novel, enzyme optical fiber-based biosensor for determination of glucose via ECL of luminol. The reaction of glucose and glucose oxidase (GOD) generated hydrogen peroxide, which reacted with luminol to produce light when the correct potential was applied to the mixture. The electrode used in this study was comprised of GOD and graphite powder impregnated in a sol–gel. The distal end of a fiber optic bundle faced the electrode surface, and the proximal end was connected to a photomultiplier tube for measurement. Using this configuration, the investigators were able to observe luminol's ECL reaction with hydrogen peroxide when the potential of the electrode was stepped to 0.7 V. The useful range of the sensor was determined to be 0.01–10 mM glucose with a limit of detection of 8.2 µM. The sensor demonstrated a rapid response as equilibration occurred in only 20 s. Polishing was used to expose fresh enzyme and renew the electrode. The response of six polishings differed by only 6.1 per cent. The sensor maintained 70 per cent of its original response after 30 days of storage at 4°C. In a separate study, Zhu *et al.* [53] developed a flow injection-based ECL enzyme optical fiber-based biosensor for glucose. The sensor employed a sol–gel to immobilize GOD on the surface of a glassy carbon electrode. Peroxide formation was detected optically via a fiber optic imaging system. Maintaining the working electrode at 0.8 V versus Ag/AgCl induced luminal ECL. Luminol concentration, pH, and flow rate were optimized, and the linear range for glucose was determined to be 50 µM – 10 mM with a detection limit of 26 µM. Interference effects were negligible at 50 times the glucose concentration for fructose, sucrose, phenylformic acid, and citric acid and ten times the glucose concentration for lactose, maltose, oxalate acid, and lactic acid. Concentrations of Cr^{3+}, Cu^{2+}, Fe^{3+}, Mn^{2+}, Ni^{2+} equal to the glucose concentration and Co^{2+} at one-tenth the glucose concentration also produced negligible interference. The system was employed for the determination of glucose in soft drinks with minimal sample preparation. The sensor retained 85 per cent of its original signal value after 90 measurements over 3 h. Marquette *et al.* [54] also employed luminol-peroxide ECL as the mechanism for detection and fabricated a series of oxidase optical fiber-based biosensors for flow injection analysis detection of glucose, lactate, choline, and cholesterol.

Immunoassay Optical Fiber-Based Biosensors

Immunoassay optical fiber-based biosensors have been the subject of a wide variety of research, ranging from the detection of specific analytes and harmful microorganisms to broad-based immunoassay systems for general use [55-56]. Immunoassays are performed in one of four modes: direct, competitive, binding inhibition and sandwich (described in detail in ref. [2]). These techniques typically require the immobilization of a primary antibody to the surface of a substrate and the use of an optical label. Several studies have focused on the immobilization of antibodies to optical fibers to examine whether the antibody retains its solution binding capability when immobilized [57]. Vijayendran *et al.* [58] examined the heterogeneity associated with the binding of a soluble, fluorescently labeled analyte, cyanine-5-ethylenediaminetrinitrobenzene, with immobilized monoclonal anti-TNT antibodies using five different immobilization strategies. Silica optical fibers were employed as the solid support for both random and structured immobilization schemes using carbohydrate, protein G, and streptavidin attachment strategies. Analysis of the binding affinity heterogeneity was performed using the Sips isotherm (a mathematical model for describing surface interactions) and was compared to binding in solution. Antibody binding affinities for each of the different immobilization strategies were at least an order of magnitude lower than the solution value. As expected, randomly immobilized antibodies gave heterogeneous binding affinities, while more uniformly aligned antibodies gave more homogeneous behavior. Antibody immobilization using binding to the antibody's carbohydrate residues was determined to be an effective binding method with the highest biological activity and most homogeneous binding affinity. It was noted that these binding strategies were not optimized, and higher antibody activities and more homogeneous binding may be achieved upon optimization.

Fry and Bobbitt [59-60] employed dynamic modification to measure the binding properties of fluorescently labeled anti-dinitrophenyl antibodies (anti-DNP) with the hapten octadecyl 6-(2,4 dinitrophenyl) aminohexanoic acid (C8-DNP) using an optical fiber [61]. The hydrophobic hapten was synthesized, characterized, and immobilized on the end of a C18 modified optical fiber. The modified fibers were used for analysis and then cleaned in an appropriate washing solution for re-use. Sensor performance for the quantification of anti-DNP was examined, and the limit of detection was determined to be 10 fmol for a 20-µL probe volume. The binding constant $K_{apparent}$ for the formation of the antibody – hapten complex was determined to be $1.0 \pm 0.2 \times 10^6$. The measurement was highly reproducible; a single fiber subjected to ten cycles gave an RSD of less than 7 per cent [62].

Marks *et al.* [63] developed a novel technique for immobilizing biotin in a polypyrrole film and used it to design an immunoassay optical fiber-based biosensor for cholera antitoxin antibodies [64]. This biosensor combined electrodeposition and optical analysis, employing ITO on the distal end of an optical fiber as the electrode material. Biotinylated polypyrrole was electrodeposited on the ITO to provide a biofunctionalized surface for modification with avidin. A sandwich immunoassay was initiated by incubating the avidinmodified sensor surface in a solution of biotinylated cholera toxin B subunit, followed by immersion in solutions of varying anti-cholera toxin B subunit concentrations, and finally by dipping it in a solution of the secondary antibody, HRP-labeled goat anti-rabbit IgG immunoglobulin. The modified end of the optical fiber was placed in a solution of oxidizing reagent, and luminal and CL light emission was monitored via a photomultiplier tube. The sensor was demonstrated to be effective for quantification of antibody titers of 1 : 3000 and lower with a proposed limit of detection of 1 : 1,200,000.

The improvement of optical fiber-based surface plasmon resonance (SPR) biosensors has been the focus of several manuscripts written in the time covered by this review. SPR biosensors may be useful for many sensing applications, and offer the possibility of generating highly selective sensors that would be broadly applicable for immunoassays [65-66]. Additionally, optical fiberbased SPR biosensors possess inherent advantages over traditional optical methods as they exhibit high sensitivity and do not require the use of a label [2].

There are two modes of SPR used in optical fiberbased SPR biosensing, localized SPR (LSPR) and propagating SPR (PSPR), both of which take advantage of the plasmon wave's sensitivity to the refractive index of the medium surrounding a noble metal. The majority of literature applications featuring optical fiber-based SPR biosensors employ PSPR with multimode optical fibers. These fibers are side-polished to expose the core, which is coated with a thin metal layer to support plasmon waves. Sensors that use multimode fibers typically suffer from limited resolution due to modal noise, so sensors based on single-mode tapered and side-polished fibers have been proposed to overcome these limitations [67-68]. Single-mode tapered fibers also have inherent limitations; they exhibit low sensitivity due to operation at shorter wavelengths or in amplitude interrogation mode. Side-polished fibers offer greater sensitivity, but are susceptible to fiber deformation. Fiber deformation results in changes in light polarization and therefore surface plasmon strength and resultant signal.

A novel fiber optic PSPR sensing device fabricated by side-polishing a single-mode optical fiber was demonstrated by Slavik *et al.* [65]. Their approach employed depolarized light, which facilitated construction of an entirely fiber optic system and reduced sensitivity to fiber deformations. The sensing element was created using a side-polished single-mode fiber optic, which was coated with 65 nm of gold, followed by 19 nm of tantalum pentoxide to adjust the refractive index operating range. The sensor was tested using solutions of diethyleneglycol and water of known refractive indices. Sensitivity of 3,500 nm/refractive index unit (RIU) and resolution as high as 5×10^{-7} RIU were achieved, comparable to much bulkier optical bench top units. The sensor was used for detecting IgG, concentrations as low as 40 ng mL^{-1} could be detected. In a continuation of this research, Slavik *et al.* [66] used the same fiber optic PSPR sensor for the detection of Staphylococcal enterotoxin B (SEB). The sensor surface was functionalized with a double-layer of anti - SEB crosslinked by glutaraldehyde. The sensor was operated using a flow injection system and mild acid was used to regenerate the surface for replicate analyses. The total time for an analysis cycle was approximately 9 min, and SEB solutions of 10, 20, 50, and 100 ng mL^{-1} were analyzed. The theoretical detection limit was calculated to be approximately 4 ng mL^{-1}, which would be comparable to conventional immunoassay methods. Two recent techniques have not been used as optical fiber-based biosensors, but warrant mention as they show promise for future biosensor applications. The first technique was an LSPR sensor fabricated by depositing self-assembled colloidal gold on the exposed core of an optical fiber [69]. This technique simplified the preparation of fiber optic SPR sensors, but maintained the advantages of high sensitivity and the ability to monitor biological binding events without the use of a label. Initial experiments for glucose/water solutions with refractive indices between 1.33 and 1.41 gave a linear response. The resolution was calculated to be 2.1 ´ 10⁻⁴ RIU. Experiments were performed to monitor the binding of streptavidin to biotinylated colloidal gold. The sensor gave a concentration-dependent response for streptavidin concentrations in the range of 0.5 – 3.0 nM, with a calculated limit of detection of 0.1 nM.

In the second technique, [70] a 300 nm thick gold film was deposited on a silica substrate, modified with a monolayer of biotin, and placed facing the distal end of an optical fiber. Light from a blue LED ($\approx$ 470 nm) was coupled into the optical fiber, reflected off the modified surface, and coupled back through the optical fiber to a photodiode which monitored the resultant intensity. Gold behaves as a dielectric for blue light, so any species adsorbed on its surface (with a refractive index different from that of the solvent or gold), will produce a change in reflectivity. Watanabe and Kajikawa [70] monitored the change in reflectivity for the binding of streptavidin to the monolayer of biotin on gold in real time. As would be expected, they observed decreases in relative reflectivity of the gold surface as layers deposited.

Conclusion

Optical fiber-based biosensors have been the focus of a large volume of research in the recent years. Enzyme optical fiber-based biosensors continue to be the most commonly reported type; however, the number of publications on immunoassay, nucleic acid, and whole cell optical fiber-based biosensors has greatly increased. Immobilization of the biological recognition element has been a common focus for optical fiber-based biosensor research, and has ultimately led to increases in sensitivity, selectivity, and in some cases, reversibility. Optical fiber-based biosensors are beginning to address issues associated with shelf life and long-term stability; some of the sensors reported in this review were stored for several months, or were used for periods up to a month while retaining over 75 per cent of their initial response. Advances have been made in lowering detection limits as well; several papers reported sensors with detection limits comparable to more sophisticated large bench-top analytical instrumentation. Despite these advances, much more work needs to be done before optical fiber-based biosensors become commercially viable options for sensing. The commercial availability of miniature fiber optic spectrophotometric instrumentation and the development of custom portable sensing instrumentation will help to further sensor research.

This article is not a research work of author. It is inspired by the article of ref. [2] & [3].

References

1. Thevenot DR, Toth K, Durst RA, Wilson GS (1999). *Pure Appl Chem* 71: 2333–2348.

2. Marazuela MD and Moreno-Bondi MC (2002). *Anal Bioanal Chem* 372: 664-682.

3. Monk DJ and Walt DR (2004). *Anal Bioanal Chem* 379: 931-945.

4. Rogers KR (2006). *Anal Chim Acta 568*: 222–231.

5. Dorst BV, Mehta J, Bekaertb K, Rouah-Martin E, Coen WD, Dubruelc P, Blusta R, Robbens J (2010). *Biosen. Bioelectron* 26: 1178–1194.

6. Ligler FS (2009). *Anal Chem* 8: 519–526.

7. Fan X, White IM, Shopova SI, Zhu H, Suter JD, Sun Y (2008). *Anal. Chim Acta* 620: 8–26.

8. Palchetti I, Mascini M (2008). *Analyst* 133: 846–854.

9. Wanekaya AK, Chen W, Mulchandani A (2008). *J Environ Monit* 10: 703–712.

10. Shankaran DR, Gobi KV, Miura N (2007). *Sens Actuators B* 121: 158–177.

11. Long F, Zhu A, Gu C, Shi H (2013). *Recent Progress in Optical Biosensors for Environmental Applications. In State of the Art in Biosensors: Environmental and Medical Applications;* Rinken T (Ed.). InTech: Rijeka, Croatia, Chapter 1, pp. 4–28.

12. Freng L, Anna Z, Hanchang S (2013). *Sensors* 13: 13928-13948.

13. Hulanicki A, Glab S, Ingman F (1991). *Pure Appl Chem* 63: 1247–1250.

14. Rogers KR, Mascini M (1998). *Field Anal Chem Technol* 2: 317–331.

15. Aga D, Thurman EM (1997). (eds.). *Environmental immunoassays: alternative techniques for soil and water analysis. In: Immunochemical technology for environmental applications,* vol 657. American Chemical Society.

16. Wang J, Rivas G, Cai X, Palecek E, Nielsen P, Shiraishi H, Dontha N, Luo D, Parrado C, Chicharro M, Farias PAM, Valera FS, Grant DH, Ozsoz M, Flair MN (1997). *Anal Chim Acta* 347: 1–8.

17. Piervincenzi RT, Reichert WM, Hellinga HW (1988). *Biosens Bioelectron* 13: 305–312.

18. Charych D, Cheng Q, Reichert A, Kuziemko G, Stroh M, Nagy JO, Spevak W, Stevens RC (1996). *Chem Biol* 3: 113–120.

19. Gauglitz G (1996). *Opto-chemical and opto-immunosensors In:* Blates H, Göpel W, Hesse J (eds.). *Sensors update,* vol 1. Verlag Chemie, Weinheim.

20. Wolfbeis OS (1991). (ed.). *Fiber optic chemical sensors and biosensors,* vols 1 and 2. CRC Press, Boca Raton, Florida.

21. Wolfbeis OS (2000). *Anal Chem* 72: 81R–89R.

22. Cámara C, Pérez-Conde C, Moreno-Bondi MC, Rivas C (1995). *Chemical sensors with fiber optic devices. In:* Quevauviller Ph, Maier EA, Griepink G (eds.). *Method validation for environmental analysis,* Elsevier, Amsterdam.

23. Rogers KM, Poziomek EJ (1996). *Chemosphere* 33: 1151–1174.

24. Squillantee E (1998). *Drug Dev Ind Pharm* 24: 1163–1175.

25. Burritt M (1998). *Lab Med* 29: 684–687.

26. Wise DL, Wingard LB (1991). (eds.). *Biosensors with fiber optics,* Humana Press,

New Jersey.

27. Cámara C, Moreno-Bondi MC, Orellana G (1991). *Chemical sensing with fiber optic devices. In*: Wise DL, Wingard LB (eds.). *Biosensors with fiber optics.* Humana Press, New Jersey.

28. Lechuga LM, Calle A, Prieto F (2000). *Quím Anal* 19: 54–60.

29. Haake HM, Schütz A, Gauglitz G (2000). Fresenius J Anal Chem 366: 576–585.

30. Earp RL, Raymond ED (1998). *Surface plasmon resonance.* In: Ramsay G (ed). Commercial biosensors, Wiley, New York.

31. Purvis DR, Pollard-Knight D, Lowe PA (1998). *Biosensors based on evanescent waves.* In Ramsay G (ed). Commercial Biosensors, Wiley, New York.

32. Homola J, Yee SS, Gauglitz G (1999). *Sensor Actuator B-Chem* 54: 3–15.

33. Borisov, S.M.; Wolfbeis, O.S. Optical biosensors (2008). *Chem. Rev., 108*: 423–461.

34. Long F, Zhu A, Gu C, Shi H (2013). Recent Progress in Optical Biosensors for Environmental Applications. In *State of the Art in Biosensors: Environmental and Medical Applications*; Rinken, T., Ed.; InTech: Rijeka, Croatia,; Chapter 1, pp. 4–28.

35. Vial L, Dumy P (2009). *New J. Chem. 33*: 939–946.

36. Eggins B (1996). *Biosensors: an introduction,* Wiley, Chichester.

37. Sotomayor MDPT, Raimundo IM, Oliveira Neto G, Kubota LT (2001). *Anal Chim Acta* 447: 33–40.

38. Andreou VG, Clonis YD (2002). *Biosens Bioelectron* 17: 61–69.

39. Choi J-W, Kim Y-K, Lee I-H, Min J, Lee WH (2001). *Biosens Bioelectron* 16: 937–943.

40. Andreou VG, Clonis YD (2002). *Anal Chim Acta* 460: 151–161.

41. Choi JW, Kim YK, Song SY, Lee IH, Lee WH (2003). *Biosens Bioelectron* 18: 1461–1466.

42. Kuswandi B (2003). *Anal Bioanal Chem* 376: 1104–1110.

43. Dutta S, Padhye S, Narayanaswamy R, Persaud KC (2001). *Biosens Bioelectron* 16: 287–294.

44. Doong RA, Tsai HC (2001). *Anal Chim Acta* 434: 239–246.

45. Issberner JP, Schauer CL, Trimmer BA, Walt DR (2002). *J Neurosci Methods* 120: 1–10.

46. Ignatov SG, Ferguson JA, Walt DR (2001). *Biosens Bioelectron* 16: 109–113.

47. Durrieu C, Tran-Minh C (2002). *Ecotox Environ Safe* 51: 206–209.

48. Fahnrich KA, Pravda M, Guilbault GG (2001). *Talanta* 54: 531–559.

49. Ramos MC, Torijas MC, Diaz AN (2001). *Sens Actuators B Chem* B73: 71–75.

50. Navas Diaz A, Sanchez AG, Garcia AG (1998). *J Biolumin Chemilum* 13: 75–84.

51. Navas Diaz A, Ramos MC, Torijas MC (1998). *Anal Chim Acta* 363: 221–227.

52. Zhu L, Li Y, Tian F, Xu B, Zhu G (2002). *Sens Actuators B Chem* B84: 265–270.

53. Zhu L, Li Y, Zhu G (2002). *Sens Actuators B Chem* B86: 209–214.

54. Marquette CA, Leca BD, Blum LJ (2001). *Luminescence* 16: 159–165.

55. Wolfbeis OS (2002). *Anal Chem* 74: 2663–2677.

56. Vo-Dinh T (2002). *J Cell Biochem* 154–161.

57. Watterson JH, Piunno PAE, Wust CC, Krull UJ (2001). *Sens Actuators B Chem* B74: 27–36.

58. Vijayendran RA, Leckband DE (2001). *Anal Chem* 73: 471–480.

59. Ogasawara FK, Wang Y, Bobbitt DR (1992). *Anal Chem* 64: 1637–1642.

60. Wang Y, Bobbitt DR (1994). *Anal Chim Acta* 298: 105–112.

61. Fry DR, Bobbitt DR (2001). *Talanta* 55: 1195–1203.

62. Wang Y, Baten JM, McMaughan SP, Bobbitt DR (1994). *Microchem J* 50: 385–396.

63. Marks RS, Novoa A, Thomassey D, Cosnier S (2002). *Anal Bioanal Chem* 374: 1056–1063.

64. Konry T, Novoa A, Cosnier S, Marks RS (2003). *Anal Chem* 75: 2633–2639.

65. Slavik R, Homola J, Ctyroky J, Brynda E (2001). *Sens Actuators B Chem* B74: 106–111.

66. Slavik R, Homola J, Brynda E (2002). *Biosens Bioelectron* 17: 591–595.

67. Slavik R, Homola J, Ctyroky J (1999). *Sens Actuators B Chem* B54: 74–79.

68. Slavik R, Homola J, Ctyroky J (1998). *Sens Actuators B Chem* B51: 311–315.

69. Cheng SF, Chau LK (2003). *Anal Chem* 75: 16–21.

70. Watanabe M, Kajikawa K (2003). *Sens Actuators B Chem* B89: 126–130.

2016, Environmental Biotechnology: A New Approach   *Pages 33–61*
Editors: Dr. Rajan Kumar Gupta and Dr. Satya Shila Singh
Published by: DAYA PUBLISHING HOUSE, NEW DELHI

3

Green Synthesis of Nanoparticles Using Higher Plants

Kikku Kunui and Ashwini Kumar Dixit*

*Department of Botany, Guru Ghasidas Vishwavidyalaya, Koni,
Bilaspur – 495 009, Chattisgarh*

ABSTRACT

Possesing various economical, ethical, social and environment prospect nanoparticles viz. biologically synthesised nanoparticles are emerging to be one of the boon factor for mankind and livelihood factor. Due to their intoxicity and cost effective mechanism biological method or green chemistry is mostly preferred. They are considered due to their easier down streaming method and involvement of non hazardous ingredient to produce non hazardous nonparticles. Consideration of plant part or whole plant extracts under biological method are intended to be safe, reliable and environment friendly route for synthesizing instant and desired definite shape and size of nanoparticles of interest.

Keywords: Green synthesis, Ecofriendly, Nanoparticles.

Introduction

Nano particle denotes the fusion of atomic or molecular particles that comprises dimension between 1 to 100 nm. The properties and capability of nanoparticle to solve the problem related to technological and environmental challenges, energy conversion, catalytic property, medicinal value and water treatment etc make it on high demand. Nanoparticles are generally classified on the basis of size, shape, surface area and dispersity and these characters are analyzed using various techniques involving spectrophotometrical and microscopical analysis. The

* *Corresponding author:* E-mail: dixitak@live.com

nanoparticles can be fabricated using physical, chemical and biological methods out of which the best suited method is biological prospect (Chandrasekharan and Kamat, 2000; Peto *et al.*, 2002; Mohanpuria *et al.*, 2008; Mittal *et al.*, 2013). On the basis of their emergence and evolution the nanoparticles are classified variedly and are capable to modify their occurance accordingly. The synthesised nanoparticles can be further used in various fields and technologies *viz.* ethical, social and environmental prospects. These but consideration of biofabricated nanoparticles are mostly preferred because they act as boon to mankind and livelihood being intoxic in nature and easily synthesisable. The food that human eat, the water that human drink, the place where human live etc is encovered with the layer of nanoparticle. One of the important reasons for synthesis of nanoparticle is due to implementation in cosmetics, clothing, cheep chips, material sciences, fuel additive, material coating and drug delivery.

The Nanoparticles

Nanoparticles are the molecules or particles whose size is ranged between 1-100 nm and possess large surface area. The prefix *nano* refers to one billionth parts The synthesis of nanoparticles is becoming an active research field nowadays. Nanoparticles were firstly characterized by Michael Faraday (1857) but recent discoveries subjected to the focused research due to its unique optical, electronic, mechanical and chemical properties (Mazur, 2004; Forough *et al.*, 2010). Metallic nanoparticles are one of the efficient nanoparticles as they posses antimicrobial property, catalytic property, electronic property and photonic property (Masciangioli and Zhang, 2003; Cheng,2004; Albrecht *et al.*, 2006; Navarro *et al.*, 2008; Forough *et al.*, 2010; Lloyd *et al.*, 2011). The mechanical, electrical and magnetic behaviour, chemical properties such as solubility, reactivity and catalytic activity is based on the mechanism of nanoparticle production *i.e.* nanotechnology (Llyod *et al.*, 2011). The physico-chemical properties of nanoparticles are different as compared to the bulk particles (Brunner *et al.*, 2006; Monica and Cremonini, 2009; Lloyd *et al.*, 2011).

Characterization of Nanoparticles

Nanoparticles are generally characterised on the basis of their size, shape, surface area and dispersity. A perfect combination of the mentioned character is termed as nanoparticle. The nanoparticles are varied in appearance and their size specifically differs.

Table 3.1: Characterization of Nanoparticle

Prefix	Meaning	Value
Deci	Ten	10^{-1}
Centi	One hundred	10^{-2}
Milli	One thousand	10^{-3}
Micro	Small	10^{-6}
Nano	Dwarf	10^{-9}
Pico	Small Quantity	10^{-12}
Femto	Fifteen	10^{-15}
Atto	Eighteen	10^{-18}

When size of a particle is ranged between 2500-100 nm or then the particles are considered as *fine particles*. Particles ranging from 1-100 nm are termed as *ultrafine particles* whereas particle belonging within the range of 1-10 nm then the term provided is *nanocluster*.

Classification of Nanoparticles

On the basis of their presence and development, nanoparticles are classified as natural, incidental and engineered.

Natural nanoparticle: The existence of natural nanoparticles had been emerged from the evolution of the earth and still exists in the earth's crust. Their emergence had been derived from volcanic eruption, lunar dust and other mineral deposits.

Incidental nanoparticles: Some anthropogenic factors such as exhausting of diesel, combustion of coal and fumigation from welding factory results into the manufacturing of incidental nanoparticles.

Engineered nanoparticles: Engineered nanoparticles are grouped under carbon based materials, metal based material, dendrimers and composite type of nanoparticles.

Carbon based nanoparticles are are further classified under single walled fullerene consisting group and multi walled nanotubes. The metal based nanoparticle comprises quantum dots, nanogold, nanozinc, nanoaluminium and nanoscales metal like TiO_2, ZnO, Al_2O_3. Dendrimers are the nano sized polymer that are derived from branched unit and are capable to perform chemical function. On other behalf the composite nanoparticles are the fused nanoparticles that forms a bulk type of material (Lin and Xing, 2007; Yu-Nam and Lead, 2008; Monica and Cremonini, 2009).

Engineered nanoparticles have recently received a special kind of attention for providing a positive determined and improving approach in sectors such as economic including consumable products, pharmaceutical products, cosmetic products, transportation goods energetic and agricultural product (Nowack and Bucheli, 2007).

Synthesis of Nanoparticles

When it comes to the synthesis of nanoparticles it is a great matter to deal with because synthesis of larger molecule is very easy as compared to nanoparticle. When condensed body formed from irregular size and shape of particle powder that leads to formation of non uniform packing morphologies due to van der Waal's forces results into variation of the particles that give rise to microstructural inhomogeneities. Usage of biological technique for synthesis of nanoparticle is the new developed strategy. These techniques are quite beneficial as compared to physical and chemical based synthesising technique.

For synthesis of nanoparticles a specified technique namingly "nanotechnology" is taken into preferences. Nanotechnology deals with the building of material starting from atomic level that ends into larger nanoscale structured material. The formation of smaller nanoparticle is termed as *"top-down mechanism"* whereas formation of atomic or molecular structure is termed as *"bottom-up mechanism"*.

Top Down Method

Top-down methods is preffered to produce nano particles using nanscale technology which is used extensively today in various fields. In this method nanoparticles synthesized from larger molecules for example photolithography. Photolithography used to create patterns on semiconductors that is capable synthesise 100 nm patterns. A silicon wafer is coated with a light-sensitive film called a photoresist that hardens during dispersion of light. After completion of the process the wafer is exposed to an acid bath, to remove unhardened areas. Thus, a layer by layer processor is obtained. In top down approach of nanoparticle synthesis an imperfection in the surface structure of product is observed. But it posses some limitation in which physical and chemical properties is dependent on (Thakkar *et al.*, 2010).

Bottom Up Method

The chemically synthesised nanoparticles are determined under the Bottom up method which involves atoms and molecules. In this technology the assemblage of smaller entities to produce final particle were observed (Mukherjee *et al.*, 2001; Thakkar *et al.*, 2010; Mittal *et al.*, 2013). Nanoparticle synthesised by chemical methods has been reported to have less stable nature and produces toxic by product that are hazardous and potentially risky to human and environment.

A clean, non-toxic and ecofriendly nanoparticle can be fabricated using *Green Chemistry* by involving various organism including bacteria, fungi, algae, lower plants and higher plant intercellularly as well as extracellularly (Mann, 1996; Sastry *et al.*, 2004; Bhattacharya and Rajinder, 2005; Mohanpuria *et al.*, 2008).

Biofabrication of Nanoparticles

The biological technique used to synthesis nanoparticle is also known as green synthesis method. Both chemical and biological method when modulated together with modern developmental technologies can more effectively reduce hazardous waste. Thus it is one of the eligible method popularly termed as *Green Chemistry* (DeSimone, 2002; Raveendran *et al.*, 2003; Sharma *et al.*, 2009).

Nanoparticles can be synthesised by reduction in solution, thermal decomposition of the compound, microwave assisted synthesis, laser mediated synthesis and biological reduction methods. (Sastry *et al.*, 2003; Navaladian *et al.*, 2007; Sreeram and Nair, 2008; Zamiri *et al.*, 2011; Kulkarni *et al.*, 2011). Due to potential risk and toxicity in relation to human life the usage of chemically synthesized nanoparticles has been diminished. The place has been captured by biologically synthesized nanoparticles. The extracellular nanoparticles are synthesized using plant leaf extract or from the whole plant due to its easier down streaming and production methodology. The extraction of nanoparticles from physical and chemical method requires high pressure, energy, temperature and toxic chemicals whereas biological technique does not involves any such ingredients. The physical and chemical methods produces harmful, toxic and less effective nanoparticles whereas the biologically synthesized nanoparticles are just viceversa of the above mentioned techniques and thus biologically synthesizing procedure is proving its importance in human mankind welfare. The biologically synthesized nanoparticle can be manufactured from microbial interaction, lower plant interaction and higher plant interaction. In various prospect, microbe mediated nanoparticle that implies desired shape is a good alternative to the chemically synthesized nanoparticles. Whereas, microbe mediated synthesizing technique is very effective as it reduces environmental toxicity. Therefore it is considered to be novel protocol as truly green chemistry which provides advancement between physical and chemical protocol. Biosynthesis of nanoparticles employed as simple and viable alternative to chemical procedure and physical methods (Casida and Quistad, 2005, Lloyd *et al.*, 2011).

It has been proved that biological methods provide a very easy, safe and reliable green route for producing instant and definite shape and size of required interest. Synthesis of nanoparticle can be done by intracellular or extracellular method such as leaf broth, sundried leaves, fruit, seed and bark. Biological synthesised several metallic nanoparticles are more eco-friendly and allows a controlled synthesis with well defined size and shape.

Some other reasons for synthesizing nanoparticles are the derived useful and medicinal character of the nanoparticles. The nanoparticle synthesized posses antitumour activity, hypotensive property, cardioprotective property, quick wound healing property, easy eye disease curing property (Prasad and Elumalai, 2011).

Some secondary metabolites such as terpenoids, alkaloids, flavonoids, ketones, aldehydes, amides, carboxylic acid, tannins, flavones, benzoquinone, catechol, protocatechaldehyde, protocatechuic acid, emodin, cyperquinone, dietchequinone and reminirin are synthesised in bulk quantity which assist majorly for photochemistry. Characters of nanoparticle are distributed diversely among the plants depending on the living habitat of the plants such as mesophytic, xerophytic and hydrophytic nature.

According to Liu *et al.*, 2010 the biological fabrication is also considered ads Bottom-up fabrication and analysed to synthesis labile components and structure which follows its own mechanism and can manage its own destruction by programmed cell death, it proves the proper developmental programme that it adapt from the environment. The biological fabrication technique uses genetic information to encode amino acid sequences which bears protein folding informations. It is carried over in the complex surroundings that controls the chemistry of components that are involved through molecular recognition both related to enzymatic kinetics generated from covalent bonding and non enzymatic kinetics generated from non-covalent bondings.

The Biofabrications of Nanoparticles using Higher Plants

The characterization of nanoparticles based on the biological synthesis is a nascent topic to deal with because of its variable usage in clinical diagnostics and therapeutic prospect (Salata *et al.*, 2004; Wagner *et al.*, 2006; Fortina *et al.*, 2007; Youns *et al.*, 2011; Larguinho and Baptista, 2012; Seil and Webster, 2012; Mittal *et al.*, 2013). Formation of nanoparticles can be procured by physical, chemical as well as biological methods, but the later one is mostly preferred due to its cost effectiveness and invasive applications involved in medical fields. Nanoparticle generation was observed from corresponding metal salts (Mukherjee *et al.*, 2001; Shankar *et al.*, 2003; Gericke and Pinches, 2006; Parsons *et al.*, 2007; Mohanpuria *et al.*, 2008; Thakkar *et al.*, 2010; Dhillon *et al.*, 2012; Gan and Li, 2012; Mittal *et al.*, 2013). Preference of biological method is also taken into consideration because when compared to other methodologies (physical and chemical) it tends to synthesis large amount of nanoparticles that are of well defined size and morphology and also found to be free of contaminations (Raveendran *et al.*, 2003; Shankar *et al.*, 2004; Anastas and Zimmerman, 2007; Mittal *et al.*, 2013).

Nanoparticles synthesised from plant extract are easy to be scaled and are comparatively cost effective as compared to microbial and whole plant involvement of nanoparticle synthesis (Armendariz *et al.*, 2004; Marshall *et al.*, 2007; Kumar and Yadav, 2009; Beattie and Haverkamp, 2011; Dhillon *et al.*, 2012; Mittal *et al.*, 2013). The plant extract used for nanoparticle synthesis act as reducing as well as stabilizing agent because of presence of different extract with different concentration and combination of reducing substrate. Reducing substrate involves aqueous extract that contains aqueous solution of corresponding metal salt (Kumar and Yadav, 2009; Mukunthan and Balaji, 2012; Mittal *et al.*, 2013). Nanoparticle synthesised from plants are considered because they are eco-friendly in nature, widely distributed, easily available, safe to handle and are supplied with higher metabolite quantity.

Plants have a tendency to accumulate 1000 times higher concentration of metal ion from outer environment and utilize it for their growth and development. Thus plants can be concisely called as hyperaccumulators. This evidence was proved using alfalfa sprouts when it was grown under gold chloride. The nanoparticle synthesized during this condition was 2-20 nm in diamenter. The same was experimented in case of silver accumulation were it was observed that plant root acquires silver from the nutrient media. The silver atoms accumulated on the cell arrange themselves in a pattern to produce nanoparticle [Kumar *et al.*, 1995; Arya *et al.*, 2010].

When we talk about the general mechanism of nanoparticle synthesis both organic and inorganic biological ingredients are used for biofabrication technique, out of which generally lipid are commonly preffered because of small molecules self assembling property that compartmentalize and organizes itself around cell organelles and vesicles through electron transport protein in respiratory chains. But recently self assembling property was observed in case of amino acid and short peptides that becomes one of the major component for biofabrications (Yemini *et al.*, 2005; Zhao *et al.*, 2006; Dickerson *et al.*, 2008; Williams *et al.*, 2009; Liu *et al.*, 2010). Collagen (protein) and cellulose (polysaccharide) are also some commonly used material in biological fabrication. Consideration of nucleic acid is also preffered because it serve as an information storage that transfers the function in biological systems and also its base-pairing provides good link of assemblage (Zheng *et al.*, 2009; Macfarlane *et al.*, 2009; Liu *et al.*, 2010). Collagen forms an organized chain of triple helices, fibrils and fibre that stabilizes the enzyme catalyzed by covalent bonding. Thus by this method a particle of definite sizes and shape and surface chemistry can be obtained with respect to nanoparticles (Rodriguez-Cabello *et al.*, 2009; Liu *et al.*, 2010).

Six higher plant species *Raphanus sativus, Brassica napus, Lolium multiforum, Lactuca sativa, Zea mays* and *Cucumis sativus* were observed by Lin and Xing (2007) on the basis of root growth with respect to multiwalled nanoparticles synthesis. It was observed that the synthesised nanoparticles inhibited the root growth. Later on the effect of nanoparticle on the cell organization was observed. It was observed that ZnO nanoparticle was responsible for upward translocation in *Lolium perenne*. Similar diminishing observation was found in case of ryegrass were shranked root tip, vacuolated root epidermal and collapsed cortical cells were reported. This physical change was observed because the ZnO particle were found adhered on the root surface and individual nanoparticle were observed in apoplast and protoplast of the root epidermal layer and on the stele portion of the root. But phytotoxicity of ZnO nanoparticles was not evidenced in the mentioned case. (Liu *et al.*, 2010).

Nanoparticle of gold, silver, zinc, palladium, can be synthesised by a general mechanism by simply mixing the plant extract with that of the required metal ion. The metal ion is reduced with the help of bioreducing agents such as enzymes, proteins, flavonoids, terpenoids and other cofactors. Which is later on tend to nucleation of the metal ion which is further allowed to stand by for stabilization. While maintaining the production, quantity and other characters of the synthesising of nanoparticles a special reference was observed with that of pH, temperature and reaction time which affects the reaction rate. Larger sized nanoparticle can be obtained at lower pH during the reaction rate. It was observed that reaction rate for complete reduction of the Au nanoparticle extracted from *Piper betel* vine took place within 120 hours, whose size ranged from 50-500 nm. This variation was observed because of change in temperature. The higher range of temperature resulted into higher yields of nanoparticles (Dwivedi and Gopal, 2010; Sneha *et al.*, 2010; Reddy *et al.*, 2010; Li *et al.*, 2011; Lloyd *et al.*, 2011; Mittal *et al.*, 2013).

Zinc nanoparticle was found to be adhered on the surface of the root and also some of the nanoparticle was found dispersed on the apoplast and protoplast of the root surface but presence of ZnO nanoparticle was found responsible for slower translocation from root to shoot (Lin and Xing, 2008; Monica and Cremonini, 2009). Titanium oxide nanoparticle when induced in the plant was responsible for increased Hill reaction and also increased chloroplast activity in *Spinacia oleracea*. Titanium oxide nanoparticle was also found responsible for oxygen evolution. Non-cyclic photophosphorylation was reported to be higher as compared with cyclic photophosphorylation. Probably titanium oxide enters into the chloroplast and oxidation reduction reaction initiate the electron transport chain that result into oxygen evolution (Hong *et al.*, 2005; Monica and Cremonini, 2009).

During experimentation on *Zea mays* water based ferrofluid was supplemented as the iron source which helped in the formation of iron based nanoparticles. Small concentration of iron based ferrofluid act as growth stimulator of new plantlet whereas higher concentration of ferrofluid act as growth inhibitor (Racuciu and Creanga, 2007; Monica and Cremonini, 2009).

Lower concentration of palladium nanoparticle synthesized from the *Hordeum vulgare* was responsible to cause stress effect on the leaves (Battke *et al.*, 2008; Monica and Cremonini, 2009). Lee *et al.*, 2008 reported the toxic effect of copper nanoparticle on *Phaseolus radiatus* and *Triticum aestivum*. In both the plants the growth rate and seedling length reduced due to exposure of nanoparticles. *Triticum aestivum* showed greater accumulation of Cu nanoparticle (Monica and Cremonini, 2009).

Biofabrication of Silver Nanoparticles using Higher Plants

Many researchers performed many experiments to synthesis nanoparticles using higher plants and succeeded in synthesising silver and gold nanoparticles variedly. *Medicago sativa* and *Brassica juncea* are among the higher plants that accumulate silver at higher concentration therefore categorised under hyperaccumulator (McGrath and Zhao, 2003; Monica and Cremonini, 2009). Richardson *et al.*, 2006 reported carbohydrate and protein serving as reducing agent against silver ion. Nanoparticle obtained from *Cinnamomum camphora* when observed under FTIR (Fourier transform Infra-Red spectroscopy) it was observed that various functional groups such as –C-O-C, -C-O-, C=C and –C=O- derived from several heterocyclic compound that act as reducing and capping agent for silver nanoparticles (Huang *et al.*, 2007; Durán *et al.*, 2011). When leaf extract of *Pelargonium graveolens* was used to synthesis silver nanoparticles by Shankar *et al.* (2003) it was observed that stable sized nanoparticles of 16-40nm was formed in the experiment. When Geraniol a monoterpene alcohol was treated with an uniform sized silver nanoparticle with 1- 10 nm range was observed. In both the cases of synthesized nanoparticle it was observed that growth of *Fibrosarcoma wehi* was inhibited. The silver nanoparticle gets entrapped in the protein surface that helps in reduction of silver ion and formation of much stable silver nanoparticle. This was observed in plant extract of *Capsicum annum* through electrostatic interaction (Li *et al.*, 2007; Durán *et al.*, 2011; Mittal *et al.*, 2013). Similar experiment was performed by Jacob *et al.* (2011) in plant *Piper longum* whose outcome was reported as formation of 18–41 nm size of silver nanoparticle. Formation of silver flakes was observed by Kaviya *et al.*, 2011 from the leaf extract of *Crossandra infundibuliformis*. *Desmodium trifolium* was found to reduce silver ion nanoparticle in presence of H^+ ions, NAD^+ and ascorbic acid estimated from the plant extract (Ahmad *et al.*, 2010; Mittal *et al.*, 2013). Kesharwani *et al.* (2009) synthesized highly stable nanoparticle (16-40 nm) from leaves of *Datura metel*. The leaf extract was reported with alkaloids, protein, enzymes, aminoacid, alcoholic compounds and polysaccharides which are the main factor responsible for silver ion reduction whereas quinol and chlorophyll pigments were found responsible for stabilization of nanoparticle (Mittal *et al.*, 2013). When *Melia azadarach* were subjected for synthesis of nanoparticles was found suitable against the HeLa cervical cancer cell. Silver nanoparticle synthesized from *Eclipta prostrate* was found responsible to control the vectors of filariasis and malaria due to its larvicidal property. From *Eucalyptus hybrida* silver nanoparticle was synthesized using methanolic extract of the plant (Jha *et al.*, 2009; Rajkumar and Abdul Rahuman, 2011; Mittal *et al.*, 2013). Silver nanoparticles particles are comparitevely less cytotoxic as compared to silver ions, but their report analyzed on onion plant suggest that they are more genotoxic. From aromatic spath of the male inflorescence from *Pandanus odorifer* Forssk silver nanoparticle was synthesized using broth (Panda *et al.*, 2011; Mittal *et al.*, 2013).

Mallikarjun *et al.* (2011) reported that 4-30 nm size of silver nanoparticle was synthesized from leaves of *Ocimum sanctum* in few minutes, these synthesized nanoparticle were found to have higher amount of ascorbic acid contained in the extract. The nanoparticles emerged from the leaf extract bears high antimicrobial activity against microorganism such as *E.coli* (Gram negative) and *Streptococcus aureus* (Gram positive) (Mallikarjun *et al.*, 2011; Singhal *et al.*, 2011; Mittal *et al.*, 2013).

Leaf extract used for nanoparticle synthesis duely contains higher amount of antioxidant, antimicrobial property because of adsorptive capabilty. This has been reported from silver nanoparticle synthesised from *Syzygium cumini* (Banerjee *et al.*, 2011; Mittal *et al.*, 2013).

Some other plants mentioned in Table 3.2 were also taken in consideration for silver nanoparticles synthesis by various researchers.

Biofabrication of Gold Nanoparticles using Higher Plants

Process of obtaining gold nanoparticles is as same as that of silver nanoparticle. According to Shankar *et al.*, 2004, synthesis of gold anoparticle was produced due to reduction of aqueous chloroaurate by reducing sugars and ketones. Vast range of higher plants can be considered for synthesis of gold anoparticles. Au nanoparticle can emerge with variable morphology from plate like form to rods and triangular. Gold nanoparticles possess an unique character of selective oxidation of CO and water gas shift reaction. Gold nanoparticle also posses photonics, optics and biomedical applications (Lloyd *et al.*, 2011). When the gold particles are accumulated in the prescribed cell it can be used as a substrate for phytomining which is purposely environmental friendly and economic method for gold recovery [Gardea-Torresdey *et al.*, 2005; Arya *et al.*, 2010].

The leaf extract of *Cassia auriculata* produced triangular gold nanoparticle of 15-25 nm size within 10 min at room temperature. Experiment followed by *Szyygium aromaticum* produced gold nanoparticles of irregular shape whereas diluted phyllatin containing plant extract of *Phyllanthus amarus* was used to synthesise hexagonal triangular gold nanoparticle from chloauric acid ($HAuCl_4$) when the concentration of $HAuCl_4$ increased it leads to the formation of spherical nanoparticles. *Coriandrum sativum* was experimented by Narayan and Sakthivel, 2010 to produce variable size of gold nanoparticle ranging from 7 -58 nm and spherical, triangular and decahedral shape of nanoparticle similar experiment when performed by Edison and Sethuraman, 2012 using *Terminalia chebula* found 6-60 nm sized gold nanoparticle bearing antibacterial activity against *S. Aureus* (gram positive) and *E. Coli* (gram negative bacteria) (Kasthuri *et al.*, 2009 a; Raghunandan *et al.*, 2010; Kumar *et al.*, 2011; Mittal *et al.*, 2013). The gold particle synthesised from *Allium cepa* was 100 nm sizes which was internalized by MCF-7 breast cancer via endocytosis (Parida *et al.*, 2011; Mittal *et al.*, 2013).

Chrysanthemum sp. and tea beverages were used to synthesise gold nanopaticle bearing antioxidant property. Extract from *Cassia fistula* is hypoglycaemic, nanoparticle synthesised from these extract was experimented on rats for controlling diabetes mellitus (Liu *et al.*, 2012; Daisy and Saipriya; 2012; Mittal *et al.*, 2013). Gold

Table 3.2: List of some Plants that are Used for Biofabrication of Silver Nanoparticles

Plant Considered	Experimental Outcome	References
Eucalyptus hybrida	Reaction of methanolic biomass of leaf with aqueous solution of $AgNO_3$ at ambient temperature	Dubey et al., 2009
Jatropha curcas	Mainly consist of latex curcain and curcacyclin due to which stable silver nanoparticle was found.	Bar et al., 2009
Pelargonium graveolens	Silver nanoparticle was successfully found dispersed	Safaepour et al., 2009
Trianthema decandra	Consist of bactericidal, wound healing property and are potentially available for large scale production	Geethalakshmi et al., 2010
Allium cepa	Consist of antibacterial property	Saxena et al., 2010
Opuntia ficus indica	Presence of long chain alcoholic compound quercitin ($C_{15}H_{10}O_7$)	Gade et al., 2010
Acanthe phylum and Alhagi maurorum	Stable biofabrication act as reducing and stabilizing agent	Forough et al., 2010
Acalypha indica	Presence of antimicrobial activity	Krishnaraj et al., 2010
Enicostoma hysopifolium and Rauwolfia tetraphylla	Stable nanoparticle obtained	Veena et al., 2011
Rhizophora mucorata	Bears larvicidal activities against *Culex quinquefasciatus* and *Aedes aegypti*	Gnanadesigan et al., 2011
Desmodium trifolium	Formation of extracellular silver nanoparticle at room temperature	Ahamed et al., 2011
Moringa olifera	Formation of stable extracellular silver nanoparticle	Prasad and Elumalai, 2011
Nelumbo nucifera	Silver nanoparticle obtained by periodic sampling of aqueous component	Santhoskumar et al., 2011
Saururus chinensis	Biosynthesis of extracellular anisotropic silver nanoparticle	Nagjyothi et al., 2011
Euphorbia hirta	Highly toxic methanolic crude extract	Priyadarshini et al., 2012
Hydrilla verticillata	Protein capped stable silver nanoparticle synthesised	Sable et al., 2012
Ellattaria cardomomum	Phytofabrication of nanoparticle without involvement of any toxic chemicals	Gnanajobhita et al., 2012
Juglans regia	Ethanolic extract from plant produces economic and industrial prespective of nanoparticle	Korbekandi et al., 2012
Ocimum sanctum	Presence of microbial activity	Ramteke et al., 2013

Contd...

Table 3.2–*Contd...*

Plant Considered	Experimental Outcome	References
Ocimum sanctum	Non toxic chemical and natural capping agent used to synthesis nanoparticle	Singhal *et al.*, 2011
Garcinia mangostana	High bactericidal activity	Veerasamy *et al.*, 2011
Boswellia serrata	Presence of gum, size of produced nanoparticle are controllable	Kora *et al.*, 2012
Pisonia grandis	Inculcation of green chemistry including effective sonication. Extreamly new technique for drug delivery	Firdouse *et al.*, 2012
Chromolaena odorata	Extraction of specific phenolic compund	Geetha *et al.*, 2012
Eucalyptus chapmaniana	Reduction of aqueous silver ion when kept in bright sunlight.	Sulaiman *et al.*, 2013
Ixora coccinea	Colour changing pattern was observed after addition of silver nitrate	Karuppiah and Rajmohan, 2013
Punica granatum	Reduction of silver ion and stabilization of silver formed from precipitation of fruit broth	Gnanjobhitha *et al.*, 2012
Myrtus communis	Biologically reduced silver nanoparticle was obtained	Sulaiman *et al.*, 2013
Riccinus communis	Nanoparticle synthesised at room temperature	Mani *et al.*, 2013
Tridex procambens *Jatropha curcas* *Calotropis giganteam* *Citrus aurantium*	Formation of < 20 nm sized spherical nanoparticles	Rajshekhareddy *et al.*, 2010

Table 3.3: List of some Plants that are Used for Biofabrication of Gold Nanoparticles

Plant Considered	Experimental Outcome	References
Magnolia kobus and Diopyros kaki	Formation of stable gold nanoparticle	Song et al., 2009
Terminalia cattappa	Nanoparticle synthesised was used as medicine in cancer therapy	Ankamwar et al., 2010
Ficus benghalensis	Plant extract bears water soluble antioxidant polyphenol compound that is highly nucleophillic that contains aromatic ring with chelating groups. Also act as an antioxidant.	Tripathi et al., 2012
Adhatoda vasica	Controlled synthesis of nanoparticle of desired shape and size. Presence of Glutathione, a capping protein	Pandey et al., 2012
Azadirachta indica	Production of crude fermentation that is used as spent free reducing agent	Verma et al., 2009
Momordica charantia	Presence of triterpenes, proteins and steroids	Pandey et al., 2012
Solanum melongena Datura metel, Carica papaya	< 20 nm sized nanoparticle was synthesized	Rajsekharareddy et al., 2010
Pelargonium graveolens	Spherical flat sheets of gold nanoparticles obtained of variable size ranging from 21-70 nm	Gardea-Torresdey et al., 1999
Cymbopogon flexuosus	Triangular nanoparticle synthesised	Shankar et al., 2005
Tamarindus indica	Triangular nanoparticle synthesised	Ankamwar et al., 2005
Sesbania sp.	Formation of 6-20 nm spherical gold nanoparticle	Sharma et al., 2007
Triticum aestivum	Tetrahedral face centered twinned irregular shaped gold nanoparticle size ranging from 10-30 nm	Gardea-Torresdey et al., 1999; Armandariz et al., 2002

nanowires and nanorods were synthesised by joining of nanoparticles in room temperature depending on reaction rate along with pH maintainance (Castro *et al.,* 2011; Mittal *et al.,* 2013).

Biofabrication of Combined Nanoparticles using Higher Plants

Many researchers proved to synthesize combined bimetallic nanoparticles from a single plant. Nanoparticle containing bi metallic center can be synthesized by reduction of aqueous ions for example the Ag-Au core shell nanoparticle can be prepared by reduction of aqueous Ag and Au ion using leaf extract or plant extract which further give rise to stabilized nanoparticle (Shankar *et al.,* 2004; Mittal *et al.,* 2013). Shankar *et al.* (2003, 2004) synthesized gold and silver nanoparticles using leaf extract of *Pelargonium graveolens* and *Azadirachta indica.* Same protocol for synthesising gold and silver nanoparticle was performed using *Aloe vera* and *Camellia sinensis* by Chandran *et al.,* 2006 and Vilchis-Nestor *et al.,* 2008. From *Camellia sinensis* Caffeine and theophylline reduces the metal ion which helps in formation of nanoparticles (Mittal *et al.,* 2013). Some of the experimented plants used to synthesis both silver and gold nanoparticle by many researchers is mentioned below.

Table 3.4: List of Plants that were Used for Synthesis of Nanoparticles by Various Researchers

Plant Considered	Reference
Szygium aromaticum	Phillip, 2010
Hibiscus-rosa-sinensis	Phillip, 2010
Rosa rugosa	Dubey *et al.,* 2010a
Tanacatum vulgare	Dubey *et al.,* 2010b
Chenopodium album	Dwivedi and Gopal 2010
Murraya koengii	Phillip *et al.,* 2011
Chenopodium album	Dwivedi and Gopal, 2010
Cinnamomum camphora	Huang *et al.,* 2007
Emblica officinalis	Ankamwar *et al.,* 2005
Memecylon edule	Elavazhagan and Arunachalam, 2011
Mentha piperita	Ali *et al.,* 2011; Parashar *et al.,* 2009
Tancetum vulgare	Dubey *et al.,* 2010 b

Not only silver and gold nanoparticles are tend to be synthesized using higher plants, but other metallic ion such as iron, copper, zinc, titanium, palladium etc are also synthesizedusing the same protocol of synthesis. Kasthuri *et al.,* 2009b demonstrated the presence of apiin in the nanoparticles synthesized from the ions of gold and silver. It was due to reduction of the ions using secondary hydroxyl and carbonyl groups of apiin. The size and shape of the formed nanoparticle can be controlled by the controlling th e concentration of apiin (Mittal *et al.,* 2013).

The quasi-spherical shape having 10-30 nm size of the nanoparticle was obtained from *Chenopodium album* by Dwivedi and Gopal, 2011. The aqueous extract of sorghum bran was considered for production of iron and silver nanoparticles in room temperature by Njagi *et al.*, 2011. Bactericidal activity against Gram-negative and Gram-positive was synthesized from silver and copper nanoparticle was performed using latex of *Euphorbiaceae* (Valodkar *et al.*, 2011). Gold and palladium nanoparticle using single step mechanism was performed by Huang *et al.*, 2011 using aqueous extract of tannin synthesised from berybery plant at room temperature. When Au $^{3+}$ and Pd^{2+} ion were mixed together the subsequencial reduction in the tannin was observed. The gold nanoparticle was found responsible for initiating the Pd shell growth. Here are some other nanoparticles that have been synthesised using higher plants by many researchers:

Table 3.5: List of Plants Used for Synthesis of Nanoparticles (Other than gold and silver) by Various Researchers

Plant	Nanoparticle Synthesized	References
Aloe vera	Cu	Chandran *et al.*, 2006
Spinacia oleracea	TiO$_2$	Gao *et al.*, 2006
Byrsonima sericea	Fe	Da Silva *et al.*, 2006
Psidium guineeense	Fe	Da Silva *et al.*, 2006
Allium porrum	H$_2$O	Eichert *et al.*, 2008
Vicia faba	H$_2$O	Eichert *et al.*, 2008
Nicotiana tabaccum	Si	Torney *et al.*, 2007
Zea mays	Si	Torney *et al.*, 2007
Capsicum annum	Se	Li *et al.*, 2007
Lolium perenne	ZnO	Lin and Zing, 2008
Zea mays	ZnO	Lin and Zing, 2008
Hordeum vulgare	Pd	Battke *et al.*, 2008
Phaseolus radiatus	Cu	Lee *et al.*, 2008
Triticum aestivum	Cu	Lee *et al.*, 2008
Medicago sativa	Ag	Harris *et al.*, 2008
Brassica juncea	Ag	Harris *et al.*, 2008
Cinnamomum zeylanicum	Pd	Satishkumar *et al.*, 2009
Diopyros kaki	Pt	Ahmad *et al.*, 2010
Annona squamosa	Pd	Roopan *et al.*, 2011
Jatropha curcas	Pb	Joglekar *et al.*, 2011
Physalis alkekengi	ZnO	Qu *et al.*, 2011
Glycine max	Pd	Kumar *et al.*, 2012
Hedera helix	TiO$_2$ and ZnO	Burris *et al.*, 2012

Positive Impact of Nanoparticles

Green synthesis of nanoparticles is becoming a boon to mankind because of its multi task usage. Being ecofriendly and cost effective biological technique they are very much preferred by many scientist. They are also taken into consideration because of their intoxic behaviour as compared to physical and chemical technologies of nanoparticle synthesis. Bionanoparticle are variedly used as water purifier and as a substrate to synthesize biomineral from waste materials.

Bionanoparticles are used in special references to bionanocatalysts with special reference to platinum and gold. Palladium based nanoparticles are known for their hydrogenation, reduction in aqueous solvents and selective dehalogenation in aqueous solvents as well as from hindered materials (Creamer *et al.*, 2008; Deplanche *et al.*, 2009; Lloyd *et al.*, 2011).Silver and gold nanoparticles are broadly synthesised due to their antimicrobial activity against pathogens due to their high toxicity against microorganism and also for medical utilization and consumeric products (Huang *et al.*, 2007;Kim *et al.*, 2007; Duran *et al.*, 2007; Jain *et al.*, 2009; Rai *et al.*, 2009; Sathishkumar *et al.*, 2009; Krishnaraj *et al.*, 2010; Ravindra *et al.*, 2010; Singh *et al.*, 2010; Arulkumar and Sabesan, 2010; Ali *et al.*, 2011; Ghosh *et al.*, 2011; Ghosh *et al.*, 2012;Lara *et al.*, 2011; Prasad and Elumalai, 2011; Singhal *et al.*, 2011; Saxena *et al.*, 2012;Amarnath *et al.*, 2012; Prasad *et al.*, 2012; Kandasamy *et al.*, 2012; Patil *et al.*, 2012a,b; Seil and Webster, 2012; Mittal *et al.*, 2013).

Silver nanoparticle are synthesised because of their antilarvicidal property against filariasis and malarial vectors. The role of silver nanoparticle is also proved against plasmodial pathogen, cancerous cell and fungal growth. Zinc oxide is also preffered because of their anti-microbial property (Fortina *et al.*, 2007; Ravindra *et al.*, 2010; Sukirtha *et al.*, 2011; Rajkumar and Abdul rahuman, 2011; Santoshkumar *et al.*, 2011; Vivek *et al.*, 2011; Subramanian, 2012; Espitia *et al.*, 2012; Mittal *et al.*, 2013). Silver nanoparticle accumulation affects the membrane permeability of the microorganism and other cellular organalles. They also inactivates the protein by interfering with DNA replication (Jung *et al.*, 2008; Chaloupka *et al.*, 2010; Li *et al.*, 2010; Cui *et al.*, 2012; Mittal *et al.*, 2013).

Nanoparticles are also becoming outstanding contributors regarding crop production and agricultiural activity. They are also used to diminish the microbial growth in wastewater effluents (Duran *et al.*, 2007; Mittal *et al.*, 2013). The reference of positive impact of nanoparticle is incomplete without the mentioning of medicinal property which has been proved in biological field by various researchers (Salata *et al.*, 2004; Wagner *et al.*, 2006; Fortina *et al.*, 2007; Thanh and Green, 2010;Subbiah *et al.*, 2010; Seil and Webster, 2012; Mittal *et al.*, 2013).

Negative Impact of Nanoparticles

On behalf of positive application of nanoparticles they also bear negative impact on the environmanetal factor. Environmental prospect is damaged by presence of particulate matter. Due to which the prone species are reported to have reduced growth, flowering and friutification that ultimately cause elimination of the species from the environment. The particulate matter settles on the surface of leaves causing blockage of stomata and respiratory pores. Transpiration rate is altered, thermal imbalance and reduction in photosynthetic rate are some of the factor that are caused due to nanoparticles. The metal ions are transported to the cell causes internal damage. Penetration of metal ion in the mesophyll cell from lower leaf and accumulationof iron on lower trichome (Da Silva *et al.*, 2006; Eichert *et al.*, 2008; Lee *et al.*, 2008; Monica and Cremonini, 2009).

Water suspended nanoparticle or aqueous solutes were responsible to cause lateral heterogeneity of the stomatal foliar uptake pathway which proves the stomatal pathway is different from that of cuticular pathway.

Formation of depositing layer of nanoparticle was observed in *Cucurbita memo* by Gonzalez-Melendi *et al.*, 2008. The nanoparticles was found to be present in extracellular space as well as within the cell, these stored nanoparticles were further transported to the cell and hinders the growth of plant, this reduced growth was observed in case of copper nanoparticle in *Phaseolus radiatus* and *Triticum aestivum* by Lee *et al.*, 2008 (Monica and Cremonini, 2009).

Conclusion

Nanoparticle synthesised from fusion of atomic or molecular particles bears dimensions between 1 to 100 nm are proving to be one of the most demanding material because of their beneficial activity toward mankind. Nanoparticle synthesised from biological method are very useful as compared to nanoparticle synthesised from physical and chemical method. Biological method sare completely environment

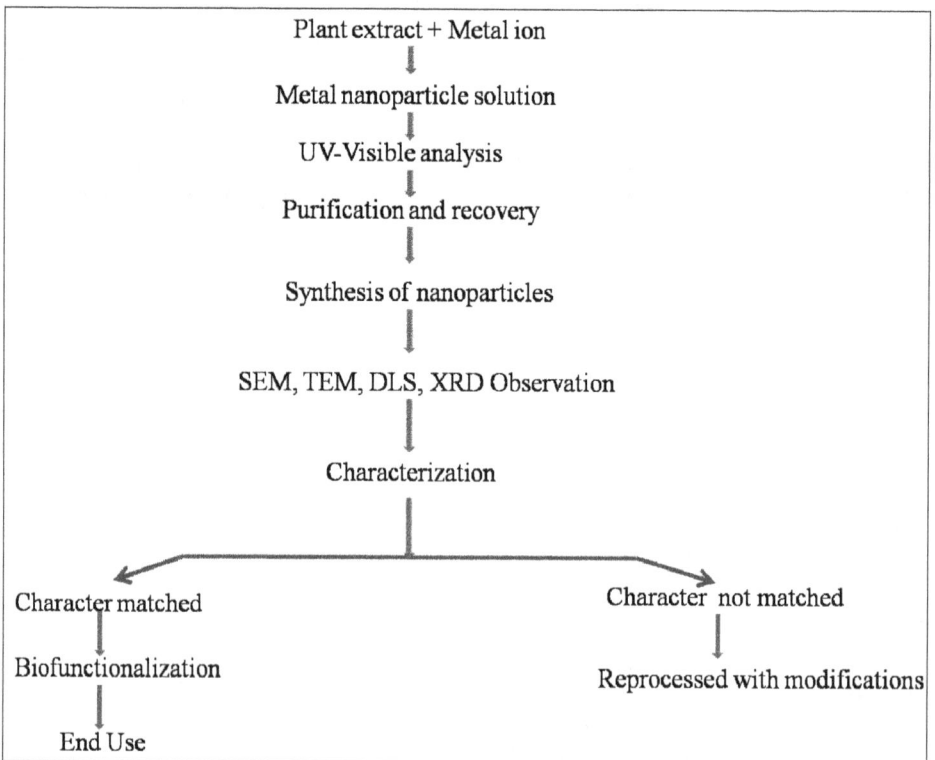

Figure 3.1: General Protocol Representing the Synthesis of Nanoparticle through Plant Extracts.

friendly and promotes Green synthesis. Nanoparticle can be synthesised using physical and chemical method are known to produces harmful, toxic and less effective nanoparticles whereas nanoparticle synthesised using biological method is preffered due to its cost effectiveness. Nanoparticles synthesised from biological method specially from higher plants uses the reactant that are easily available. Nanoparticle synthesised are one of the simple and viable alternative for physical and chemical method proving to be advanced criteria.

References

1. Ahamed M, Khan M, Siddiqui M, AlSalhi MS, Alrokayan SA (2011). Green synthesis, characterization and evaluation of biocompatibility of silver nanoparticles. *Physica E: Low Dimensional Systems and Nanostructure* 43: 1266–1271.

2. Ahmad N, Sharma S, Singh V, Shamsi S, Fatma A, Mehta B (2010). Biosynthesis of silver nanoparticles from *Desmodium triflorum*, a novel approach towards weed utilization, *Biotechnology Research International*.

3. Albrecht MA, Evans CW, Raston CL (2006). Green chemistry and the health implication of nanoparticles. *Green Chemistry* 8: 417.

4. Ali DM, Thajuddin N, Jaganathan K, Gunasekaran M (2011). Plant extract mediated synthesis of silver and gold nanoparticles and its antibacterial activity against clinically isolated pathogens. *Colloids and Surfaces B: Biointerfaces* 85: 360–365.

5. Amarnath K, Kumar J, Reddy T, Mahesh V, Ayyappan SR, Nellore J (2012). Synthesis and characterization of chitosan and grape polyphenols stabilized palladium nanoparticles and their antibacterial activity. *Colloids and Surfaces B: Biointerfaces* 94: 254–261.

6. Anastas PT, Zimmerman JB (2007). Green nanotechnology, Why we need a green nano award and how to make it happen. Washington DC, Woodrow Wilson International Center for Scholars.

7. Ankamwar B (2010). Biosynthesis of gold nanoparticles (green-gold). using leaf extract of *Terminalia catappa. European Journal of Chemistry* 7: 1334–1339.

8. Ankamwar B, Chaudhary M, Sastry M (2005). Gold nanotriangles biologically synthesised using Tamarind leaf extract and potential application in vapour sensing synthesis in reactivity. In: *Inorganic, Metal-Organic and Nanometal Chemistry* 35: 19-26.

9. Armanderiz V, Jose YM, Moller DA, Peralta-Videa RJ, Troiani H, Henera I (2002). Characterization of gold nanoparticle formation by *Avena sativa* biomass use of plant in biotechnology. *Journal of Nanoparticle Research* 6: 377-382.

10. Armendariz V, Herrera I, Peralta-Videa JR, Jose-Yacaman M, Troiani H, Santiago P, (2004). Size controlled gold nanoparticles formation by *Avena sativa* biomass, use of plants in nanobiotechnology. *Journal of Nanoparticle Research* 6: 377-382.

11. Arulkumar S, Sabesan M (2010). Biosynthesis and characterization of gold nanoparticle using antiparkinsonian drug *Mucuna pruriens* plant extract. *International Journal of Research in Pharmaceutical Science* 1: 417–420.

12. Arya V(2010). Living systems, Eco-Friendly Nanofactories. *Digest Journal of Nanomaterials and Biostructures* 5(1): 9–21.

13. Banerjee J, Narendhirakannan R (2011). Biosynthesis of silver nanoparticles from *Syzygium cumini* (L.). seed extract and evaluation of their *in vitro* antioxidant activities. *Digest Journal of Nanomaterial Biostructure* 6: 961–968.

14. Bar H, Bhui DK, Sahoo G, SarkarP, De SP,Misra A (2009). Green synthesis of silver nanoparticles using latex of *Jatropha curcas*. Colloids and Surface A: *Physico-chemical and Engeneering Aspects* 339: 134-139.

15. Battke F, Leopold K, Maier M, Schidhalter U, Schuster M (2008). Palladium exposure of Barley uptake and effects. *Plant Biology* 10: 272-276.

16. Beattie IR, Haverkamp RG (2011). Silver and gold nanoparticles in plants, sites for the reduction to metal. *Metallomics* 3: 628–32.

17. Bhattacharya D, Rajinder G (2005). Nanotechnology and potential of microorganisms. *Critical Review in Biotechnology* 25: 199– 204.

18. Brunner TI, Wick P, Manser P, Spohn P, Grass RN, Limbach LK, Bruinink A, Stark W (2006). *In vitro* cytotoxicity of oxide nanoparticles, comparison to asbestos, silica and effect of particle solubility. *Environmental Science and Technology* 40: 4378-4381.

19. Burris JN, Lenaghan SC, Zhang M, Stewart CN (2012). Nanoparticle biofabrication using English ivy (*Hedera helix*). *Journal of Nanobiotechnology* 10: 41.

20. Casida JE, Quistad GB (2005). Insecticide targets, learning to keep up with resistance and changing concepts of safety. *Agricultural Chemistry and Biotechnology* 43: 185-191.

21. Castro L, Blázquez ML, Muñoz JA, González F, García-Balboa C, Ballester A (2011). Biosynthesis of gold nanowires using sugar beet pulp. *Process Biochemistry* 46: 1076–82.

22. Chaloupka K, Malam Y, Seifalian AM (2010). Nanosilver as a new generation nanoproduct in biomedical applications. *Trends in Biotechnology* 28: 580–588.

23. Chandran SP, Chaudhary M, Pasricha R, Ahmad A, Sastry M (2006). Synthesis of gold nanotriangles and silver nanoparticles using *Aloe-vera* plant extract. *Biotechnology Progress* 22: 577–83.

24. Chandrasekharan N, Kamat PV (2000). Improving the photo-electrochemical performance of nanostructured TiO_2 films by adsorbtion of gold nanoparticles. *Journal of Physical Chemistry B* 10851-10857.

25. Cheng M-D (2004). Effects of nanophase materials(< or = 20 nm). on biological responses. *Journal of Environmental Science and Health Part A* 39: 2691–2705.

26. Creamer NJ, Deplanche K, Snape TJ, Mikheenko IP, Yong P, Samyahumbi D, Wood J, Pollmann K, Selenska-Pobell S, Macaskie LE (2008). A biogenic catalyst for hydrogenation, reduction and selective dehalogenation in non-aqueous solvents. *Hydrometallurgy* 94: 138-143.

27. Cui Y, Zhao Y, Tian Y, Zhang W, Lü X, Jiang X (2012). The molecular mechanism of action of bactericidal gold nanoparticles on *Escherichia coli*. *Biomaterials* 33: 2327–2333.

28. Da Silva LC, Oliva MA, Azevedo AA, De Araujo MJ (2006). Response of resting a plant species to pollution from an iron pelletization factory. *Water, Air and Soil Pollution* 175: 241-256.

29. Daisy P, Saipriya K (2012). Biochemical analysis of *Cassia fistula* aqueous extract and phytochemically synthesized gold nanoparticles as hypoglycemic treatment for diabetes mellitus. *International Journal of Nanomedicine* 7: 1189–1202.

30. Deplanche K, Snape TJ, Hazrati S, Harrad S, Macaskie LE (2009). Versatility of a new bioinorganic catalyst, palladized cells of *Desulfovibrio desulfuricans* and application to dehalogenation of flame retardant materials. *Environmental Technology* 30: 681-692.

31. DeSimone JM (2002). Practical approaches to green solvents. *Science* 297: 799-803.

32. Dhillon GS, Brar SK, Kaur S, Verma M (2012). Green approach for nanoparticle biosynthesis by fungi, current trends and applications. *Critical Review in Biotechnology* 32: 49–73.

33. Dickerson M B, Sandhage K H, Naik R R (2008). Proteinand peptide-directed syntheses of inorganic materials. *Chemical Reviews* 108: 4935.

34. Dubey M, Bhadauria S, Kushwah B (2009). Green synthesis of nanosilver particles from extract of *Eucalyptus hybrida* (safeda) leaf. *Digest Journal of Nanomaterial Biostructures* 4: 537–543.

35. Dubey SP, Lahtinen M, Sillanpaa M (2010a). Green synthesis and characterizations of silver and gold nanoparticles using leaf extract of *Rosa rugosa*. *Colloids Surf A* 364: 34–41.

36. Dubey SP, Lahtinen M, Sillanpää M (2010b). Tansy fruit mediated greener synthesis of silver and gold nanoparticles. *Process Biochemistry* 45: 1065–71.

37. Durán N, Marcato P, Durán M, Yadav A, Gade A, Rai M (2011). Mechanistic aspects in the biogenic synthesis of extracellular metal nanoparticles by peptides, bacteria, fungi and plants. *Applied Microbiology and Biotechnology* 90: 1609-1624.

38. Durán N, Marcato PD, De Souza GIH, Esposito E (2007). Antibacterial effect of silver nanoparticle produced by fungal process on textile febrics and their effluent treatment. *Journal of Biomedicine and Nanotechnology* 3: 203-208.

39. Dwivedi AD, Gopal K (2010). Biosynthesis of silver and gold nanoparticles using Chenopodium album leaf extract. *Colloids and Surface A: Physico-chemical and Engineer Aspects* 369: 27-33.

40. Edison T, Sethuraman M (2012). Instant green synthesis of silver nanoparticles using *Terminalia chebula* fruit extract and evaluation of their catalytic activity on reduction of Methylene Blue. *Process Biochemistry* 47: 1351–7.

41. Eichert T, Kurtz A, Steiner U, Goldbach HE (2008). Size exclusion limits and lateral heterogeneity of the stomatal foliar uptake pathway for aqueous solutes and water suspended nanoparticles. *Physiologia Plantarum* 134: 151-160.

42. Elavazhagan T, Arunachalam KD (2011). *Memecylon edule* leaf extract mediated green synthesis of silver and gold nanoparticles. *International Journal of Nanomedicine* 6: 1265–78.

43. Espitia PJP, Soares NFF, Coimbra JSRC, Andrade NJ, Cruz RS, Eber Antonio Alves Medeiros EAA (2012). Zinc Oxide Nanoparticles Synthesis. *Antimicrobial Activity and Food Packaging Applications Food Bioprocess Technology* 5: 1447–1464.

44. Firdhouse MJ, Lalitha P, Sripathi SK (2012). Novel synthesis of nanoparticle using leaf ethanol extract of *Pisonia grandis* (R,Br,). *Der Pharma Chemica* 4 (6): 2320-2326.

45. Forough M, Farhadi K (2010). Biological and green synthesis of Silver nanoparticles. *Turkish Journal of Engineering and Environmental Science* 34: 281-287.

46. Fortina P, Kricka LJ, Graves DJ, Park J, Hyslop T, Tam F (2007). Applications of nanoparticles to diagnostics and therapeutics in colorectal cancer. *Trends In Biotechnology* 25: 145–152.

47. Gade S, Gaikwad A, Tiwari V, Yadav A, Ingle A, Rai M (2010). Biofabrication of silver nanoparticles by *Opuntia ficus-indica*, *In vitro* antibacterial activity and study of Mechanism involved in synthesis. *Current Nanoscience* 6: 370-375.

48. Gan PP, Li SFY(2012). Potential of plant as a biological factory to synthesize gold and silver nanoparticles and their applications. *Reviews in Environmental Science and Biotechnology* 11: 169–206.

49. Gao F, Hong F, Liu C, Zheng L, Su M, Wu X, Yang F, Wu C, Yang P (2006). Mechanism of nano-anatase TiO_2 on promoting photosynthetic carbon reaction of spinach. *Biological Traces in Element Research* 111: 239-253.

50. Gardea-Torresdey JL, Rodriguez E, Parsons JG, Peralta-Videa JR, Meitzner G, Cruz-Jimenez GC (2005). Analytical and Bioanalytical Chemistry 382: 347-352.

51. Gardea-Torresdey JL, Tiemann KJ, Gomez E, Dokken K, Tehuacanero S, Yacaman MJ (1999). *Journal of Nanoparticle Research* 1: 397.

52. Geetha N, Harini K, Showmya JJ, Selva P (2012). Biofabrication of silver nanoparticle using leaf extract of *Chromolaena odorata* (L,). King and Robinson, International conference on Nuclear Energy. *Environmental and Biological Sciences* 56-59.

53. Geethalakshmi R, Sarada DV(2010). Synthesis of plant mediated silver nanoparticle using *Trianthema decandra* extract and evaluation of their

antimicrobial activities. *Int Journal of Engeneering Science and Technology* 2(5): 970-975.

54. Gericke M, Pinches A (2006). Biological synthesis of metal nanoparticles. *Hydrometallurgy* 83: 132-140.

55. Ghosh S, Patil S, Ahire M, Kitture R, Jabgunde A, Kale S (2011). Synthesis of gold nano-anisotrops using *Dioscorea bulbifera* tuber extract. *Journal of Nanomaterial* http.//dx.doi.org/10.1155/2011/354793. [Article ID 354793 2011].

56. Ghosh S, Patil S, Ahire M, Kitture R, Kale S, Pardesi K (2012). Synthesis of silver nanoparticles using *Dioscorea bulbifera* tuber extract and evaluation of its synergistic potential in combination with antimicrobial agents. *International Journal of Nanomedicine* 7: 483–96.

57. Gnanadesigan M, Anand M, Ravikumar S, Maruthupandy M, Vijayakumar V, Selvam S, Dhineshkumar M, Kumaraguru AK (2011). Biosynthesis of silver nanoparticles by using mangrove plant extract and their potential mosquito larvicidal property. *Asian Pacific Journal of Tropical Medicine* 799-803.

58. Gnanajobhitha GD, Rajesh KS, Annadurai G, Kannan C (2012). Green Synthesis of nanoparticles using *Ellataria cardomomum* and assessment of antimicrobial activity. *International Journal of Pharmaceutical Science and Research* 3 (3): 323-330.

59. Gonzales-Melendi P, Fernandez-Pacheco R, Coro-nado M, Corredor JE, Testillano PS, Risueno MC, Marquina C, Ibarra M R, Rubiales D, Perez-de-Luque A (2008). Nanoparticles as smart treatment-delivery systems in plants, assessment of different techniques of microscopy for their visualization in plant tissues. *Annals of Botany* 101: 187-195.

60. Harris AT, Bali R (2008). On the formation and extent of uptake of silver nanoparticles by live plants. *Journal of Nanoparticles Research* 10: 691-695.

61. Hong F, Zhou J, Liu C, Yang F, Wu C, Zheng L, Yang P (2005). Effect of Nano TiO_2 on photochemical on photochemical reaction of chloroplast of Spinach. *Biological Trace Animal Research* 104: 249-260.

62. Huang J, Chen C, He N, Hong J, Lu Y, Qingbiao L, Shao W, Sun D, Wang XH, Wang Y, Yiang X (2007). Biosynthesis of silver and gold nanoparticles by novel sundried *Cinnamomum camphora* leaf. *Nanotechnology* 18: 105-106.

63. Huang X, Wu H, Pu S, Zhang W, Liao X, Shi B (2011). One-step room-temperature synthesis of Au-Pd core–shell nanoparticles with tunable structure using plant tannin as reductant and stabilizer. *Green Chemistry* 13: 950–957.

64. Jacob S, Finub J, Narayanan A (2011). Synthesis of silver nanoparticles using *Piper longum* leaf extracts and its cytotoxic activity against Hep-2 cell line. *Colloids and Surfaces B: Biointerfaces* 91: 212–214.

65. Jain D, Daima HK, Kachhwaha S, Kothari S (2009). Synthesis of plant-mediated silver nanoparticles using papaya fruit extract and evaluation of their antimicrobial activities. *Digest Journal of Nanomaterial Biostructure* 4: 557–563.

66. Jha AK, Prasad K, Kumar V (2009). Biosynthesis of silver nanoparticles using Eclipta leaf. *Biotechnology Progress* 25: 1476–1479.

67. Joglekar S, Kodam K, Dhaygude M, Hudlikar M (2011). Novel route for rapid biosynthesis of lead nanoparticles using aqueous extract of *Jatropha curcas* L. Latex. *Material Letters* 65: 3170–3172.

68. Jung WK, Koo HC, Kim KW, Shin S, Kim SH, Park YH (2008). Antibacterial activity and mechanism of action of the silver ion in *Staphylococcus aureus* and *Escherichia coli. Applied Environmental and Microbiology* 74: 2171–2178.

69. Kandasamy K, Alikunhi NM, Manickaswami G, Nabikhan A, Ayyavu G (2012). Synthesis of silver nanoparticles by coastal plant *Prosopis chilensis* (L,). and their efficacy in controlling vibriosis in shrimp *Penaeus monodon*, *Applied Nanoscience* http.//dx.doi.org/ 10.1007/s13204-012-0064-1.

70. Karruppiah M, Rajmohan R (2013). Green synthesis of silver nanoparticle using *Ixora coccinea* leaves extract. *Material Letters* 97: 141-143.

71. Kasthuri J, Kathiravan K, Rajendiran N (2009a). Phyllanthin-assisted biosynthesis of silver and gold nanoparticles, a novel biological approach. *Journal of Nanoparticle Research* 11: 1075–1085.

72. Kasthuri J, Veerapandian S, Rajendiran N (2009b). Biological synthesis of silver and gold nanoparticles using apiin as reducing agent. *Colloids and Surfaces B: Biointerfaces* 68: 55–60.

73. Kaviya S, Santhanalakshmi J, Viswanathan B (2011). Biosynthesis of silver nano-flakes by *Crossandra infundibuliformis* leaf extract. *Material Letters* 67: 64–6.

74. Kesharwani J, Yoon KY, Hwang J, Rai M (2009). Phytofabrication of silver nanoparticles by leaf extract of *Datura metel*: hypothetical mechanism involved in synthesis. *Journal of Bionanoscience* 3: 39–44.

75. Kim JS, Kuk E, Yu KN, Kim JH, Park SJ, Lee HJ (2007). Antimicrobial effects of silver nanoparticles. *Nanomedicine: Nanotechnology, Biology and Medicine* 3: 95-101.

76. Kora AJ, Sashidhar RB, Arunachalam J (2012). Aqueous extract of Gum olibanum (*Boswellia serrata*). A reductant and stabilizer for biosynthesis of antibacterial silver nanoparticles. *Process Biochemistry of Anti Bacterial Silver Nanoparticle* 47: 1516-1520.

77. Korbekandi H, Asghari G, Jalayer SS, Jalayer MS, Bendagani M (2012). Nanosilver particle production using *Juglans regia* (Walnut). leaf extract. *Jundishapur Journal of Nantional Pharmaceutical Product.* 8(1): 20-26.

78. Krishnaraj C, Jagan E, Rajasekar S, Selvakumar P, Kalaichelvan P, Mohan N (2010). Synthesis of silver nanoparticles using *Acalypha indica* leaf extracts and its antibacterial activity against water borne pathogens. *Colloids and Surfaces B: Biointerfaces* 76: 50–56.

79. Kulkarni AP, Srivastava AA, Harpale PM, Zunjarro RS (2011). Plant mediated synthesis of silver nanoparticles-tapping the unexploited sources. *Journal of Natural Product and Plant Resources* 1(4): 100-107.

80. Kumar PBAN, Dushenkov V, Motto H, Raskin I (1995). Phytoextraction: the use of plants to remove heavy metals from soil. *Environmental Science and Technology* 29: 1232-1238.

81. Kumar PR, Vivekanandhan S, Misra M, Mohanty A, Satyanarayana N (2012). Soybean (*Glycine max*). leaf extract based green synthesis of palladium nanoparticles. *Journal of Biomaterial and Nanobiotechnology* 3: 14–19.

82. Kumar V, Gokavarapu S, Rajeswari A, Dhas T, Karthick V, Kapadia Z (2011). Facile green synthesis of gold nanoparticles using leaf extract of antidiabetic potent *Cassia auriculata*. *Colloids and Surfaces B: Biointerfaces* 87: 159–63.

83. Kumar V, Yadav SK (2009). Plantmediated synthesis of silver and gold nanoparticles and their applications. *Journal of Chemical Technology and Biotechnology* 84: 151–7.

84. Lara HH, Garza-Treviño EN, Ixtepan-Turrent L, Singh DK (2011). Silver nanoparticles are road-spectrum bactericidal and virucidal compounds. *Journal of Nanobiotechnology* 9: 30.

85. Larguinho M, Baptista PV (2012). Gold and silver nanoparticles for clinical diagnostics from genomics to proteomics. *Journal of Proteomics* 75: 2811–23.

86. Lee F, Chung J E,Kurisawa M (2008). An injectable enzymatically crosslinked hyaluronic acid-tyramine hydrogel system with independent tuning of mechanical strength and gelation rate. *Soft Matter* 4: 880.

87. Li S, Qui L, Shen Y, Xie A, Yu X, Zhang Q (2007). Green synthesis of silver nanoparticles using *Capsicum annum* L extract. *Green Chemistry* 9: 852-858.

88. Li WR, Xie XB, Shi QS, Zeng HY, Ou-Yang YS, Chen YB (2010). Antibacterial activity and mechanism of silver nanoparticles on *Escherichia coli*. *Applied Microbiology and Biotechnology* 85: 1115–1122.

89. Li X, Xu H, Chen ZS, Chen G (2011). Biosynthesis of nanoparticles by microorganisms and their applications. *Journal of Nanomaterial*. [Article 270974].

90. Lin D, Xing B (2007). Phytotoxicity of nanoparticle, inhibition of seed germination and root growth. *Environmental Pollution* 150: 243-250.

91. Lin D, Xing B (2008). Root uptake and Phytotoxicity of ZnO nanoparticles. *Environmental Science and Technology* 5580-5585.

92. Liu Q, Liu H, Yuan Z, Wei D, Ye Y (2012). Evaluation of antioxidant activity of chrysanthemum extracts and tea beverages by gold nanoparticles-based assay. *Colloids Surf B Biointerfaces* 92: 348–352.

93. Liu Y, Kim E, Ghodssi R, Rubloff GW, Culver JN, Bentley WE, Payne GF (2010). Biofabrication to build the biology-device interface. *Biofabrication* 2: 022002.

94. Lloyd JR, Byrne JM, Coker VS (2011). Biotechnological synthesis of functional nanomaterials. *Current Opinion in Biotechnology* 22: 509–515.

95. Macfarlane R J, Lee B, Hill H D, Senesi A J, Seifert S, Mirkin C A (2009). Assembly and organization processes in DNA-directed colloidal crystallization. *Proceedings of National Academy of Science*. USA 106: 10493.

96. Mallikarjun K, Narsimha G, Dillip G, Praveen B, Shreedhar B, Lakshmi S (2011). Green synthesis of silver nanoparticles using Ocimum leaf extract and their characterization. *Digest Journal of Nanomaterial Biostructure* 6: 181–186.

97. Mani U, Dhansingh S, Arunachalam R, Paul E, Shanmugam P, Rose C, Mandal AB (2013). A simple and green method for synthesis of silver anaoparticle using *Ricinus communis* leaf extract. *Progress in Nanotechnology and Nanomaterials* 2(1): 21-25.

98. Mann S (1996). *Biomimetic materials chemistry* (Ed). VCH Publishers, New York.

99. Marshall AT, Haverkamp RG, Davies CE, Parsons JG, Gardea-Torresdey JL, van Agterveld D (2007). Accumulation of gold nanoparticles in *Brassic juncea*. *International Journal of Phytoremediation* 9: 197–206.

100. Masciangioli T, Zhang WX (2003). Environmental technologies at the nanoscale. *Environmental Science and Technology* 37: 102a-108a.

101. Mazur M (2004). Electrochemically prepared silver nanoflakes and nanowires. *Electrochemistry Communications* 6: 400-403.

102. McGrath SP, Zhao FJ (2003). Phytoextraction of metals and metalloids from contaminated soils. *Current Opinion in Biotechnology* 14: 277-282.

103. Mittal AK, Chisti Y, Banerjee UC (2013). Synthesis of metallic nanoparticles using plant extract. *Biotechnology Advances* 31: 346-356.

104. Mohanpuria P, Rana NK, Yadav SK (2008). Biosynthesis of nanoparticles, technological concepts and future applications. *Journal of Nanoparticle Research* 10: 507–517.

105. Monica RC, Cremonini R (2009). Nanoparticles and Higher plants. *Caryologia* 62 (2): 161-165.

106. Mukherjee P, Ahmad A, Mandal D, Senapati S, Sainkar SR, Khan MI (2001). Fungus mediated synthesis of silver nanoparticles and their immobilization in the mycelia matrix, a novel biological approach to nanoparticle synthesis. *Nano Letters* 1: 515–519.

107. Mukunthan K, Balaji S (2012). Cashew apple juice (*Anacardium occidentale* L,). speeds up the synthesis of silver nanoparticles. *International Journal of Green Nanotechnology* 4: 71–79.

108. Nagajyothi P, Lee K (2011)Synthesis of plant-mediated silver nanoparticles using *Dioscorea batatas* rhizome extract and evaluation of their antimicrobial activities. *J. Nanomaterial*. [Article 573429].

109. Narayanan KB, Sakthivel N (2010). Phytosynthesis of gold nanoparticles using leaf extract of *Coleus amboinicus* Lour. *Material Characterization* 61: 1232–1238.

110. Navaladian S, Vishwanathan B, Vishwanath RP, Varadarajan TK (2007). Thermal decomposition as route for silver nanoparticles. *Nanoscale Research Letters* 2: 44-48.

111. Navarro E, Baun A,Behra R, Hartmann NB, Filser J, Miao AJ, Quigg A, Santschi PH, Sigg L (2008). Environmental behaviour and ecotoxicity of engineered nanoparticles to algae, plants and fungi. *Ecotoxicology* 17: 372-386.

112. Njagi EC, Huang H, Stafford L, Genuino H, Galindo HM, Collins JB (2011). Biosynthesis of iron and silver nanoparticles at room temperature using aqueous sorghum bran extracts. *Langmuir* 27: 264–271.

113. Nowack B, Bucheli TD (2007). Occurence, behaviour and effects of nanoparticles in the environment. *Environmental Pollution* 150: 5-22.

114. Panda KK, Achary VMM, Krishnaveni R, Padhi BK, Sarangi SN, Sahu SN (2011). *In vitro* biosynthesis and genotoxicity bioassay of silver nanoparticles using plants. *Toxicology in vitro* 25: 1097–1105.

115. Pandey S, Oza G, Kalita G, Sharon M (2012). *Adathoda vasica* - an Intelligent Fabricator of Gold Nanoparticles. *Pelagia Research Library* 468-474.

116. Parashar UK, Saxena P, Srivastava A (2009). Bioinspired synthesis of silver nanoparticles. *Digest Journal of Nanomaterial Biostructure* 4: 159–66.

117. Parsons J, Peralta-Videa J, Gardea-Torresdey J (2007). Use of plants in biotechnology, synthesis of metal nanoparticles by inactivated plant tissues, plant extracts, and living plants. *Dev Environment Science* 5: 463–485.

118. Patil R, KokateM, Kolekar S (2012b). Bioinspired synthesis of highly stabilized silver nanoparticles using *Ocimum tenuiflorum* leaf extract and their antibacterial activity. *Spectrochimica Acta Part A : Molecular and Biomolecular Spectroscopy* 91: 234–8.

119. Patil SV, Borase HP, Patil CD, Salunke BK (2012a). Biosynthesis of silver nanoparticles using latex from few euphorbian plants and their antimicrobial potential. *Applied Biochemical and Biotechnology* 167: 776–790.

120. Peto G, Molnar GL, Paszti Z, Geszti O, Beck A, Guczi L (2002). Electronic structure of gold nanoparticles deposited on SiOx/Si. *Material Science and Engeneering* C 19: 95–99.

121. Phillip D (2010). Green synthesis of gold and silver nanoparticle using *Hibiscus rosa sinensis*. *Physica E: Low Dimensional Systems and Nanostructure* 42: 1417-1424.

122. Phillip D, Unni C, Aromal Sa, Vidhu VK (2011). *Murraya koenigii* leaf assisted rapid green synthesis of silver nanoparticle and gold nanoparticles. *Spectrochimica Acta Part A: Molecular and Biomolecular Spectroscopy* 78: 899-904.

123. Prasad KS, Patel H, Patel T, Patel K, Selvaraj K (2012). Biosynthesis of Se Nanoparticles and its Effect on UV-Induced DNA Damage. *Colloids and Surfaces B, Biointerfaces* 103: 261–266.

124. Prasad TNVKV, Elumalai EK (2011). Biofabrication of Ag nanoparticles using *Moringa oleifera* leaf extract and their antimicrobial activity. *Asian Pacific Journal of Tropical Biomedicine* 1: 439-442.

125. Priyadrashini KA, Murugan K, Pannerselvam C, Ponnarulselvam S, Hwang JS, Nicolletti M (2012). Biolarvicidal and pupicidal potential of silver nanoparticle synthesised using *Euphorbia hirta* against *Anopheles stephensi* Liston (Diptera : Culicidae). *Parasito Res* 111: 997-1006.

126. Qu J, Yuan X, Wang X, Shao P (2011). Zinc accumulation and synthesis of ZnO nanoparticles using *Physalis alkekengi* L. *Environmental Pollution* 159: 1783–1788.

127. Racuciu M and Creanga DE (2007). TMA-OH coated magnetic nanoparticles internalized in vegetal tissues. *Romania Journal of Physics* 52: 395.

128. Raghunandan D, Bedre MD, Basavaraja S, Sawle B, Manjunath S, Venkataraman A (2010). Rapid biosynthesis of irregular shaped gold nanoparticles from macerated aqueous extracellular dried clove buds (*Syzygium aromaticum*). solution. *Colloids and Surfaces B: Biointerfaces79*: 235–40.

129. Rai M, Yadav A, Gade A (2009). Silver nanoparticles as a new generation of antimicrobials. *Biotechnology Advance* 27: 76–83.

130. Rajakumar G, Abdul Rahuman A (2011). Larvicidal activity of synthesized silver nanoparticles using *Eclipta prostrata* leaf extract against filariasis and malaria vectors. *Acta Tropica* 118: 196–203.

131. Rajasekharreddy P, Rani PU, Sreedhar B (2010). Qualitative assessment of silver and gold nanoparticle synthesis in various plants, a photobiological approach. *Journal of Nanoparticle Research* 12: 1711-1721.

132. Ramteke C, Chakrabarti T, Sarangi BK, Pandey RA (2013). Synthesis of silver nanoparticles from aqueous extract of leaves of *Ocimum sanctum* for enhanced antibacterial activity. *Gen of Chem*: 7.

133. Raveendran P, Fu J, Wallen SL (2003). Completely "green" synthesis and stabilization of metal nanoparticles. *Journal of American Chemical Society* 125: 13940–13941.

134. Ravindra S, Mohan YM, Reddy NN, Raju KM (2010). Fabrication of antibacterial cotton fibres loaded with silver nanoparticles via "green approach". *Colloids and Surface A: Physicochemical and Engineering* A 367: 31–40.

135. Reddy AS, Chen CY, Chen CC, Jean JS, Chen HR, Tseng MJ, Fan CW, Wang JC (2010). Biological synthesis of gold and silver nanoparticles mediated by the bacteria *Bacillus subtilis*. *Journal of Nanoscience and Nanotechnology* 10: 6567-6574.

136. Richardson A, Chan BC, Crouch RD, Janiec A, Chan BC, Crouch RD (2006). Synthesis of silver nanoparticles, an undergraduate laboratory using green approach. *Journal of Chemical Education* 11: 331-333.

137. Rodriguez-Cabello J C, Martin L, Alonso M, Arias F J, Testera A M (2009). 'Recombinamers' as advanced materials for the post-oil age. *Polymer* 50: 5159.

138. Roopan SM, Bharathi A, Kumar R, Khanna VG, Prabhakarn A (2011). Acaricidal, insecticidal, and larvicidal efficacy of aqueous extract of *Annona squamosa* L peel as biomaterial for the reduction of palladium salts into nanoparticles. *Colloids and Surfaces B: Biointerfaces* 92: 209–12.

139. Sable N, Gaikwad S, Bonde S, Gade A, Rai M (2012). Phytofabrication of silver nanoparticles using aquatic plant *Hydrilla verticillata*. *Nunastra Bioscience* 4 (2): 45-49.

140. Safaepour M, Shahverdi AR, Shahverdi HR, Khorramizadeh MR, Gohari AR (2009). Green synthesis of small silver nanoparticles using geraniol and its cytotoxicity against Fibrosarcoma–Wehi 164. *Avicenna Journal of Medical Biotechnology* 1: 111–115.

141. Salata O (2004). Applications of nanoparticles in biology and medicine. *Journal of Nanobiotechnology* : 2. http://dx.doi.org/10.1186/1477-3155-2-3.

142. Santhoshkumar T, Rahuman AA, Rajakumar G, Marimuthu S, Bagavan A, Jayaseelan C (2011). Synthesis of silver nanoparticles using *Nelumbo nucifera* leaf extract and its larvicidal activity against malaria and filariasis vectors. *Parasitol Res* 108: 693–702.

143. Sastry M, Ahmad A, Islam Khan M, Kumar R (2003). Biosynthesis of metal nanoparticles using fungi and actinomycete. *Current Sciences* 85: 162–70.

144. Sastry M, Ahmad A, Khan MI, Kumar R (2004). Microbial nanoparticle production *In*: Niemeyer CM, Mirkin CA (Eds). *Nanobiotechnology*. Wiley-VCH, Weinheim, Germany, 126–135.

145. Sathishkumar M, Sneha K, Won S, Cho CW, Kim S, Yun YS (2009). *Cinnamon zeylanicum* bark extract and powder mediated green synthesis of nano-crystalline silver particles and its bactericidal activity. *Colloids and Surfaces B: Biointerfaces* 73: 332–338.

146. Saxena A, Tripathi RM. Singh RP (2010). Biological Synthesis of silver nanoparticles by using Onion (*Allium cepa*). extract and their antibacterial activity. *Digest J. Nanomater. Biostruct* 5: 427-432.

147. Saxena A, Tripathi R, Zafar F, Singh P (2012). Green synthesis of silver nanoparticles using aqueous solution of *Ficus benghalensis* leaf extract and characterization of their antibacterial activity. *Material Letters* 67: 91–94.

148. Seil JT, Webster TJ (2012). Antimicrobial applications of nanotechnology, methods and literature. *International Journal of Nanomedicine* 7: 2767–2781.

149. Shankar SS, Ahmad A, Sastry M (2003). Geranium leaf assisted biosynthesis of silver nanoparticles. *Biotechnology Progress* 19: 1627-1631.

150. Shankar SS, Rai A, Ahmad A, Sastry M (2005). Controlling the optical properties of lemongrass extract synthesized gold nanotriangles and potential application in infrared-absorbing optical coatings. *Chemistry of Materials* 17: 566–572.

151. Shankar SS, Rai A, Ahmad A, Sastry M (2004). Rapid synthesis of Au, Ag, and bimetallic Au core–Ag shell nanoparticles using Neem (*Azadirachta indica*). leaf broth. *Journal of Colloid and Interface Science* 275: 496–502.

152. Sharma NC, Sahi SV, Nath S, Parson JG, Gardea-Torresdey JL, Pal T (2007). Synthesis of plant mediated gold nanoparticle and catalytic role of biometrics embedded nanoparticles. *Environmental Science and Technology* 41: 5137-5142.

153. Sharma VK, Yngard RA, Lin Y (2009). Silver nanoparticles, Green synthesis and their antimicrobial activities. *Advances in Colloid and Interface Science* 145: 83–96.

154. Singh A, Jain D, Upadhyay M, Khandelwal N, Verma H (2010). Green synthesis of silver nanoparticles using *Argemone mexicana* leaf extract and evaluation of their antimicrobial activities. *Digest Journal of Nanomaterials and Biostructures* 5: 483–489.

155. Singhal G, Bhavesh R, Kasariya K, Sharma AR, Singh RP (2011). Biosynthesis of silver nanoparticles using *Ocimum sanctum* (Tulsi). leaf extract and screening its antimicrobial activity. *Journal of Nanoparticle Research* 13: 2981–2988.

156. Sneha K, Sathishkumar M, Kim S, Yun YS (2010). Counter ions and temperature incorporated tailoring of biogenic gold nanoparticles. *Process Biochemistry* 45: 1450-1458.

157. Song JY, Jang HK, Kim BS (2009). Biological synthesis of gold nanoparticles using *Magnolia kobus* and *Diopyros kaki* leaf extracts. *Process Biochemistry* 44: 1133–1138.

158. Sreeram KJ, NIdin M, Nair BU (2008). Microwave assisted template synthesis of silver nanoparticles. *Bulletin of Material Sciences* 31(7): 937-942.

159. Subbiah R, VeerapandianM, Yun KS (2010). Nanoparticles, functionalization and multifunctional applications in biomedical sciences. *Current Medicinal Chemistry* 17: 4559–4577.

160. Subramanian V (2012). Green synthesis of silver nanoparticles using *Coleus amboinicus* lour, antioxitant activity and invitro cytotoxicity against Ehrlich's *Ascite carcinoma*. *Journal of Pharmaceutical Research* 5: 1268–72.

161. Sukirtha R, Priyanka KM, Antony JJ, Kamalakkannan S, Ramar T, Palani G (2011). Cytotoxic effect of green synthesized silver nanoparticles using *Melia azedarach* against *in vitro* HeLa cell lines and lymphoma mice model. *Process Biochemistry* 47: 273–279.

162. Sulaiman GM, Wahed HEA, Waheb AI, Mohammad AA (2013). Anti microbial activity of silver nanoparticle synthesised by *Myrtus communis* extract. *Journal of Engeneering and Technology* 31(3): 400-408.

163. Thakkar KN, Mhatre SS, Parikh RY (2010). Biological synthesis of metallic nanoparticles. *Nanomedicine: Nanotechnology, Biology and Medicine* 6: 257–262.

164. Thanh NTK, Green LAW (2010). Functionalisation of nanoparticles for biomedical applications. *Nano Today* 5: 213–230.

165. Torney F, Trewyn, B, G, Lin, Y, Wang, K (2007). Mesoporous silica nanoparticles deliver DNA and chemicals into plants. *Nature Nanotechnology* 2: 295-300.

166. Tripathi RM, Shrivastava A, Shrivastav BR (2012). Biofabrication of Gold nanoparticles using leaf extract of *Ficus benghalensis* and their characterization. *International Journal of Pharmaceutical and Biological Sciences* 3(3): 551-558.

167. Valodkar M, Nagar PS, Jadeja RN, Thounaojam MC, Devkar RV, Thakore S (2011). *Euphorbia ceaelatex* induced green synthesis of non-cytotoxic metallic nanoparticle solutions, a rational approach to antimicrobial applications. *Colloids Surf A* 384: 337–344.

168. Veena K, Nima P, Ganesan V (2011). Green synthesis of silver nanoparticles using two different higher plants. *Advanced Biotech* 1(2).

169. Veerasamy R, Xin TZ, Gunasagaran S, Xiang TFM, Yang EFC, Jayakumar N, Dhanaraj SA (2011). Biosynthesis of silver nanoparticle using Mangosteen leaf extract and evaluation of their anti microbial activities. *Journal of Saudi Chemical Society* 15(2): 113-120.

170. Verma VC, Gond SK, Mishra A, Kumar A, Kharwar RN, Gange AC (2009). Endophytic actinomycetes from *Azadirachta indica* A. Juss.: isolation, diversity, and antimicrobial activity. *Microbial Ecology* 57: 749–756.

171. Vilchis-Nestor AR, Sánchez-Mendieta V, Camacho-López MA, Gómez-Espinosa RM, Arenas-Alatorre JA (2008). Solvent less synthesis and optical properties of Au and Ag nanoparticles using *Camellia sinensis* extract. *Material Letters* 62: 3103–3105.

172. Vivek M, Kumar PS, Steffi S, Sudha S (2011). Biogenic silver nanoparticles by *Gelidiella acerosa* extract and their antifungal effects. *Avicenna Journal of Medical Biotechnology* 3: 143–8.

173. Wagner V, Dullaart A, Bock AK, Zweck A (2006). The emerging nanomedicine landscape. *Nature Biotechnology* 24: 1211–1218.

174. Williams R J, Smith A M, Collins R, Hodson N, Das A K, Ulijn R V(2009). Enzyme-assisted self-assembly under thermodynamic control. *Nature Nanotechnology* 4: 19.

175. Yemini M, Reches M, Rishpon J, Gazit E (2005). Novel electrochemical biosensing platform using self-assembled peptide nanotubes. *Nano Letters* 5: 183.

176. Youns M, Hoheisel JD, Efferth T (2011). Therapeutic and diagnostic applications of nanoparticles. *Current Drug Targets* 12: 357–365.

177. Yu-Nam Y, Lead R (2008). Manufactured nanoparticles, an overview of their chemistry, interactions and potential environmental implications. Science of the *Total Environment* 400: 396-414.

178. Zamiri R, Zakaria A, Abbastabar H, Darroudi M, Husin MS, Mahdi MA (2011). *International Journal of Nanomedicine* 6: 565-568.

179. Zhao XJ and Zhang S G (2006). Molecular designer self-assembling peptides. *Chemical Society of Review* 35: 1105.

180. Zheng JP, Birktoft JJ, Chen Y, Wang T, Sha RJ, Constantinou PE, Ginell SL, Mao CD and Seeman NC (2009). From molecular to macroscopic via the rational design of a self-assembled 3D DNA crystal. *Nature* 461: 74.

2016, Environmental Biotechnology: A New Approach
Editors: Dr. Rajan Kumar Gupta and Dr. Satya Shila Singh
Published by: DAYA PUBLISHING HOUSE, NEW DELHI

Pages 63–82

Chapter 4

Calcium Ion Regulation in Cyanobacteria

Savita Singh, Balkrishna Tiwari and A.K. Mishra*

Laboratory of Microbial Genetics,
Department of Botany, Banaras Hindu University,
Varanasi – 221 005, India

ABSTRACT

Calcium is a well known second messenger both in eukaryotes as well as prokaryotes owing to its favourable chemical nature such as binding strength, ionization potential and kinetic parameters. New roles played by calcium in biological systems are constantly been identified. Although a number of reviews are available pertaining to role of calcium in eukaryotes however, knowledge about prokaryotes particularly the cyanobacteria in this regard is scanty. Recently, a number of processes have been described in cyanobacteria which appear to be calcium-dependent viz. photosystem II, heterocyst differentiation and N_2-fixation, inorganic phosphate assimilation, trichome motility etc. The aim of this review is to identify those lines of thought and research that traced Ca^{2+} ion regulation, its transport in the cyanobacterial system beginning from influx, efflux, to its regulation through calmodulins and cross-talk among Ca^{2+} and other nutrients. This review strives to identify the roots of modern Ca^{2+} research and to chart its prospects in the near future.

Keywords: Calcium, Cyanobacteria, Regulation, Calmodulins, Cross talk.

Introduction

Calcium is a soft gray alkaline earth metal, and is the fifth-most-abundant element by mass in the earth's crust. Calcium is also the fifth-most-abundant dissolved ion in

---

* *Corresponding author.* E-mail: akmishraau@rediffmail.com; akmishraau@hotmail.com

seawater by both molarity and mass, after sodium, chloride, magnesium, and sulphate (Dickson and Goyet 1994). With due course of time, however, calcium emerged as a major control because calcium ions interact eagerly with biological molecules due to numerous specific properties (flexible coordination chemistry, high affinity for carboxylate oxygen, which is the most frequent motif in amino acids, rapid binding kinetics, etc) (Case *et al.*, 2007).

Calcium is essential for living organisms, in particular in cell physiology, where movement of the calcium ion, Ca^{2+} into and out of the cytoplasm functions as a signal for many cellular processes (Berridge, 1997). Cytosolic free calcium in the eukaryotic cells regulate processes pertaining to chromosome segregation, chemotaxis, fertilization, muscle contraction, cell cycle transition and many more (Campbell, 1983). As a major material used in mineralization of bone, teeth and shells, calcium is the most abundant metal by mass in many animals. The intracellular level of calcium ions is tightly controlled by means of specific pumps and channels occurring in plasma membrane and sub cellular organelles (Carafoli, 1987; Tsien and Tsein, 1990). Evidence for an equally important role in prokaryotes is accumulating day by day. There are a myriad of processes of which calcium is directly or indirectly involved starting from the most primitive life forms such as cyanobacteria to the most complex life forms of higher plants and animals. It acts as a universal second messenger in signal transduction pathway.

Cyanobacteria are among the most widely distributed micro-organisms in the biosphere known to survive a wide spectrum of environmental stresses (Tandeau de Marsac and Houmard 1993) and playing a dominant role in the global nitrogen and carbon cycles (Schwarz and Forchhammer 2005). Certain filamentous forms evolved mechanisms for adapting themselves both structurally and functionally, by differentiating 5-10 per cent of their vegetative cells into heterocysts (Wolk, 2000) which provide anaerobic site for the synthesis of nitrogen by nitrogenase enzyme (Zhang *et al.*, 2006). Calcium (Ca^{2+}) plays a pivotal role in the physiology and biochemistry of organisms ranging from prokaryots to eukaryots. Considering the importance of cellular processes in cyanobacteria mediated by calcium (Torrecilla *et al.*, 2004; Smith, 1988; Rodriguez *et al.*, 1990; Norris *et al.*, 1996, Legane´ s *et al.*, 2009) in the present review, recent developments related to calcium regulation particularly in cyanobacteria has been compiled as very few reports are available throwing light on this phenomenon. Calcium ion regulation in terms of influx, efflux and its regulation through Calcium binding proteins (CaM) have been discussed in detail. Further, a glimpse of cross talk among Ca^{2+} and other ions has also been presented.

Thus, this review is an eclectic mixture of events and processes involved in calcium ion regulation along with emphasis on the mechanism of transport. More precisely the role played by calcium ions and the effect of calcium on heterocyst development (if any) have been discussed and also tried to summarize the progress made during the recent past in the identification of several key signals involved in heterocyst development and to provide a basis for future research.

Calcium in Cyanobacteria

Cyanobacteria (blue green algae) group of oxygenic photosynthetic prokaryotes, possessing chlorophyll *a* (Whitton and Potts 2000; Castenholz 2001a) owe their existence on this earth approximately billions of years ago and are responsible for making an oxygen-rich atmosphere on earth about 2,300 million years ago (Blankenship 1992). These organisms are important in the biosphere, being among the main group of primary producers and the largest suppliers of reduced nitrogen.

Probably owing to their physiological flexibility and long evolutionary history, cyanobacteria are known to survive through a range of environmental stresses and consequently inhabiting a large variety of terrestrial and aquatic habitats from deserts to lakes as well as hot springs and glaciers (Mur *et al.*, 1999). To survive in extreme or variable environments, cyanobacteria have developed specific regulatory systems, in addition to more general mechanisms equivalent to those of other prokaryotes or photosynthesis eukaryotes (Tandeau de Marsac and Houmard, 1993).

Anabaena sp. strain PCC 7120, a filamentous cyanobacterium is capable of both nitrogen fixation and oxygenic photosynthesis. When deprived of combined nitrogen approximately every 10th cell along the filament differentiates into specialized cells called heterocyst (Fay *et al.*, 1968; Tyagi, 1975; Wolk, 2000; Mishra, 2003). Heterocyst provides anaerobic environment in which nitrogenase can function (Ernst *et al.*, 1992; Black *et al.*, 1993). Detailed coverage of heterocyst development and function can be found in several comprehensive reviews (Meeks and Elhai 2002; Wolk, 2000; Zhang *et al.*, 2006)

The role of calcium as a second messenger in the regulation of cellular events is well established in both plants and animals (Dieter and Marmé 1983). In contrast, there is little evidence in support of a similar diverse regulatory role for calcium in prokaryotic organisms. Recently, several processes have been described in cyanobacteria which appear to be calcium-dependent such as photosystem II (Becker and Brand 1985), N_2-fixation (Gallon and Hamadi, 1984), heterocyst differentiation (Smith *et al.*, 1987; Torrecilla *et al.*, 2004), inorganic phosphate assimilation (Kerson *et al.*, 1984) and trichome motility (Abeliovitch and Gan, 1982). The role of Ca^{2+} seems to be related to protection of nitrogenase from inactivation, by conferring heterocyst resistance to oxygen (Rodriguez *et al.*, 1990)

Involvement of Ca^{2+} in the regulation of cellular processes has been described essentially in terms of an influx or efflux of Ca^{2+} from the cytosol or methods based on the calcium-binding luminescent photoprotein aequorin (Knight *et al.*, 1991) which has allowed the monitoring of intracellular free-calcium concentration ($[Ca^{2+}]i$) in such a way that fluxes of free calcium in prokaryotic cells can be precisely portrayed.

Artificial means of altering the intracellular concentration of Ca^{2+}, including Ca^{2+} depletion and inhibitors and enhancers of Ca^{2+} uptake and binding have been shown to affect the various cell processes (Becker and Brand 1985; Gallon and Hamadi 1984; Smith *et al.*, 1987).

Role of Calcium Ion in Heterocyst Development

Under nitrogen starvation some filamentous cyanobacteria form specialised terminally differentiated structures called heterocysts having ability to reduce atmospheric dinitrogen to ammonia. Heterocyst development is influenced by numerous factors such as nitrogen deprivation, light quality, cell physiology and intercellular communication (Golden and Yoon, 1998). Smith *et al.*, 1987 presented evidence for calcium mediated regulation of heterocyst frequency and nitrogenase activity in *Nostoc* 6720. The proportion of heterocysts (heterocyst frequency) in cultures of *Nostoc* 6720 varied with the extracellular Ca^{2+} concentration present in the growth medium. Some workers have also speculated Ca^{2+} to be related to protection of nitrogenase from inactivation, by conferring heterocysts resistance to oxygen (Rodriguez *et al.*, 1990). Most recently, Torrecilla *et al.*, 2004 have reported that a specific calcium spiking is triggered exclusively when cells are deprived of combined nitrogen and generated by intracellular calcium stores. This group manipulated intracellular calcium signals by treatment with specific calcium drugs, and the assessed the effect of such manipulations on the process of heterocyst differentiation. Suppression, magnification or poor regulation of this signal prevented the process of heterocyst differentiation, thereby suggesting that a calcium signal with a defined set of kinetic parameters may be required for differentiation. NtcA, a global transcriptional regulator exerts global nitrogen control and is thought to activate the expression of genes involved in heterocyst differentiation in response to ammonium withdrawal via HetR. PatS seem to play a key role in heterocyst pattern formation by inhibiting the formation of adjacent heterocyst and maintaining a minimum number of vegetative cells between heterocysts (Golden and Yoon, 1998). The expression of hetC is controlled by NtcA (Muro-Pastor *et al.*, 1999). It has been observed that the hetC mutant of *Anabaena* 7120 was able to initiate heterocyst differentiation but stopped at an early stage of the process, and that the developing cells fail during a transition step to reach a non-dividing state (Khudyakov and Wolk 1997; Xu and Wolk 2001). So, although heterocyst differentiation have been reviewed by a number of workers (Wolk *et al.*, 1994; Wolk, 1996, 2000, Golden and Yoon, 1998, 2003) but an overall model of the regulatory networks controlling development remains elusive. Zhang *et al.*, 2006 have described a Ca^{2+} binding protein, CcbP, from *Anabaena* sp. PCC 7120 and shown that free calcium accumulates in differentiating cells and mature heterocyst can be correlated with changes in the level of the Ca^{2+}-binding protein. However, where and how free Ca^{2+} acts is still obscure, and any of the steps involved in the auto- and mutual regulation of hetR-ntcA could be potential targets of this regulation process (Zhang *et al.*, 2006) (Figure 4.1). So, the observations provide evidence that calcium ions regulate the development of heterocyst, the site of nitrogen fixation are fairly countable and considering the importance of cyanobacteria having a great ecological impact as primary producers, contributing substantially to the nitrogen economy (Singh,1961), it becomes inevitable to understand the physiology of cyanobacterial cells in context to calcium.

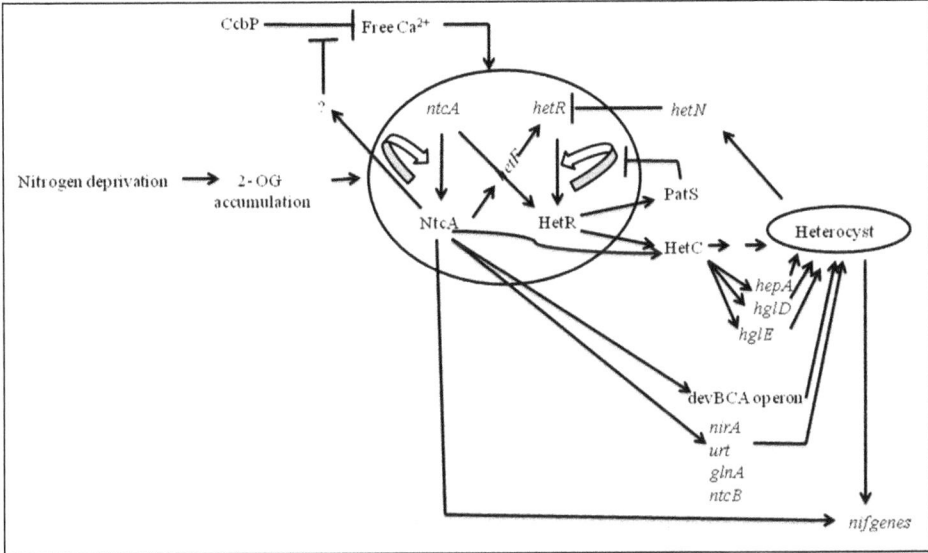

Figure 4.1: Modified Regulatory Circuit Possibly Involved in Heterocyst Differentiation (Zhang *et al.*, 2006).

Ion Channels in Prokaryotes

There are two distinctive features of ion channels that differentiate them from other types of ion transporter proteins:(1) The rate of ion transport through the channel is very high (often 10^6 ions per second or greater), (2) Ions pass through channels down their electrochemical gradient, which is a function of ion concentration and membrane potential, "downhill", without the input (or help) of metabolic energy (*e.g.* ATP, co-transport mechanisms, or active transport mechanisms).

Ion channels are located within the plasma membrane of nearly all cells and many intracellular organelles (Gouaux and MacKinnon 2005). They are often described as narrow, water-filled tunnels that allow only ions of a certain size and/ or charge to pass through. This characteristic is referred as selective permeability. The archetypal channel pore is just one or two atoms wide at its narrowest point and is selective for specific species of ion, such as sodium or potassium (Hurtley 2005). However, some channels may be permeable to the passage of more than one type of ion, typically sharing a common charge: positive (cations) or negative (anions). Ions often move through the segments of the channel pore in single file nearly as quickly as the ions move through free solution. In many ion channels, passage through the pore is governed by a "gate", which may be opened or closed in response to chemical or electrical signals, temperature, or mechanical force.

Most of the literatures give extensive coverage of various types of ion channels in eukaryotes, prokaryotes particularly cyanobacteria suffer negligence and have not been discussed in detail. Here, we present a rather clearer picture of ion channels in cyanobacteria with an emphasis on Calcium ion. In prokaryotes two types of channels are reported (i) Large-Conductance Mechanosensitive Channels of Prokaryotes (A)

Mechanosensitive channels of bacteria (B) Mechanosensitive channels of Archaea and (ii) Ion-Specific Channels of Prokaryotes that includes (A) K^+ channels and Ca^{2+} channels (B) Na^+ channels (NaChBac) (C) Ammonium channel (Amt, Mep) and (D) Cl "channels" (ClC-ec1) (Figure 4.2).

Large-Conductance Mechanosensitive Channels of Prokaryotes

MS channels have evolved specifically to serve as emergency valves for fast release of osmolytes in osmotically challenged bacterial cells. The opening of MS channels upon a hyposmotic shock enables rapid loss of cytoplasmic solutes and excess water, which helps to restore a normal cellular turgor. Mechanosensitive channels of bacteria (Martinac *et al.*, 1987, Betanzos *et al.*, 2002, Blount *et al.*, 1999) primarily serve to transport cellular osmoprotectants other than ions by acting as osmosensors that regulate the cellular turgor such as MscL (Chang *et al.*, 1998; Ajouz, *et al.*, 2000). MscS (Bass *et al.*, 2002), and MscK (Li *et al.*, 2002). Whereas Mechanosensitive channels of Archaea are gated by tension in the lipid bilayer and are activated by amphipaths (Dain *et al.*, 1998). They have large conductance and low selectivity for ions and are weakly voltage dependent. Number of MS channel have been reported in archaea namely MscMJ (Kloda and Martinac, 2001a), MscMJLR (Kloda and Martinac 2001b), and MscTA (Kloda *et al.*, 2001).

Ion-Specific Channels of Prokaryotes

K^+ Channels and Ca^{2+} Channels

K^+ along with Ca^{2+} are major cations in any cytoplasm and therefore K^+ and Ca^{2+} channels genes are found in nearly all the three superkingdom of life. Historically, studies of prokaryotic ion specific channels are largely the extensions of crystallographic studies and basically not related to microbial-physiological concern (Boris *et al.*, 2008). Most of the ion channel studies uptill root to neurophysiology where changes in Na^+ and K^+ permeability is considered. In this regard fine literature is almost lacking pertaining to regulation of these ions in cyanobacteria. It has been deduced that most cation specific channels K^+, Na^+ and Ca^{2+} and cation specific TRP channels have a similar structural motif (Boris *et al.*, 2008). A tetrameric assembly of subunits each with an inner transmembrane helix, a short pore helix, filter sequence and an outer transmembrane helix is apparently the core of all K^+ and Ca^{2+} channels besides Na^+, cyclic nucleotide gated (CNG) and TRP channels. The Ca^{2+} ATPase pump has to discriminate between Ca^{2+} and the other dominant cations in biology: Na^+, K^+, and Mg^{2+} (Toyoshima and Inesi 2004). Atomic structures in numerous conformations have revealed how this pump creates high-affinity Ca^{2+} binding sites in one conformation and then reorganizes the sites in another to change their Ca^{2+} affinity (Toyoshima *et al.*, 2000; Toyoshima and Nomura 2002). In cyanobacteria several K^+ channel cores trailed by sequences of CNBD (cyclic nucleotide binding domain) can be found (Kuo *et al.*, 2005). One important fact has also been realized that channels that open by Ca^{2+} seem to require much higher concentration of Ca^{2+} compared to eukaryotic channels.

Na⁺ Channels (NaChBac)

In prokaryotes, Na⁺ channels have been proposed to play a central role in Na⁺-dependent flagellar mobility in some prokaryotes (Koishi *et al.*, 2004). In bacteria, the Na⁺/H⁺ exchanger prevents cytotoxic Na⁺ accumulation and also supports pH homeostasis at elevated pH (Krulwich*et al.*, 2001, Sugiyama *et al.*, 1985, Booth *et al.*, 2003). In low [Na⁺]o environments or in the absence of solutes to support Na⁺ uptake through Na⁺-coupled solute transporters, the pH homeostasis function may rely on a Na⁺ channel (Krulwich *et al.*,2001, Sugiyama *et al.*, 1985, Booth *et al.*, 2003). The selectivity of bacterial sodium channel is converted from Na⁺ to Ca²⁺ by replacing an amino acid adjacent to glutamic acid in pore domain by aspartate (Gouaux and MacKinnon 2005). The animal depolarization-activated Na⁺ or Ca²⁺ channels are similar to K⁺ channels in structure. However, each Na⁺ or Ca²⁺ channel comprises 24 TM domains that are covalently linked, with each 6-TM repeat being comparable to a Shaker-type 6TM K-channel subunit (Boris *et al.*, 2008) (Figure 4.2).

Ammonium Channel (Amt, Mep)

Ammonium channels (Amt proteins) are transmembrane proteins that are ubiquitous in living organisms. These proteins are known as AMT proteins in higher plants and as MEP proteins in fungi, and the mammalian Rh proteins also belong to the same protein family. These proteins are involved in ammonia translocation across the biological membranes. However, whether they mediate transport of the ammonium

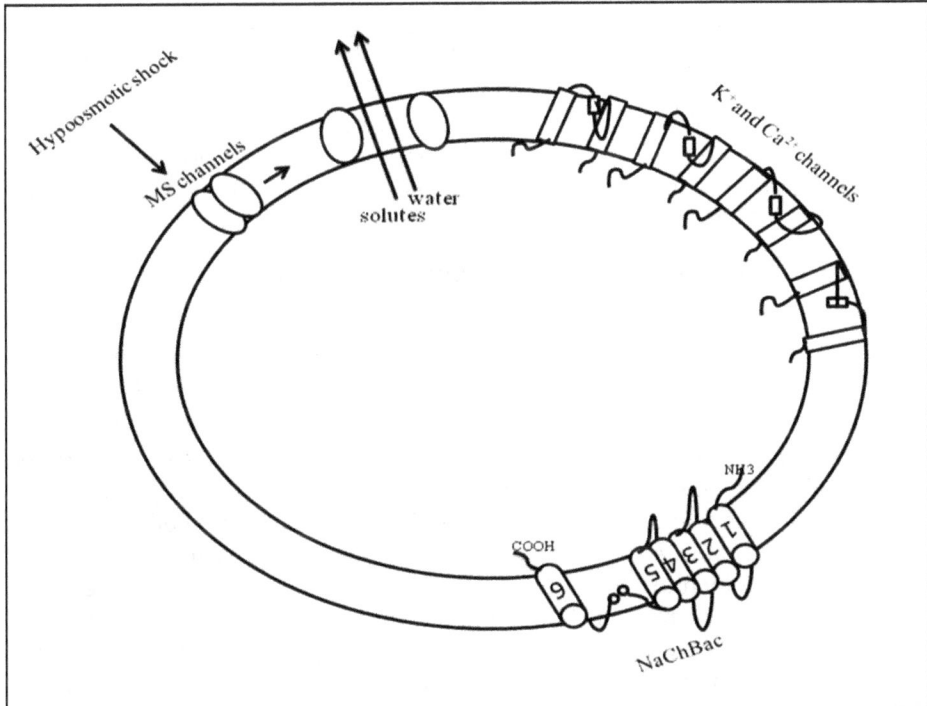

Figure 4.2: Schematic Representation of Ion Channels in Prokaryotes.

ion or translocation of the nonionic species ammonia (pK_a [NH_4^+/NH_3], 9.25) has been the subject of debate (Ninnemann *et al.*, 1994, Soupene *et al.*, 1998). It has recently been shown that plant AMT proteins mediate electrogenic transport of ammonium, whereas the Rh proteins appear to mediate a nonelectrogenic translocation of ammonia (Mayer at al., 2006). The crystal structures of the Amt proteins from *Escherichia coli* (Khademi *et al.*, 2004, Zheng *et al.*, 2004) and *Archaeoglobus fulgidus* (Andrade *et al.*, 2005) have been determined. Inspection of these structures indicates that these proteins have a channel through which only ammonia can pass; however, the proteins initially bind ammonium, which is deprotonated before translocation. Coordinated transport of a proton could take place in some systems but not in others (Ludewig, 2006).

Cl "Channels" (ClC-ec1)

The exact mechanism(s) of permeation through ClCs are still under infancy. They are also very different from all other channels discussed above except some physiochemical characteristics. Channels from *Salmonella typhimurium* and *E. coli* (ClC-ec1) have been crystallized and their structures solved at high resolution (Dutzler *et al.*, 2002).Crystal structures of two prokaryotic members of the ClC Cl⁻ channel/Cl⁻ H⁺ exchanger family provide a first view of the chemistry underlying Cl⁻ selectivity in transport proteins (Dutzler *et al.*, 2002; Dutzler *et al.*, 2003). These proteins blur the distinction between channels and pumps: Some members of the ClC family are channels (such as ClC⁻0 from the torpedo fish) (White and Miller 1981) whereas others are Cl⁻/H⁺ exchangers (such as ClC from *E.coli*) (Accardi and Miller 2004), and yet their related amino acid sequences indicate that they have essentially the same protein structure. Accardi *et al.*, 2004 reported these channels to be acting like exchanger rather than channel. Current literature now divides the animal ClCs into two classes those in plasma membrane (ClC-0, -1, etc.) acting channels and those in vesicular membranes(ClC-4, -5, etc.), being more homologous to bacterial ClCs, evidence support the view that they are employed in the extreme acid-resistance response.

Calcium Ion Transport and Regulation in Cyanobacteria

The exact time when calcium was used as regulatory ion firstly is not known however it is sure that it happened at a very early time. This could be exemplified by the fact that a calcium gradient exists in the most primitive prokaryote (cytosolic calcium at about 80-100 nM) (Gandola and Rosen1987, Watkins *et al.*, 1995).

Most of the studies pertaining to calcium ions have taken into consideration mainly the specialized eukaryote or prokaryote particularly *E.coli*, cyanobacterial prospective in this regard is at utmost negligience. In *E.coli* Ca^{2+} homeostasis has been studied in details. A Ca^{2+} concentration in bacterial cells is maintained at low concentrations which are brought about by three important factors: Selective permeability of cell membrane, high buffering capacity of cytosol, and presence of Ca^{2+} with-drawing system. Further, recent evidences have accumulated concerned with calcium regulation. These basically comprise of voltage gated channel for influx, primary and secondary transporters for efflux and calmodulin like proteins for mediating response to calcium. Changes in nucleoid structure, phosphorylation and proteolysis of target proteins are the methods through which calcium dependent regulation is brought about in prokaryots. Since for any study relating to transport there are basically two important aspects such as influx and the other is efflux, therefore we have discussed the calcium transport under these heads.

Calcium Influx

Considering calcium influx in bacteria it has been mainly by non proteinacious calcium channels which are a complex of Polyhydroxy butarate and calcium polyphosphate These have been supposed to be quite similar to VOCCs in terms of voltage dependence, selectivity for divalent ions and blockage by Lanthanum. In *Bacillus subtilis* Ca^{2+} uptake has been reported by L type Calcium channels.

Ca^{2+} dependent ATPase associated with AF1 resembaling CF1 particle of higher plant chloroplasts have been reported. AF1 isolated from thermophilic cyanobacterium *Mastiglcladus laminosus* and reconstructed into organism's thylakoid membranes (Wolff *et al.*, 1980). Similar latent Ca^{2+}-ATPase has been isolated from alkalinophile *Spirulina platensis*. A passive Ca^{2+} influx associated with Ca^{2+} channels has been observed with inverted vesicles of plasmalemma prepared from *A. variabilis* (Lockau and Pfeffer 1983) while *Streptococcus faecalis* happens to be the only prokaryote in which $Mg-ATP^{2-}$-dependent membrane transport of Ca^{2+} has been detected (Kobayashi *et al.*, 2006).

In cyanobacteria, Ca^{2+} influx associated with trichome motility suggest that cyanobacteria maintain a low intracellular concentration of Ca^{2+} and it is the calcium gradient across cell membrane which facilitate Ca^{2+} influx. Influx may have gated channels as it is blocked by known inhibitors of eukaryote Ca^{2+} channels like lanthanum and ruthenium red.

Calcium Efflux

In *E.coli*, the calcium export system consists of Primary exchangers (ATP hydrolysis) and secondary exchangers (utilizing electrochemical gradient of Na^+ and H^+). A number of secondary exchangers have been reported for example Cha A from bacteria. In *S. pneumonia* Na^+/calcium exchangers are known to be involved in Ca efflux and influx, lysis and competence. P- type ATPases are characteristic of eukaryotes however, in the recent period Ca^{2+} pump have been cloned from cyanobacterium *Synechococcus* sp. strain PCC 7942 which are P-type ATPase similar to eukaryotic P-type ATPases. Energy-dependent Ca^{2+} efflux and its regulation from the diazotrophic cyanobacterium *Nostoc calcicola* Bre'b has been investigated. Ca^{2+} followed hyperbolic efflux kinetics. Further strong inhibition in Ca^{2+} efflux by addition of metabolic inhibitors like carbonyl cyanide-p-nitrofluoromethoxyl-phenyl hydrazone(FCCP) and N, N, -dicyclohexylcarbo-diimide (DCCD) suggested the vital role of membrane potential and ATP hydrolysis in driving the efflux process (Pandey *et al.*, 1999).

Calcium Regulation through Calmodulins

In Prokaryots Calcium action can be met at three major levels first is calcium acting as major primary trigger, Secondly it acts as a modulator and lastly a structural element. Calmodulin, a prototypical EF-hand protein, is a small (15–22 kDa), acidic (pI 3.9–4.3) Ca^{2+}-binding protein that can interact with >25 distinct target proteins, thereby regulating the activity of many vital enzymes, including kinases, phosphatases, nitric oxide synthases, phosphodiesterases and ion channels (Crivici and Ikura 1995). Calmodulins essentially expressed in all eukaryotic cells where it participates in signalling pathways that regulate many crucial processes such as growth, proliferation and movement. These Calmodulins behave as Ca^{2+} sensors by being able to detect and respond to a biologically tolerant range of intracellular free Ca^{2+} concentration of 10^{-7} to 10^{-6} M. Additional discrimination for Ca^{2+} comes from the C-terminal pair of E-F hands which has 3 to 5 fold higher affinity for Ca^{2+} than the N-terminal pair of sites. Calmodulins constitute atleast 0.1 per cent of the total protein present on cells (10^{-6}- 10^{-5}M) and is expressed at even higher levels in rapidly growing cells undergoing cell division and differentiation. The local intracellular availability of CaM is likely to be significant because of various CaM-dependent effectors regulated over a wide range of free CaM concentration (10^{-12}M -10^{-6}M) (Chin and Means, 2000).

Ca^{2+} binding motifs in bacteria tend to be diverse in nature. Besides motifs, protein can also bind Ca^{2+} through oxygen atoms provided by several charged glutamate or aspartate residues (Michiels *et al.*, 2001). Yang 2001 proposed the idea that Calcium-binding protein originated in high G+C gram positive bacteria and were later acquired by eukaryotes.

Prokaryotic "Calmodulin-like" proteins present a complex picture in which no single protein is readily identified as a CaM. Fry *et al.*, 1986 purified Calmodulin like protein from *Bacillus subtilis*. Further, CAMLPs (Calmodulin like protein) have been demonstrated to be present in heat treated extracts of five strains of four species of the genus *Mycobacterium* by monitoring the stimulation of bovine phosphodiesterase (PDE) (Reddy *et al.*, 2003). In *E.coli*, calmodulin like proteins have been reported on the basis of change in stability and Ca binding as well as Ca dependent activation and inhibition of Phosphodiesterase enzyme (PDE) (Norris *et al.*, 1996). Calmodulin like proteins have been discovered in several cyanobacteria as well. Kerson *et al.*, 1984 reported CAMLP in cyanobacteria, he found that stimulation observed in cultures of *Oscillatoria limnetica* was inhibited by Fluphenazine, an anti-psychosis drug known to inhibit Ca^{2+}/CaM dependent processes in eukaryotes and also that an antigen to a commercially prepared anti-CaM antibody was preaent in the cell free extracts of *O. limnetica*.

Eldik and Wolchok 1984 detected a protein of about 17 kDa by using a polyclonal antibody raised against spinach calmodulin. An important aspect of the report was the localization of protein by electron microscopy to both vegetative and heterocyst of *Anabaena variabilis*. Using a combination of ion-exchange (DEAE-cellulose) and hydrophobic interaction (phenyl-sepharose) column chromatography, Bunchingi *et at.*, 1990 achieved a partial purification of putative cyanobacterial CaM from *Anabaena* spp. In all three polypeptides, one major (58 kDa) and two minor (40 and 16 kDa) all were heat stable and capable of activating PDE in the presence of Ca^{2+}.

Physiological as well as biochemical and immunological evidences have been reported for the occurrence of calmodulins in cyanobacteria. Kerson *et al.*, 1984 demonstrated that Calmodulin antagonists 4-(3-[2(trifluoromethyl) phenyl thiazin-10ylpropyl)-1-piperazine ethanol-HCl and its monofluoro-analog inhibited orthophosphate uptake of *O. limnetica* by over 97 per cent implying involvement of CaM in this process. CAMLP was quantified in cell free extract from *O.limnetica* by radioimmunoassay. Petterson and Bergman 1989 documented the occurrence, activity and localization of CaM in three heterocystous cyanobacteria of the genus *Anabaena*. Boiled crude extracts caused a Ca^{2+} dependent stimulation of NAD kinase. SDS-PAGE and Western Blot analyses using antiserum against eukaryotic spinach CaM, revealing a polypeptide of about 17 kDa. Immunogold localization of CaM gave a dense gold label both in vegetative cells and heterocysts. The label being mainly confined to the centroplasm of vegetative cells while it was evenly distributed in the cytoplasm of mature heterocysts.

Thus, the presence of proteins in different cyanobacterial species in the range of 16-18 kDa has been observed. Further work to resolve whether one or more calmodulin like proteins exist in cyanobacteria having same isoelectric point as the eukaryotes. More data such as amino acid sequence and *in vivo* physiological functions of the polypeptides are required before a definitive conclusion be made.

Calcium Homeostasis and Cross Talk between Ca^{2+} and Other Nutrients

As cell has to maintain a definite concentration of intracellular calcium inside it there has to be a fine mechanism that contributes to its realisation inside the cellular system. Controlled ion homeostasis is critical to all organisms. This is certainly the case for alkali metals (Na^+, K^+) and alkaline earth metals (Ca^{2+}, Mg^{2+}) that play essential roles in many organisms but that must be carefully regulated with regard to their cellular concentrations to prevent osmotic and metal stress. So, is the case with calcium ion where a definite level is responsible for cell signalling while a higher level may prove toxic to the cell survival. Most of the studies related to calcium homeostasis has been initiated in *E. coli*. Using Fura-2 and aequorin, $[Ca^{2+}]$ in cells of this microorganism was found to be maintained at a constant level around 90 nM (Gandola and Rosen 1987; Watkins *et al.*, 1995). The low Ca^{2+} concentration in the cytoplasm of bacterial cells is provided by three factors: selective permeability of the cell membrane regulated by fine mechanisms of control of Ca^{2+} entry into the cell, high buffering capacity of cytosol, and efficient Ca^{2+}-withdrawing system. Calcium concentration both intracellular and extracellular happen to affect the regimes of Na^+ ion, H^+ ions, K^+ ions, Mg^{2+}, phosphate ions as well as changes in pH (Figure 4.3). Recently studies relating to Ca^{2+} and phosphate induced modulation of nitrogen assimilatory enzymes under high stress have been investigated. Nazarenko *et al.*, 2003 have suggested the involvement of MscL an outward channel in regulation of Ca^{2+} homeostasis in *Synechocystis* cells under temperature-stress conditions. One of the recent studies by Singh and Mishra 2014 have demonstrated the interactive effect of calcium influx/ efflux rate, remaining Ca^{2+} and intracellular levels of Na^+ and K^+ when *Anabaena* sp. PCC 7120 cells subjected to varied levels of calcium chloride. A biphasic calcium uptake with definite K_s 0.5 and V_{max} values have been obtained for wild type and mutant strains of *Anabaena* sp. PCC 7120.

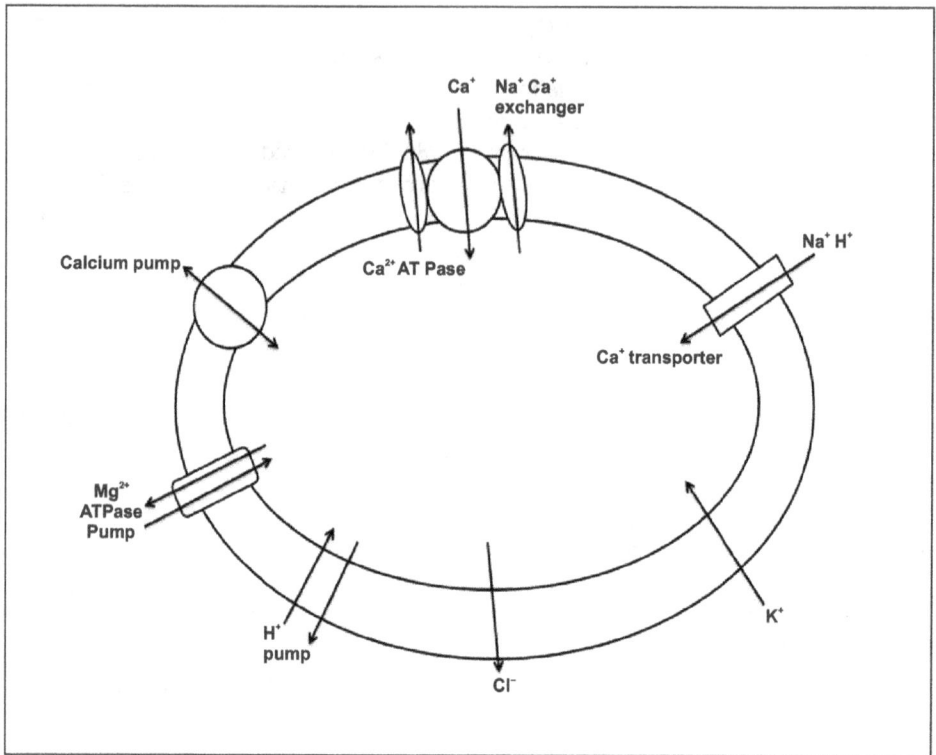

Figure 4.3: Diagrammatic Sketch of Calcium Ion Regulation in Cyanobacteria.

Role of Calcium Ions in Maintenance of Cell Structure, Trichome Motility and Photosynthesis

Besides role of calcium ion in nitrogen metabolism some more phenomenons affected by calcium such as cell structure maintenance, trichome motility and photosynthesis have been undertaken here to get a clearer picture of the range of processes being affected by Ca^{2+} levels. Giraldez-ruiz et al., 1997 studied the relationship between intracellular pH, growth characteristics and calcium in Anabaena sp. PCC 7120. Increasing concentrations of Ca allowed Anabaena sp. strain PCC 7120 to perform better at lower pH. Similarly, Carneiro et al., 2011 studied the inhibitory effect of calcium on Cylindrospermopsis raciborskii metabolism. He found that continued exposure to calcium limited growth, decreased trichome length, increased chl-a content. Role of Ca^{2+} in water oxidation became an issue only after investigations began focusing on inhibitory treatments of chloroplast-derived thylakoids (Barr et al., 1983) and of PS II membranes (Ghanotakis et al., 1984; Miyao and Murata 1984a) that could be reversed quite specifically by Ca^{2+} ions. Involvement of voltage-gated calcium channels was related with motility in cyanobacteria (Abeliovich and Gan, 1982).

Conclusions and Future Prospects

The importance of calcium as a cell regulator is well established in eukaryotes however, the role of calcium ion in prokaryotes is still elusive particularly in the cyanobacteria. It has been found that calcium ions are involved in the maintenance of cell structure, trichome motility, heterocyst formation, photosynthesis and nitrogen fixation in cyanobacteria. Therefore, in the present review a comprehensive outlook has been presented to support the cause. Carefully regulated Ca^{2+} partitioning is an essential requirement in all cells to prevent toxic accumulation of Ca^{2+} in the cytosol but also to allow Ca^{2+} to fulfill a role as a cellular signal.

Ca^{2+} influx channels, and Ca^{2+} efflux transporters comprising of Ca^{2+} ATPases and ion-coupled Ca^{2+} exchangers, along with changes in pH, allow cytosolic Ca^{2+} levels to be dynamically shaped. Further, Calmodulin, the calcium binding proteins have also been known to participate in regulating the regime of calcium ion concentration inside the cells of cyanobacteria. The growing evidence for calmodulin like protein similar to higher eukaryotes in cyanobacteria suggest that Ca^{2+} and calmodulins mediate the regulation of variety of cellular processes including cell motility, exocytosis, cytoskeletal assembly, and the intracellular modulation of Ca^{2+} concentration and Ca^{2+} ion channels (Klee and Vanaman 1982, Allan and Helper 1989, Hetherington *et al.*, 1992). Future studies should aim to dissect the specific roles of these calmodulins, their subcellular distribution and the physiological relevance of their interaction with target proteins to accelerate our understanding of Ca^{2+}- and calmodulin mediated signal transduction pathways. Pertaining to heterocyst development though a wave of explorations have begin correlating with Ca^{2+} transients more efforts are needed to unleash the exact mechanisms and the whole lot of different genes involved at the molecular level.

Acknowledgements

The Head, Department of Botany, Banaras Hindu University, Varanasi, India is gratefully acknowledged for providing laboratory facilities. One of us (Savita Singh) is also thankful to UGC, New Delhi for financial support in form of JRF.

References

1. Abeliovich A, Gan J (1982). Site of Ca^{2+} action in triggering motility in the cyanobacterium *Spirulina subsalsa*. *Cell. Motil.* 2(4): 393–403.

2. Abeliovitch A, Gan, J (1983). Factors affecting motility of *Spirulina subsala*, Photosynthetic Prokaryotes. In: Papogeorgiou GC and Packer L, *Cell Differentiation and Function* (Eds.). Elsevier Biomedical, New York, USA : 354-361.

3. Accardi A, Kolmakova-Partensky L, Williams C, Miller C (2004). Ionic currents mediated by a prokaryotic homologue of CLC Cl– channels. *J Gen Physiol* 123: 109–119.

4. Accardi A, Miller C (2004). Secondary active transport mediated by a prokaryotic homologue of ClC Cl- channels. *Nature* 427(6977): 803-807.

5. Ajouz B, Berrier C, Besnard M, Martinac B, Ghazi A (2000). Contributions of the different extramembraneous domains of the mechanosensitive ion channel MscL to its response to membrane tension. *J. Biol. Chem.* 275: 1015–1022.

6. Allan E, Helper PK (1989). Calmodulin and calcium-binding proteins: In: Marcus A (Eds.). *The biochemistry of plants. A comprehensive treatise.* Academic Press, San Diego 455-484.

7. Andrade S. L, Dickmanns A, Ficner R, Einsle O (2005). Crystal structure of the archaeal ammonium transporter Amt-1 from *Archaeoglobus fulgidus. Proc. Natl. Acad. Sci. USA* 102: 14994–14999.

8. Barr R, Troxel KS, Crane FL (1983). A calcium-selective site in Photosystem II of spinach chloroplasts. *Plant Physiol.* 73(2): 309-15.

9. Bass RB, Strop P, Barclay M, Rees D (2002). Crystal structure of *Escherichia coli* MscS, a voltage-modulated and mechanosensitive channel. *Science* 298: 1582–1587.

10. Becker DW, Brand JJ (1985). *Anacystis nidulans* demonstrates a photosystem II cation requirement satisfied only by Ca^{2+} or Na^+. *Plant Physiol.* 79 (2): 552-558.

11. Berridge M J (1997). Elementary and global aspects of calcium signalling. *Journal of Physiology* 499(2): 291-306.

12. Betanzos M, Chiang CS, Guy HR, Sukharev S (2002). A large iris-like expansion of a mechanosensitive channel protein induced by membrane tension. *Nat Struct Biol* 9: 704–710.

13. Black TA, Cai Y, Wolk CP (1993). Spatial expression and autoregulation of *hetR*, a gene involved in the control of heterocyst development in *Anabaena. Mol. Microbiol.* 9(1): 77–84.

14. Blankenship RE (1992). Origin and early evolution of photosynthesis. Pho*tosynth Res.* 33: 91-111.

15. Blount PSS, Moe P, Martinac B, Kung C (1999). Mechanosensitive channels in bacteria. *Meths Enzymol.* 294: 458–482.

16. Booth IR, Edwards MD, Miller S (2003). Bacterial ion channels. *Biochemistry* 42: 10045–10053.

17. Boris Martinac, Yoshiro Saimi, Ching Kung (2008). Ion channels in microbes. *Physiol Rev.* 88 (4): 1449–1490.

18. Bunchingi M, Pastini AC, Muschiettj IP, T~llez-Inon MT, Martinettho E, Torresh N, Flawiam M (1990). Adenylate cyclase activity in cyanobacteria : activation by Ca^{2+}- calmodulin and calmodulin-like activity. *Biochimica et Biophysica Acta* 1055: 75-81.

19. Campbell, A. K. 1983. Intracellular calcium: its universal role as regulator. *Monographs in Molecular Biophysics and Biochemistry*, John Wiley and Sons, Chichester, England.

20. Carafoli, E. (1987). Intracellular calcium homeostasis. *Annu. Rev. Biochem.* 56: 395–433.

21. Carneiro RL, Alípio ACN, Bisch PM, Azevedo SMFO, Pacheco ABF (2011). The inhibitory effect of calcium on *Cylindrospermopsis raciborskii* (cyanobacteria). metabolism. Brazilian *J. Microbiol.* 42: 1547-1559.

22. Case RM, Eisner D, Gurney A, Jones O, Muallemd S, Verkhratsky A (2007). Evolution of calcium homeostasis: From birth of the first cell to an omnipresent signalling system. *Cell Calcium* 42: 345–350.

23. Castenholz RW (2001a). Oxygenic photosynthetic bacteria. In: Boone DR, Castenholz RW (Eds). *Bergey's manual of systematic bacteriology*, New York, USA, Springer-Verlag, 474.

24. Chang G, Spencer RH, Lee AT, Barclay MT, Rees DC (1998). Structure of the MscL homolog from Mycobacterium tuberculosis: a gated mechanosensitive ion channel. *Sci.* 282: 2220–2226.

25. Chin D, Means AR (2000). Calmodulin: a prototypical calcium sensor. *Trends in Cell Biology* 10(8): 322-328.

26. Crivici A, Ikura M (1995). Molecular and structural basis of target recognition by calmodulin. *Ann. Rev. Biophys. Biomol. Struct.* 24: 85-116.

27. Dickson AG, Goyet C (1994). Handbook of method for the analysis of the various parameters of the carbon dioxide system in sea water, version 2. *ORNL/CDIAC-74.*

28. Dieter P, Marmé D (1983). The effect of calmodulin and far-red light on the kinetic properties of the mitochondrial and microsomal calcium-ion transport system from corn. *Planta,* 159 (3): 277-81.

29. Dutzler R, Campbell EB, Cadene M, Chait BT, MacKinnon R (2002). X-ray structure of a ClC chloride channel at 3.0 A reveals the molecular basis of anion selectivity. *Nature* 415(6869): 287-294.

30. Dutzler R, Campbell EB, MacKinnon R (2003). Gating the selectivity filter in ClC chloride channels. *Science* 300(5616): 108-112.

31. Eldik LJV, SR (1984). Conditions for reproducible detection of calmodulin and Sl00 beta in immunoblots. *Biochem. Biophys. Res. Comm.*124: 752-759.

32. Ernst A, Black T, Cai Y, Panoff JM, Tiwari DN, Wolk CP (1992). Synthesis of nitrogenase in mutants of the cyanobacterium Anabaena sp. strain PCC 7120 affected in heterocyst development or metabolism. *J. Bacteriol.*174(19): 6025.

33. Fay P, Stewarwt DP, Walsbya E, Fogg GE (1968). Is the heterocyst the site of nitrogen fixation in blue-green algae? *Nature,* London 220: 810-812.

34. Fry IJ, Villa L, Kuehn GD, Hagemanj H (1986). Calmodulin-like protein from Bacillus subtilis. *Biochemical and Biophysical Research Communications* 134: 212-217.

35. Gandola P, Rosen BP (1987). Maintenance of intracellular calcium in *Escherichia coli. J. Biol. Chem.* 262: 12570–12574.

36. Gallon JR, Hamadi AF(1984). Studies on the effects of oxygen on acetylene reduction (nitrogen fixation). in *Gloeothece* sp, ATCC27152. J. *Gen. Microbiol.* 130: 295- 503.

37. Giraldez-Ruiz N, Mateo P, Bonilla I, Fernandez-Pinas F (1997). The relationship between intracellular pH, growth characteristics and calcium in the cyanobacterium *Anabaena* sp. strain PCC 7120 exposed to low pH. *New Phytol.* 137(4): 599-605.

38. Ghanotakis DF, Babcock GT, Yocum CF (1984). Calcium reconstitutes high rates of oxygen evolution in polypeptide depleted Photosystem II preparations. *FEBS Lett* 167: 127–130.

39. Golden JW, Yoon HS (1998). Heterocyst formation in *Anabaena*. *Curr Opin Microbiol* 1: 623–629.

40. Golden JW, Yoon HS (2003). Heterocyst development in *Anabaena*. *Curr Opin Microbiol* 6: 557–563.

41. Gouaux E, MacKinnon R (2005). Principles of Selective Ion Transport in Channels and Pumps. *Science* 310 (5753): 1461-1465.

42. Hetherinton AM, Graziana A, Mazars C, Thuleau P, Ranjeava R (1992). The biochemistry and pharmacology of plasma-membrane calcium channels in plants. *Proc R Soc Lond Ser B* 338: 91-96.

43. Hurtley SM, (2005). Crossing the bilayer, *Science*, 310: 1451.

44. Legane´ s F, Forchhammer K, Ferna´ ndez-Pin˜ as F (2009). Role of calcium in acclimation of the cyanobacterium *Synechococcus elongatus* PCC 7942 to nitrogen starvation. *Microbiol.* 155: 25–34.

45. Li Y, Moe PC, Chandrasekaran S, Booth IR, Blount P (2002). Ionic regulation of MscK, a mechanosensitive channel from *Escherichia coli*. *EMBO J* 21: 5323–5330.

46. Lockau W, Pfeffer S (1983). ATP-dependent calcium transport in membrane vesicles of the cyanobacterium *Anabaena variabilis*. *Biochim. Biophys. Acta* 733(1): 124-132.

47. Ludewig U (2006). Ion transport versus gas conduction: function of AMT/Rh-type proteins. *Transfus. Clin. Biol.* 13: 111–116.

48. Kerson GW, Miernyk JA, Budd K (1984). Evidence for the occurrence and possible physiological role for cyanobacterial calmodulin. *Plant Physiol* 75: 222-224.

49. Khademi S, O'Connell J, Remis J, Robles-Colmenares Y, Miercke L J, Stroud RM (2004). Mechanism of ammonia transport by Amt/MEP/Rh: structure of AmtB at 1.35 A°. *Sci.* 305: 1587–1594.

50. Khudyakov I, Wolk CP (1997). hetC, a Gene Coding for a Protein Similar to Bacterial ABC Protein Exporters, Is Involved in Early Regulation of Heterocyst Differentiation in *Anabaena* sp. Strain PCC 7120. *J. Bacteriol.* 179(22): 6971–6978.

51. Klee CB, Vanaman TC (1982). Calmodulin. *Adv Protein Cem* 35: 213-321.

52. Kloda A,. Martinac B (2001a). Molecular identification of a mechanosensitive ion channel in archaea. *Biophys. J.* 80: 229–240.

53. Kloda A, Martinac B (2001b). Structural and functional similarities and differences between MscMJLR and MscMJ, the two homologous MS channels of *M.jannaschii. EMBO J.* 20: 1888-1896.

54. Kloda A, Martinac B (2001). Mechanosensitive channel of Thermoplasma, the cell wall-less archaea: cloning and molecular characterization. *Cell Biochem Biophys* 34: 321–347.

55. Knight MR, Campbell AK, Smith SM, Trewavas AJ (1991). Transgenic plant aequorin reports the effects of touch and cold –shock and elicitors on cytoplasmic calcium. *Nature* 352: 524-526.

56. Kobayashi M, Katoh H, Ikeuchi M (2006). Mutations in a putative chloride efflux transporter gene suppress the chloride requirement of photosystemII in the cytochrome c550-deficient mutant. *Plant Cell Physiol.* 47: 799–804

57. Koishi R, Xu H, Ren D, Navarro B, Spiller BW, Shi Q, Clapham DE (2004). A Superfamily of Voltage-gated Sodium Channels in Bacteria. *J. Bio. Chem.* 279(10): 9532–9538.

58. Krulwich TA, Ito M, Guffanti AA (2001). The Na^+-dependence of alkaliphily in *Bacillus. Biochim. Biophys. Acta* 1505, 158–168.

59. Kumar K, Mella-Herrera RA, Golden JW (2010). Cyanobacterial Heterocysts, Cold Spring Harb Perspect Biol doi: 10.1101/cshperspect.a000315.

60. Kuo, MM-C, Kung C, Saimi Y (2005). K^+ channels: a survey and a case study of Kch of *Escherichia coli.* In: Kubalski A, Martinac B (Eds.). *Bacterial Ion Channels and their Eukaryotic Homologs.* Washington, D.C.: ASM Press, 1-20.

61. Kuo MM, Haynes WJ, Loukin SH, Kung C, Saimi Y(2005). Prokaryotic K(+). channels: from crystal structures to diversity. *FEMS Microbiol Rev* 29: 961–985.

62. Le Dain AC, Saint N, Kloda A, Ghazi A, Martinac B (1998). Mechanosensitive ion channels in the archaeaon *Haloferax volcanii. J. Biol. Chem.* 273: 12116-12119.

63. Martinac, B., Buechner, M., Delcour, A.H., Adler, J. and Kung. C. (1987). Pressure-sensitive ion channel in *Escherichia coli. Proc. Natl. Acad. Sci. USA* 84, 2297-2301.

64. Mayer M, Schaaf G, Mouro I, Lo´pez C, Colin Y, Neumann P, Cartron JP, Ludewig U (2006). Different transport mechanisms in plant and human AMT/Rh-type ammonium transporters. *J. Gen. Physiol.* 127: 133–144.

65. Meeks JC, Elhai J (2002). Regulation of cellular differentiation in filamentous cyanobacteria in free-living and plant-associated symbiotic growth states. *Microbiol Mol Biol Rev* 66: 94–121.

66. Michiels J, Xi C, Verhaert J, Vanderleyden J (2002). The functions of Ca^{2+} in bacteria: a role for EF-hand proteins? *Trends Microbiol.* 10(2): 87-93.

67. Miyao M and Murata N (1984a). Calcium ions can be substituted for the 24 kDa polypeptide in photosynthetic oxygen evolution. *FEBS Lett* 168: 118–120.

68. Mishra AK (2003). MSX-resistant mutants of *Anabaena* 7120 with depressed heterocyst development and nitrogen fixation. *W. J. Microbiol. Biotechnol.,*19: 675-680.

69. Mur LR, Skulberg OM, Utkilen H (1999). Cyanobacteria in the Environment. In: Chorus I, Bartram J (Eds.). *Toxic Cyanobacteria in Water: A Guide to their Public Health Consequences, Monitoring, and Management*. London and New York, 416.

70. Muro-Pastor AM, Valladares A, Flores E and Herrero A (2002). Mutual dependence of the expression of the cell differentiation regulatory protein HetR and the global nitrogen regulator NtcA during heterocyst development. *Molecular Microbiology* 44(5): 1377–1385.

71. Ninnemann O, Jauniaux JC, Frommer WB (1994). Identification of a high affinity NH4 transporter from plants. *EMBO J.* 13: 3464–3471.

72. Norris V, Grant S, Freestone P, Canvin J, Sheikh FN, Toth I, Trinei M, Modha K, Norman RI (1996). *Calcium Signalling in Bacteria Journal of Bacteriology. American Society for Microbiology* 178(13): 3677–3682.

73. Pandey PK, Gour RK, Bisen PS (1999). Energy-dependent Ca^{2+} efflux from the cells of *Nostoc calcicola Bréb*: role of modifying factors. *Curr. Microbiol.* 39: 254–258.

74. Reddy PT, Prasad CR, Reddy PH, Reeder D, McKenney K, Jaffe H, Dimitrova MN, Ginsburg A, Perkofsky A, Murthy PS (2003). Cloning and Expression of the Gene for a Novel Protein from *Mycobacterium smegmatis* with Functional Similarity to *Eukaryotic calmodulin*. *J. Bacteriol.*185(17): 5263-5268.

75. Rodriguez H, Rivas J, Guerrero MG, Losada M (1990). Calcium requirement for aerobic nitrogen fixation by heterocystous blue-green algae. *Plant Physiol.* 92(4): 886-90.

76. Schwarz R, Forchhammer K (2005). Acclimation of unicellular cyanobacteria to macronutrient deficiency: emergence of a complex network of cellular responses. *Microbiology* 151(8). 2503-2514.

77. Singh RN (1961). Role of blue green algae in nitrogen economy of Indian Agriculture, *Indian Council of Agricultural Research*, New Delhi, 175.

78. Singh S, Mishra AK (2014). Regulation of Calcium Ion and Its Effect on Growth and Developmental Behavior in Wild Type and *ntc*A Mutant of *Anabaena* sp. PCC 7120 under Varied Levels of $CaCl_2$. *Microbiol.* 83(3): 235-246.

79. Smith RJ, Hobson S, Ellis IR (1987). Evidence for calcium mediated regulation of heterocyst frequency and nitrogenase activity in *Nostoc* 6720. *New Phytol* 105: 531–541.

80. Smith RJ (1988). Calcium mediated regulation in the cyanobacteria. In: Rogers LJ, Gallon JR (Eds.). *Oxford, Biochemistry of the Algae and Cyanobacteria* Clarendon Press 185-199.

81. Sugiyama S, Matsukura H, Imae Y (1985). A specific inhibitor for the Na⁺-driven flagellar motors of alkalophilic *Bacillus*. *FEBS Lett*. 182, 265–268.

82. Soupene E, He L, Yan D, Kustu S (1998). Ammonia acquisition in enteric bacteria: physiological role of the ammonium/methylammonium transport B (AmtB). protein. *Proc. Natl. Acad. Sci. USA* 95: 7030–7034.

83. Tandeau-de-Marsac N, Houmard J (1993). Adaptation of cyanobacteria to environmental stimuli, new steps towards molecular mechanisms. *FEMS Microbiol Rev* 104: 119–189.

84. Torrecilla I, Legane's F, Bonilla I, Ferna'ndez-Pin˜as F (2004). Light-to-dark transitions trigger a transient increase in intracellular Ca²⁺ modulated by the redox state of the photosynthetic electron transport chain in the cyanobacterium *Anabaena* sp. PCC7120. *Plant Cell Environ* 27: 810–819.

85. Toyoshima C, Inesi G (2004). Structural basis of ion pumping by Ca2+-ATPase of the sarcoplasmic reticulum. *Annu. Rev. Biochem* 73: 269-292.

86. Toyoshima C, Nakasako M, Nomura H, Oqawa H (2000). Crystal structure of the calcium pump of sarcoplasmic reticulum at 2.6 A resolution. *Nature* 405(6787): 647-655.

87. Toyoshima C, Nomura H (2002). Structural changes in the calcium pump accompanying the dissociation of calcium. *Nature* 418(6898): 605-11.

88. Tsien RW, Tsien RY(1990). Calcium channels, stores, and oscillations. *Annu. Rev. Cell Biol*. 6: 715–760.

89. Tyagi VVS (1975). The heterocysts of blue-green algae (Myxophyceae). *Biol. Rev*. 50 (3): 247-284.

90. Watkins NJ, Knight MR, Trewalas AJ, Campbell AK (1995). Free calcium transients in chemotactic and non-chemotactic strains of *Escherichia coli* determined by using recombinant aequorin. *Biochem. J*. 306: 865–869.

91. White MM, Miller C (1981). Chloride permeability of membrane vesicles isolated from Torpedo californica electroplax. *Biol. Chem*. 35(2): 455–462.

92. Wolff J, Cook H, Goldhammae RR, Berkowitsz SA (1980). Calmodulin activates prokaryotic adenylate cyclase. *Proc.Natl. Acad. Sci. USA* 77: 3841-3844.

93. Wolk CP (1996). Heterocyst formation. *Annu Rev Genet* 30 : 59-78.

94. Wolk CP (2000). Heterocyst formation in *Anabaena*. In: Brun YV, Shimkets LJ, (Eds.). *Prokaryotic Development*. American Society for Microbiology, Washington DC: 83-104.

95. Wolk CP (1975). *Differentiation and pattern formation in filamentous blue-green algae*. In: Gerhardt P, Sadoff HL, Costilow RN (Eds.). American Society for Microbiology, Washington: 85-96.

96. Wolk CP, Ernst A, Elhai J (1994). Heterocyst metabolism and development. In: Bryant DE, Dordrecht(Eds.). Kluwer Academic Publisher 769-823.

97. Xu X, Wolk CP (2001). Role for hetC in the transition to a nondividing state during heterocyst differentiation in *Anabaena* sp. *J Bacteriol*. 183(1): 393-396.

98. Yang K (2001). Prokaryotic Calmodulins: Recent Developments and Evolutionary Implications. *J. Mol. Microbiol. Biotechnol*. 3(3): 457-459.

99. Zhang CC, Laurent S, Sakr S, Peng L, Bédu S (2006). Heterocyst differentiation and pattern formation in cyanobacteria: a chorus of signals. *Mol. Microbiol*. 59(2). 367-375.

100. Zheng L, Kostrewa D, Berneche S, Winkler FK, Li XD (2004). The mechanism of ammonia transport based on the crystal structure of AmtB of *Escherichia coli*. *Proc. Natl. Acad. Sci. USA* 101: 17090–17095.

2016, Environmental Biotechnology: A New Approach   *Pages 83–100*
Editors: Dr. Rajan Kumar Gupta and Dr. Satya Shila Singh
Published by: DAYA PUBLISHING HOUSE, NEW DELHI

Chapter 5

Cyanobacteria: Diversity and Applications in Biotechnology

Prashant Singh[1] and Satya Shila Singh[2]

[1]*Microbial Culture Collection (MCC),*
National Centre for Cell Science (NCCS), Pune
E-mail: sps.bhu@gmail.com
[2]*Department of Botany, Guru Ghasidas Vishwavidyalaya,*
Bilaspur, Chhattisgarh
E-mail: satyashila@rediffmail.com

ABSTRACT

Apart from having evolutionary importance, cyanobacteria are an equally important group of economically beneficial organisms. Their adaptive capacity in virtually all realms of earth has aloe enabled them to develop substantial characteristics for various beneficial aspects for many of the living forms of the planet earth. Due to their economic benefits only, much of the science fraternity has paid attention to the proper identification, cultivation and mass culturing of cyanobacteria. Strains of Arthrospira, Nostoc and Aphanizomenon are some of the most commonly used forms that have reaped immense economic benefits. The use of cyanobacteria as food supplements, in the animal feed industry, as cosmetics, as biofertilizers and in wastewater treatment are only some of the fields where active work is being pursued by workers all across the globe. The usage of cyanobacteria in terms of biotechnology and economic aspects has been discussed in detail in this chapter along with discussing the intricacies involved in mass culturing of cyanobacteria for the above mentioned purposes.

Cyanobacteria

Cyanobacteria are oxygenic, gram negative, photosynthetic bacteria found from the Arctic Zones (Mataloni and Komárek, 2004) to the hot springs of Yellowstone National Park, thus representing life in a wide range of aquatic and terrestrial environments at the "limits of life" (Whitton and Potts, 2000). They are common in lakes, ponds, springs, wetlands, streams, and rivers, and they play a major role in the nitrogen, carbon and oxygen dynamics of many aquatic environments. Their success in such diverse habitats is attributed to a very long standing evolutionary history. The exact timing of appearance of the first cyanobacteria like microbes on Earth is still unclear because of controversy over the interpretation of Precambrian fossils; however, much of their present diversity was achieved more than 2 billion years ago, and they likely played a major role in the accumulation of oxygen in the Earth's early atmosphere. Very ancient cyanobacteria, having a great resemblance to their modern-day counterparts, have been discovered in 3.5 billion year old conglomerate Apex Chert (Castenholz, 1992), and they have dominated the Proterozoic Era (2500-570 Ma), thus being titled appropriately, the "Age of Cyanobacteria" (Whitton and Potts 2000). Some of the modern and well known physiological and anatomical features of cyanobacteria that support a long history on Earth include: a high tolerance level of low oxygen and free sulfide, the utilization of even H_2S as a photoreductant in place of H_2O and the tolerance to ultraviolet B and C radiation (Whitton and Potts, 2000). These features have allowed cyanobacteria to grow and dominate in areas of the world where many other organisms are virtually unknown.

The origin of cyanobacteria and their basic anatomical architecture is of typical bacterial type but their ecological, biological and morphological features are very specific and sometimes quite different from their bacterial counterparts (Flores *et al.*, 2006; Kalaitzis *et al.*,2009; Flores and Herrero2010). They possess certain fascinating features such as the capacity of buoyancy, performing oxygenic photosynthesis and very importantly, fixation of atmospheric nitrogen (Walsby, 1994; Castenholz, 2001; Burja *et al.*, 2001; Berman-Frank *et al.*, 2003). They play a significant role as phototrophic primary producers in natural ecosystems (Field *et al.*, 1998; Bryant, 2003). Cyanobacteria also contain chlorophyll *a*, carotenes, phycobillins and a wide variety of important bioactive compounds (Bryant, 1991; Stevens and Nierzwicki-Bauer 1991; Gantt, 1994; Burja *et al.*, 2001; Six *et al.*, 2004).

Different kinds of organism have different kinds of adaptive capacities and also, different kinds of nutritional demands and ecological tendencies. This fact holds true for cyanobacteria also. For example, *Prochlorococcus* is a cyanobacterium that dwells comfortably in the oligotrophic deep oceans along with co-existing in peace with the cyanophages. The first reports of *Prochlorococcus* came in 1988 by Chisholm *et al.*, and it is now recognized to be of major importance in the oceans (Zwirglmaier *et al.*, 2007). This association is responsible for the events of lateral gene transfer amongst the two biological systems which has been held responsible for the huge amount of studies on molecular evolution through gene transfer events (Johnson *et al.*, 2006). One of the commonly found examples, *Prochlorococcus marinus*, has been reported to be the tiniest photosynthetic organism along with being the most abundant and frequently found photoautotroph in an oligotrophic deep ocean habitats. The density of this organism has been estimated to be around 105 cells per ml of sea water. Due to their high density and perhaps incredible pace of division, these organisms are regarded as an extremely important reservoir of carbon and primary production. This capacity is thus also an indirect measure of the global warming that can be enhanced by the activities of these organisms.

On having a close look at the cyanobacterium *Chroococcidiopsis* that is possibly the only inhabitant that can not only survive but also flourish in the highly inhospitable Atacama deserts habitats has been found to have importance in astrobiological contexts (Friedmann, 1982; Wierzchos *et al.*, 2006).

Modern day rRNA sequencing approaches have, in a way, changed the course and approach of environmental biotechnology and microbiology. The advent of molecular microbial ecology with fast development of sequencing approaches has served well to delineate closely related cyanobacteria of even the same genera. Apart from the 16S rRNA sequencing, assortments of molecular markers are being used regularly for understanding the diversity and phylogeny of cyanobacteria. But again, sequencing based approaches along with the classical approaches are the need of the hour; thus in a nutshell, vehemently arguing for the development of polyphasic approaches for studying cyanobacteria (Komarek, 2006).

Evolution of Cyanobacteria

In terms of their evolutionary lineage, the cyanobacteria have also played a major role in the evolution of other forms of life on Earth. Earth was nearly anaerobic until about 2 billion years ago; the conversion of its anaerobic atmosphere to an aerobic one was initiated by cyanobacteria, thus supplying oxygen to the atmosphere through oxygenic photosynthesis for approximately 1.5 billion years (Berman-Frank *et al.*, 2003; Schopf *et al.*, 1983). This occurrence of oxygenation of the earth's atmosphere allowed for a greater broadening of life on the Earth. Cyanobacteria have also influenced life on earth by being the progenitors of chloroplasts (Giovannoni *et al.*, 1988; Sergeev *et al.*, 2002), ultimately leading to diversification of algae and land plants.

Cyanobacteria continue to produce their impact on life on Earth, as major oxygen producers on this planet. Many of them can help in fixing nitrogen, making them an important member of nutrient poor environments in the extremes terrains of the earth's wide geographical and geological contours. Cyanobacteria are also known to form a range of important nitrogen-fixing symbiotic relationships with representatives from almost all plant groups such as *Azolla-Anabaena*, *Cycas-Nostoc*, fungi, bryophytes, pteridophytes, gymnosperms and angiosperms thus, supplying the host organism with fixed nitrogen and acquiring carbohydrates from the host (Morot-Gaudry and Touraine 2001) and allowing the host organisms to grow in regions where they may otherwise not survive.

Cyanobacterial Diversity

If the diversity of cyanobacteria could be described in one word then at the simplest forms also, it could be described as "immense". Along with having a universal trademark, omnipresent occurrence with potential ecological benefits, cyanobacteria are without doubt one of the most important contributors in the global nitrogen fixing and carbon fixation chains (Karl *et al.*, 2002). The physiological plasticity is of scales that are colossal in magnitude thus enabling them to be representatives of virtually all geographical regions on the planet earth (Kol, 1968; Castenholz, 1973; Whitton, 1973; Van Landingham, 1982; Reed *et al.*, 1984; Kann,

1988; Dor and Danin 1996; Laamanen, 1996; Skulberg, 1996; Weber *et al.*, 1996). Many of them are freshwater dwelling forms, but some also thrive very well in marine environments. Terrestrial habitat is common for the majority of the strains of the cyanobacteria while they are also present in the benthic habitats as planktonic forms. Due to their high physiologically adaptive capacity, cyanobacteria are also known to form frequent endophytic and symbiotic associations (Rai, 1990; Adams, 2000; Bergman *et al.*, 2007; Thajuddin *et al.*, 2010). Sometimes they also form biofilms and microbial mats on shores, surface of the stones, plants and artificial objects (Stal, 2000). Common and heavy occurrence of cyanobacteria in fresh water blooms, rocky crevices, bark of trees, marine ecosystems, rice fields, within limestone, salt subjugated lands, deserts, polar environments and in symbiotic associations thus, highlights their ability to survive in the above niches (Elster *et al.*, 1999; Whitton and Potts 2000).

Importance of Cyanobacteria

On taking an overall view of the beneficial and economic perspectives of cyanobacteria, three cyanobacterial genera stand out of the rest because of the commercially relevant products that they bring out as a by-product of their metabolism: *Arthrospira, Nostoc* and *Aphanizomenon*. These are being produced and/or collected for different purposes, mostly as health food and dietary supplement all across the globe. The ability of some cyanobacteria to fix atmospheric nitrogen through various mechanisms makes them unique in their ability to independently secure their carbon and nitrogen requirements (Kulasooriya, 2008). Some of these organisms and symbiotic systems like *Azolla* are used as biofertilizers, particularly in the paddy fields (Venkataraman, 1972, Tiwari *et al.*, 1991). They can also be utilized for application in oxidation ponds and sewage and sludge treatment plants (Lincoln *et al.*, 1996). Recently, some species have been investigated for biofuel production because of their capacity to convert solar energy, which has been found to be the most efficient amongst all living organisms. Furthermore, their modest genomic structure has enabled the production of biofuel secreting strains through bio-engineering approaches (Lane, 2010). Many cyanobacteria are also known to form blooms in eutrophied water bodies and most of such bloom forming strains are capable of producing cyanotoxins that are harmful and sometimes even lethal to the animals and humans (Carmichael, 1994; Carmichael *et al.*, 2001). Some species of cyanobacteria are also utilized for the production of highly nutritious food supplements (Kulshreshtha *et al.*, 2008).

Arthrospira which is marketed as "*Spirulina*" is well known to be one of the most popular microalga as well as also the most extensively studied one in various economical perspectives. *A. platensis* has been grown worldwide as a food supplement (Gershwin and Belay 2007) due to its quality of being one of the best protein rich diets known worldwide. *Arthrospira* is also very well known as a rich and in fact, very efficient source of vitamins, essential amino acids, minerals, essential fatty acids like - linolenic acid (GLA, and w6-polyunsaturated fatty acid) and antioxidant pigments like phycobiliproteins and carotenoids. *A. platensis* and *A. maxima* are seemingly the only producers of GLA so far reported for any cyanobacteria. This fatty acid is known to be a precursor of arachidonic acid, which is required for the synthesis of very important metabolic mediators. The proportion of linoleic, linolenic acid and palmitic acid are known to be species specific within the genus *Arthrospira* (Mühling *et al.,* 2005) and thus can also aid in strain identification.

Nostoc has been used as a food supplement since times immemorial. The use of *Nostoc* as a food or as a supplement has been described in many parts of Asia, especially China for more than 2000 years (Gao,1998; Qiu *et al.,*2002). This economic importance is a result of both its herbal and the pharmaceutical value (Khaing, 2004). The most commonly used ones are *Nostoc commune, Nostoc sphaeroides, Nostoc verrucosum* and *Nostoc flagelliforme* with the last one being probably the most commonly used and also the best known. Owing to its morphological appearance, *Nostoc flagelliforme* is also known as 'Facai' (hair vegetable) and is grown prominently on the soils in the China and Myanmar throughout the year.

Aphanizomenon flos-aquae (AFA) is a very recent introduction in the food market as a health food supplement. It was first marketed as a health food supplement in the early 1980s in the US market, owing to its quality of possessing a very similar chemical composition as *Arthrospira*. The complete biomass was taken from the Upper Klamath Lake, a shallow lake system in Oregon, USA. The production of AFA differs from that of *Arthrospira* in being harvested solely from a natural lake rather than from artificially constructed ponds. It has been reported by Carmichael *et al.* (2000) that the AFA blooms occur between late May and October or November with biomass concentrations of approximately 3–50 mgL^{-1}. This bloom also has its own dynamics where the blooms of AFA would occur with biomass less than 1 per cent during the above mentioned period. The occurrence of blooms of *Microcystis aeruginosa* and *Coelosphaerium* (probably *Woronichinia*) has been reported in July and known to be persistent.

Mass Culture of Cyanobacteria

The mass cultivation of cyanobacteria has always been a technique that has developed slowly and slowly with the gain of expertise and experience. In contrast to fermentation through the use of heterotrophs, in the photobioreactors the innate ability of cyanobacteria to use photosynthesis as a source of energy is employed. Hence, the important conditions for algal growth in the suspension like light and nutrients are optimized with precision for maintenance of these photobioreactors. The sufficient light supply is achieved by very thin layers of the culture suspension, given that by the algal growth and by the increased clouding of the suspension a rapid decrease of the available light for the algal cells occurs. On the other hand, it is the high cell densities in the culture, which assures high growth rates per volume unit. The culture area obligatory to produce biomass also plays a very significant role. Microalgae, including the cyanobacteria, can be cultivated in open, closed or ultrathin layer systems. The open cultivation systems are basically natural or artificial ponds, raceway ponds, or the so-called inclined surface systems. They are representatives of the classic methods for the mass production of algal biomass and require large areas. Thus, if the land utilization costs are low and climatic conditions are favorable, then the investment costs for sizes up to 100 ha are also relatively low. One of the paramount advantages of this approach to production is that the investment costs are low and which can be almost none if the pond construction occurs in natural waters. For example, Lake Chad is the best known natural system for

Arthrospira production. However, with the advent of science and more knowledge on the mass cultivation, during the past few decades a considerable *Arthrospira* production has also been developed in several alkaline crater lakes in central Myanmar. Natural and artificial ponds are usually used for the mass cultivation of marine, naturally predominant or extremophilic species, where there is a relatively low risk of contamination and sometimes even better efficiencies. Thus, the open pond systems dominate the industrial scale algal biotechnology landscape in general (Grobbelaar, 2009) and cyanobacterial biotechnology cultivating *Arthrospira* in particular, both in terms of the annual output and their distribution worldwide.

The open ponds typically comprise of cement or plastic basins, with depth not more than 0.2–0.35 m, in order to maintain light conditions for optimal growth. The overall area of the pond may range from 25 to 5,000 m². The problem of stirring is of fundamental importance in the open ponds because a very large amount of energy is required to prevent the formation of concentration gradients and algal sedimentation. For this purpose, paddle wheels combined with aerating units are incorporated into parallel, loop-like channels that may be several km in total length thus, resulting in suspension velocities of 0.01–0.3 ms⁻¹ thereby, requiring relatively low operating expenses. The growth of biomass in raceway ponds is also dependent a lot on the prevailing regional climate with the mean growth rates for *Arthrospira platensis* ranging from 5 to 20 gm⁻¹ day⁻¹. Also, the local weather conditions like heavy rain, aridity and thunderstorms can influence the morphology and productivity of the cultures pretty frequently. Due to the above mentioned conditions, this cultivation method is restricted to tropical or subtropical climate zones where the light input is directly linked to the prevailing temperatures and thus is sufficient for meeting the ecological demands of the algae. The semiarid and arid climates pose lesser danger of flooding, but in the other way, pose a higher water demand.

The significantly low productivity and the sensitivity of open systems to contamination and climatic changes of various kinds has now led to the development of closed reactors in which photo-biological processes are less dependent on the interfering environmental influences.

Such closed reactors have a range of primary advantages when compared to the open systems: (1) significantly low CO_2 -loss; (2) low water losses; (3) abridged contamination risk; (4) prime temperature regulation; (5) much efficient controllable hydrodynamics; (6) reproducible cultivation conditions; (7) substantial flexibility in tackling environmental influences; (8) significantly low space requirements. These closed photo bioreactors allow the introduction of the much needed light into the culture suspension by their light transparent reactor walls (tubes, plates); roughly, about 90 per cent of the incident light can reach the cells in this system. In general, these closed photo-bioreactors consist of photosynthetically active modules (glass or plastic tubes or extruded profile plates), a compensation tank, distribution pipes and pumps. The reactor is also characterized by its narrow gas exchange with the ambient conditions outside. Thus, this limited exchange ensures relatively low contamination risk, making the process technically well controllable.

Cyanobacteria as a Source of Food

The use of algae as a food or food supplement has been reported for times immemorial. In the non -western civilizations, algae and cyanobacteria have been used as food supplement in the human diet for millennia. The first records report about the consumption of macroalgae in coastal regions dates back to around 6,000 years ago. Cyanobacteria, with the first records from *Arthrospira*, have been utilized as a food supplement by the Aztecs (LakeTexcoco) in Mexico by the Aztek population since 1300 AD (Pulz and Gross 2004) and in Africa (Lake Chad) even earlier (Abdulqader *et al.*, 2000). *Arthrospira* is commercially traded in Africa as "dihé" after collecting and sun drying (Ciferri, 1983) and is provided in soups at up to 60 g per meal (Delpeuch *et al.*,1975). In Mexico, the dried cake of *Arthrospira* called "tecuitlatl" was commonly consumed with maize, different types of cereals or in a sauce of tomatoes and spices. *Aphanothece sacrum* (Fujishiro *et al.*, 2004), *Nostoc muscorum* and *Nostoc commune* have all been used as food supplements or side dishes in Japan since very ancient times (Lee,2008). The consumption of *Nostoc* by humans has been reported from Japan, Peru and Mongolia (China) along with many other Asian countries like Myanmar. Dried and stir-fried *Nostoc* balls are available in many Asian markets and are used mainly as soup ingredients or soup delicacies. Interestingly, rising demands of these products led to rise in the prices and on the other hand, massive collection of *Nostoc flagelliforme* led to grassland degradation, desertification and social problems, all additional collection, sale and exportation of these products was banned by the Chinese authorities (Roney *et al.*, 2009).

Cyanobacteria in Animal Feed Industry

Animal feed industry is another place where microalgae and cyanobacteria have been found to be having a lot of use. It is known that the survival, growth, development, productivity and fertility of animals are essentially reflected by a good health. Amongst the many other factors that may affect animal growth and development, the quality of fodder and feed that they get, is one of the most important exogenous factor influencing them, particularly in connection with intensive breeding conditions and the trend to reduce or to avoid antibiotics. When the use of the microalgae started for the animal feed industry the proportion of the algal biomass used was almost 50 per cent of the common feed with an aim to replace the protein source. But with more and more trials, advanced and more scientifically planned research projects, it has come to notice that even very small amounts of microlagal biomass could have significantly positive effects on the physiology of the animals. Normally 01-10 per cent of *Arthorpsira* are being used as animal feed. On this kind of diet what is most noticeable is the positive effects of vitamins, minerals and PUFAs, an unspecific but advantageous immune response of the animals (Khan *et al.*, 2005). The augmentation of growth, fertility rate, rate of survival, live weight, feed conversion efficiency, resistance to bacterial and viral infections and enhanced color have all been reported in chickens, buffalos, prawns, salmon, carp and tilapia (Belay *et al.*, 1996). *Arthrospira* has been proven very effective in the poultry industry for the proper coloration of egg yolk (Vonshak, 1997), and it has been reported that this coloration reaches a maximum after a 7 days diet. Studies have very precisely indicated that a content of 21 per cent *Arthrospira* in the diet led to the color score being 37 per cent higher than the indigenous eggs and also 2.6 times more effective than dehydrated berseem (*Trifolium alexandrinum*) meal and 1.9 times more than 40 per cent yellow maize. Research has indicated that at all used quantities and ratios, *Arthrospira* -fed birds produced egg yolks with a deeper and better color than the two conventional carotenoid sources, clover and maize. The examination of diverse processing methodologies in the preparation of quail feed has revealed that freeze drying of the biomass is preferable to an extrusion process, since the latter process degrades the carotenoid content of the cyanobacterial biomass by heat. But, the drawback of this process is that economically it is not a very feasible process as processing costs in freeze drying are comparatively higher and thus not being preferred as of now.

There are a large number of reports of increased animal production in addition to improved quality. The supplementation of corn-based diet for hens with 1 per cent *A. platensis* (1 g biomass per hen per day) has been reported to have supported an average increase of 51.7 per cent in eggs laid over a trial period of 30 days) compared to the control groups (Storandt *et al.*, 2000), whereas the egg weight was found to have increased by 2.4 per cent. Very positively, the firmness of the egg-shell, the appearance of the plumage and general health were all influenced with increasing trends. The positive effect on egg productivity was still quantifiable in the subsequent period, when feeding was carried out without *A. platensis* addition. In a feeding trial on broiler hens, a diet containing 0.1 per cent *A. platensis* was shown to result in an increased nutrient uptake and final slaughter weight (Pulz *et al.*, 2008).

Aquaculture is an important industry that holds a special position in the use of algal biomass as feed, since algae are the basis of the natural food chain in almost all aquatic systems. Beside the direct nutritional use for various marine organisms like the molluscs, zooplankton, crustaceans and fish larvae, they are also being used as an addition for the enhancement of coloration, growth and immunity. More than 40 species of microalgae are currently being used in aquaculture worldwide depending on the special requirements for production. Key features needed for standardizing the use of algae in aquaculture are tolerable cell size for ingestion by the particular animal, the absence of toxins and the nutritional profile, including fatty acids. Marine aquaculture is a vital economic sector worldwide with a predicted growth trend of approximately 8 per cent p.a., which will lead to an increase in growth systems using artificial ponds. As feed production signifies one of the key input cost in aquacultural production, biotechnological production of algae is increasingly the focus of interest for aquaculturists. The worldwide production of long chain polyunsaturated fatty acid containing eukaryotic algae, suchas *Nannochloropsis*, *Tetraselmis* and *Isochrysis*, was roughly estimated to be 1,000 ton (Muller Feuga *et al.*, 2003).

Among the cyanobacteria *Arthrospira* is currently being used mainly for aquaculture feed (Muller Feuga, 2004), with its maximum use being in the Japanese region. Its benefits have also been investigated by several authors. El-Sayed (1994) reported that silver sea bream utilized *Arthrospira* biomass in a more efficient way as compared to either soybean meal or chicken meal. Some other cyanobacteria have also been tested and are also being used for their suitability as the sole sourceof nutrition. Cultured *Phormidium valderianum* has been used successfully for tilapia production in India (Thajuddin and Subramanian, 2005). The utilized strain had excellent tolerance for salinities of high order and the biomass was incorporated into feed pellets in order to reduce handling at the production site. Taking into context, for water quality control in aquaculture, *Arthrospira* has also been efficaciously co-cultured with black tiger shrimp (*Penaeus monodon*), resulting in reduced N concentrations in the tanks and thus, enhanced shrimp survival rate (Chuntapa *et al.,*2003).

Cyanobacteria in Cosmetics

With a world that is aging fast and more importantly being affected by global climate change and pollution, cosmetics from natural resources have been continuously having an increasing demand. Anti-aging products and cosmetics market has increased manifolds in particular in markets of Europe, USA and Asia primarily due to two reasons: increased awareness and increased per capita income. Along with their treasured nourishing constituents, many algae are known to contain active dermogenic substances. Nowadays, numerous products having active cyanobacterial extracts have been formulated by the biochemical industries and are being marketed, such Protulines® and Aquaflor®, both of which contain *Arthrospira* extracts with well proven moisturizing and anti-wrinkling effects. Water extracts of *Arthrospira* with a high concentration of magnesium salt content have been found to facilitate both ATP and matrix protein synthesis, that ultimately results in a stimulation of keratinocyte differentiation process (Schlotmann *et al.*, 2005). The percentage of those extracts is typically in the range of 2–10 per cent of the final formulation. Amino acids also, represent about 40 per cent of the group of natural moisturizing substances and thus, contribute to the hydration of the corneous layer cells by holding water. In cosmetic products, they are also being used for maintaining and also improving the softness, flexibility and elasticity of our skin. Investigations have in fact shown that hot-water extracts of *Arthrospira* contain predominantly high amounts of amino acids, since during extraction the water soluble proteins get degraded. These extracts are thereafter, frequently used as moisture regulating products. Recently, mainly aqueous and aqueous-ethanolic extracts are being used frequently in different cosmetic products such as creams, lotions, sun and hair care. For lipid-based cosmetics, like creams or lotions, supercritical CO_2-extracts have started gaining more commercial importance, so that toxic solvents could be avoided. Mendiola *et al.* (2007) described very effective antibacterial and antifungal CO_2 extracts from *Arthospira platensis*, and also a tocopherol- enriched extract (Mendiola *et al.,*2008), both of which may be of possible interest for cosmetic preparations. *Aphanizomenon* and *Arthrospira* are cyanobacteria which are known to produce high molecular weight polysaccharides and which have been reported to have *in vitro* higher immune-stimulatory activities than the commercially available immune-therapeutics (Pugh *et al.*, 2001). For example, calcium spirulan (Ca-Sp) is of immense interest for its use in cosmeceuticals. A 3-step purification process for this polysaccharide from *Arthrospira* has been developed and high anti-HSV activities have been detected (Sandau and Pulz, 2009). Cyanobacteria are routinely exposed to high oxygen and radical stresses, especially in extreme environments. In the long term, this has resulted in the improvement and development of numerous efficient protective systems against oxidative and radical stressors (Whitton and Potts 2000). These protective mechanisms are able to prevent the accumulation of free radicals and reactive oxygen species and thus help the cyanobacteria to counteract cell damaging activities. Because these antioxidative components have their origin from a natural source, their application in cosmetics for preserving and protecting purposes is hence, developing rapidly. It is known that the exposure of the skin to UV light is one of the main reasons for premature skin aging and also for skin cancer; hence, sun-protecting cosmetics represent an area of high demand. Many cyanobacteria are adopt of overcoming the toxic effects of ultraviolet radiation by synthesizing UV-absorbing compounds. These naturally synthesized compounds are biotechnologically being exploited by the cosmetic industry for the development of sunscreens and various other cosmetic products. In general, the cosmetic market segment is changing at a breathtaking pace and new products with advanced skin- protecting characteristics are being added to the industry with an envious rate. The almost untapped and truly excellent potential of the cyanobacteria with their vast adaptative mechanisms is of precise interest in this connection.

Biofertilizers

The use of algae and cyanobacteria as biofertlizer is a technology that holds immense promises for the future. Macroalgae have been used as soil fertilizers in many coastal regions all over the world (Critchley and Ohno 1998; Zemke-White and Ohno 1999). But, the beneficial role of cyanobacteria in the soil ecosystem has often been neglected. If we look at the beneficial effects of cyanobacteria as biofertilizers they are in fact numerous: amplified water-binding capacity and water storage, particle adherence and reduction of soil erosion, improvement of mineral composition of the soil, the production and secretion of bioactive compounds such as phytohormones, which ultimately stimulate the growth of agricultural crops (Stirk *et al.*,2002). These same properties can also be utilized in liquid fertilizers produced from the macroalgae, such as in the improvement of cover for abandoned mining lands to avoid erosion and also to initiate floral succession. For the past few years, several studies have been carried out in order to include better strain identification, isolation and culture of algae, analyzing their N_2- fixing activity and related physiological processes, energetic and biochemistry, and elucidating the structure and regulation of nitrogen fixing (*nif*) genes and the nitrogenase enzyme (Vaishampayan *et al.*, 2001). As a consequence of their wide tolerance of adverse environmental conditions like desiccation, hot temperatures etc. Cyanobacteria appear to be particularly suitable for the use as excellent site specific biofertilizers. Also, diverse potentially other useful effects have been shown for the cyanobacteria, such as their role as antifungal substances (Kim, 2006). The pre-soaking of paddy seeds with cyanobacterial cultures or extracts has also been reported to enhance germination and growth, although very consistent investigations are still lacking (Sharma *et al.*, 2010). However, it is mainly the ability of some cyanobacteria to fix N_2 through various physiological mechanisms which makes them excellent biofertilizers. The study of N_2 fixation in rice fields of north-east Spain by Quesada *et al.* (1997) gives an example of the importance of cyanobacterial fixation; it was estimated that N_2 fixed on a per crop basis was in the range of 5–80 kg ha^{-1}N, the efficiency being strongly subject to environmental conditions. This high rate of nitrogen supplementation is indeed very helpful for maintaining a sustained development of the agricultural fields. It has been estimated that using *Aulosira fertilissima* and *Anabaena doliolum* with or without the combination of urea, the actual nitrogen demand for a rice field in north India could be reduced by an amount as low as 25 per cent (Dubey and Rai 1995). While cyanobacterial N_2 fixation can be used to enhance crop yields and reduce the use of chemical N fertilizers, the cyanobacteria can also be used in the more arid regions to reduce erosion processes, because of their ability to form EPS, which is known to improve the water-binding capacity and soil structure. A point of caution for the use of cyanobacteria as biofertilizers is that they should be site specific and thus indigenous to the area where they are to be used. An experiment in Chile by Pereira *et al.* (2009) used *Anabaena iyenganii* and *Nostoc* spp. indigenous to the region in their biofertilizer fortrials on local rice fields. As a consequence, the use of the biofertilizer decreased the amount of synthetic nitrogen fertilizer (50 kg N ha$^{-1)}$ required by as much as 50 per cent, while still maintaining the same yield of 7.4 tha^{-1} rice. Another different method is to use *Azolla* with its symbiotic cyanobacteria *Anabaena*. This strategy has been applied for green manuring rice fields in China and Vietnam for centuries (Watanabe, 1982). There are primarily two standard approaches used in different situations for manuring with *Azolla* (Sharma *et al.*, 2010). Either it can be grown as a monocrop and applied to the paddy prior to the rice being planted or alternatively, it can also be grown together with the rice. The rice yield thereby was found to be enhanced by about 0.5–0.75 tha^{-1}. However, it has been shown that free living cyanobacteria release ammonium into the water, where it can be utilized directly by the crop, whereas with *Azolla* the ammonium is not as directly available as compared to the free living forms of cyanobacteria.

Wastewater Treatment

Sustained development of any ecosystem which may be as big as the earth requires proper mitigation of all the potential hazards through an approach that has longevity and that supports life eventually. Thus, the protection and preservation of the natural and continuous basis of life are not only ethical demands, but also essential for long-lasting economic and social development. This approach is also expected to create technological progress and jobs. Further development and perfection of the present existing systems for wastewater treatment, the feasible reduction of problematic emissions and the need for water recycling are important objectives that hold the hey for a better future. Micro- and macroalgae, sometimes in combination with other microorganisms, can be utilized to treat wastewater and other harmful effluents. As of now, the current uses include: removal of nutrients from circulating process water, use of CO_2 from industrial exhaust gases; disposal of deleterious contaminants from agricultural waste water; purification of wastewater from biogas production and tertiary wastewater purification. Governments all across the world and energy based companies worldwide are showing an ever-increasing interest in CO_2 - fixation biotechnology, especially due to the introduction of CO_2 certificates and the concepts of carbon credit. Numerous cyanobacteria have been used in the past decade for the remediation of agricultural or industrial wastewaters. *Arthrospira* has been used for pig wastewater treatment in Mexico (Olguín *et al.*, 2003) after dilution with sea water, while *Nostoc muscorum* and *Anabaena subcylindrica* were extensively used for industrial wastewater effluents in Egypt (El-Sheekh *et al.*, 2005).

Conclusion

It is evident that cyanobacteria have significant economic uses and the field of biotechnology can be used for proper facilitation of these resources (Figure 5.1). The use of cyanobacteria as biofertilizers has long been a topic of interest for the

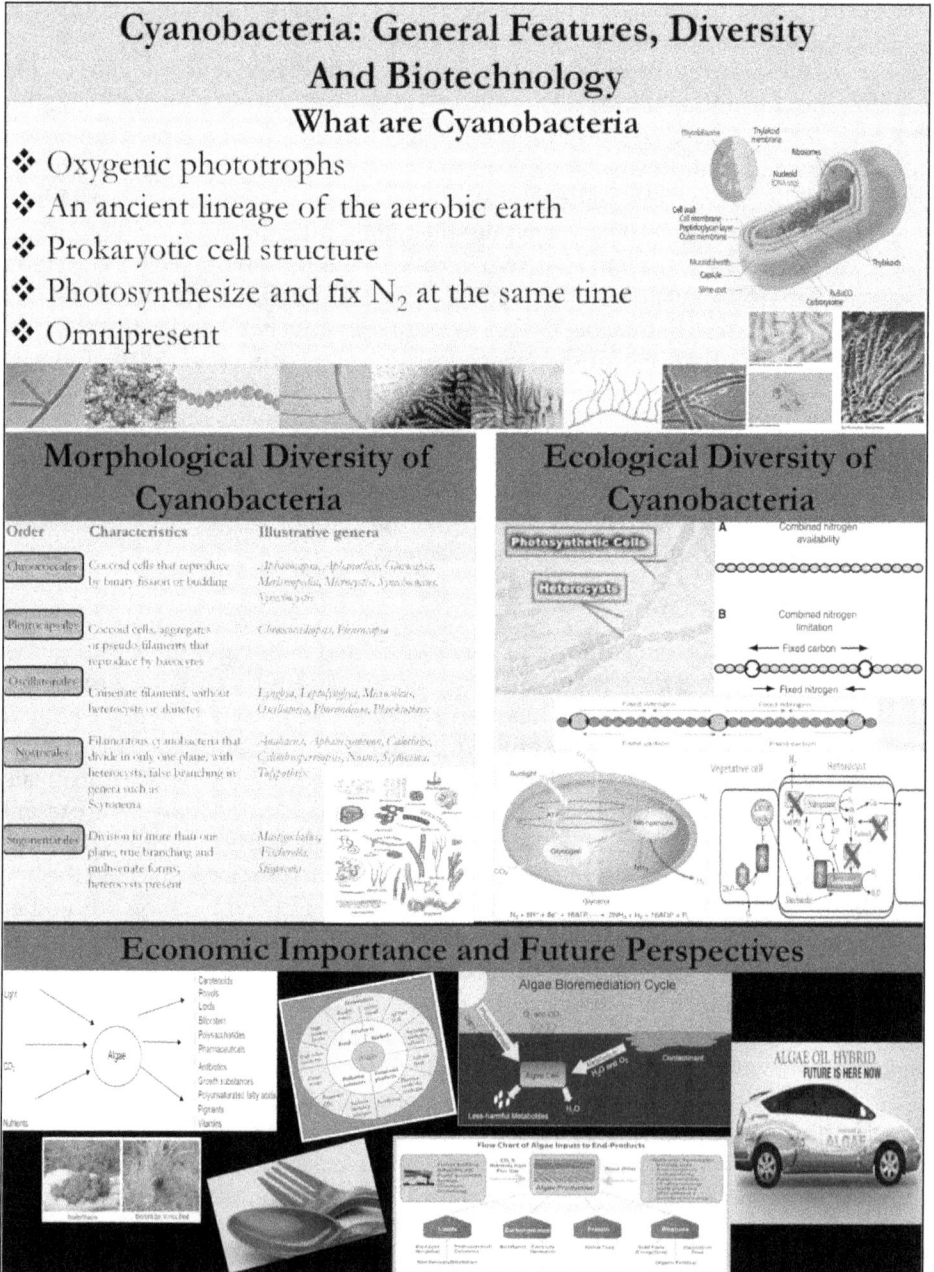

Figure 5.1: Cyanobacterial Diversity and its Applications in Biotechnology.

agriculturist and their inherent property of nitrogen fixation and photosynthesis must be studied with greater detail so that proper usage could be warranted. In the same way, their use of food supplements both for human consumption and animal feed industry also still has some lacunas which need to be filled up. The use of cyanobacteria as cosmetics and in bioremediation of waste water systems is still at a nascent stage and these fields need to be developed in such a way that the laboratory to field transition is equally beneficial. Future prospects of cyanobacterial usage in plastic industry and as biofuels are also important ventures that need to be tackled properly and through scientific methodologies.

Acknowledgement

Dr. Satya Shila Singh is thankful to Department of Botany, Guru Ghasidas Vishwavidyala for providing necessary facilities. She is also thankful to funding agencies UGC and CSIR, New Delhi for financial assistance. Dr. Prashant Singh is grateful to MCC, NCCS, Pune for providing necessary facilities.

References

1. Abdulqader GL, Barsanti L, Tredici MR (2000). Harvest of *Arthrospira platensis* from Lake Kossorom (Chad). and its household usage among the Kanembu. *J. Appl. Phycol.* 12(3): 493–498.

2. Adams DG (2000). Heterocyst formation in cyanobacteria. *Curr. Opin. Microbiol.* 3: 618-624.

3. Belay A (1997). Mass culture of *Spirulina* outdoors: the Earthrise Farms experience. In: Vonshak A (ed). *Spirulina platensis (Arthrospira): Physiology, cell-biology and biotechnology.* Taylor and Francis, London, pp. 131–142, 233 pp.

4. Bergman B, Rasmussen U RaiAN (2007). Cyanobacterial associations, In: Elmerich, C. and Newton, W.E. (Eds.), *Associative and Endophytic Nitrogen-Fixing Bacteria and Cyanobacterial Associations.* vol. 5. Kluwer Acad., Dordrecht, pp. 257-301.

5. Berman-Frank I, Lundgren P, Falkowski P (2003). Nitrogen fixation and photosynthetic oxygen evolution in cyanobacteria. *Res. Microbiol.* 154: 157-164.

6. Bryant DA (1991). Cyanobacterial phycobilisomes: progress towards complete structural and functional analysis via molecular genetics, In: Bogorad, L. and Vasil, I.K. (Eds.), *Cell Culture and Somatic Cell Genetics of Plants*, Vol. 7B. Academic Press., New York, pp. 257-300.

7. Bryant DA (2003). The beauty in small things revealed. *Proc. Nat. Acad. Sci. USA.* 100: 9647-9649.

8. Burja AM, Banaigs B, Abou-Mansour E, Burgess JG Wright PC (2001). Marine cyanobacteria- a prolific source of natural products. *Tetrahedron.* 57: 9347-9377.

9. Carmichael WW (1994). The toxins of Cyanobacteria. *Scientific American* 270 (1): 78-86.

10. Carmichael WW, Drapeau C, Anderson DM (2000). Harvesting of *Aphanizomenon flos-aquae* Ralfs ex Born. and Flah. var. *flos-aquae* (Cyanobacteria). from Klamath Lake for human dietary use. *J. Appl. Phycol.* 12(6): 585–595.

11. Carmichael WW, Azevedo SM, An JS, Molica RJ, Jochimsen EM, Lau S, Rinehart KL, Shaw G R, Eaglesham GK (2001). Human fatalities from cyanobacteria: chemical and biologicalevidence for cyanotoxins. *Environ. Health Perspect.* 109 (7): 663 – 668.

12. Castenholz RW (1973). Ecology of blue-green algae in hot springs, In: Carr, N.G. and Whitton, B.A. (Eds.), *The Biology of Blue-Green Algae.* Blackwell Scientific Publications. Oxford, pp. 379-414.

13. Castenholz RW (2001). Phylum BX. Cyanobacteria, Oxygenic Photosynthetic Bacteria, In: Boone, D.R. and Castenholz, R.W. (Eds.). *Bergey's Manual of Systematic Bacteriology*, Vol. 1. Springer-Verlag., New York-Berlin-Heidelberg. pp. 473-597.

14. Castenholz RW (1992). Species usage, concept, and evolution in the cyanobacteria (blue-green algae). *J. Phycol.* 28: 737-745.

15. Chuntapa B, Powtongsook S, Menasveta P (2003). Water quality control using *Spirulina platensis* in shrimp culture tanks. *Aquacult.* 220(1–4): 355–366.

16. Ciferri O (1983). *Spirulina*, the edible microorganism. *Microbiol. Mol. Biol. Rev.* 47(4): 551–578.

17. Delpeuch F, Joseph A, Cavelier C (1975). Consumption as food and nutritional composition of blue-green algae among populations in the Kanem region of Chad. *Ann. Nutr. Aliment* 29: 497–516.

18. Dor I, Danin A (1996). Cyanobacterial desert crusts in the Dead Sea Valley, Israel. *Arch. Hydrobiol./Algol. Stud.* 83: 197-206.

19. Dubey A, Rai A (1995). Application of algal biofertilizers (*Aulosira fertilissima* Tenuis and *Anabaena doliolum* Bhardawaja). for sustained paddy cultivation in northern India. *Isr. J. Plant Sci.* 43(1): 41–52.

20. El-Sayed AFM (1994). Evaluation of soybean meal, *Spirulina* meal and chicken offal meal as protein sources for silver seabream (*Rhabdosargus sarba*). fingerlings. *Aquacult.* 127(2–3): 169–176.

21. El-Sheekh MM, El-Shouny WA, Osman MEH, El-Gammal EWE (2005). Growth and heavy metals removal efficiency of *Nostoc muscorum* and *Anabaena subcylindrica* in sewage and industrial wastewater effluents. *Environ. Toxicol. Pharmacol.* 19(2): 357–365.

22. Elster J, Lukešová A, Svoboda J, Kopecký J, Kanda, H (1999). Diversity and abundance of soil algae in the polar desert, Sverdrup Pass central Ellesmere Island. *Polar Record* 35: 231-254.

23. Field CB, Behrenfeld MJ, Randerson JT, Falkowski P (1998). Primary production of the biosphere: integrating terrestrial and oceanic components. *Science.* 281: 237-240.

24. Flores E, Herrero A (2010). Compartmentalized function through cell differentiation in filamentous cyanobacteria. *Nature Rev. Microbiol.* 8: 39-50.

25. Flores E, Herrero A, Wolk CP, Maldener I (2006). Is the periplasm continuous in filamentous multicellular cyanobacteria? *Trends Microbiol.* 14: 439-443.

26. Fujishiro T, Ogawa T, Matsuoka M, Nagahama K, Takeshima Y, Hagiwara H (2004). Establishment of a pure culture of the hitherto uncultured unicellular cyanobacterium *Aphanothece sacrum*, and phylogenetic position of the organism. *Appl. Environ. Microbiol.* 70(6): 3338–3345.

27. Gantt E (1994). Supramolecular Membrane Organization, In: Bryant, D.A. (Ed.), *The Molecular Biology of Cyanobacteria*, Kluwer., Dordrecht, pp 119- 138.

28. Gao K (1998). Chinese studies on the edible blue-green alga, *Nostoc flagelliforme*: a review. *J. Appl. Phycol* 10(1): 37–49.

29. Gershwin ME, Belay A (2007). *Spirulina in human nutrition and health*. CRC Press, Boca Raton, 328 pp.

30. Giovannoni SJ, Turner S, Olsen GJ, Barns S, Lane DJ, Pace NR (1988). Evolutionary relationships among cyanobacteria and green chloroplasts. *J. Bacteriol.* 170: 3584-3592.

31. Grobbelaar JU (2009). Factors governing algal growth in photobioreactors: the "open" versus "closed" debate. *J. Appl. Phycol.* 21(5): 489–492.

32. Kalaitzis JA, Lauro FM, Neilan BA (2009). Mining cyanobacterial genomes for genes encoding complex biosynthetic pathways. *Nat. Prod. Rep.* 26: 1447-1465.

33. Kann E (1988). Zur Autøkologie benthischer Cyanophyten in reinen europäischen Seenund Fliessgewässern. *Arch. Hydrobiol./Algol. Stud.* 50/53: 473-495.

34. Karl D, Michaels A, Bergman B, Capone D, Carpenter E (2002). Dinitrogen fixation in the world's oceans. *Biogeochem.* 87/88: 47-98.

35. Khaing MK (2004). A study on the edible cyanobacteria (blue green algae). *Nostoc* species in Upper Myanmar. *Ph.D. Thesis*, Department of Botany, Mandalay University, Myanmar.

36. Khan Z, Bhadouria P, Bisen P (2005). Nutritional and therapeutic potential of *Spirulina. Curr. Pharm. Biotechnol.* 6(5): 373–379.

37. Kim JD (2006). Screening of cyanobacteria (blue-green algae). from rice paddy soil for anti-fungal activity against plant pathogenic fungi. *Korean J. Mycol.* 34(3): 138–142.

38. Kol E (1968). Kryobiologie. I. Kryovegetation, In: Elster, H.J. and Ohle, W. (Eds.), Die Binnengewässer, Band XXIV. E. *Schweizerbart'sche Verlagsbuchhandlung, Stuttgart*, 216 pp.

39. Kulasooriya SA (2008). *Biological Nitrogen Fixation: Fundamentals and Utilization*. Science Education Unit, Faculty of Science, University of Peradeniya. Pp. 143.

40. Kulshreshtha AJ, Ansih Z, Jarouliya U, Bhadauriya P, Prasad GBKS, Bisen PS (2008). Spirulina a wonder alga full of nutrition. *Curr. Pharmaceut. Biotech.* 9 (5): 400 –405.

41. Laamanen M (1996). Cyanoprokaryotes in the Baltic Sea ice and winter plankton. *Arch. Hydrobiol. Algol. Stud.* 83: 423-433.

42. Lane J (2010). Joule wins key patent for GMO cyanobacteria that creates fuel from sunlight, CO2 and water. *Biofuels Digest* 14(September).

43. Lee RE (2008). *Phycology*, 4th edn. Cambridge University Press, Cambridge, UK, 560 pp.

44. Lincoln EP, Wilkie AC, French BT(1996). Cyanobacteria process for renovating dairy waste water. *Biomass Bioen.* 10 (1): 63 – 68.

45. Mataloni G, Komarek J (2004). *Gloeocapsopsis aurea*, a new subaerophytic cyanobacterium from maritime *Antarctica. Polar Biol.* 27: 623-628.

46. Mendiola JA, García-Martínez D, Rupérez FJ, Martín-Álvarez PJ, Reglero G, Cifuentes A, Barbas C, Ibañez E, Señoráns FJ (2008). Enrichment of vitamin E from *Spirulina platensis* microalga by SFE. *J Supercrit Fluid* 43(3): 484–489.

47. Mendiola JA, Jaime L, Santoyo S, Reglero G, Cifuentes A, Ibanez E, Señoráns F (2007). Screening of functional compounds in supercritical fl uid extracts from *Spirulina platensis. Food Chem.* 102(4): 1357–1367.

48. Morot-Gaudry JF,Touraine B (2001)Sources of nitrogen, nitrogen cycle, root structure and nitrogen assimilation, In: Morot-Gaudry JF (Ed.), *Nitrogen Assimilation by Plants: Physiological, Biochemical and Molecular Aspects*, Science Publishers Inc., NH, pp. 5-14.

49. Mühling M, Belay A, Whitton BA (2005). Variation in fatty acid composition of *Arthrospira* (Spirulina). strains. *J. Appl. Phycol.* 17(2): 137–146.

50. Muller Feuga A (2004). Microalgae for aquaculture: the current global situation future trends. In: Richmond A (ed). *Handbook of Microalgal Culture: Biotechnology and Applied Phycology*. Blackwell Publishing, Oxford, pp 352–364, 566 pp.

51. Muller Feuga A, Robert R, Cahu C, Robin J, Divanach P (2003). Uses of microalgae in aquaculture. In: Stottrup J, McEvoy L (eds). *Live feeds in marine aquaculture*. Blackwell Publishing Co, Oxford, pp 253–299, 336 pp.

52. Olguín EJ, Galicia S, Mercado G, Pérez T (2003). Annual productivity of *Spirulina* (*Arthrospira*). and nutrient removal in a pig wastewater recycling process under tropical conditions. *J. Appl. Phycol.* 15(2): 249–257.

53. Pereira I, Ortega R, Barrientos L, Moya M, Reyes G, Kramm V (2009). Development of a biofertilizer based on fi lamentous nitrogen- fi xing cyanobacteria for rice crops in Chile. *J. Appl. Phycol.* 21(1): 135–144.

54. Pugh N, Ross SA, ElSohly HN, ElSohly MA, Pasco DS (2001). Isolation of three high molecular weight polysaccharide preparations with potent immunostimulatory activity from *Spirulina platensis, Aphanizomenon fl os-aquae* and *Chlorella pyrenoidosa. Planta Med.* 67(8): 737–742.

55. Pulz O, Gross W (2004). Valuable products from biotechnology of microalgae. *Appl. Microbiol. Biotechnol.* 65(6): 635–648.

56. Pulz O, Storandt R, Boback A (2008). Mikroalgen in der Broileraufzucht. *DGS* 9: 15–18.

57. Qiu B, Liu J, Liu Z, Liu S (2002). Distribution and ecology of the edible cyanobacterium Ge-Xian-Mi (*Nostoc*). in rice fields of Hefeng County in China. *J. Appl. Phycol.* 14(5): 423–429.

58. Quesada A, Leganés F, Fernández-Valiente E (1997). Environmental factors controlling N 2 fi xation in Mediterranean rice fields. *Microb. Ecol.* 34(1): 39–48.

59. Rai AN (1990). *CRC Handbook of Symbiotic Cyanobacteria.* CRC Press, Boca Raton, 253 pp.

60. Reed RH, Chudek JA, Foster R, Stewart WDP (1984). Osmotic adjustment in cyanobacteria. *Arch. Microbiol.* 138: 333-337.

61. Roney BR, Renhui L, Banack SA, Murch S, Honegger R, Cox PA (2009). Consumption of facai *Nostoc* soup: a potential for BMAA exposure from *Nostoc* cyanobacteria in China? *Amyotroph Lateral Scler* 10(S2): 44–49.

62. Sandau P, Pulz O (2009). Untersuchungen zu bioaktiven Wirkungen des Algenpolysaacharids Calcium-Spirulan aus *Arthrospira platensis. OM and Ernährung* 131: F40–F45.

63. Schlotmann K, Waldmann-laue M, Jassoy C, Kaeten M, Koehler E, Pulz O, Kurth E (2005). Extract of blue-green alga *Spirulina* and the use thereof in cosmetic skin-care and body care agents, *EP Patent* 1,239,813.

64. Schopf JW, Hayes JM, Walter MR (1983). Evolution of Earth's earliest ecosystems: recent progress and unsolved problems, In: Schopf J. (Ed.), *Earth's Earliest Biosphere: its Origin and Evolution,* Princeton University Press., Princeton, pp.361-384.

65. Sergeev VN, Gerasimenko LM, Zavarzin GA (2002). The proterozoic history and present state of cyanobacteria. *Mikrobiol.* 71: 623-637.

66. Sharma NK, Tiwari SP, Tripathi K, Rai AK (2010). Sustainability and cyanobacteria (blue-green algae): facts and challenges. *J. Appl. Phycol.* 23(6): 1059–1081.

67. Stal LJ (2000). Cyanobacterial mats and stromatolites. In: Whitton BA, Potts M. (Eds.), *The Ecology of Cyanobacteria,* Kluver Academic Publishers, Dordrecht, pp 61-120.

68. Stevens SE Jr,Nierzwicki-BauerS(1991). Thecyanobacteria, In: Stolz JF(Ed.), *Structure of Phototrophic Prokaryotes,* CRCPress Inc, Boca Raton, pp. 15-47.

69. Stirk W, Ördög V, Van Staden J, Jäger K (2002). Cytokinin-and auxin-like activity in Cyanophyta and microalgae. *J. Appl. Phycol.* 14(3): 215–221.

70. Storandt R, Franke H, Pulz O, Loest K, Ecke M, Steinberg KH (2000). *Algae in animal nutrition.* Tierernährung – Ressourcen und neue Aufgaben, Hannover, Germany, Landbau forschung Völkenrode.

71. Thajuddin N, Subramanian G (2005). Cyanobacterial biodiversity and potential applications in biotechnology. *Curr. Sci.* 89(1): 47–57.

72. Thajuddin N, Muralitharan G, Sundaramoorthy M, Ramamoorthy R, Ramachandran S, Akbarsha MA, Gunasekaran M (2010). Morphological and genetic diversity of symbiotic cyanobacteria from cycads. *J. Basic Microbio.*50: 254-265.

73. Tiwari DN, Kumar A, Mishra AK (1991). Use of cyanobacterial diazotrophic technology in rice agriculture. *Appl. Biochem. Biotech.* 28/29: 387-396.

74. Vaishampayan A, Sinha R, Hader DP, Dey T, Gupta A, Bhan U, Rao A (2001). Cyanobacterial biofertilizers in rice agriculture. *Bot. Rev.* 67(4): 453–516.

75. Van Landingham SL (1982). *Guide to the Identification, Environmental Requirements and Pollution Tolerance of Freshwater Blue-Green Algae (Cyanophyta)*. United States Environmental Protection Agency, Cincinnati, Ohio, pp 341.

76. Venkataraman GS (1972). *Algal Biofertilizers and Rice Cultivation*. Today and Tomorrows Printers and Publishers, Calcutta: Pp 81.

77. Vonshak A (1997). *Spirulina platensis (Arthrospira)*: physiology, cellbiology and biotechnology. Taylor and Francis, London/Bristol, 233 pp.

78. WalsbyAE (1994). Gas vesicles. Microbiol. rev. pp.94-144.

79. Watanabe I (1982). *Azolla -Anabaena* symbiosis – its physiology and use in tropical agriculture. In: Dommergues YE (ed). *Microbiology of tropical soils and plant productivity*. Kluwer Academic Publishers, Dordrecht, p 169.

80. Weber B, Wessels DCJ, Büdel B (1996). Biology and ecology of cryptoendolithic cyanobacteria of a sandstone outcrop in the Northern Province, South Africa. *Arch.Hydrobiol./Algol. Stud.* 83: 565-579.

81. Whitton BA (1973). Freshwater plankton, In: Fogg GE, Stewart WDP, Fay P, Walsby AE (Eds.), *The Blue-Green Algae*. Academic Press., London, pp 353-367.

82. Whitton BA, Potts M (2000). Introduction of cyanobacteria. In: Whitton BA, Potts M (Eds.), *The Ecology of Cyanobacteria. their Diversity in Timeand Space*. Kluwer Academic, Dordrecht, Netherlands, pp 1-10.

83. Zemke-White W, Ohno M (1999). World seaweed utilisation: an end-ofcentury summary. *J. Appl. Phycol.* 11(4): 369–376.

2016, Environmental Biotechnology: A New Approach   *Pages 101–110*
Editors: Dr. Rajan Kumar Gupta and Dr. Satya Shila Singh
Published by: DAYA PUBLISHING HOUSE, NEW DELHI

Chapter 6

Eutrophication, Algal Bloom and their Toxicity on Human and Animals

Kaushal Kishore Choudhary[1]* and Raisana Kumari[2]

*[1]Department of Botany,
Dr. Jagannath Mishra College,
(Affiliated to B.R. Ambedkar Bihar University, Muzaffarpur)
Muzaffarpur – 842 001, Bihar
[2]Department of Botany, R.D.S. College,
Muzaffarpur, Bihar*

ABSTRACT

Cyanobacteria or blue green algae are the most vital biological system on the earth surface that play significant role in nutrient cycling particularly nitrogen fixation and phosphorus mobilization. However, recent findings have revealed the harmful effects of cyanobacteria forming bloom in estuaries, ponds, reservoirs etc. Study suggests that rapid industrialization and mismanaged disposal of industrial, agricultural and domestic wastes may lead to development of harmful cyanobacteria producing toxins deleterious to animal and human health. The inputs of nutrient form various sources potentially stimulate the primary productivity in eutrophic water bodies resulting into the development of algal blooms development. Algal bloom creates hypoxia/anoxia in benthic and metalimnetic strata of the water and generates harmful or toxic effects on the coexisting organisms especially on fisheries resources, ecosystems and human health or recreation. The many phytoplankton communities are capable of forming blooms, however cyanobacteria are more prominent in such waters. Nevertheless, the production of toxins by bloom forming organisms is the increasing concern worldwide. They

* *Corresponding author.* E-mail: kkc1970@gmail.com

produce various kinds of toxins of neurotoxic, hepatotoxic and cytotoxic nature. The current chapter aimed towards introduction to cyanobacteria, causes of bloom formation i.e. eutrophication and nature of toxicity to humans and animals.

Keywords: Algal bloom, Cyanobacteria, Distribution, Nutrients inputs, Toxins.

Introduction

The distribution, diversity and dominance of the organisms in a particular ecosystem is the function of the availability of nutrient supply, climatic conditions, altitudes, latitudes, pH and the ability of the organisms to utilize the resources efficiently. Most of the plants particularly photosynthetic plants are characteristics of a certain environmental condition. But cyanobacteria are photosynthetic oxygenic group of microorganisms that inhabit all kinds of possible biomes. They are endowed with the ability to store nitrogen, phosphorus and carbon in their cells. Some cyanobacteria are equipped with a specialized structure called 'heterocyst' that enable them to convert elementary nitrogen into soluble form of nitrogen with the help of enzyme "nitrogenase complex" contained within the boundary of heterocyst (Adams and Duggan, 1999). Few cyanobacteria, unlike other photosynthetic organisms, utilize other sources of substrate *i.e.* bicarbonate for their photosynthetic activity. Cyanobacterial doubling time ranges from ten thousands years as shown by cyanobacterium of dry desert of Antarctic (Nienow and Friedlmann, 1993) to fastest growth rate of 2.1h exhibited by a cyanobacterium *Anacystis nidulans* (*Synechococcus* PCC 6803) isolated from a freshwater site in Texas (Kratz and Myers, 1955). This suggested that cyanobacterial dominance depends on the climatic conditions. Such tropic independence of cyanobacteria for carbon and nitrogen together with a great adaptability to variations in environmental factors enables them to be unique creatures of the nature. Cyanobacteria are the common features of many aquatic environments including lakes, rivers and estuaries but their dominance, diversity and distribution differ. The shift in nutrient status and the climatic condition including light, temperature and pH of the system witnesses the change in the existence of cyanobacteria with time and space (Granhall, 1975).

The recent unidirectional development in terms of industrialization and modernization to support the human population is leading into the complete destruction of natural ecosystem and the loss of many important floras and faunas. Release of agricultural, domestic and industrial wastes into rivers, ponds and lakes are causing the enrichment of nutrient status of the system leading to eutrophication (Paerl, 1997; Nixon, 1995). Not only that, the upgradation of aquaculture at industrial scale is also fuelling the eutrophication of aquatic bodies. The change in water bodies natural color to blue-green is primarily referred as 'bloom'. Bloom is generally defined in terms of phytoplankton biomass higher than the lake average. Bloom affected water bodies becomes soupy, more often turquoise, bright blue, grey or even red in color. And the organisms responsible for all these events are predominantly belong to cyanobacteria.

The rapid increase in eutrophication is the increasing concern for the ecologists globally because they are posing a threat to water quality. Additionally, the decomposition of cell pigments causes water to turn purple, red or turquoise, whereas the decomposition of cyanobacterial cells causes loss of oxygen from water and increase in level of ammonia commonly resulting into die-off of fish and other aquatic organisms. Nevertheless, the production of toxins by bloom forming cyanobacteria are most important aspects of present day environmental programme. This chapter has been aimed to describe some aspects of bloom formation and ecological crisis in terms of diversity of aquatic organisms and toxins produced and their influence on humans and animals health.

Cyanobacteria and Bloom Formation

Cyanobacterial evolution in "proterozoic era" (Schopf and Walter, 1982) in nutrient limiting conditions and their distribution in wide range of habitats suggest their potential of adaptability in different ecological niches. The advancement in their morphological and cellular characterization confirmed their adaptability in diverse habitats with presence of spore formation and nitrogen-fixing potential in adverse and nitrogen-limiting condition. Their ability to co-exist in mutualistic and symbiotic associations with other microbes, higher plants and animals provide additional support for their existence under adverse conditions. Additionally, they can utilize other sources of carbon *i.e.* bicarbonate, as a substrate for photosynthesis. Moreover, some cyanobacteria developed gas-vacuole that helps in buoyancy and vertical migration in the deep water. And this is the most important characteristics of the cyanobacteria that help in formation of surface blooms. The adaptation to wide range of habitats along with possession of gas-vacuole is believed to largely responsible for the success of cyanobacteria in nutrient-rich conditions of ponds, lakes and rivers.

There are commonly three kinds of trophic state in the natural system particularly aquatic bodies. These are: oligotrophic, mesotrophic and eutrophic. Oligotrophic state describes the nutrient poor state of the fertility whereas mesotrophic state is nutrients rich and provides optimal condition for growth and development of plants. The eutrophic state of aquatic bodies is the resultant of nutrient pollution emerged due to industrialization. In other words, the eutrophic state or eutrophication is a process rather than a trophic state. The eutrophication is not a permanent state of the estuaries rather it is a temporary state of the system and demolishes with the utilization of nutrients by phytoplankton leading to the increased productivity. With the establishment of excess of nutrients in the estuaries, the primary producers particularly algae witness a population increase in comparison above the average of the system to other living organisms. This is due to the fact that they can utilize the nutrients more efficiently and can build up their populations above lake average in prevailing nutrient rich condition. This abnormal increase in population size of algae in aquatic bodies is called 'Algal Bloom' (Reynolds and Walsby, 1975). The occurrence of bloom is the function of physico-chemical properties to water and ability of organisms to utilize the resources efficiently. It may occur due to single species or a group of species. The bloom has been termed variously on the basis of dominance of the phytoplankton communities *e.g.* cyanobacterial bloom, Diatom bloom or *Anabaena* bloom. In recreational water, a bloom is defined as the development of phytoplankton structure that causes nuisance to humus. The bloom formation *i.e.* over growth of algae, is commonly associated with the eutrophic state of the system *i.e.* eutrophication. In other words, eutrophication is the results of addition of effluent from agriculture industrial and blooms are the result of eutrophication.

Causes of Eutrophication

The eutrophication is the natural aging process of aquatic ecosystems. According to Burkholder (2000), the eutrophication of lakes, rivers, estuaries, and marine waters is the resultant of hundreds or thousands of years of human activities that added nutrients to them. Nixon (1995) has defined eutrophication in terms of "the process of increased organic enrichment of an ecosystem generally through increased nutrients inputs".

The ultimate causes of eutrophication is an increase in nutrients particularly nitrogen (N) and phosphorus (P) and primary productivity of an ecosystem. And this is achieved due to regular increase in industry and application of excess chemical fertilizer in the cultivable land and ultimate release of effluents into the aquatic bodies directly or indirectly. However, the eutrophication may also occur naturally in depositional environments where nutrients accumulate regularly and is beyond the control of humans. According to Odum *et al.* (1995), all estuaries undergo eutrophication periodically. The most striking observation of the eutrophication is the addition of nutrients from outside the system. The increase in nutritional content of the system may occur due to either recycling of nutrients or may enter from the other sources in the system. The former is called allochthonous and later is termed as autochthonous. There are mainly two sources of addition of nutrients and organic matter in the estuaries:

(I) Point Sources

Those nutritional pollutants which enter to the estuaries from source of origin to estuaries is referred as point sources of nutrient pollutants. These include industrial and domestic wastes, wastewater treatment plants and storm water drains (Pinckney *et al.*, 2001). Point sources are less important because it may be controlled and regulated with little efforts. Additionally, they contribute a little in eutrophication. Howarth *et al.* (1996) have demonstrated that the nonpoint sources of nitrogen are much more than those contributed from point sources of sewage inputs in Europe and North America. Further it was shown that only 12 per cent of N is contributed by sewage in North America. The total contribution of N and P in Missisipi River has been observed to be 20 per cent and 40 per cent respectively through point sources (Goolsby *et al.*, 2000).

(II) Nonpoint Sources

Nonpoint sources are also called as diffused or runoff pollution. These are the sources of nutrients which are gathered at one site and carried away to the estuaries with the rain water. These include soil retention, groundwater run-off and atmospheric deposition (acid rain). This kind of nutrient pollution is difficult to regulate and vary with season, precipitation and other irregular events. Nixon (1995) has proposed that the nonpoint pollution is the dominant sources of N and other nutrient inputs in many water estuaries. The survey of different bloom affected water bodies suggest that the nonpoint sources of nutritional loading is the main source of eutrophication. Carpenter *et al.* (1998) have observed that nonpoint sources are also responsible for maximum contribution of phosphorus (P) in estuaries. There are three common sources of nonpoint nutrient pollution. These are:

(i) Soil Retention

The excess application of chemical fertilizer to agricultural field are resulting into their accumulation in soils. And these chemicals are entering into the estuaries along with rain water. The observation with many water bodies suggest that the amount of phosphorus lost to surface waters increases linearly with the amount of phosphorus in the soil (Sharpley *et al.*, 1996). Thus, much of the nutrients loading in soil eventually make its way to water thereby causing eutrophication.

(ii) Runoff to Surface Water and Leaching to Groundwater

Nutrients from human activities tend to travel from land to either surface or ground water. Nitrogen in particular is removed through storm drains, sewage pipes and other forms of surface runoff. Nutrient loss in runoff and leachate are often associated with agriculture. Modern agriculture often involves the application of excess fertilizer into fields in order to maximize production. However, farmers frequently apply more nutrients than are taken up by crops (Buol, 1995) or pastures. Regulations aimed at minimizing nutrient exports from agriculture are typically far less stringent than those placed on sewage treatment plants (Carpenter *et al.*, 1998) and other point source polluters.

(iii) Atmospheric Deposition

The different compounds is released into the atmosphere because of combustion of fossil fuels. The accumulation of nitrogen is more important in this context. Nitrogen is released into air due to volatilization of ammonia and nitrous oxide production. These compounds deposited in the air eventually make it way to estuaries along with rain (acid rain). Gathering of literature on causes of eutrophication suggests that the regular increase in industrialization and urbanization is leading to the deposition of different substances and fine particulate in the atmosphere. Consequently, they are joining to the earth surface in the form of acid rains thereby increasing the nutrient status of the estuaries (Duce, 1986; Paerl, 1995). Thus, atmospheric deposition (*e.g.* in the form of acid rain) effect nutrient concentration in water (Paerl, 1997) and more prominent in highly industrialized regions.

(iv) Other Sources

Besides the above discussed causes of eutrophication, the rate of water renewal plays a critical role in eutrophication. Additionally, stagnant water is more sensitive to eutrophication than running water because former is more susceptible to collection of nutrients from other sources than replenished water supplies. It has also been shown that the drying of wetlands causes an increase in nutrient concentration and subsequent eutrophication booms (Mungall and McLaren, 1991).

Toxicity and Harmful Algal Blooms

The last several decades has witnessed an escalation in the incidence of bloom formation that is toxic or harmful. The bloom forming phytoplankton community has been grouped under a single umbrella that is called Harmful Algal Blooms or HABs. The bloom forming cyanobacteria cause harm either by producing toxins or by altering the food web of the system by depleting the oxygen and light energy supply to the co-existing organisms. There are commonly two types of HABs occurring in the estuaries. One that produces toxins that cause death of wildlife and other that harms in other way. The progress in relationship between nutrients supply and toxin producing cyanobacteria has variously been studied. Smith (1983) has proposed that the noxious blooms of toxic or harmful cyanobacteria are stimulated by P enrichment. Many bloom forming cyanobacteria produce bioactive compounds that include cytotoxins, hepatotoxins and neurotoxins that cause death of wildlife around the world (Skulberg *et al.*, 1993; Codd *et al.*, 1997) and rarely death of humans as well (Chorus and Bartram, 1999). Boyer *et al.* (1987) has reported the 5-10 fold increase in toxin production in P-limited versus N-limited conditions. The production of domoic acid has been found to be inversely correlated to the Si concentration in batch culture (Pan *et al.*, 1996). Moreover, not all species of cyanobacteria dominating the earth surface are toxic. Even within a single species, some strains are toxic while others are not. In a eutrophic water bodies, the toxic and non-toxic strains of a species occur simultaneously. Toxicity in a bloom-affected lakes, ponds and rivers are not a permanent function rather it degrades with depletion in nutrients and collapse of bloom occurrence. Neurotoxins degrade rapidly in the environment whereas microcystins are more persistent and may thrive in the system upto two weeks.

Cyanobacterial toxicity has been reported since long time over a century ago. The first literature on toxin production by cyanobacteria accumulated in the Journal 'Nature' by Francis in 1878 with cyanobacterium *Nodularia*. The cyanotoxins have variously been reported to affect the humans as well as other animals (Carmichael, 1992; Sivonen and Jones, 1999). The neurotoxins that include anatoxin-a, homoanatoxin-a, anatoxin-a(s) and saxitoxin are produced by different genera of cyanobacteria (*Anabaena, Aphanizomenon, Oscillatoria* and *Trichodesmium*). The anatoxin-a has been reported to block the postsynaptic cholinergic transmission that causes fatigue and paralysis whereas homoanatoxin-a (*Oscillatoria formosa*) and anatoxin-a(s) (*Anabaena flos-aquae*) are responsible for neuromuscular blocking (Skulberg *et al.*, 1992) and salivation in vertebrates (Mahmood and Carmichael, 1986). Another kind of toxins produced by cyanobacteria is hepatotoxins. They have been categorized into microcystins; and the most common species producing are *Microcystis* sp, *Anabaena flos-aquae, Nostoc rivulare,* and *Oscillatoria agardhii* (Codd and Beattie, 1991); Nodularin produced by *Nodularia spumigena* (Codd and Beattie, 1991) and Cylindrospermopsin from *Cylindrospermopsis reciborskii* (Harada *et al.*, 1994). They all cause acute liver necrosis those results into the death of affected organisms. The cyanotoxins produced by cyanobacteria fall primarily into scytophycins, cyanobacterin, hepalindole A, acutiphycins and lyngbyatoxins. The scytophycins have been reported from *Scytonema pseudohofmanii* and have been found to affect the human epidermoid carcinoma (Moore *et al.*, 1986). The cyanobacterin (*Scytonema hofmanii*) is algicide (Mason *et al.*, 1982) whereas hepalindole A (*Hapalosiphon fontinalis*) showed antialgal and antimycotic property (Moore *et al.*, 1984). The acutiphycin (*Oscillatoria acutissima*) and lyngbyatoxins (*Lyngbya majuscula*) have been reported to cause lung carcinoma (Barchi *et al.*, 1984) and promote tumour in organisms (Moore, 1981) respectively. In this way cyanobacteria not only affect the aquatic organisms where they develop but also terrestrial organisms through drinking water.

Conclusion

Eutrophication of ponds, rivers and lakes are increasing rapidly throughout the WHO and now a global concern. The nutrient loading into the aquatic bodies is the main reason for the development of algal bloom and associated consequences of disturbance of food web as well as death of fishes and animals. The significant knowledge has been accumulated regarding regulation of phytoplankton community composition and their dominance and distribution in the eutrophic water bodies. The information describes that the how nutrients input and their proportions govern some species or group of species. Now, it is well established that impacts of nutrient loading is governed by the species composition, nutritional state of the organisms at the time of input of new nutrients supply and the environmental conditions at the time of nutrient input. The similar nutrient status may not alter the food web of the system in different environment. Thus it is appealing to control the outbreaks of the frequent algal bloom occurrence by developing the suitable strategies to control the rate of nutrient supply into the aquatic bodies.

References

1. Adams DG and Duggan PS (1999). Heterocyst and akinete differentiation in cyanobacteria. *New Phytol* 144: 3-33.

2. Barchi JJ, Moore RE, Furusawa E and Patterson GML (1984). Acutiphycin and 20,21-didehydroacutiphycin, new antineoplastic agents from the cyanophyte *Oscillatoria acutissima*. *J Am Chem Soc* 106: 8193-8197.

3. Boyer GL, Sullivan JJ, Andersen RJ, Harrison PJ and Taylor FJR (1987). Effects of nutrient limitation on toxin production and composition in the marine dinoflagellate *Protogonyaulax tamarensis*. *Mar Biol* 96: 123–128.

4. Buol SW (1995). Sustainability of Soil Use. *Annual Rev. Ecol. Syst.* 26: 25-44.

5. Burkholder JM (2000). Eutrophication and oligotrophication. In: *Encyclopedia of Biodiversity* (ed. S. Levin), Academic Press, New York. Volume 2. pp. 649–670.

6. Carmichael WW (1992). A Review: Cyanobacteria secondary metabolites – the cyanotoxin. *J. Appl. Bact.* 72: 445-459.

7. Carpenter SW, Caraco NF and Smith VH (1998). Nonpoint pollution of surface waters with phosphorus and nitrogen. *Ecol Appl* 8: 559-568.

8. Chorus I and Bartram J (1999). *Toxic Cyanobacteria in Water - A Guide to their Public Health Consequences, Monitoring, and Management* (Eds. Chorus I and Bartram J). E and FN Spon, published on behalf of the World Health Organization, New York.

9. Codd GA and Beattie KA (1991). Cyanobacteria (blue-green alga) and their toxins: awareness and action in the United Kingdom. *PHLS Microbiology Digest Supplement*, London 8: 82-86.

10. Codd GA, Ward CJ and Bell SG (1997). Cyanobacterial toxins: Occurrence, modes of action, health effects and exposure routes. In J. P. Seiler and E. Vilanova (eds.), *Applied Toxicology: Approaches through Basic Science. Archives of Toxicology Supplement* 19. Springer-Verlag, Berlin, Germany p. 399–410.

11. Duce RA (1986). The impact atmospheric nitrogen, phosphorus and iron species on marine biological productivity. In: *The role of Air-Sea Exchange in Geochemical Cycling* (ed. Buat-Menard) Norwell, MA: D. Riedel: 497-529.

12. Francis G (1878). Poisonous Australian Lake. *Nature* (London) 18: 11-12.

13. Goolsby DA, Bettaglin WA, Aulenbach BT and Hooper RP (2000). Nitrogen flux and sources in Mississipi River Basin. *Sci Total Environ* 248: 75-86.

14. Granhall U (1975). Nitrogen fixation by blue-green algae in temperate soils. In: *Nitrogen-Fixation by Free-living Microorganisms* (ed Stewart WDP), Cambridge University Press, Cambridge, UK, 189-198.

15. Harada K-I, Ohtani I, Iwamoto K, Sazuki M, Watanable MF and Terao K (1994). Iasolation of cylindrospermopsin from a cyanobacterium *Cylindrosprmopsis raciborskii* and its screening method. *Toxicon* 32: 73-84.

16. Howarth, RW, Billen G, Swaney D, Townsend A, Jaworski N, Lajtha K, Downing JA, Elmgren, R, Caraco N, Jordan T, Berendse F, Freney J, Kudeyarov V, Murdoch P and Zhao-Liang Zhu (1996). Regional nitrogen budgets and riverine inputs of N and P for the drainages to the North Atlantic Ocean: natural and human influences. *Biogeochemistry* 35: 75-139.

17. Kratz WA and Myers J (1955). Nutrition and growth of several blue-green algae. *Am J Bot* 42: 275-280.

18. Mahmood NA and Carmichael WW (1986). Paralytic shellfish poisons produced by the freshwater cyanobacterium *Anabaena flos-aquae* NH-5. *Toxicon* 24: 175-186.

19. Mason CP, Edwards KR, Carlson RE, Pignatello J, Gleason FK and Wood JM (1982). Isolation of chlorine-containing antibiotic from the freshwater cyanobacterium, *Scytonema hofmanni. Science* 215: 400-402.

20. Moore RE (1981). Constituents of the blue-green algae. In: *Marine Natural Products: Chemical and Biological Perspectives* (ed Scheuer PJ), Academic Press, 4: 1-52.

21. Moore RE, Cheuk C, Yang XG and Patterson GML (1984). Hapalindoles: new alkaloids from the blue-green alga *Hapalosiphon fontinalis. J. Amm Chem Soc* 106: 6456-6457.

22. Moore RE, Patterson GML, Entzeroth M, Morimoto H, Saganuma M, Hakii H, Fujiki H and Sugimura T (1986). Binding studies of [H-3] lyngbyatoxin A and [H-3] debromoplysiatoxin to the phorbol ester receptor in a mouse epidermal particulate fraction. *Carcinogenesis* 7: 641-644.

23. Mungal C and McLaren DJ (1991). *Planet under stress: The challenge of global change*. Oxford University Press, New York, USA.

24. Nienow JA and Friedmann EI (1993). Terrestrial lithophytic (rock) communities. In: Friedmann EI (ed), *Antarctic Microbiology*, Willey-Liss, New York, 353-412.

25. Nixon SW (1995). Coastal marine eutrophication: a definition, social causes and future concerns. *Ophelia* 41: 199-219.

26. Odum WE, Odum EP and Odum HT (1995). Nature's pulsing paradigm. *Estuaries* 18: 547-555.

27. Paerl HW (1995). Coastal eutrophication in relation to atmospheric nitrogen deposition: current perspectives. *Ophelia* 41: 237-259.

28. Paerl HW (1997). Coastal eutrophication and harmful algal blooms: importance of atmospheric deposition and groundwater as "New" nitrogen and other nutrient sources. *Limnol Oceanogr* 42: 1154-1165.

29. Pan Y, Subba Rao DV, Mann KH, Brown RG and R Pocklington (1996). Effects of silicate limitation on production of domoic acid, a neurotoxin, by the diatom Pseudo-nitzschia multiseries. II. Continuous culture studies. *Mar Ecol Prog Ser* 131: 235–243.

30. Pinckney JL, Paerl HW, Tester PA and Richardson T (2001). The role of nutrient loading and eutrophication in estuarine ecology. *Environment Health Perspectives* 109: 699-706.

31. Reynolds CS and Walsby AE (1975). Water-blooms. *Biological Reviews* 50: 437-481.

32. Schopf JW and Walter MR (1982). Origin and early evolution of cyanobacteria. The geological evidence. In: *The Biology of Cyanobacteria* (eds Carr NG and Whitton BA), Blackwell, Oxford and University of California Press, Berkeley, 543-564.

33. Sharpley AN, Daniel TC, Sims JT and Pote DH (1996). Determining environmentally sound soil phosphorus levels. *J Soil Water Conserv* 51: 160-166.

34. Sivonen K and Jones G (1999). Cyanobacterial toxins. In: *Toxic Cyanobacteria in Water. A Guide to their Public Health Consequences. Monitoring and Management* (eds. Chorus I and Bortram J). E and FN, London and New York 41-111.

35. Skulberg OM, Carmichael WW, Codd GA, and Skulberg R (1993). Taxonomy of toxic Cyanophyceae (cyanobacteria). In: *Algal Toxins in Seafood and Drinking Water* (ed Falconer IR), Academic Press, New York, pp. 1–28, 145–164.

36. Skulberg OM, Carmichael WW, Anderson RA, Matsunaga, Moore RE and Skulberg R (1992). Investigation of a neurotoxic Oscillatorian strain (cyanophyceae) and its toxin. Isolation and characterization of homoanatoxin-a. *Env Toxicol Chem* 11: 321-329.

37. Smith VH (1983). Low nitrogen to phosphorus ratios favors dominance by blue-green algae in lake phytolankton. *Science* 221: 669-671.

2016, Environmental Biotechnology: A New Approach *Pages 101–129*
Editors: Dr. Rajan Kumar Gupta and Dr. Satya Shila Singh
Published by: DAYA PUBLISHING HOUSE, NEW DELHI

Chapter 7

Azolla and Blue Green Algae Supplementation Affecting Mycorrhization and Growth of Rice

Deepak Vyas[1], Ashok Shukla[2], Anuradha Jha[2] and R.K. Gupta[3]

[1]*Lab Microbial Technology and Plant Pathology, Department of Botany, Dr. Hari Singh Gour University, Sagar, Madhya Pradesh*
[2]*National Research Centre for Agroforestry, Jhansi, Uttar Pradesh*
[3]*Department of Botany, Govt P.G. College, Kotdwar, Uttarakhand*

ABSTRACT

Aim of the present study was to determine the mycorrhization and growth of rice under the influence of Azolla and blue green algae (BGA), grown on two potting substrates viz., rehli soil (paddy field) and garden soil (natural soil). Threshold concentrations of Azolla (10, 20 and 40 tone ha^{-1}) and BGA (10, 20 and 40 kg ha^{-1}) were determined. Results indicate that higher concentrations of Azolla and BGA were responsible for reduction in mycorrhization. At maturation stage of rice, reduction in mycorrhization was observed, but to our surprise mycorrhization increased at harvesting stage. However, in terms of growth and biomass, maximum values were noticed at 20 tone Azolla ha^{-1} and 20 kg BGA ha^{-1} in garden soil, whereas in rehli soil, 10 tone Azolla ha^{-1} and 20 kg BGA ha^{-1} gave the best results. Thus, increased growth under the influence of Azolla and BGA reflected the importance of mycorrhizae. Therefore, mycorrhizal fungi can be proposed to mix with Azolla and BGA to get maximum benefits of the synergistic effects on rice. However, extrapolation of the results to real conditions of rice cropping should be done with precaution because of differences in growth conditions (greenhouse) and substrates used in the present study.

Keywords: Azolla, Blue green algae, Mycorrhizal fungi, Rice.

Introduction

In India, rice (*Oryza sativa* L.) are being cultivated in various ecological environments *viz.*, humid/low lying/waterlogged areas of Assam, Manipur, West Bengal, Orissa, Bihar, Eastern Uttar Pradesh and Southern parts of the India [1]. Though, Sagar (Madhya Pradesh) is predominantly known as wheat and soybean growing area, but in its some pockets (village rehli), farmers are cultivating rice [2]. In recent past, its cultivation has become less profitable due to increased production cost. But its demand is increasing every year and estimated that in 2025 AD the requirement would be 140 M tones [3]. In order to keep pace with the food requirement of ever increasing population of India, there is a constant pressure on crop production from the available cultivable land. To sustain present food self sufficiency and to meet future requirements, India has to increase its production by 3 per cent per annum [4].

Moreover, intensive agricultural practices exploiting the lands and other natural resources increased manifolds and simultaneously the use of chemical fertilizers are also increased [5]. Use of fertilizers helps to make country self dependent in terms of food production, but its injudicious and untimely applications adversely affecting the natural balance of soil ecosystems and consequently corresponds to widespread decline in crop yields [6]. Due to greater demand of crops in low fertility soils and high cost of inorganic fertilizers, supplementary cheaper sources of nutrients are needed [6]. This demand can be reduced to some extent by adopting new technology including biofertilizers [7].

Biofertilizers serves as a substitute of chemical fertilizer and improve plant growth. These are non bulky and low cost agricultural inputs [9]. *Azolla*, blue green algae (BGA), arbuscular mycorrhizal fungi (AMF) etc. are important constituents of biofertilizer [10]. *Azolla* and BGA are the symbiotic nitrogen-fixers, found in aquatic environment of rice fields [11]. AM fungi are known to improve plant growth, yield and nutrient uptake [12], and its presence in aquatic/wetland habitat has also documented [13]. *Azolla* and BGA play an important role in maintaining soil fertility [14]. The important factor in using *Azolla* as a biofertilizer is its quick decomposition in soil [17].

Most paddy soils have natural population of *Azolla* and BGA [18]. However, there is no information regarding its influence on mycorrhization and growth of rice plant. Therefore, considering the above mentioned fact, authors undertook the present study to identify the threshold concentrations of *Azolla* and BGA that support optimal AMF association.

Materials and Methods

Description of Site

Present study was carried out at Dr HS Gour Central University, Sagar, India. The region (23°10'-24°27' N latitude and 78°4'-79°21' E longitude) is situated between North of Tropic of Cancer on an average altitude of 580 m above mean sea level. Sagar lies in Agro-ecoregion 10; Central Highlands, hot sub-humid ecoregion with medium and deep black soils (I5C3), and has two distinct cropping seasons *viz.*, *kharif* (rainy season, July to October) and *rabi* (winter season, November to February).

Biological Materials

Rice seeds (var. IR-36; procured from Indian Agriculture Research Institute, New Delhi) were used. These were surface sterilized with 0.01 per cent (w/v) $HgCl_2$, washed several times (five to six) with distilled water and germinated on sterile sand.

Naturally occurring *Azolla pinnata* was collected from ponds at university botanic garden, and its large quantity was cultured in plastic trays containing modified medium [19].

BGA was collected from paddy fields (rehli) and rose in unialgal culture. Four strains (*Anabaena variabilis, Nostoc* sp 1, *Nostoc* sp 2 and *Nostoc* sp 3) were selected and grown on modified Chu-10 medium [20]. The composition of the medium was: K_2HPO_4 0.01 g/L, $MgSO_4.7H_2O$ 0.025 mg/L, $CaCl_2.2H_2O$ 0.04 mg/L, Na_2CO_3 0.02 mg/L, $Na_2SiO_3.9H_2O$ 0.025 mg/L, ferric citrate 0.003 mg/L, citric acid 0.003 mg/L, $MnCl_2$ 0.5 mg/L, $Na_2MoO_4.2H_2O$ 0.01 mg/L, boric acid 0.05 mg/L, $CuSO_4.5H_2O$ 0.02 mg/L, cobalt chloride 0.04 mg/L, $ZnSO_4$ 0.05 mg/L and distilled water 1 L.

Potting Substrates and Biological Treatments

Experiments were carried out in two potting substrates, collected from botanic garden (natural soil) and village rehli (paddy field). Soils are generally derived from basalts, black in color (vertisol), clayey and slightly basic in nature. Some physico-chemical characteristics of soils are given in Table 7.1.

Table 7.1: Physico-chemical Properties of Potting Substrates *i.e.* Soils of Garden and Rehli

Potting Substrate	Soil pH (1:2.5 H_2O)	Organic Carbon	Available Phosphorus (kg ha⁻¹)	Available Nitrogen (kg ha⁻¹)	Available Potassium (kg ha⁻¹)
Garden soil	7.1	1.14	7.25	210.85	190.00
Rehli soil	7.2	1.23	9.52	267.57	224.13

The following treatments with four replications each were included in this study: unsterilized soil (control; neither *Azolla* nor BGA); unsterilized soil + 10 tone *Azolla* ha⁻¹; unsterilized soil + 20 tone *Azolla* ha⁻¹; unsterilized soil + 40 tone *Azolla* ha⁻¹; unsterilized soil + 10 kg BGA ha⁻¹; unsterilized soil + 20 kg BGA ha⁻¹; and unsterilized soil + 40 kg BGA ha⁻¹. Plastic pots (size: 24 x 16 cm) filled with three kg unsterilized soils were used where pre-germinated seedlings were transplanted. Pots were kept on separate benches under greenhouse and water logging was maintained by blocking the holes and/or by daily watering. Viability of *Azolla* and BGA were checked at every stages of crop growth.

Greenhouse Experiments

To study the threshold concentrations of *Azolla* and BGA, and their effects on mycorrhization (colonization index, spore population/50 g soil and distribution of AMF species), separate experiments were carried out with both the substrates. As above mentioned, seven treatments were made which were replicated four times. Thus, a total of 28 pots per substrate were used where six rice seedlings per pot were transplanted. Two plants per pot were harvested at seedling (30 days), maturation (60 days) and harvesting stages (90 days); then observations on colonization index, spore population and distribution of AMF species were recorded. Fine roots were cleared with 10 per cent KOH and stained with acid fuschin (0.01 per cent in lacto-glycerol) following the method of Phillips and Hayman [21]. Colonization in cleared root parts was determined with a microscope at 100 X using grid line intersect method [22]. AMF spores were counted and isolated according to the method of Gerdemann and Nicolson [23]. The isolated spores were mounted in polyvinyl-lactoglycerol (PVLG) or Melzer's reagent and examined microscopically; then identified according to taxonomic criteria and by using information from the INVAM website on the internet (http://invam.caf.wvu.edu).

To study the effect of *Azolla* and BGA supplementations on growth of rice plant, separate experiments were laid down for each substrate, and aforesaid treatments (28) were maintained. Three pre-germinated healthy seedlings were transplanted to each pot; then thinning was carried out leaving one plant/pot. Experiment was terminated after 90 days and seedlings were harvested (one per replicate; 28 per potting substrate). Measurement of shoot length (cm) was done by ruler. Dry mass, expressed in gram was obtained after drying shoot and root tissues in an oven at 70 °C, until constant weight was achieved.

Statistical Analysis

Mycorrhizal colonization data were arcsine transformed to normalize their distribution and subjected to analysis of variance in which effect of treatments were tested. For each factor analyzed, the means of different treatments were compared using Fischer F-test (P<0.05). Mean of the experiment was analyzed statistically using a general linear model for analysis of variance in completely randomized design. Least significant difference (LSD) was used to compare treatment differences. The ANOVA was performed by using the statistical package SYSTAT version 12.

Results

Effect of *Azolla* and BGA Supplementations on Mycorrhization

The data presented in the Table 7.2 shows the effect of *Azolla* and BGA on mycorrhization in pots filled with garden soil. Since, amendment was done during transplantation of seedlings hence the data on spore population, colonization index and number of AMF species were recorded before amendment also. At seedling stage, differences were non-significant among the treatments. As plants grew, changes in degree of mycorrhization were noticed. Even though, 10 tone *Azolla* ha^{-1} was enough to decrease the spore population. Mycorrhization and number of AMF species reduced with the increase in concentrations of *Azolla*. Supplementation of 20 tone *Azolla* ha^{-1} found to be responsible for disappearance of one AMF species. While, at 40 tone *Azolla* ha^{-1}, reduction in spore population, colonization index and number of AMF species was more severe. But to our surprise, mycorrhization and number of AMF species increased with the age of the plant *i.e.* at harvesting stage. Maximum

Table 7.2: Spore Density (50 g^{-1} Soil), Root Colonization Index and Number of AM Species during Growth of Rice in Pots Supplemented with Azolla and Blue Green Algae (BGA) Filled with Garden Soil

Treatment	Seedling Stage			Maturation Stage			Harvesting Stage		
	Spore Density	TCI[a]	Number of AMF Species	Spore Density	TCI	Number of AMF Species	Spore Density	TCI	Number of AMF Species
Control (unsterilized soil)	299[b]	37.46	7	217	30.94	4	326	38.45	8
Unsterilized soil + *Azolla*									
+ 10 tone ha^{-1}	296	37.72	7	194	35.57	5	341	41.80	9
+ 20 tone ha^{-1}	284	37.42	7	174	29.65	4	359	44.92	9
+ 40 tone ha^{-1}	288	37.29	7	158	27.00	3	314	34.21	6
Unsterilized soil + BGA									
+ 10 kg ha^{-1}	298	37.96	7	187	32.11	6	330	40.65	10
+ 20 kg ha^{-1}	298	37.59	7	209	36.10	6	377	47.60	11
+ 40 kg ha^{-1}	293	37.41	7	170	28.82	5	246	37.37	8
LSD$_{0.05}$	ns	ns		17.5	3.40		13.1	3.83	

a: Arcsine transformed value of colonization index; b: Average of four replication.

mycorrhization and number of AMF species were observed in pots supplemented with 20 tone *Azolla* ha^{-1}, and interesting to note that 40 tone *Azolla* ha^{-1} was not so detrimental as its effect was seen at maturation stage. The data suggested that it was almost comparable to control, except number of AMF species. The data on the effect of BGA on mycorrhization and AMF diversity are also presented in the same table. Almost similar trends were observed with BGA, except types and number of AMF species at maturation and harvesting stages, respectively.

The data presented in the Table 7.3 shows the effect of *Azolla* and BGA on mycorrhization in pots filled with rehli soil. During maturation stage, more or less similar trend was observed as observed in garden soil. Here, 10 tone *Azolla* ha^{-1} significantly enhance spore population and root colonization. Whereas, supplementation of 40 tone *Azolla* ha^{-1} significantly reduce the spore population as well as number of AMF species, but colonization index remain almost equal to control. Similarly, at harvesting stage, 10 tone *Azolla* ha^{-1} gave best results, while 40 tone *Azolla* ha^{-1} significantly reduce the mycorrhization and number of AMF species. On the contrary, BGA supplemented pots showed little different results. Twenty kg BGA ha^{-1} gave best results at maturation as well as harvesting stage. Interestingly, colonization index remain lesser at both the stages as compared to control.

Effect of *Azolla* and BGA Amendments on AMF Distribution

Distributions of AMF species at different cropping stages of rice are presented in Tables 7.4 and 7.5. In control pots, ASCB, LAGR, LDPH, LFSC, LHOI, LINR and LMSS were recorded at seedling stage. ASCB, LDPH and LMSS were disappeared at maturation stage, and LAGR, LFSC, LHOI and LINR remain as such. AT harvesting stage, ASCB, LDPH and LMSS reappeared along with two new species (LOCT and LPST), while LAGR disappeared, and other species like LFSC, LHOI, LINR and LMSS were remained as such. Pots amended with 10 tone *Azolla* ha^{-1} were analyzed and similar AMF species were recorded as observed in control pots, at seedling stage. At maturation, here also ASCB and LMSS were disappeared and all other species remain same, but at harvesting stage ASCB and LMSS reappeared along with three new species (LETC, LSGT and LPST) whereas LAGR was not observed. In 20 tone *Azolla* ha^{-1} amended pots, almost similar AMF species were observed as recorded in 10 tone *Azolla* ha^{-1} at all stages. Further addition *i.e.* 40 tone *Azolla* ha^{-1} did not influence AMF species composition at seedling stage, but at maturation stage only LAGR, LFSC and LHOI were observed. Loss of AMF species were also distinguished at harvesting stage, here ASCB, LETC and LPST were not found which were present in 20 tone *Azolla* ha^{-1} treated pots at harvesting stage. Data on distribution of AMF species in BGA amended pots are also presented in the Table 7.4. Results revealed that at seedling stage, amendment of BGA did not influence AMF species distribution at all, however changes were observed at maturation and harvesting stages. At maturation, in 10 kg BGA ha^{-1} amendment pots, the only difference was observed that ASCB was replaced by GABD, whereas LMSS was disappeared. At harvesting stage, in 20 kg BGA ha^{-1} treated pots, ADTC, LSGT, LPST and LVSF were new records, and ASCB and LMSS reappeared, and GABD was disappeared. Similarly, in 40 kg BGA ha^{-1} amended pots, GABD disappeared at maturation stage and at harvesting

Table 7.3: Spore Density (50 g⁻¹ Soil), Root Colonization Index and Number of AM Species during Growth of Rice in Pots Supplemented with *Azolla* and Blue Green Algae (BGA) Filled with Rehli Soil

Treatment	Seedling Stage			Maturation Stage			Harvesting Stage		
	Spore Density	TCI[a]	Number of AMF Species	Spore Density	TCI	Number of AMF Species	Spore Density	TCI	Number of AMF Species
Control (unsterilized soil)	319[b]	37.99	6	296	36.63	5	394	43.03	8
Unsterilized soil + *Azolla*									
+ 10 tone ha⁻¹	314	38.43	6	327	41.72	5	353	48.24	9
+ 20 tone ha⁻¹	318	38.12	6	276	39.54	4	322	41.82	9
+ 40 tone ha⁻¹	320	38.38	6	244	36.69	3	283	40.36	6
Unsterilized soil+BGA									
+ 10 kg ha⁻¹	319	37.86	6	289	33.68	4	406	37.24	8
+ 20 kg ha⁻¹	321	33.40	6	325	35.41	5	442	41.11	10
+ 40 kg ha⁻¹	320	33.34	6	269	34.09	3	297	39.16	7
LSD₀.₀₅	ns	ns	ns	16.9	3.96		20.6	2.85	

a: Arcsine transformed value of colonization index; b: Average of four replication.

Table 7.4: Distribution of AM Species at different Growth Stages of Rice in Pots Supplemented with *Azolla* and Blue Green Algae (BGA) Filled with Garden Soil

Treatment	Distribution of AM Fungal Species at		
	Seedling Stage	Maturation Stage	Harvesting Stage
Unsterilized soil (control)	ASCB[a], LAGR, LDPH, LFSC, LHOI, LINR, LMSS	LAGR, LFSC, LHOI, LINR	ASCB, LDPH, LFSC, LHOI, LINR, LMSS, LOCT, LPST
Unsterilized soil + *Azolla*			
+ 10 tone ha⁻¹	ASCB, LAGR, LDPH, LFSC, LHOI, LINR, LMSS	LAGR, LDPH, LFSC, LHOI, LINR	ASCB, LDPH, LETC, LFSC, LHOI, LINR, LMSS, LPST, LSGT
+ 20 tone ha⁻¹	ASCB, LAGR, LDPH, LFSC, LHOI, LINR, LMSS	LAGR, LFSC, LHOI, LINR	ASCB, LDPH, LETC, LFSC, LHOI, LINR, LMSS, LPST, LSGT
+ 40 tone ha⁻¹	ASCB, LAGR, LDPH, LFSC, LHOI, LINR, LMSS	LAGR, LFSC, LHOI	LDPH, LFSC, LHOI, LINR, LMSS, LSGT
Unsterilized soil + BGA			
+ 10 kg ha⁻¹	ASCB, LAGR, LDPH, LFSC, LHOI, LINR, LMSS	GABD, LAGR, LDPH, LFSC, LHOI, LINR	ADTC, ASCB, LDPH, LETC, LFSC, LHOI, LINR, LMSS, LPST, LSGT,
+ 20 kg ha⁻¹	ASCB, LAGR, LDPH, LFSC, LHOI, LINR, LMSS	GABD, LAGR, LDPH, LFSC, LHOI, LINR	ADTC, ASCB, LDPH, LETC, LFSC, LHOI, LINR, LMSS, LPST, LSGT, LVSF
+ 40 kg ha⁻¹	ASCB, LAGR, LDPH, LFSC, LHOI, LINR, LMSS	LAGR, LDPH, LFSC, LHOI, LINR	ASCB, LDPH, LFSC, LHOI, LINR, LMSS, LPST, LSGT

a: Codes as per Perez and Schenck [54]; ADTC: *Acaulospora denticulata*; ASCB: *A. scrobiculata*; GABD: *Gigaspora albida*; LAGR: *Glomus aggregatum*; LDPH: *G. diaphanum*; LETC: *G. etunicatum*; LFSC: *G. fasciculatum*; LHOI: *G. hoi*; LINR: *G. intraradices*; LMSS: *G. mosseae*; LOCT: *G. occultum*; LPST: *G. pustulatum*; LSGT: *G. segmentatum*; LVSF: *G. vesiculiferum*.

Table 7.5: Distribution of AM Species at different Growth Stages of Rice in Pots Supplemented with *Azolla* and Blue Green Algae (BGA) Filled with Rehli Soil

Treatment	Distribution of AM Fungal Species at		
	Seedling Stage	Maturation Stage	Harvesting Stage
Unsterilized soil	AMLL[a], LAGR, LDMR, LFSC, LMRC, LMSS	LAGR, LDMR, LFSC, LMRC, LMSS	GABD, *Glomus* sp, LDPH, LFSC, LHOI, LHTS, LINR, LMSS
Unsterilized soil + *Azolla*			
+ 10 tone ha⁻¹	AMLL, LAGR, LDMR, LFSC, LMRC, LMSS	LAGR, LDMR, LFSC, LMRC, LMSS	AMLL, GABD, *Glomus* sp, LDPH, LFSC, LHOI, LHTS, LINR, LMSS
+ 20 tone ha⁻¹	AMLL, LAGR, LDMR, LFSC, LMRC, LMSS	LDMR, LFSC, LMRC, LMSS	AMLL, GABD, *Glomus* sp, LDPH, LFSC, LHOI, LHTS, LINR, LMSS
+ 40 tone ha⁻¹	AMLL, LAGR, LDMR, LFSC, LMRC, LMSS	LFSC, LMRC, LMSS	*Glomus* sp, LDPH, LFSC, LHOI, LINR, LMSS
Unsterilized soil + BGA			
+ 10 kg ha⁻¹	AMLL, LAGR, LDMR, LFSC, LMRC, LMSS	LDMR, LFSC, LMRC, LMSS	GABD, *Glomus* sp, LDPH, LFSC, LHOI, LHTS, LINR, LMSS
+ 20 kg ha⁻¹	AMLL, LAGR, LDMR, LFSC, LMRC, LMSS	LAGR, LDMR, LFSC, LMRC, LMSS	ADTC, AMLL, GABD, *Glomus* sp, LDPH, LFSC, LHOI, LHTS, LINR, LMSS
+ 40 kg ha⁻¹	AMLL, LAGR, LDMR, LFSC, LMRC, LMSS	LFSC, LMRC, LMSS	GABD, LDPH, LFSC, LHOI, LHTS, LINR, LMSS

a: Codes as per Perez and Schenck [54]; ADTC: *Acaulospora denticulata*; AMLL: *A. mellea*; GABD: *Gigaspora albida*; *Glomus* sp (unidentified *Glomus*); LAGR: *Glomus aggregatum*; LDPH: *G. diaphanum*; LDMR: *G. dimorphicum*; LFSC: *G. fasciculatum*; LHTS: *G. heterosporum*; LHOI: *G. hoi*; LINR: *G. intraradices*; LMRC: *G. microcarpum*; LMSS: *G. mosseae*.

stage ASCB and LMSS were reappeared, and LAGR disappeared. Here, LPST and LSGT were new AMF species.

Results showed that soil of Rehli harbor different AMF species then recorded with garden soil (Table 7.5). In control pots, AMLL, LAGR, LDMR, LFSC, LMRC and LMSS were recorded. At maturation stage, AMLL disappeared, while all other species were remain same. However at harvesting stage, *Glomus* sp, GABD, LDPH, LFSC, LHOI, LHTS, LINR and LMSS were recorded, where *Glomus* sp, GABD, LDPH, LHOI, LHTS and LINR were new records. LMSS was found at all the three stages, while LAGR, LDMR and LMRC disappeared at this stage. Supplementation of 10 tone *Azolla* ha⁻¹ did not influence the AMF diversity at seedling and maturation stages hence we recorded same species at both stages. At harvesting stage, similar species were recorded as observed in control pots, except reappearance of AMLL. In 20 tone *Azolla* ha⁻¹ supplemented pots, AMF diversity remained same at seedling stage, but at maturation stage AMLL and LAGR disappeared. However, at harvesting stage, *Glomus* sp., GABD, LDPH, LHOI, LHTS and LINR were the new species. Here also, LMSS continued their presence and AMLL reappeared. Addition of 40 tone *Azolla* ha⁻¹ again did not cause any change in AMF diversity at seedling stage, however at maturation stage, only LFSC, LMRC and LMSS were recorded, and at harvesting stage *Glomus* sp., LDPH, LHOI and LINR were new records. At this stage, LFSC and LMSS continued their presence, while AMLL, LAGR and LDMR were absent. On the other hand, addition of BGA did not influence the distribution of AMF species, at seedling stage. However, at maturation stage, AMLL and LAGR were absent in 10 kg BGA ha⁻¹ treated pots, while at 20 kg BGA ha⁻¹, AMLL was absent. 40 kg BGA ha⁻¹ cause further reduction in species composition and only three species *i.e.* LFSC, LMRC and LMSS were observed at same stage. Whereas, at harvesting stage there was considerable variation in rhizosphere of 10, 20, and 40 kg BGA ha⁻¹ amended pots. In 10 kg BGA ha⁻¹, *Glomus* sp, GABD, LDPH, LHOI, LHTS and LINR appeared, and LDMR and LMRC disappeared. In 20 kg BGA ha⁻¹, ADTC, AMLL, *Glomus* sp, GABD, LDPH, LHOI, LHTS and LINR were appeared and here again LDMR and LMRC were not observed. In 40 kg BGA ha⁻¹ amended pots, GABD, LDPH, LHOI, LHTS and LINR were appeared and LDMR and LMRC were disappeared.

Effect of *Azolla* and BGA Amendments on Plant Growth

In garden soil, amendment of *Azolla i.e.* 10 and 20 tone ha⁻¹ gave better results as compared to control pots, while significantly higher shoot length, and shoot and root dry weights were recorded in 20 tone *Azolla* ha⁻¹ treated pots (Figure 7.1). When its concentration was doubled *i.e.* 40 tone ha⁻¹, it cause adverse effects on plant growth. Similarly, supplementation of 10 and 20 kg BGA ha⁻¹ increased the plant growth. Here, also 20 kg BGA ha⁻¹ gave significantly better results, whereas 40 kg BGA ha⁻¹ found to be responsible for reduced growth. Same set of experiment was also carried out in rehli soil. Plants grown in control pots or amended with 10 and 20 tone *Azolla* ha⁻¹ showed better results where significantly greater growth and biomass were recorded in 10 tone *Azolla* ha⁻¹ amended pots. Here, 40 tone *Azolla* ha⁻¹ become detrimental and observed values were less then control plants. It is deduced that higher the concentration greater the loss in terms of growth. Results obtained with

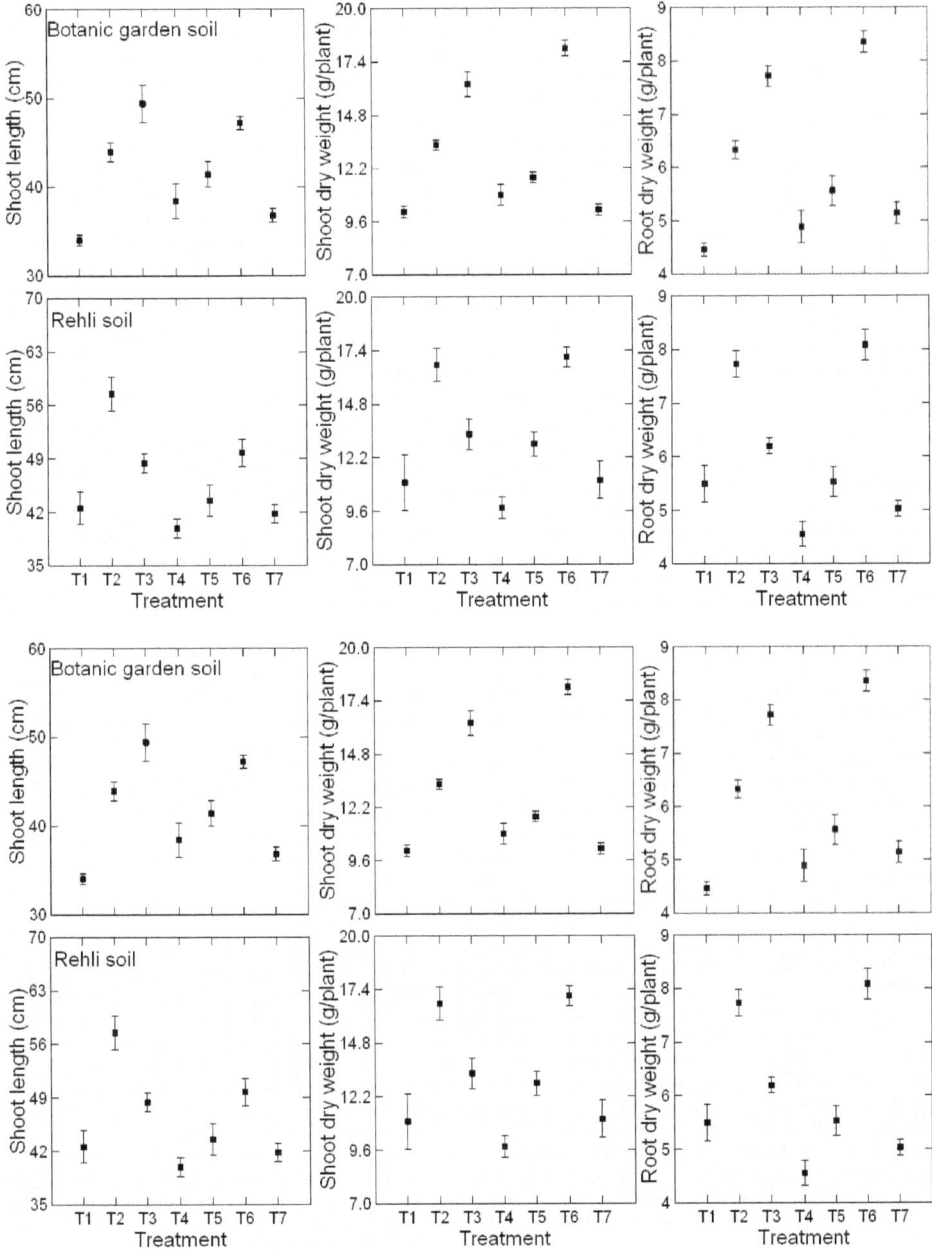

Figure 7.1: Effect of *Azolla* and BGA on Shoot Length (cm), Shoot Dry Weight (g) and Root Dry Weight (g) of Rice in Soils of Garden and Rehli. T1: control (unsterilized soil); T2: unsterilized soil + 10 tone *Azolla* ha^{-1}; T3: unsterilized soil + 20 tone *Azolla* ha^{-1}; T4: unsterilized soil + 40 tone *Azolla* ha^{-1}; T5: unsterilized soil + 10 kg BGA ha^{-1}; T6: unsterilized soil + 20 kg BGA ha^{-1}; T7: unsterilized soil + 40 kg BGA ha^{-1}.

the BGA showed that 20 kg BGA ha^{-1} amended pots gave best results and its higher concentration *i.e.* 40 kg BGA ha^{-1} was found to be responsible for negative results, which were at par with 10 kg BGA ha^{-1}.

Discussion

Effect of *Azolla* and BGA Supplementations on Mycorrhization

Results of the present study suggested that *Azolla* and BGA influenced the mycorrhization grown in two different soils *i.e.* garden and rehli soil (Tables 7.2 and 7.3). It is clearly evident from the results that at harvesting stage, 20 tone *Azolla* ha^{-1} positively influences mycorrhization in the garden soil, whereas same amount of *Azolla* found to be unfavorable for the mycorrhization in rehli soil. This might be due to that garden soil was not treated with chemical fertilizers (natural soil) therefore, primary requirement of nitrogen was higher to the plants and that required amount of nitrogen was provided by inoculated *Azolla* having *A. azollae* symbiont which fixes nitrogen. Several workers have reported that higher concentration of nitrogen affect mycorrhization [25, 26, 27, 28]. At maturation stage, BGA supplementation did not reduce mycorrhization as much as observed with *Azolla*, while mycorrhization and number of AMF species were found higher and different than *Azolla* treated pots. Similarly, at harvesting stage, BGA treated pots showed more mycorrhization then *Azolla* treated pots. This might be due to carbon and nitrogen provided by dried cells of BGA which influenced the rhizosphere of the test plants positively [29, 30, 11]. According to Choudhury and Kennedy [31], higher growth of BGA increases the biomass reserve of soil. Singh *et al.* [32] and Jha *et al.* [33] reported gradual accumulation of organic matter due to BGA, which directly influenced the availability of nitrogen. However, supplementation of higher amount of BGA did not support the growth [34].

Potting substrate used in the present study *i.e.* rehli soil was already saturated with chemicals therefore, need of nitrogen and other chemical substances in this soil was less, hence reduced mycorrhization was observed at 20 tone *Azolla* ha^{-1} and 40 kg BGA ha^{-1} treated pots. On the other hand, elevated carbon in rhizosphere of rice reduces the mycorrhization [35, 36]. During maturation stage, water logging was maintained in pots, which might be responsible for reduced mycorrhization. Our results are consistent with previous results [37, 38]. Occurrence of lesser propagules of AM fungi during wet conditions might be due to changes in soil pH [39] because in such conditions excess of CO_2 combine with water and forms carbonic acid (HCO_3^-) which works as a neutralizing source of pH. An increase in mycorrhization towards the end of the cropping season *i.e.* harvesting stage might be due to drying of the soil [40], which provide conducive environment for the exchange of the gases [41].

Effect of *Azolla* and BGA Amendments on AMF Distribution

Results on effect of *Azolla* and BGA on distribution of AM fungi reveal that AMF species composition did not influenced by supplementation of *Azolla* or BGA at seedling stage (Tables 7.4 and 7.5). However, AMF species composition showed some variations in garden and Rehli soil, which might be not much influenced by the presence of either *Azolla* or BGA at maturation stage. However, in comparison to seedling stage, loss of species was noticed which was reappeared at harvesting stage and it might be due to the anaerobiosis created in the pots [40, 2]. This suggests that *Azolla* or BGA and water logging conditions cause reduction in AMF distribution at maturation stage and reappearance suggested that effects of *Azolla* or BGA were temporary, and secondary dryness provides conducive environment to flourish the AM fungi.

One of the unexpected but important finding of the present experiment was the emergence of new AMF species at harvesting stage (Table 7.6) which were not observed at seedling and maturation stages. Emergence of GABD, *Glomus* sp., LDPH, LHOI, LHTS and LINR at almost all the concentrations indicates that these species remain in the Rehli soil and as the dryness comes they emerges, which were remain suppressed under wet conditions. On the contrary, few AMF species were disappeared at higher concentrations of *Azolla* and BGA and were found at its lower concentrations. This suggested that these species were not influenced by wet conditions, but when concentration of *Azolla* and BGA increases they disappeared. Our results are consistent with Dubey *et al.* [2] who reported the appearance and disappearance of certain AMF species at different growth stages of rice. Bohrer *et al.* [37] reported that the infectivity of AMF propagules decreases in wet and anoxic soil conditions. Another important feature of waterlogged soil with high organic content is its attraction for wide array of soil organism enhancing the parasitic pressure on AMF spores. Our results also provide a clue that probably soil works as a store house. The appearance and disappearance of AMF species probably influenced by root exudation, as it is well documented that root exudation plays important role in maintaining AMF diversity [42, 43, 44].

But what was the actual reason of emergence and disappearance, is the subject of further exploration. Since, the experiments were carried out in pots which were kept under identical conditions of green house, except the potting substrates (garden and Rehli soil). Therefore, we negate the possibility of grazing and assume that probably root exudation might have played the role in emergence and disappearance of certain AMF species.

Effect of *Azolla* and BGA Amendments on Plant Growth

Our results suggested that in garden soil, best results were obtained in pots either treated with 20 tone *Azolla* ha^{-1} or 20 kg BGA ha^{-1}. But, when concentration was doubled *i.e.* 40 tone *Azolla* ha^{-1} and 40 kg BGA ha^{-1}, it reduce the plant growth. On the other hand, 10 tone *Azolla* ha^{-1} and 20 kg BGA ha^{-1} yielded better results in rehli soil. There have been reported cases on the beneficial effects of *Azolla* and BGA on rice growth [45, 11]. The importance of *Azolla* as an organic fertilizer in rice cultivation is well appreciated and practiced in several countries [45]. Similarly, role of BGA in maintenance of fertility of rice field has been well substantiated and documented [46, 47]. Glick [48] reported that BGA liberate wide array of secondary metabolites, which have direct or indirect effect on plant growth. In India alone, the beneficial effects of BGA on yield of many rice varieties have been demonstrated in number of field locations [49]. Beneficial effects of BGA inoculation have also been reported in different varieties of tomato [50]. According to Venkataraman [51], successful establishment of

Table 7.6: AMF Species Emerges at Harvesting Stage of Rice in Garden and Rehli Soil

Treatment	Garden Soil	Rehli Soil
Control (unsterilized soil)	LOCT[a], LPST	GABD, *Glomus* sp, LDPH, LHOI, LHTS, LINR
Unsterilized soil + *Azolla*		
+ 10 tone ha^{-1}	LETC, LPST, LSGT	AMLL, GABD, *Glomus* sp, LDPH, LHOI, LHTS, LINR
+ 20 tone ha^{-1}	LETC, LPST, LSGT	AMLL, GABD, *Glomus* sp, LDPH, LHOI, LHTS, LINR
+ 40 tone ha^{-1}	LSGT	*Glomus* sp, LDPH, LHOI, LINR
Unsterilized soil + BGA		
+ 10 kg ha^{-1}	ADTC, LETC, LPST, LSGT	GABD, *Glomus* sp, LDPH, LHOI, LHTS, LINR
+ 20 kg ha^{-1}	LETC, LPST, LSGT, LVSF	ADTC, AMLL, GABD, *Glomus* sp, LDPH, LHOI, LHTS, LINR
+ 40 kg ha^{-1}	LPST, LSGT	GABD, LDPH, LHOI, LHTS, LINR

a: Codes as per Perez and Schenck [54]; ADTC: *Acaulospora denticulata*; AMLL: *A. mellea*; GABD: *Gigaspora albida*; *Glomus* sp (unidentified *Glomus*); LDPH: *G. diaphanum*; LETC: *G. etunicatum*; LHOI: *G. hoi*; LHTS: *G. heterosporum*; LINR: *G. intraradices*; LOCT: *G. occultum*; LPST: *G. pustulatum*; LSGT: *G. segmentatum*; LVSF: *G. vesiculiferum*.

promising strains of BGA enhances the crop yield up to 30 percent, and also provide 15-25 kg ha^{-1} biologically fixed nitrogen/season [52, 53]. Wilson [16] postulated that *A. azollae* fixes atmospheric nitrogen estimated between 120 to 312 kg N$_2$ ha^{-1}.

Conclusions

☆ Supplementation of 20 and 10 tone *Azolla* ha^{-1} positively influenced the mycorrhization in garden and rehli soils, respectively whereas 20 kg ha^{-1} of BGA was found responsible for the greater mycorrhization in both the soils.

☆ Results also suggested that certain AMF species emerges at harvesting stage, while few were disappeared at this stage, but what should be the actual reason of emergence and disappearance is the matter of further exploration. Since, experiments were carried out in identical conditions, except potting substrates (garden and rehli soils), therefore assume that root exudation played an important role in AMF distribution. Our findings are of great importance, therefore it is important to look into the matter more thoroughly. Thus, to understand the actual cause of emergence and disappearance of AMF species, it needed further exploration with sincere investigations.

☆ However, in terms of plant growth, 20 and 10 tone *Azolla* ha^{-1} supplementation gave best results in garden and rehli soils, respectively whereas maximum growth was obtained in 20 kg BGA ha^{-1} amended pots filled with rehli soil. Increased growth of rice under the influence of *Azolla* and BGA reflect the importance of indigenous AM fungi. Therefore, it can be proposed to mix *Azolla* and BGA with AMF consortia to get maximum benefits of the synergistic effects on growth of rice. However, extrapolation of the results to the real conditions of rice cropping systems should be done with precaution because of differences in the growth conditions (greenhouse) and substrate used in the present study.

Acknowledgements

Authors are thankful to head department of botany, Dr HS Gour Central University, Sagar for facilitating the research program and constant encouragement during the period of study. Ashok Shukla acknowledges funding through University Grants Commission's Dr DS Kothari Post Doctoral Fellowship Scheme, New Delhi, India during the preparation of this paper. Thanks are also due to anonymous reviewers for critical comments and suggestions.

References

1. R. Prasanna, L. Nain, A.K. Pandey, S. Nayak, Exploring the ecological significance of microbial diversity and networking in the rice ecosystem, in: P. Dion (Ed.), *Soil Biology and Agriculture in the Tropics, Soil Biology* vol. 21. Springer-Verlag, Berlin Heidelberg, 2010, pp. 139-161.

2. A. Dubey, P.K. Singh, M.K. Mishra, D. Vyas, Occurrence of AM fungi at varying stages of growth of rice plants. *Proc. Ind. Nat. Acad. Sci. Part B Biol. Sci.,* 78 (2008): 45-48.

3. G. Khush, Productivity improvements in rice. *Nutr. Rev.* 61 (2003): S114–116.

4. M.N. Pandey, S.B. Verulkar, D. Sharma, Rice research: past achievement, present scenario and future thrust. *Ind. J. Agric. Sci.* 80 (2010): 447-469.

5. A. Shukla, A. Kumar, A. Jha, Ajit, D.V.K.N. Rao, Phosphorus threshold for arbuscular mycorrhizal colonization of crops and tree seedlings. *Biol. Fertil. Soil* (2011): (DOI: 10.1007/s00374-011-0576-y)

6. M. Vosatka, J. Albrechtova, Benefits of arbuscular mycorrhizal fungi to sustainable crop production, in: M.S. Khan, A. Zaidi, J. Mussarat (Eds.), *Microbial strategies for crop improvement.* Springer-Verlag, Berlin Heidelberg, 2009, pp. 205-225.

7. J.S. Singh, V.C. Pandey, D.P. Singh, Efficient soil microorganisms: A new dimension for sustainable agriculture and environmental development. *Agric. Ecosyst. Environ.* 140 (2011): 339-353.

8. B. Wu, S.C. Cao, Z.H. Li, Z.G. Cheung, K.C. Wong, Effects of biofertilizer containing N-fixer, P and K solubilizers and AM fungi on maize growth. *Geoderma* 125 (2005): 155-162.

9. D. Uribe, J.S. Nieves, J. Venegas, Role of microbial biofertilizers in the development of a sustainable agriculture in the tropics, in: P. Dion (Ed.), *Soil Biology and Agriculture in the Tropics, Soil Biology,* vol. 21. Springer-Verlag, Heidelberg, 2010, pp. 235-250.

10. A. Vaishampayan, R.P. Sinha, D.P. Hader, T. Dey, A.K. Gupta, U. Bhan, A.L. Rao, Cyanobacterial biofertilizers in rice agriculture. *Bot. Rev.* 67 (2001): 453–516.

11. S. Nayak, R. Prasanna, A. Pabby, T.K. Dominic, P.K. Singh, Effect of urea and BGA-*Azolla* biofertilizers on nitrogen fixation and chlorophyll accumulation in soil cores from rice fields. *Biol. Fertil. Soil* 40 (2004): 67–72.

12. A. Shukla, A. Kumar, A. Jha, O.P. Chaturvedi, R. Prasad, A. Gupta, Effects of shade on arbuscular mycorrhizal colonization and growth of crops and tree seedlings in Central India. *Agroforest. Syst.* 76 (2009): 95-109.

13. D. Beck-Nielsen, T.V. Madsen, Occurrence of vesicular arbuscular mycorrhiza in aquatic macrophytes from lakes and streams. *Aquat. Bot.* 71 (2001): 141-148.

14. E. Fernandez-Valiente, A. Ucha, A. Quesada, F. Leganes, R. Carreres, Contribution of N_2 fixing cyanobacteria to rice production: availability of nitrogen from ^{15}N-labelled cyanobacteria and ammonium sulphate to rice. *Plant Soil* 221 (2000): 107–112.

15. H. Saadatnia, H. Riahi, Cyanobacteria from paddy fields in Iran as a biofertilizer in rice plants. *Plant Soil Env.* 55 (2009): 207–212.

16. L.T. Wilson, Cyanobacteria: A potential nitrogen source in rice fields. *Texas Rice* 6 (2006): 9–10.

17. S. Kannaiyan,. Nitrogen contribution by *Azolla* to rice crop. *Proc. Ind. Nat. Acad Sci Part B Biol. Sci.* 59 (1993): 309–314.

18. W. Liesack, S. Schnell, N.P. Revsbech, Microbiology of flooded rice paddies. *FEMS Microbiol. Rev.* 24 (2000): 625-645.

19. I. Watanabe, C.R. Espinas, N.C. Berja, V.B. Alimagno, Utilization of the *Azolla-Anabaena* complex as a nitrogen fertilizer for rice. *Int. Rice Res. Inst. Paper Ser No.* 11, 1977.

20. R.S. Safferman, M.E. Morris, Growth characteristics of the blue green algae virus LPP-1. *J. Bacteriol.* 88 (1964): 771-773.

21. J.M. Phillips, D.S. Hayman, Improved procedures for clearing roots and staining parasitic and vesicular arbuscular mycorrhizal fungi for rapid assessment of infection. *Trans. Br. Mycol. Soc.* 55 (1970): 158-161.

22. M. Giovannetti, B. Mosse, An evaluation of techniques for measuring vesicular arbuscular mycorrhizal infection in roots. *New Phytol.* 84 (1980): 489–500.

23. J.W. Gerdemann, T.H. Nicolson, Spores of mycorrhizal *Endogone* species extracted from soil by wet sieving and decanting. *Trans. Br. Mycol. Soc.* 46 (1963): 235–244.

24. L. Wilkinson, M. Coward, Linear models III-general linear models. In: *SYSTAT II statistics II.* SYSTAT software Inc., Richmond, 2004, pp. 139.

25. L. Corkidi, D.L. Rowland, N.C. Johnson, E.B. Allen, Nitrogen fertilization alters the functioning of arbuscular mycorrhizae at two semiarid grasslands. *Plant Soil* 240 (2002): 299–310.

26. N.C. Johnson, D.L. Rowland, L. Corkidi, L.M. Egerton-Warburton, E.B. Allen, Nitrogen enrichment alters mycorrhizal allocation at five mesic to semiarid grasslands. *Ecology* 84 (2003): 1895-1908.

27. K.K. Treseder, K.M. Turner, M.C. Mack, Mycorrhizal responses to nitrogen fertilization in boreal ecosystems: potential consequences for soil carbon storage. *Global Change Biol.* 13 (2007): 78–88.

28. M.O. Garcia, T. Ovasapyan, M. Greas, K.K. Treseder, Mycorrhizal dynamics under elevated CO_2 and nitrogen fertilization in a warm temperate forest. *Plant Soil* 303 (2008): 301-310.

29. B. Bergman, J.R. Gallon, A.N. Rai, L.J. Stal, N_2 fixation by non-heterocystous cyanobacteria. *FEMS Microbiol. Rev.* 19 (1997): 139-185.

30. M.M. Hoque, K. Inubushi, S. Miura, Nitrogen dynamics in paddy fields as influenced by free-air CO_2 enrichment (FACE): at three levels of nitrogen fertilization. *Nutr. Cycl. Agroecosyst.* 63 (2002): 301–308.

31. A.T.M.A. Choudhury, I.R. Kennedy, Prospects and potentials for systems of biological nitrogen fixation in sustainable rice production. *Biol. Fertil. Soil* 39 (2004): 219-227.

32. P.K. Singh, B.C. Panigrahi, K.B. Satpathy, Coperative efficiency of *Azolla*, blue green algae and other organic manures in relation to N and P availability in flooded rice soil. Plant Soil 62 (1981): 35-41.

33. M.N. Jha, A.N. Prasad, S.K. Misra, Influence of source of organics and soil organic matter content on cyanobacterial nitrogen fixation and distributional pattern under different water regimes. *World J. Microbiol. Biotechnol.* 20 (2004): 673–677.

34. M.N. Jha, M.K. Mallik, N. Ahmad, Response of superimposed cyanobacterial inoculation of different varieties of paddy grown in summer and kharif seasons, in: B.D. Kaushik (Ed.), *Proceedings of National Symposium on Cyanobacterial Nitrogen Fixation.* IARI, New Delhi, India, 1990.

35. P.L. Staddon, A.H. Fitter, Does elevated atmospheric carbon dioxide affect arbuscular mycorrhizas? *Trends Ecol. Evol.* 13 (1998): 455-458.

36. M.C. Rillig, G.Y. Hernandez, P.C.D. Newton, Arbuscular mycorrhizae respond to elevated atmospheric CO_2 after long-term exposure: evidence from a CO_2 spring in New Zealand supports the resources balance model. *Ecol. Let.* 3 (2000): 475-478.

37. K.E. Bohrer, C.F. Friese, J.P. Amon, Seasonal dynamics of arbuscular mycorrhizal fungi in differing wetland habitats. *Mycorrhiza* 14 (2004): 329-337.

38. V. Escudero, R. Mendoza, Seasonal variation of arbuscular mycorrhizal fungi in temperate grasslands along a wide hydrologic gradient. *Mycorrhiza* 15 (2005): 291-299.

39. A. Rohyadi, F.A. Smith, R.S. Murray, S.E. Smith, Effects of pH on mycorrhizal colonisation and nutrient uptake in cowpea under conditions that minimise confounding effects of elevated available aluminium. *Plant Soil* 260 (2004): 283-290.

40. L.L. Ilag, A.M. Rosales, F.V. Elazegvi, T.W. Mew, Changes in the population of infective endomycorrhizal fungi in a rice based cropping system. *Plant Soil* 103 (1987): 67-73.

41. L.K. Abbott, A.D. Robson, Factors influencing the occurrence of vesicular arbuscular mycorrhizae. *Agric. Ecosyst. Environ.* 35 (1991): 121-150.

42. G. Becard, Y. Piche, Fungal growth stimulation by CO_2 and root exudates in vesicular-arbuscular mycorrhizal symbiosis. *App. Environ. Microbiol.* 55 (1989): 2320-2325.

43. D.L. Jones, A. Hodge, Y. Kuzyakov, Plant and mycorrhizal regulation of rhizodeposition. *New Phytol.* 163 (2004): 459–480.

44. C.D. Broeckling, A.K. Broz, J. Bergelson, D.K. Manter, J.M. Vivanco, Root exudates regulate soil fungal community composition and diversity. *App. Environ. Microbiol* (2008): 74, 738–744.

45. B. Mandal, P.L.G. Vlek, L.N. Mandal, Beneficial effects of blue-green algae and *Azolla*, excluding supplying nitrogen, on wetland rice fields: a review. *Biol. Fertil. Soil* 28 (1999): 329-342.

46. P.K. De, The role of blue green algae in nitrogen fixation in rice fields. *Proc. R. Soc. Lond. B* 127 (1939): 121-139.

47. D.L.N. Rao, R.G. Burns, The influence of blue-green algae on the biological amelioration of alkali soils. *Biol. Fertil. Soil* 11 (1991): 306-312.

48. B.R. Glick, The enhancement of plant growth by free-living bacteria. *Can. J. Microbiol.* 41 (1995): 109–117.

49. G.S. Venkataraman, Blue-green algae for rice production. A manual for its promotion. *FAO Soils Bull.* 46 (1981): 120.

50. B.D. Kaushik, G.S. Venkataraman, Effect of algal inoculation on the yield and vitamin C content of two varieties of tomato. *Plant Soil* 52 (1979), 135-137.

51. G.S. Venkataraman, *Algal Biofertilizers and Rice Cultivation.* Today and Tomorrow Printers and Publishers, New Delhi, India, 1972.

52. S. Nayak, R. Prasanna, Soil pH and its role in cyanobacterial abundance and diversity in rice field soils. *Appl. Ecol. Environ. Res.* 5 (2007): 103-113.

53. R. Prasanna, S. Nayak, Influence of diverse rice soil ecologies on cyanobacterial diversity and abundance. *Wetlands Ecol. Manag.* 15 (2007): 127-134.

54. Y. Perez, N.C. Schenck, A unique code for each species of VA mycorrhizal fungi. *Mycologia* 82 (1990): 256-260.

2016, Environmental Biotechnology: A New Approach Pages 131–145
Editors: Dr. Rajan Kumar Gupta and Dr. Satya Shila Singh
Published by: DAYA PUBLISHING HOUSE, NEW DELHI

Chapter 8

Application of Biotechnology in Agricultural Production

Fouzia Ishaq[1], Santosh Kumar Singh[2]
and Amir Khan[2]*

*[1]Department of Zoology and Environmental Science,
Gurukula Kangri University, Haridwar, Uttarakhand
[2]Glocal School of Life Sciences, The Glocal Univesrity,
Mirzapur Pole, Saharanpur, Uttar Pradesh*

ABSTRACT

Humans have always relied on plants and animalsfor food, shelter, clothing and fuel, and forthousands of years farmers have been changingthem to better meet our evolving needs. Society's demandfor resources provided by plants and animals will increaseas the world's population grows. The global population,which numbered approximately 1.6 billion in 1900, hassurged to more than 6 billion and is expected to reach 10billion by 2030. The United Nations Food and Agriculture Organization estimates world food production will have todouble on existing farmland if it is to keep pace with theanticipated population growth.Biotechnology can help meet the ever-increasing need byincreasing yields, decreasing crop inputs such as waterand fertilizer, and providing pest control methods that aremore compatible with the environment.Biotechnology can help meet the ever-increasingneed by increasing yields, decreasing crop inputssuch as water and fertilizer, and providing pestcontrol methods that are more compatible with the environment.

In its simplest form, biotechnology involves inserting, changing, ordeleting genetic information within a host organism to give it newcharacteristics. This technology will likely bring great benefits to agriculture,just as breeding has over several thousand years of human history. The development and use of new techniques is allowing researchers to manipulatethe genetic character of organisms while

* Corresponding author: E-mail: amiramu@gmail.com

overcoming the complications andlimitations of sexual gene exchange. Genetic engineering is reducing theamount of time needed to analyze genetic information and transfer genes. Bothgenetic engineering and monoclonal antibody technology, another majordevelopment in biotechnology, greatly increase the specificity and accuracy of analytical research methods. Further, these new technologies are permittinghighly specific molecular analyses to be done and are opening new areas ofinquiry. The tools of biotechnology, combined with traditional techniques in biology and chemistry, increase enormously both the power and the pace of discoveries in biological investigation.

Keywords: Genetic engineering, Biotechnology, Fertilizer and agriculture.

Introduction

The tools of biotechnology offer both a challenge and tremendous opportunity. They do not change the purpose of agriculture to produce needed food, fiber, timber, and chemical feed stocks efficiently. Instead, they offer new techniques for manipulating the genes of plants, animals, and microorganisms. Biotechnology tools complement, rather than replace, the traditional methods used to enhance agricultural productivity and build on a base of understanding derived from traditional studies in biology, genetics, physiology, and biochemistry. Biotechnology has opened an exciting frontier in agriculture. The new techniques provided by biotechnology are relatively fast, highly specific, and resource efficient. It is a great advantage that a common set of techniques; gene identification and cloning, for example are broadly applicable. Not only can we improve on past, traditional methods with the more precise modern methods, but we can explore new areas as well. We can seek answers toquestions that only a few years ago we never thought to ask.The power of biotechnology is no longer fantasy. In the last few years, we have begun to transform ideas into practical applications. For instance, scientists have learned to genetically alter certain crops to increase their tolerance to certain herbicides. Biotechnology has been used to design and develop safer and more effective vaccines against viral and bacterial diseases such aspseudorabies, enteric colibacillosis (scours), and foot-and-mouth disease.Yet we have barely scratched the surface of the potential benefits. This topic briefly reviews the major uses of biotechnology in agriculture. It looks specifically at the progress and potentials of genetic engineering and other new biotechnologies in plant and animal agriculture and bioprocessing. These sections review traditional approaches, discuss examples of progress using biotechnology, and highlight opportunities on the horizon.

Crop Biotechnology

Farmers and plant breeders have relied for centuries on crossbreeding, hybridization and other genetic modification techniques to improve the yield and quality of food and fiber crops and to provide crops with built-in protection against insect pests, disease-causing organisms and harsh environmental conditions. Stone Age farmers selected plants with the best characteristics and saved their seeds for the next year's crops. By selectively sowing seeds from plants with preferred characteristics, the earliest agriculturists performed genetic modification to convert wild plants into domesticated crops long before the science of genetics was understood. As our knowledge of plant genetics improved, we purposefully crossbred plants with desirable traits (or lacking undesirable characteristics) to produce offspring that combine the best traits of both parents. In today's world, virtually every crop plant grown commercially for food or fiber is a product of cross breeding, hybridization or both. Unfortunately, these processes are often costly, time consuming, inefficient and subject to significant practical limitations. For example, producing corn with higher yields or natural resistance to certain insects takes dozens of generations of traditional crossbreeding, if it is possible at all.

The tools of biotechnology allow plant breeders to select single genes that produce desired traits and move them from one plant to another. The process is far more precise and selective than traditional breeding in which thousands of genes of unknown function are moved into our crops. Biotechnology also removes the technical obstacles to moving genetic traits between plants and other organisms. This opens up a world of genetic traits to benefit food production. We can, for example, take a bacterium gene that yields a protein toxic to a disease-causing fungus and transfer it to a plant. The plant then produces the protein and is protected from the disease without the help of externally applied fungicides.

Improving Crop Production

The crop production and protection traits agricultural scientists are incorporating with biotechnology are the same traits they have incorporated through decades of crossbreeding and other genetic modification techniques: increased yields; resistance to diseases caused by bacteria, fungi and viruses; the ability to withstand harsh environmental conditions such as freezes and droughts; and resistance to pests such as insects, weeds and nematodes. Natural Protection for Plants just as biotechnology allows us to make better use of the natural therapeutic compounds our bodies produce; italso provides us with more opportunities to partner with nature in plant agriculture. Through science, we have discovered that plants, like animals, have built-in defense systems against insects and diseases, and we are searching for environmentally benign chemicals that trigger these natural defense mechanisms so plants can better protect themselves. Biotechnology will also open up new avenues for working with nature by providing new biopesticides, such as microorganisms and fatty acid compounds, that are toxic to targeted crop pests but do not harm humans, animals, fish, birds or beneficial insects. Because bio pesticides act in unique ways, they can control pest populations that have developed resistance to conventional pesticides. Bio pesticide farmers (including organic farmers) have used since the 1930s is the microorganism *Bacillus thuringiensis*, or *Bt*, which occurs naturally in soil. Several of the proteins the *Bt* bacterium produces are lethal to certain insects, such as the European corn borer, a prevalent pest that costs the United States $1.2 billion in crop damage each year. *Bt* bacteria used as a biopesticidal spray can eliminate target insects without relying on chemically based pesticides.

Using the flexibility provided by biotechnology, we can transplant the genetic information that makes the *Bt* bacterium lethal to certain insects (but not to humans, animals or other insects) into plants on which that insect feeds. The plant that once was a food source for the insect now kills it, lessening the need to spray crops with chemical pesticides to control infestations. The plant that once was a food source for the insectnow kills it, lessening the need to spray crops withchemical pesticides to control infestations.

Herbicide Tolerance

Good planting conditions for crops will also sustain weeds that can reduce crop productivity as they compete for the same nutrients the desired plant needs. To prevent this, herbicides are sprayed over crops to eliminate the undesirable weeds. Often, herbicides must be applied several times during the growing cycle, at great expense to the farmer and possible harm to the environment. Using biotechnology, it is possible to make crop plants tolerant of specific herbicides. When the herbicide is sprayed, it will kill the weeds but have no effect on the crop plants. This lets farmers reduce the number of times herbicides have to be applied and reduces the cost of producing crops and damage to the environment.

Resistance to Environmental Stresses

In addition to the biological challenges to plant growth and development just described, crop plants must contend with abiotic stresses nature dispenses regularly: drought, cold, heat and soils that are too acidic or salty to support plant growth. While plant breeders have successfully incorporated genetic resistance to biotic stresses into many crop plants through crossbreeding, their success at creating crops resistant to abiotic stresses has been more limited, largely because few crops have close relatives with genes for resistance to these stresses. The crossbreeding limitation posed by reproductive compatibility does not impede crop biotechnology; genes found in any organism can be used to improve crop production. As a result, scientists are making great strides in developing crops that can tolerate difficult growing conditions. For example, researchers have genetically modified tomato and canola plants that tolerate salt levels 300 percent greater than non-genetically modified varieties. Other researchers have identified many genes involved in cold, heat and drought tolerancefound naturally in some plants and bacteria. Scientists in Mexico have produced maize and papaya that are tolerant to the high levels of aluminium that significantly impede crop plant productivity in many developing countries.

Increasing Yields

In addition to increasing crop productivity by using built-in protection against diseases, pests, environmental stresses and weeds to minimize losses, scientists use biotechnology to improve crop yields directly. Researchers at Japan's National Institute of Agrobiological Resources added maize photosynthesis genes to rice to increase its efficiency at converting sunlight to plant starch and increased yields by 30 percent. Other scientists are altering plant metabolism by blocking gene action in order to shunt nutrients to certain plant parts. Yields increase as starch accumulates in potato tubers and not leaves, or as oil-seed crops, such as canola, allocate most fatty acids to the seeds. Biotechnology also allows scientists to develop crops that are better at accessing the micronutrients they need. Mexican scientists have genetically modified plants to secrete citric acid, a naturally occurring compound, from their roots. In response to the slight increase in acidity, minerals bound to soil particles, such as calcium, phosphorous and potassium, are released and made available to the plant. Nitrogen is the critical limiting element for plant growth and, step-by-step, researchers from many scientific disciplines are teasing apart the details of the symbiotic relationship that allows nitrogen-fixing bacteria to capture atmospheric nitrogen and provide it to the plants that harbour them in root nodules.Plant geneticists in Hungary and England have identified the plant gene and protein that enable the plant to establish a relationship with nitrogen-fixing bacteria in the surrounding soil. Microbial geneticists at the University of Queensland have identified the bacterial gene that stimulates root nodule formation. Collaboration among molecular biologists in the European Union, United States and Canada yielded the complete genome sequence of one of the nitrogen-fixing bacteria species. Protein chemists have documented the precise structure of the bacterial enzyme that converts atmospheric nitrogen into a form the plant can use.

Crop Biotechnology in Developing Countries

Today, 70 percent of the people on the planet grow what they eat, and, despite the remarkable successes of the Green Revolution in the 1960s, millions of them suffer from hunger and malnutrition. Continuing population growth, urbanization, poverty, inadequate food distribution systems and high food costs impede universal access to the higher yields provided by technological advances in agriculture. In addition, the crops genetically improved by plant breeders who enabled the Green Revolution were large-volume commodity crops, not crops grown solely by small-scale subsistence farmers. For many farmers in developing countries, especially those in sub-Saharan Africa, the Green Revolution never materialized because its agricultural practices required upfront investments; irrigation systems, machinery, fuel, chemical fertilizers and pesticides beyond the financial reach of small-scale farmers. Today's biological agricultural revolution is knowledgeintensive, not capital intensive, because its technologicaladvances are incorporated into the crop seed.Today's biological agricultural revolution is knowledge intensive, not capital intensive, because its technological advances are incorporated into the crop seed. As a result, small-scale farmers with limited resources should benefit.In addition, because of the remarkable flexibility provided by crop biotechnology, crop improvement through genetic modification need no longer be restricted to the large-volume commodity crops that provide a return on industrial R&D investments. A beneficial gene that is incorporated into maize or rice can also be provided to crops grown by subsistence farmers in developing countries because the requirement for plant reproductive compatibility can be circumvented. Realizing biotechnology's extraordinary capacity for improving the health, economies and living conditions of people in developing countries, many universities, research institutions, government agencies and companies in the industrialized world have developed relationships for transferring various biotechnologies to developing countries. The nature of the relationship varies, depending on the needs and resources of the partners involved. For example: Cornell University donated transgenic technology for controlling the papaya ring spot virus to research institutions in Brazil, Thailand and Venezuela and provided their scientists with training in transgenic techniques. Japan's International Cooperation Agency built tissue culture facilities at an Indonesian research institution so that scientists there could develop disease-free potato materials for planting. The Indonesian researchers are also working with scientists at Michigan State University to develop insect-resistant potatoes and sweet potatoes. An Australian agricultural research centre collaborated with Indonesian researchers on studies of nitrogen fixation and development of disease-resistant peanuts. Seiberdorf Laboratories (Austria) worked with the Kenyan Agricultural Research Institute to transfer technology for cassava mutagenesis and breeding. Monsanto has donated virus resistance technologies to Kenya for sweet potatoes, Mexico for potatoes and Southeast Asia for papaya and technology for pro-vitamin-A production in oilseed crops to India. Pioneer Hi-Bred and the Egyptian Agricultural Genetic Engineering Research Institute (AGERI) collaborated to discover potentially novel strains of Bt in Egypt. Pioneer trained AGERI scientists in methods for characterizing Bt strains and transgenic techniques. Patents are owned by AGERI and licensed to Pioneer. AstraZeneca trained scientists from Indonesia's Central Research Institute for Food Crops in the use of proprietary technologies for creating insect-resistant maize. The Malaysian palm oil research institute has collaborated with Unilever and universities in England, the United States and the Netherlands on research to change the nutritional value of palm oil and find new uses for it, such as lubricants, fuels, a vitamin-E precursor, natural polyester and biodegradable plastics.

While technology transfer has been and, no doubt, will continue to be an essential mechanism for sharing the benefits of crop biotechnology, many developing countries are taking the next step: investing resources to build their own capacity for biotechnology research, development and commercialization. The leaders in these countries recognize the potential of crop biotechnology to provide agricultural self-sufficiency, preserve their natural resources, lower food prices for consumers and provide income to their small farmers. Even more important, they understand that biotechnology has the potential to improve existing exports and create new ones, leading to amore diversified economy and increased independence. But they also know that many of their agricultural problems are unique and can best be solved by local scientists who are familiar with the intricacies of the problems, local traditions, and applicability or lack of it of technologies that were developed to solve agricultural problems in industrialized countries. To move their countries forward, they are investing human and financial resources in developinglocal strength in crop biotechnology. For example: The Malacca government in Malaysia formed a unit in the Chief Minister's Office to promote research and development in biotechnology and established the Sarawak Biodiversity Centre to ensure sustainable use of genetic resources and to build a strong database for bioresources. Taiwan opened an extension of the Hsinchu industrial park devoted exclusively to biotechnology. Companies in the park will have access to $850 million in government research and development funds and $4 billion in state and private venture capital, plus a wide range of support services including marketing and global patent applications. Pakistan's Ministry of Science and Technology prepared a biotechnology action plan and funded a three year program to promote biotechnology research and development. Uganda's National Council of Science and Technology established its first commercial agricultural biotechnology lab to produce disease-free coffee and banana plantlets. Egypt's government, a long-time supporter of agricultural biotechnology, released a report encouragingfarmers to plant genetically modified crops to benefit from reduced pesticide applications, lower production costs, higher yields and increased income.

Environmental and Economic Benefits

Beyond agricultural benefits, products of crop biotechnology offer many environmental and economic benefits. As described above, biotech crops allow us to increase crop yields by providing natural mechanisms of pest control in place of chemical pesticides. These increased yields can occur without clearing additional land, which is especially important in developing countries. In addition, because biotechnology provides pest-specific control, beneficial insects that assist in pest control will not be affected, facilitating the use of integrated pest management. Herbicidetolerant crops decrease soil erosion by permitting farmers to use conservation tillage. Because farmers in many countries have grown biotech crops for years, data are now available for assessing the magnitude of the environmental and economic benefits provided by biotechnology. In the past few years, a number of independent researchers have produced reports documenting these benefits. According to the National Centre for Food and Agricultural Policy's (NCFAP) 2004 report, in 2003 the 11 biotech crop varieties adopted by U.S. growers increased crop yields by 5.3 billion pounds, saved growers $1.5 billion by lowering production costs, and reduced pesticide use by 46.4 million pounds. Based on increased yields and reduced production costs, growers realized a net economic impact or savings of $1.9 billion. Three new traits for corn and cotton were introduced in 2003, and the NCFAP study takes into account six biotech crops—canola, corn, cotton, papaya, soybean and squash. In its report "Conservation Tillage and Plant Biotechnology," the Conservation Tillage Information Centre (CTIC) at Purdue University attributes the recent improvements in tillage reduction to the increased use of the herbicide tolerant varieties produced through biotechnology. CTIC concludes that the increase in conservation tillage associated with herbicide-tolerant crops decreases soil erosion by 1 billion tons of soil material per year, saves $3.5 billion per year in sedimentations costs and decreases fuel use by 3.9 gallons per acre. According to the International Service for the Acquisition of Agri-Biotech Applications, a single biotech crop, Bt cotton, has led to the following environmental and economic benefits for farmers in developing countries: From 1999 to 2000 in China, insecticide usage decreased by 67 percent and yields increased by 10 percent, leading to income gains of $500 per hectare.Extensive field trials in India from 1998 to 2001 demonstrated a 50 percent reduction in insecticide spraying and a 40 percent increase in yields, which equals an increase in income from $75 to $200 per hectare. Small farmers in South Africa gained through a 25 percent yield increase and decreased number of insecticide sprays from 11 to four, reducing pesticide costs by $45 per acre. The higher cost of Btseed (up to $15 per hectare for small farmers) resulted in an average economic advantage of $35 per hectare.

Using Gene Transfer to Enhance Agriculture

Throughout the history of agriculture, humans have taken advantage of the natural process of genetic exchange through breeding that creates variation in biological traits. This fact underlies all attempts to improve agricultural species, whether through traditional breeding or through techniques of molecular biology. In both cases, people manipulate a natural process to produce varieties of organisms that display desired characteristics or traits, such as disease resistant crops or food animals with a higher proportion of muscle to fat. The major differences between traditional breeding and molecular biological methods of gene transfer lie neither in goals nor processes, but rather in speed, precision, reliability, and scope. When traditional breeders cross two sexually reproducing plants or animals, tens of thousands of genes are mixed. Each parent, through the fusion of sperm and egg, contributes half of its genome (an organism's entire repertoire of genes) to the offspring, but the composition of that half varies in each parental sex cell and hence in each cross. Many crosses are necessary before the "right" chance recombination of genes results in offspring with the desired combination of traits. Molecular biological methods alleviate some of these problems by allowing the process to be manipulated one gene at a time. Instead of depending on the recombination of large numbers of genes, scientists can insert individual genes for specific traits directly into an established genome. They can also control the way these genes express themselves in the new variety of plant or animal. In short, by focusing specifically on a desired trait, molecular gene transfer can shorten the time required to develop new varieties and give greater precision. It also can be used to exchange genes between organisms that cannot be crossed sexually. Gene transfer techniques are key to many applications of biotechnology.The essence of genetic engineering is the ability to identify a particular gene; one that encodes a desired trait in an organism, isolate the gene, study its function and regulation, modify the gene, and reintroduce it into its natural host or another organism. These techniques are tools, not ends in themselves. They can be used to understand the nature and function of genes, unlock secrets of disease resistance, regulate growth and development, or manipulate communication among cells and among organisms.

Isolation of Important Genes

The first step in an effort to genetically engineer an organism is to locate the relevant gene(s) among the tens of thousands that make up the genome. Perhaps the researcher is searching for genes to improve tolerance to some environmental stress or to increase disease resistance. This can be a difficult task similar to trying to find a citation in a book without an index. This task is made easier with restriction enzymes that can cut complex, double-stranded macromolecules of DNA into manageable pieces. A restriction enzyme recognizes a unique sequence in the DNA, where it snips the strands.By using a series of different restriction enzymes, an organism's genomic DNA can be reduced to lengths equivalent to one or several genes. These smaller segments can be sorted and then cloned to produce a quantity of genetic material for further analysis. The collection of DNA segments from one genotype, a gene library can be searched to locate a desired gene. Patterns can also be analysed to link a particular sequence, a marker to a particular trait or disease, even though the specific gene responsible is still unknown. Restriction enzymes are also used in cloning genes. To clone a gene, a small circle of DNA that exists separate from an organism's main chromosomal complement, a plasmid cut open using the same restriction enzyme that was used to isolate a desired gene. When the cut plasmid and the isolated gene are mixed together with an enzyme that rejoins the cut ends of DNA molecules, the isolated gene fragment is incorporated into the plasmid ring. As the repaired plasmid replicates, the cloned gene is also replicated. In this way, numerous reproductions of the cloned gene are produced within the host cell, usually a bacterium. After replication, the same restriction enzyme is used to snip out the cloned gene, allowing numerous copies of that gene to be isolated. The ability to isolate and clone individual genes has played a critical role in the development of biotechnology. Cloned genes are necessary research tools for studies of the structure, function, and expression of genes. Further, specific gene traits could not be transferred into new organisms unless numerous gene copies were available. Cloned genes also are used as diagnostic test probes in medicine and agriculture to detect specific diseases.

Gene Transfer Technology

To transfer genes from one organism to another, molecular biologists use vectors. Vectors are the "carriers" used to pass genes to a new host, and they can mediate the entry, maintenance, and expression of foreign genes in cells. Vectors used to transfer genes include viruses, plasmids, and mobile segments of DNA called transposable elements. Genes can also be introduced by laboratory means, such as chemical treatments, electrical pulses, and physical treatments including injection with microneedles. The basic principles behind these technologies are the same for animals, plants, and microbes, although specific modifications may be necessary. Vectors based on viruses, plasmids, and transposable elements have been adapted from naturally occurring systems and engineered to transfer desired genes into animals, plants, and microbes. For plants, the classic example is the Ti plasmid from the soil bacterium *Agrobacterium tumefaciens*, which in nature transfers a segment of DNA into plant cells, causing the recipient cells to grow into a tumor. Scientists have adapted this plasmid by eliminating its tumor-causing properties to create a versatile vector that can transfer foreign genes into many types of plants. Similarly, the transposable P-element of the fruit fly *Drosophila melanogaster* is an effective vector for gene transfer into *Drosophila*. This or similar transportable elements should prove to be adaptable to insects of agricultural importance. Animal viruses such as simian virus 40 (SV40), adeno, papilloma, herpes, vaccinia, and the retroviruses, all originally studied because of their role in disease, are now being engineered as vectors for gene transfer into animal cells and embryos. Plant viruses such as cauliflower mosaic virus, brome mosaic virus, and Gemini viruses are similarly being exploited for their abilities to transfer genes.

Cell Culture and Regeneration Techniques

The ability to regenerate plants from single cells is important for progress with gene transfer into plants. Animals cannot be regenerated asexually, so the only way to introduce a foreign gene into all cells of an animal is to insert it into the sperm, egg, or zygote. Cell culture techniques are important for the regeneration of plants. They are also critical for fundamental studies on both plant and animal cells, and for the manipulation of microorganisms. The vegetative propagation of stem cuttings or other growing plant parts to produce genetic clones is common for some agricultural crops. Potatoes, sugarcane, bananas, and some horticultural species, for example, are cultivated by vegetative propagation. Techniques exist to propagate and regenerate whole plants from tissues, isolated plant cells, or even protoplasts (plant cells from which the cell wall has been enzymatically removed) in culture. This set of techniques is complete for some agricultural species, such as alfalfa, carrots, oilseed rape, soybeans, tobacco, tomatoes, and turnips. Progress on other crops, including major food species such as many cereals and legumes, has been slower. Cell culture techniques have taken on added importance as biotechnology has progressed. Genetic engineering requires an ability to manipulate individual cells as recipients of isolated genes. Cell culture techniques allow scientists to maintain and grow cells outside the organism and thus expand their ability to perform gene transfer and study the results. In addition, cell culture allows scientists to regenerate numerous copies (clones) of the manipulated varieties, which is easier, more efficient, and more convenient, especially for producing significant quantities of stock plants. A third use of cell culture is to regenerate "somaclonal variants," plants with altered genetic traits that can prove useful as new or improved crops. Thus, cell culture techniques are important to increasing the productivity and versatility of agriculture. However, there are some important limitations. Chromosomal abnormalities appear as cultures age. These changes are related to the phenomenon of somaclonal variation, which may prove useful to agriculture, but in many instances the changes are undesirable. Therefore, scientists must learn how to prevent chromosomal changes in cell cultures. Second, long-term cultures lose regenerative potential. As biotechnology expands, it will be critical to understand why different species have differing abilities to regenerate from cell cultures into plants and how factors such as the genetic or physiological origin of the cells and the culture conditions affect growth. Most plant cells appear to be totipotent, that is, they are in a reversible differentiated state that will permit them to regenerate into a whole plant under appropriate conditions. Understanding what these appropriate conditions are remains a fundamental question in the study of plant development and its genetic control.

Monoclonal Antibody Technologies

The development of monoclonal antibody technology is based on advances in our ability to culture cells. Antibodies are the protein components of the immune system found in the blood of mammals. They have a unique ability to identify particular molecules and select them out. When a foreign substance (an antigen) enters the body, specialized cells called B lymphocytes produce a protein (an antibody) to combat it. To envision how antibodies work, think of a lock and key: The antibody key "fits" only the specific antigen lock. This marks the antigen for destruction. Each of the specialized B lymphocyte cells produces only a single type of antibody and thus recognizes only one antigen. Apart from their natural role in protecting organisms via the immune response, antibodies are important scientific tools. They are used to detect the presence and level of drugs, bacterial and viral products, hormones, and even other antibodies in the blood. The conventional method of producing antibodies is to inject an antigen into a laboratory animal to evoke an immune response. Antiserum (blood serum containing antibodies) is then collected from the animal. However, antiserum collected in this way contains many types of antibodies, and the amount that can be collected is limited. Modern biotechnology has opened a door to a more efficient, more specific, and more productive way of producing antibodies. By fusing two types of cells, antibody-producing B lymphocytes and quasi-immortal cancer cells from mice, scientists found that the resulting hybrid cells, called hybridomas, secreted large amounts of homogeneous antibodies. Each hybridoma has the ability to grow indefinitely in cell culture and thus can produce an almost unlimited supply of a specific "monoclonal" antibody. By immunizing mice with specific antigens, researchers can create and select hybridomas that produce a culture of specific, desired monoclonal antibodies. Thus, biotechnology has produced a way of creating pure lines of antibodies that can be used to identify complex proteins and macromolecules. Monoclonal antibodies are powerful tools in molecular analyses, and their uses in detecting low levels of disease agents such as bacteria and viruses are rapidly expanding. Beyond many diagnostic uses, hybridoma technology shows promise for immunopurification of substances, imaging, and therapy. Immunopurification is a powerful technique to separate large, complex molecules from a mixture of either unrelated or closely related molecules. For imaging, easily visualized tags can be attached to monoclonal antibodies to provide images of organs and to locate tumors to which the antibody will specifically bind. Finally, new therapeutic methods have been developed that use monoclonal antibodies to inactivate certain kinds of immunological cells and tumor cells or to prevent infection by certain microorganisms. Although many applications of this technology are still in the experimental stages, the commercial agricultural use of monoclonal antibodies has begun. For example, monoclonal antibodies are now on the market as therapeutics against calf and pig enteric colibacillosis, which causes neonatal diarrhoea (scours). This approach is often more effective than conventional vaccines, and it supplements genetically engineered vaccines. Monoclonal antibody-based diagnostic kits can detect whether scouring animals are infected with a particular strain of an *Escherichia coli* bacterium that causes scours, and thus help veterinarians determine the appropriate therapeutic monoclonal antibody to use on an infected herd.

New Approaches to Crop Production

In the past 50 years, agricultural production in the United States has more than doubled while the amount of land under cultivation has actually declined slightly. This impressive agricultural success is the result of many factors: an abundance of fertile land and water, a favourable climate, a history of innovative farmers, and a series of advances in the science and technology of agriculture that have made possible more intensive use of yield-enhancing inputs such as fertilizer and pesticides. Yet the productivity successes brought about by farm mechanization, improved plant varieties, and the development of agricultural chemicals may be harder to repeat in the future unless new approaches are pursued. Biotechnology offers vast potential for improving the efficiency of crop production, thereby lowering the cost and increasing the quality of food. The tools of biotechnology can provide scientists with new approaches to develop higher yielding and more nutritious crop varieties, to improve resistance to disease and adverse conditions, or to reduce the need for fertilizers and other expensive agricultural chemicals. The following paragraphs highlight some examples of how genetic engineering can be used to enhance crop production.

The Genetic Engineering of Plants

Perhaps the most direct way to use biotechnology to improve crop agriculture is to genetically engineer plants, that is, alter their basic genetic structure so they have new characteristics that improve the efficiency of crop production. The traditional goal of crop production remains unchanged: to produce more and better crops at lower cost. However, the tools of biotechnology can speed up the process by helping researchers screengenerations of plants for a specific trait or work more quickly and precisely to transfer a trait. These tools give breeders and genetic engineers access to a wider universe of traits from which to select. Although powerful, the process is not simple. Typically, researchers must be able to isolate the gene of interest, insert it into a plant cell, induce the transformed cell to grow into an entire plant, and then make sure the gene is appropriately expressed. If scientists were introducing a gene coding for a plant storage protein containing a better balance of essential amino acids for human or animal nutrition, for example, it would need to be expressed in the seeds of corn or soybeans, in the tubers of potatoes, and in the leaves and stems of alfalfa. In other words, the expression of such a gene would need to be directed to different organs in different crops.

Putting the New Technologies to Work

There are already successes that demonstrate how plants can be genetically engineered to benefit agriculture. Herbicide resistance traits are being transferred to increase options for controlling weeds. Soon, the composition of storage proteins, oils, and starches in plants may be altered to increase their value. One plant gene that has been isolated, cloned, and transferred is for the sulphur-rich protein found in the Brazil nut, *Berthalletiaexcelsa*. This protein contains large amounts of two nutritionally important sulphur-containing amino acids: methionine and cysteine. These are the very nutrients in which legumes, such as soybeans, are deficient. If the sulphur-rich protein gene were transferred into soybeans, it might enhance this legume's role as a protein source throughout the world. By purifying the Brazil nut protein and determining the order and kind of amino acids in the protein, scientists were able to synthesize an artificial segment of DNA coding for a section of this protein. This DNA "probe" was used to find and pull out the natural gene from the Brazil nut. Researchers then transferred the gene into tomato and tobacco plants, which were chosen because they are easier to manipulate than soybeans. Researchers have also transferred the gene into yeast cells. Early results show that the genetically engineered yeast does produce the sulphur-rich protein. Similar work is being done to improve oil crops. Oil crops produced in the United States in 1984 were worth $11.8 billion. Depending on their chemical composition, oils and waxes from plants have uses in feed, food, and industrial products such as paints and plastics. Chemical properties, and thereby the uses of plant oils, vary depending on the length of the fatty acid chains that composethe oil and their degree of saturation. Many of the enzymes controlling the biochemical pathways that regulate molecular chain length and degree of saturation have been well studied, and this reservoir of knowledge now makes it possible to genetically engineer the type of oil a crop produces. Although traditional breeding methods have succeeded in modifying the oil compositionof some crops, genetic engineering opens a broader range of possibilities. Scientists have taken another important step in using genetic engineering to improve crop production: They have for the first time engineered plants to be resistant to powerful herbicides. One example is glyphosate (trade name: "Roundup"), a common, effective, and environmentally safe herbicide. However, glyphosate indiscriminately kills crops as well as weeds. Thus, it must usually be used before crop plants germinate. Yet by engineering crops to be resistant to glyphosate, scientists hope to expand the range of the herbicide's applications. Scientists have isolated a glyphosate-resistance gene and successfully transferred it into cotton, poplar trees, soybeans, tobacco, and tomatoes. The gene was derived from the bacterium *Salmonella typhimurium*. Similarly to other accomplishments in biotechnology, this success depended on extensive prior basic research on biochemical pathways in bacteria and plants, and sophisticated gene cloning and transfer techniques. Field testing and commercialization of glyphosate-resistant crops should follow soon. Analysis of tomato growers' costs in California predicts that farmers could save up to $100 per acre in weed control costs if they used glyphosate in place of current herbicides, with concomitant reductions in labour, equipment, and environmental damage. This advance would also give farmers improved flexibility, yield, quality, and spectrum of weed control.

Looking to the Future

With such promising examples already being realized, it is interesting to speculate about other possibilities. For instance, could scientists take naturally occurring chemicals that hinder plant growth such as the compound crabgrass releases that prevents other grasses from invading its territory and engineer crop plants with their own ability to control weeds? Scientists have long known that some plants produce chemicals that affect the growth of other plants; by studying these allelopaths, scientists may be able to engineer or breed plants that would give farmers new biological tools to fight weeds, in addition to mechanical cultivation and other cultural tools, and chemical herbicides. The potential value of research on biological methods of weed control is great, but the work is very complicated and significant advances are not expected quickly. One of the complicating factors that must be understood is how certain plants produce allelopathic molecules and at the same time protect themselves against these chemicals.

Observations of nature combined with abilities to engineer plants might also provide opportunities to manipulate plant growth and development. Through research, scientists have determined that flowering, dormancy, fruit ripening, and a host of other growth and developmental processes come under the influence of a relatively few plant hormones or growth regulating substances. Agricultural chemists have already discovered a number of inhibitors and mimics of these regulating compounds, and these have readily found commercial applications. For example, they are used to induce and synchronize flowering and fruit production in pineapple fields, to control ripening and premature dropping of fruit from trees and vines, and to block elongation growth to create more compact and attractive potted plants, such

144 | 144 | 144 | 144

aschrysanthemums and poinsettias.Because the natural growth regulators are active in very small amounts, it has been difficult to study their synthesis and mode of action. However, the availability of new techniques and genetic probes to locate the genes responsible for their synthesis is giving researchers new tools to study these chemicals. As our understanding grows, we will likely discover additional ways to regulate and control plant growth and development. For example, perhaps scientists can improve on ways to control fruit ripening, so ripening can be delayed until the fruit is en route to market. Scientists may also develop ways to increase flowering, fruiting, seed set, or other growth habits of plants to improve efficiency of production.

References

1. Abdullah, R., E. C. Cocking, and J. A. Thompson., 1986. Efficient plant regeneration from rice photoplasts through somatic embryogenesis. *Bio/Technology* 4: 1087–1090.

2. Abel, P. P., R. S. Nelson, B. De, N. Hoffmann, S. G. Rogers, R. T. Fraley, and R. N. Beachy., 1986. Delay of disease development in transgenic plants that express the tobacco mosaic virus coat protein gene. *Science* 232: 738–743.

3. An, G., B. D. Watson, S. Stachel, M. P. Gordon, and E. W. Nester., 1985. New cloning vehicles for transformation of higher plants. *EMBO J.* 4: 277–284.

4. Austin, S., M. A. Baer, and J. P. Helgeson., 1985. Transfer of resistance to potato leaf roll virus from *Solanumbrevidens*into *Solanumtuberosum*by somatic fusion. *Plant Sci.* 39: 75–82.

5. Baulcombe, D. C., G. R. Saunders, M. W. Bevan, M. A. Mayo, and B. D. Harrison., 1986. Expression of biologically active viral satellite RNA from the nuclear genome of transformed plants. *Nature* 321: 446–449.

6. Bevan, M., 1984. Binary *Agrobacterium* vectors for plant transformation. *Nucleic Acids Res.* 12: 8711–8721.

7. Bravo, J. E., and D. A. Evans., 1985. Protoplast fusion for crop improvement. *Plant Breeding Rev.* 3: 193–218.

8. Broglie, R., G. Coruzzi, R. T. Fraley, S. G. Rogers, R. B. Horsch, J. G. Niedermeyer, C. L. Fink, J. S. Flick, and N.-H. Chua., 1984. Light-regulated expression of a pea ribulose-1,5- bisphosphate carboxylase small subunit gene in transformed plant cells. *Science* 224: 838– 843.

9. Crossway, A., H. Hauptli, C. M. Houck, J. M. Irvine, J. V. Oakes, and L. A. Perani., 1986. Micromanipulation techniques in plant biotechnology. *BioTechniques* 4: 320–334.

10. David, C., M.-D. Chilton, and J. Tempé., 1984. Conservation of T-DNA in plants regenerated from hairy root cultures. *Bio/Technology* 2: 73–76.

11. de Block, M., J. Schell, and M. van Montagu., 1985. Chloroplast transformation by *Agrobacterium tumefaciens. EMBO J.* 4: 1367–1372.

12. French, R., M. Janda, and P. Ahlquist., 1986. Bacterial gene inserted in an engineered RNA virus: efficient expression in monocotyledonous plant cells. *Science* 231: 1294–1297.

13. Fujimura, T., M. Sakurai, H. Akagi, T. Negishi, and A. Hirose., 1985. Regeneration of rice plants from protoplasts. *Plant Tissue Culture Letters* 2: 74–75.

14. Gamble, H. R., 1986. Applications of hybridoma technology to problems in the agricultural sciences. In: *Biotechnology for Solving Agricultural Problems* (pp. 39–52), P. C. Augustine, H. D. Danforth, and M. R. Bakst, eds. Dordrecht, the Netherlands: MartinusNijhoff.

15. Graves, A. C. F., and S. L. Goldman., 1986. The transformation of *Zea mays* seedlings with *Agrobacterium tumefaciens*–Detection of T-DNA specific enzyme activities. *Plant Mol. Biol. Rep.* 7: 43–50.

16. Hernalsteens, J.-P., L. Thia-Toong, J. Schell, and M. van Montagu., 1984. An *Agrobacterium* transformed cell culture from the monocot *Asparagus officinalis*. *EMBO J.* 3: 3039–3041.

17. Horsch, R. B., J. E. Fry, N. L. Hoffmann, D. Eichholtz, S. G. Rogers, and R. T. Fraley., 1985. A simple and general method for transferring genes into plants. *Science* 227: 1229–1231.

18. Potrykus, I., M. Saul, J. Petruska, J. Paszkowski, and R. D. Shillito., 1985a. Direct gene transfer to cells of a graminaceous monocot. *Mol. Gen. Genet.* 199: 183–188.

19. Potrykus, I., R. D. Shillito, M. W. Saul, and J. Paszkowski., 1985b. Direct gene transfer state of the art and future potential. *Plant Mol. Biol. Rep.* 3: 117–128.

20. Schocher, R. J., R. D. Shillito, M. W. Saul, J. Paszkowski, and I. Potrykus., 1986. Co-transformation of unlinked foreign genes into plants by direct gene transfer. *Bio/Technology* 4: 1093–1096.

21. Spradling, A. C., and G. M. Rubin., 1982. Transposition of cloned P elements into *Drosophila* germ line chromosomes. *Science* 218: 341–347.

22. Tepfer, D., 1984. Transformation of several species of higher plants by *Agrobacterium rhizogenes*: sexual transmission of the transformed genotype and phenotype. *Cell* 37: 959–967.

23. Turgeon, B. G., R. C. Garber, and O. C. Yoder., 1985. Transformation of the fungal maize pathogen *Cochliobolusheterostrophus*using the *Aspergillusnidulans*and S gene. *Mol. Gen. Genet.* 201: 450–453.

24. Vaeck, M., A. Reynaerts, H. Hofte, M. van Montagu, and J. Leemans., 1987. New developments in the engineering of insect resistant plants. *J. Cell. Biochem. Suppl.* 11B: 13.

25. Yamada, Y., Z. Q. Yang, and D. T. Tang., 1986. Plant regeneration from protoplast-derived callus of rice (*Oryza sativa* L.). *Plant Cell Rep.* 5: 85–88.

26. Yelton, M. M., J. E. Hamer, and W. E. Timberlake., 1984. Transformation of *Aspergillus nidulans* by using a *trp* C plasmid. *Proc. Natl. Acad. Sci. USA.* 81: 1470–1474.

2016, Environmental Biotechnology: A New Approach   *Pages 147–177*
Editors: Dr. Rajan Kumar Gupta and Dr. Satya Shila Singh
Published by: DAYA PUBLISHING HOUSE, NEW DELHI

Chapter 9

Prospects of Bio-composting in Organic Farming and Environment Management

Anchal Kumar Srivastava[1], Madhumita Srivastava[2],
Prem Lal Kashyap[1] and Alok Kumar Srivastava[1]*

[1]*ICAR–National Bureau of Agriculturally Important Microorganisms,*
Kushmaur, Mau – 275 103, Uttar Pradesh
[2]*Sunbeam College for Women, Varanasi – 221 005, Uttar Pradesh*

ABSTRACT

Bio-composting is a novel way to imitate the natural system occurring in ecosystems. It is a biological process in which various organic biodegradable wastes are converted into hygienic, humus rich products (compost) which could be used as a soil conditioner. Bio-compost provides a dwelling place for millions of microbes due to which biological activities are markedly enhanced in the plant rhizosphere. The use of bio-compost in agricultural production systems started long time ago, and there are several evidences that compost can enhance plants' tolerance to biotic and adverse environmental stresses, which include salt stress, drought stress, weed infestation, nutrient deficiency, and heavy metal contaminations etc. Addition of compost corroborates nutrients, micronutrients and organic matter availability to soil and favor colonization of symbiotic bacteria resulting into improved biomass. Moreover, microbial communities alter the composting process due to variation in temperature, moisture, nutrient levels etc. The initial decomposition accelerates with due heat generation through oxidation and respiratory activity that leads to further increase in decomposition process of compost formation resulting production of bio-compost. Many benefits from bio-compost are long term and are

---

* *Corresponding author.*

found from repeated applications, improving crop production, soil health, nutrient levels, organic matter, crop growth and soil-borne disease suppression. Briefly, this chapter discusses the microbiology and benefits of bio-compost in broad terms, but attempts were also made to present specifics about the role of bio-compost in organic farming and environment management.

Keywords: Agriculture, Compost, Fertility, Microbes, Stress.

Introduction

Agriculture faces the unprecedented challenge of securing food supplies for a rapidly growing human population, while seeking to minimize adverse impacts on the environment (Loos *et al.*, 2014). Recent reports indicate that crop growth and yield are adversely affected by abiotic and biotic factors including weather (rain, heat and temperature), soil conditions (water, pH and nutrients), insect populations, disease incidence and management practices (cultivar, irrigation, fertilization and rotation). These factors represent the principal cause of crop failure, decreasing average yields for major crops by more than 50 per cent (Mehta *et al.*, 2012). A study on global land use pattern revealed that 7 per cent of the world's land area, amounting to 1,000 million hectares, has become saline (Tester and Davenport, 2003). Abiotic stress affects plant growth, as well as development processes such as seed germination, seedling growth and vigour, vegetative growth, flowering and fruit set (Sairam and Tyagi, 2004). Similarly, biotic injury on crop yield impacts population dynamics, plant stress or co-evolution, and ecosystem nutrient cycling. The major biotic stresses affecting crops are fungal diseases although insects, viruses, bacteria and parasitic weeds can also drastically decrease crop production. The excessive use of agro-chemicals (fertilizers and pesticides) to manage these stresses is posing serious threats on the environment. Moreover, the production of urban and industrial organic wastes is increasing worldwide, and strategies must be developed for disposal in such a way that these do not further degrade soil. In addition, there are several environmental issues which need to be addressed such as the climate change, eutrophication and contamination of natural waters, land degradation and desertification, and loss of biodiversity (Lal, 2008). To a greater extent, the gradual reduction in the use of chemicals in agriculture without affecting yield or quality of the crop produce and waste management can only be possible with effective handling of the indigenous resources of agro-ecosystems. In this context, bio-composting is considered a promising alternative to reduce pesticide use and sustain agricultural productivity.

Bio-composting is a controlled biological decomposition process by which organic materials are degraded through the activities of successive groups of microorganisms (Dees and Ghiorse, 2001). It transforms raw organic waste materials into biologically stable, humic substances that make excellent soil amendments (Adani *et al.*, 1995). Composting has been used in farming to improve soil fertility and crop health for centuries, however the process was somewhat modernized in the nineteenth century in Europe, with the onset of organic farming (Heckman, 2006). In bio-composting processes, the most important step is the decomposition of organic matter, and this occurs via mostly aerobic decomposition, although some anaerobic decomposition also occurs (Cooperband, 2002). Bio-compost application can help to reduce pathogen attacks, in addition to the improvement of soil health and its nutrient levels. Most of the literature on the role of compost, its mechanism of action, its microbial structure and the possibilities to improve compost quality for biotic and abiotic stress management is scattered. Therefore, in this chapter pioneering as well as recent works pertaining to the role of compost in disease suppression, abiotic stress management as well as the major factors and mechanisms contributing to compost quality were discussed.

Historical Developments in Compost Technology

The art of agriculture dates back to prehistoric times. The dawn of human civilization is marked by the realization of the importance of agriculture by mankind. It brought about self-sustenance and self-reliance in the prehistoric man as regards his food requirements and livelihood. Since then, the exploitation of arable land for crop production became a necessity for the survival of mankind. In India, the use of cattle dung as manure and stalks and stems of sesame as green manure appears to have been practiced since the vedic age (2500-1000 B.C.). At different periods people used different fertilizer sources as manures to ensure better yield from the field soil. Thus, while organic manures probably found their use in field from the time man began cultivating the soils, use of inorganic fertilizers is of comparatively recent origin. It is also believed that onset of waste generation and its management begins with human civilization and urbanization. As evidenced in Neolithic period, it appears that man began to live in urban settlements, changing their habits from essentially hunters and gatherers to farmers and breeders. Since the establishment of these settlements, waste pits became commonly used. The first waste pits made out of stone and built outside the houses were found in Sumerian cities about 6000 years ago. In these pits, organic urban waste was stored for eventual application on agricultural fields (Diaz *et al.*, 2007). The early civilizations in South America, India, China, and Japan practiced intensive agriculture and it is believed that they utilized agricultural, animal, and human residues as fertilizers (Food and Agriculture Organization, 1978). Many of these residues either were placed in pits or placed in heaps and allowed to rot for long periods of time for the production of a soil amendment.

The earliest known reference to composting is found in clay tablets dated to Akkadian empire, approximately 2300 B. C. (Rodale *et al.*, 1960), it was not until early in the 20th Century that the results of controlled studies on both compost making and compost use to increase crop productivity were first published. The devolvement of composting system can be traced to the work of Sir Albert Howard. The composting procedure developed by him initially known as Indore method and later, after additional improvements, the Bangalore method employed layered mixtures of high carbon to nitrogen ration (C:N), feedstock's like leafy plant materials with low C:N materials like animal manure (3:1). VAM composting systems is also one of the earliest composting systems formed by the Netherland Government in 1932 (Gotaas, 1956). This system was an adaptation of Indore systems, with modifications including the mechanized equipments, such a strains that delivered the composting material and grappling hooks that removed finished compost material from open compost bins. This systems composted municipal solid waste rather than just agricultural crop residues.

The first rotary drum composting systems was developed in the U.S. by Eric Eweson in 1940, after got influenced by Sir Albert Howard's belief that anaerobic fermentation at initial phase could materially accelerate the composting process. This system has been very successful and there are many such systems operating throughout the world at the current time.

Composting system that operated in the second half of 20th century were of two types; i) systems in which composting took place in the open air and; ii) systems in which composting took place inside enclosures or reactors systems. The common type of open air composting systems includes the window and aerated stated pile systems. Window composting is an example of open composting systems that use relatively low level technology and is most popular technology for yard trimming composting facility. Moreover, this system generally does not required dedicated specialized equipments. Framers and others who have large amounts of organic material suitable for composting typically have general purpose farm equipments that can be used for compost turning and handlings. A large number of reactor systems were introduced in the later 20th century. The rotary drum systems developed by Ewson in the 1940s remains popular. Other noteworthy contemporary reactor systems include rectangular agitated beds system and *in-silo* systems. Rectangular agitated bed system have user friendly feed and withdrawal systems and most are constructed with aeration pipes recessed within the floor to enable maximum air movements through the compost mass (Haug 1993). On the other hand, *in silo* systems, the compost material is added to a large cylindrical shaped vessel. In the system, air blow up from the base through the composting material. In currently used silo systems, initial composting is followed by a curing periods to allow the compost to achieve the desired level of maturity.

Microbiology of Bio-Composting

Millions of microbes representing different groups act on compost substrates in succession (Suárez-Estrella *et al.*, 2013). The bio-composting process comprised of three different phase (Figure 9.1). Firstly, a mesophilic or moderate-temperature phase ($\leq 40°C$), followed by a thermophilic or high temperature phase ($\geq40°C$) and finally a mesophilic curing or maturation phase ($\leq 40°C$). Different microbial communities play an important role during the various temperature phases. The initial decomposition is carried out by mesophilic microorganisms, which are responsible for a rapid degradation of the soluble and readily degradable compounds. During this degradation process, temperatures are initially mesophilic, but start to rise. Once temperatures exceed 40°C, thermophilic microorganisms become more active and replace mesophilic microorganisms in the composting process. The thermophilic stage is one of the most important phases in composting, and the high temperatures allow microbes to break down proteins, fats, and complex carbohydrates like cellulose and hemicellulose, the major structural molecules in plants. This phase is also important because the high temperatures in the compost heap kill weed seeds and pathogenic organisms. As the availability of major structural compounds becomes exhausted, the compost temperature gradually decreases and mesophilic microorganisms once again enter the final phase of maturation of the remaining

Composting phases	Composting phases vs. Temperature	Major microbial communities during different composting phases
Mesophilic Phase (Early)		**Bacteria:** *Pseudomonas, Bacillus, Flavobacterium, Clostridium, Serratia, Enterobacter* and *Klebsiella* **Fungi:** *Alternaria, Cladosporium, Mucor, Aspergillus, Humicola,* and *Penicillium*
Thermophilic Phase		**Bacteria:** *Bacillus* and *Thermus* **Fungi:** *Aspergillus, Mucor, Chaetomium, Humicola, Absidia, Sporotrichum, Thermoascus* and yeast. **Actinomycetes:** *Streptomyces, Thermoactinomyces,* and *Thermomonospora*
Cooling/ Maturation Phase (Late Mesophilic phase)		**Bacteria:** *Bacillus, Flavobacterium, Pseudomonas* and *Cellulomonas* **Fungi:** *Alternaria, Aspergillus, Bipolaris* and *Fusarium* **Actinomycetes:** *Streptomyces* and *Thermopolyspora*

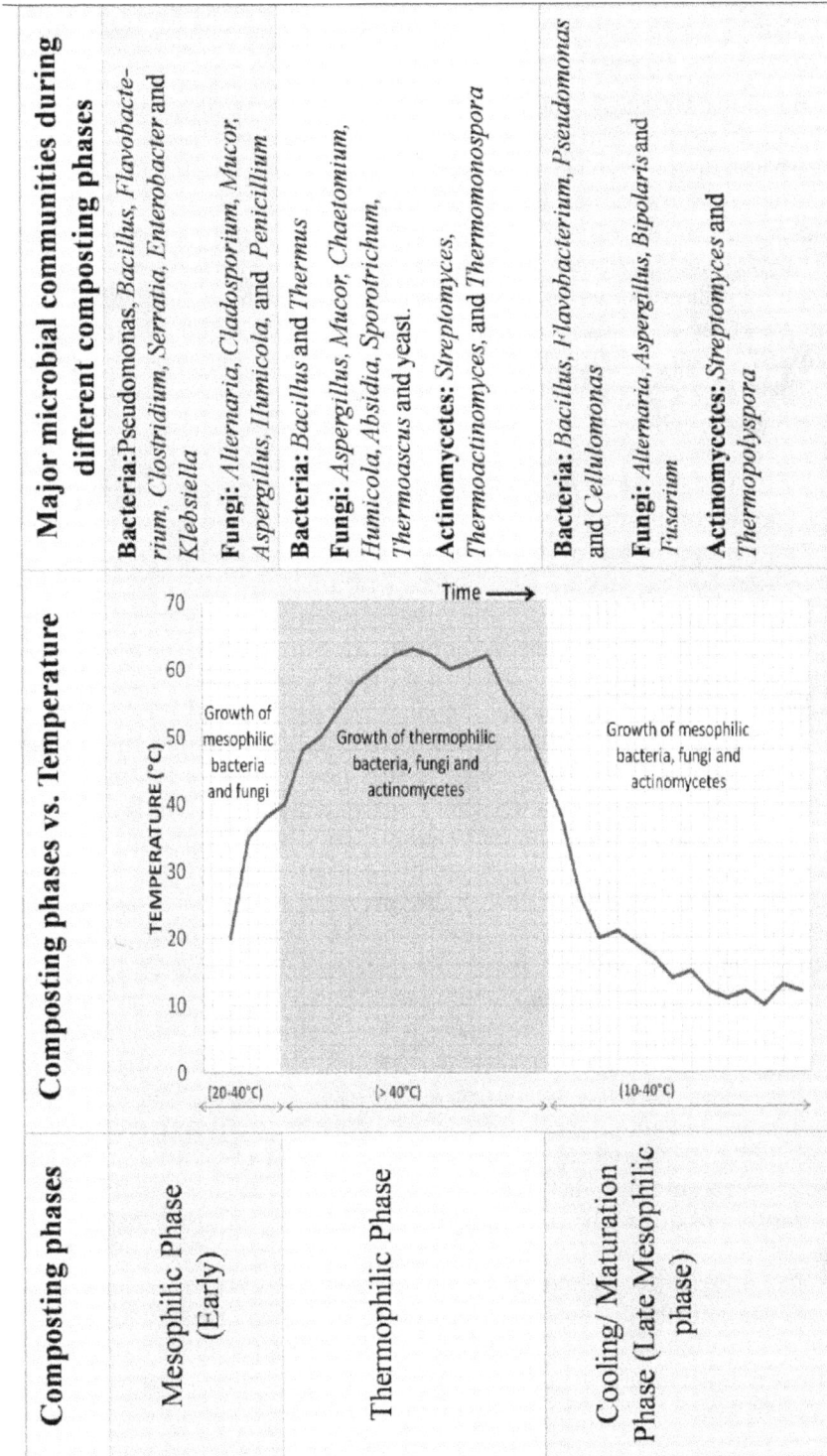

Figure 9.1: Different Phases of the Composting Process and the Microbial Communities Involved in These Phases (*Source:* Mehta *et al.,* 2013).

organic matter (Insam and de Bertoldi, 2007). The ambient temperature of the compost during the curing phase helps to make the remaining organic matter more stable and suitable for agricultural use (Epstein, 1997).

During bio-composting, the bacterial population is mainly responsible for substrate decomposition and heat generation. Bacteria constitute the majority of microorganisms in composting piles, with eubacteria and actinomycetes usually present in higher number than fungi (Rebollido *et al.*, 2008). Usually, mesophilic bacteria observed in the initial stages of bio-composting and include *Escherichia*, *Klebsiella*, *Aeromonas*, *Alcaligenes*, *Enterococcus* and *Bacillus* species (Ghazifard *et al.*, 2001). Their populations increase exponentially during the initial stages of bio-composting as they take advantage of the readily available sugars and starches. (Golueke 1972) found actinomycetes *viz.*, *Thermophilus*, *Streptomyces* and *Micromonospora* spp. to be common in compost. These microorganisms were found to be ineffective competitors when nutrient levels are high because of their slow development compared to bacteria or fungi, but become more competitive as nutrient levels decreased (Nakasaki *et al.*, 1985). Fungi are present in higher numbers when moderate temperatures and low moisture levels prevail in the bio-compost piles. Nonetheless, mesophilic fungi, yeasts and molds have been observed in the initial stages of the composting process, despite temperature of up to 60–68°C (Beffa *et al.*, 1996). Rawat *et al.* (2005) observed that diverse populations of mesophilic fungi existed throughout the bio-composting process. Several *Bacillus* species have been recorded during the thermophilic stage of the composting process including the thermotolerant *B subtilis*, *B. polymyxa*, *B. pumilus*, *B. sphaericus*, and *B. licheniformis*, as well as thermophilic species like *B. stearothermophilus*, *B. acidocaldarius*, and *B. schleglii* (Ghazifard *et al.*, 2001). As the activity of thermophilic bacteria decrease, the temperatures also decrease and both Gram positive and Gram-negative mesophilic bacteria again predominate and are involved in compost curing and maturation. A high diversity of thermotolerant fungi including *Thermomyces* spp., *Penicillium duponti*, *Geotrichum candidum* (Le Goff *et al.*, 2010), *Cladosporium*, *Aspergillus*, *Mucor*, *Rhizopus* and *Absidia* spp. have been observed during the later stages of composting where temperatures between 40 and 60°C occur (Rawat *et al.*, 2005). The optimum range for the survival of thermotolerent fungi is between 45 and 50°C, and the disappearance of viable fungi in composts is well advanced before temperatures reach 60°C, and is essentially completed by 65°C (Mehta *et al.*, 2013). It has also been reported that a consortium of microorganisms may be necessary to degrade lignocellulosic materials, however, interactions among species are not well documented (Davis *et al.*, 1992). At the late mesophilic stage where temperatures decrease and the activity of thermophilic fungi also decrease, mesophilic fungi again begin to recolonise. Several studies on fungal communities during the later stages of composting reported that species of *Alternaria*, *Aspergillus*, *Bipolaris*, *Fusarium*, *Mucor*, *Rhizopus*, *Peziza*, *Phoma* and *Trichoderma* dominate (Ryckeboer *et al.*, 2003). Recently, Steel *et al.* (2009) reported that immediately after the thermophilic phase peak, the compost nematode population (members of Rhabditidae, Panagrolaimidae and Diplogastridae) was comprised of opportunists that fed solely on bacteria. Afterwards, general opportunists who fed on either bacteria (members of Cephalobidae), or fungi (members of Aphelenchoididae) could be found. During the maturation phase, the bacterial feeding predator nematodes (*Mononchoides* sp.) became dominant and finally, in the most mature stage, the fungal feeding Anguinidae (mainly *Ditylenchus filimus*) dominated. Besides this, protozoa also represent a small fraction of microbial biomass in compost. They are found in water droplets in compost and feed on bacteria and fungi (Epstein, 1997). In bio-composting processes, Protozoa play an important role in the decomposition of organic matter, in disease suppression and in nutrient cycling. Protozoa feed on bacteria which have high nitrogen contents, thus these organisms can have a significant effect on the nitrogen cycle in compost (Hoorman and Islam, 2010). Therefore, it is essential to understand the diversity and distribution of these microbial components in compost ecosystems.

Factors Affecting Bio-composting Process

Bio-composting process is affected by several biotic and abiotic factors, which maintain its integrity on optimum and these factors are quite important for effective and efficient bio-composting process. Temperature is the main factor that regulates microbial activity during composting. Heating is essential to enable the development of a thermophilic population of microorganisms which is capable of degrading the more recalcitrant compounds (natural and anthropogenic), and to kill pathogens and weed seeds (Boulter *et al.*, 2000). The pH is a good indication showing the development in different stages of composting. The pH dropped slightly at the beginning of the composting process due to production of organic acids. Generally, a pH of 6.7–9.0 is required for good microbial activity during bio-composting, although, optimum value lies between 5.5 and 8.0 (Miller, 1992). The optimization of pH factor is very important for controlling N-losses by ammonia volatilization, which must be higher than 7.5. Aeration is another factor which has an indirect effect on temperature by speeding the rate of decomposition and therefore, the rate of heat production. The air requirement depends upon the type of waste (type of material, particle size), the temperature of the compost and the stage of the process. Air supply can be controlled to some extent by the use of a system of aeration. Under natural conditions warm air diffuses from the top of the windrow drawing fresh air into the base and sides (Hellmann *et al.*, 1997). Forced aeration has also been used successfully on static piles giving a high degree of process control (Sasaay *et al.*, 1997).

Microbial activity is strongly influenced by moisture content. Usually, microbial activity decreases under dry conditions, and aerobic activity decreases under water-logged conditions due to the resulting decrease in air supply. The recommended optimum water content is 40–60 per cent on a mass basis (Epstein, 1997). The microorganisms involved in composting develop according to the temperature of the mass, which defines the different steps of the process (Keener *et al.*, 2000). Bacteria predominate early in composting, fungi are present during all the process but predominate at water levels below 35 per cent and are not active at temperatures >60°C. Actinomycetes predominate during stabilization and curing, and together with fungi are able to degrade resistant polymers.

During bio-composting process, moisture is essential but if the compost is too wet then anaerobic conditions develop. Anaerobic conditions are also undesirable because of the loss of N by denitrification. There may also be build-up of organic acids, such as acetic acid, which can be toxic to plants. Alteration in moisture content is directly related to aeration and temperature and it is observed that in an aerated static pile system ~90 per cent of the heat loss is due to evaporation of water (Sasaay *et al.*, 1997). Systems which actively encourage aeration can lead to desiccation and result in a decrease in the rate of decomposition in windrow composting (Itavaara *et al.*, 1997). Hence, bio-compost must be kept aerobic to avoid the production of odours. The carbon nitrogen ratio (C: N) is another factor which is important in determining the rate of decomposition of organic materials. High C/N ratios make the process very slow as there is an excess of degradable substrate for the microorganisms. But with a low C/N ratio there is an excess of N per degradable C and inorganic N is produced in excess and can be lost by ammonia volatilization or by leaching from the composting mass. Then, low C/N ratios can be corrected by adding a bulking agent to provide degradable organic-carbon.

Enzymes and Enzymatic Processes Involved in Bio-composting

Bio-composting is a complex process that mainly depends on the metabolic machinery of the microbes and involves different enzymes and enzymatic processes. The enzymes involved in bio-composting process vary depending on the composition of the organic matter in the compost, the physico-chemical conditions that are prevalent during the composting and the changing dynamics of microbial community succession. A pictorial representation of the enzymes and microbial community prevailing during bio-composting process has been shown in Figure 9.2.

The role and significance of various enzymes *viz.* avicelase, cellobiase, cellulases, CMCase, endoglucanases, hemicelluases, mannanase, phosphatase, proteases, urease, xylanases, β-glucosidase during bio-composting process has been reported by several workers. Various microbial strains and their hydrolytic enzymes produced during different stages of bio-composting processes have been described in Table 9.1. Most common microbial enzymes produced during the initial phase of the composting process is the proteases, and generally associated with the nitrogen cycle and hydrolysis of proteins (Castaldi *et al.*, 2008). The protease activity was associated with a number of *Bacillus* sp. during the early phase of animal manure composting. Proteases from *B. stearothermophillus* were found to be activated in the presence of calcium and ferrous ions during the aerobic degradation of sludge compost (Kim *et al.*, 2002).

Another category of enzymes is cellulases and hemicellulases which include a wide array of enzymatic reactions that plays significant roles in the degradation of organic matter, particularly plant matter and lignocellulosic residues. These enzymes have decisive influence on the carbon cycle and also affect the overall nutrient cycle during composting. He *et al.* (2013) reported the highest cellulase and β-glucosidase activity was associated with the thermophilic Bacilli. Similarly, Goyal *et al.* (2005) also highlighted that alterations in the cellulase and xylanase activity were associated with the breakdown of various compost types and includes sugarcane trash + cattle dung (4:1), sugarcane trash + cattle dung (1:1), press mud, poultry waste and water

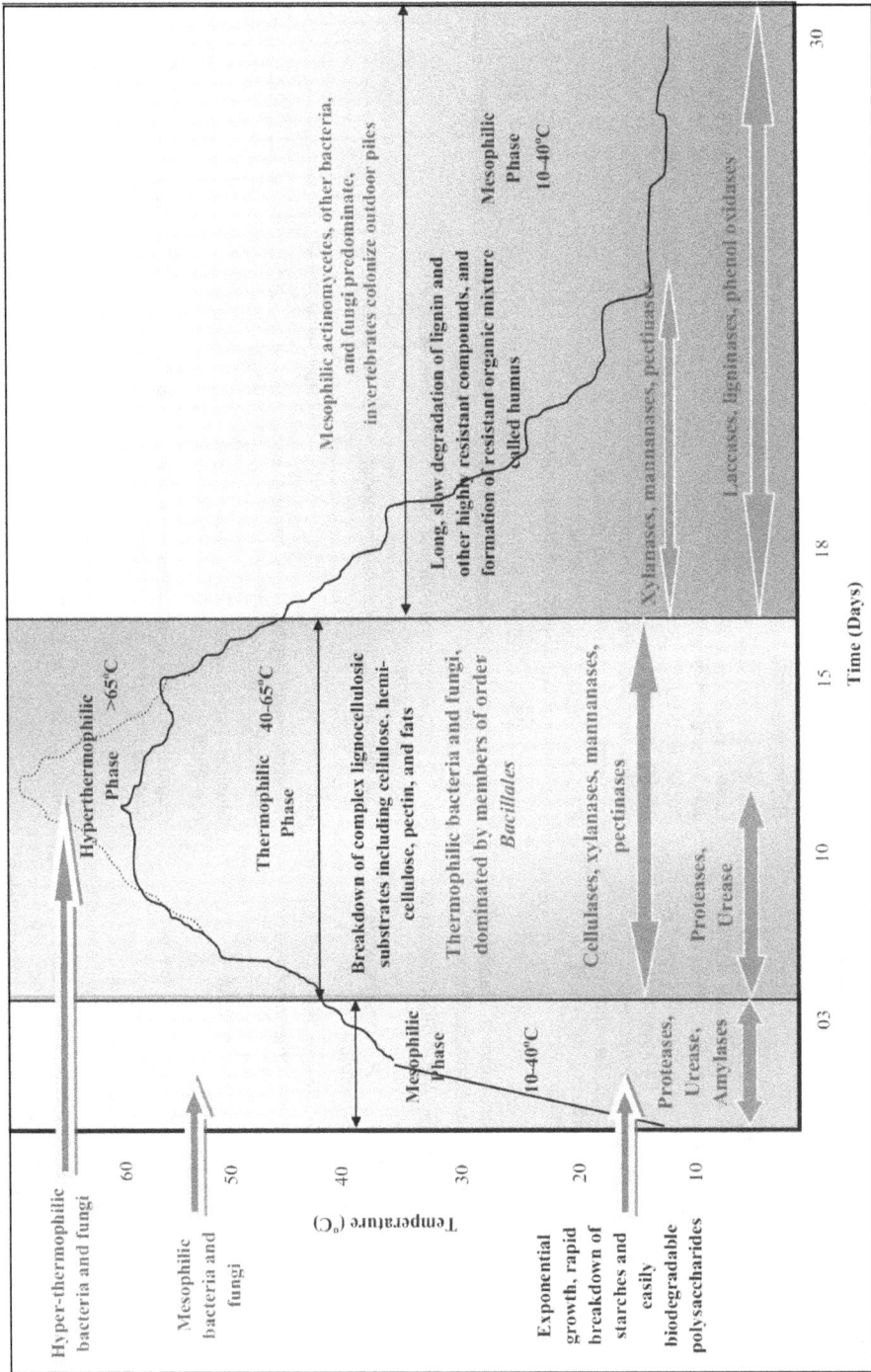

Figure 9.2: Enzymes and Microbial Activities Associated with Bio-composting Process (*Source:* Bhattacharya and Pletschke, 2014).

Table 9.1: Enzymes Secreted by Microbes during various Bio-composting Processes

Enzyme	Microbe	Bio-composting Process	Reference(s)
Avicelase	*B. subtilis*	Hot compost	Kim *et al.,* 2002
cellobiase	*B. subtilis*	Rotting rice straw	Shu-bin *et al.,* 2012
Cellulases	*Bacillus* sp., *B. stearothermophillus, B. licheniformis, B.subtilis* sub *subtilis, B.subtilis* sub *zizenii, B.amyloliquefaciens; Paenibacillus cookie,Paenibacillus* spp., *Geobacillus* spp., *Paenibacillus* spp. (*P.woosongensis) Cohnella* sp.,, *Streptomyces* spp., Thermophilic microbialconsortium (TMC) and other thermophiles	Municipal solid waste, Sugarcane trash + cattle mud, press mud poultry waste, water hyacinth; Mix of yellow poplar wood chips and mown lawn grass; Industrial waste compost; Garden compost; Poultry manure compost; Rice straw with pig manure; thermal compost' Lignocellulosic substrates (corn stover and prairie grass)	Amore *et al.,* 2013 Wei *et al.,* 2012; Eida *et al.,* 2012; Zambare *et al.,* 2011; Yang *et al.,* 2010; Ng *et al.,* 2009; Wang *et al.,* 2008; Raut *et al.,* 2008; Mayende *et al.,* 2006; Goyal *et al.,* 2005
CMCase	*B. subtilis*	Hot compost, rotting rice straw	Shu-bin *et al.,* 2012; Kim *et al.,* 2002
Endoglucanases	*B. subtilis, B. licheniformis, Bacillus* spp., *Geobacillus* spp. *Paenibacillus* spp.	Hot compost; Lignocellulosic biomass	Acharya *et al.,* 2012; Nizamudeen and Bajaj 2009
Hemicelluases	Thermophilic *Bacilli* and other groups of thermophiles	Mix of yellow poplar woodchips and mown lawn grass	Wei *et al.,* 2012
Mannanase	*Paenibacillus* spp. (*P. woosongensis) Cohnella* sp., *Streptomyces* spp.	Saw dust and coffee residue compost	Eida *et al.,* 2012
Phosphatase	Majority mesophiles and some thermophiles	Municipal solid waste	Raut *et al.,* 2008
Proteases	*Bacillus* sp., *B. stearothermophillus* and other thermophiles	Municipal solid waste, Diary manure with rice, sludge compost	Liu *et al.,* 2011; Raut *et al.,* 2008;Kim *et al.,* 2002
Urease	Majority mesophiles and some thermophiles	Municipal solid waste, Daily manure and rice chaff	Liu *et al.,* 2011; Castaldi *et al.,* 2008
Xylanases	Mesophiles belonging to *Bacillus* and *Paenibacillus, Paenibacillus* spp. (*P. woosongensis) Cohnella* sp., *Streptomyces* spp. and some thermophiles	Sugarcane trash + cattlemud, press mud poultrywaste, water hyacinth	Eida *et al.,* 2012; Goyal *et al.,* 2005
β-glucosidase	*Bacillus* sp. *B. subtilis, Paenibacillus* spp. (*P. woosongensis) Cohnella* sp., *Streptomyces* spp. and other thermophiles	Municipal solid waste, Hot compost	He *et al.,* 2013, Eida *et al.,* 2012; Castaldi *et al.,* 2008; Kim *et al.,* 2002

hyacinth. Raut *et al.* (2008) reported an increase in cellulose activity that reached a maximum on day 12 of municipal sludge waste composting (aerated with added glucose) followed by a decline in activity until the end of the composting period.

Urease is another type of enzyme which involved in the hydrolysis of urea to ammonium and carbon dioxide and is closely related to the nitrogen cycle. Castaldi *et al.* (2008) observed a strong correlation between urease, protease and β-glucosidase activity. Urease activity increased during the first 21 days of composting followed by a steep decline in activity in the fourth week and then stabilized after 70 days of composting. The initial increase in urease activity could be attributed to the availability of water soluble nitrogen due to the action of proteases. The sharp decrease in urease activity may be due to the accumulation of nitrates in composting mixture. Cayuela *et al.* (2008) reported very low urease activity during the entire duration of semi-solid olive mill waste composting, which was attributed to the low concentration of nitrogen and also to the low levels of available substrates released during mineralization. Liu *et al.* (2011) reported the presence of significant urease activity during the second week of composting of daily manure with rice chaff.

Alkaline phosphatase is a relevant enzyme for characterization of the composting process. Phosphatase is a key enzyme in the phosphorus cycle and is induced by the presence of carbohydrate derived structures that are degraded to glucose by enzymatic action. This enzyme is only synthesized by microbes, and is not released by plants and/or organic matter residues. The activity of phosphate enzyme has been reported during the early phase of composting followed by decline in activity (Raut *et al.*, 2008); however, Ros *et al.* (2006) reported an increase in alkaline phosphatase activity during the beginning of composting, reaching a maximum by the end of the process. Phosphatase activity could be related to the amount and availability of organic phosphate compounds present in the composting mixture. This enzyme has high agronomic value because it hydrolyses compounds of organic phosphorous and transforms them into different forms of inorganic phosphorus which is assimilated by plants.

Recently, Eida *et al.* (2012) reported the isolation and characterization of cellulose decomposing bacteria belonging to the genus *Paenibacillus*, *Cohnella*, *Streptomyces* and *Microbiospora* from saw dust and coffee residue composts. The two novel strains belonging to *P. woosongensis* producing several hydrolytic enzymes, including cellulases, xylanases, β-glucanase, and mannanase were also identified. Yang *et al.* (2010) reported the isolation and identification of a novel cellulase producing *Bacillus subtilis* strain 115 from long-term thermal compost containing rich cellulose materials that showed optimum activity at pH 6.0 and 60 °C. Shu-bin *et al.* (2012) reported the isolation and identification of *B. subtilis* strain Pa5 from soil samples rich in rotting rice straw. The optimal pH and temperature for both CMCase and cellobiase were observed at pH 7.0 and 50 °C. The enzyme exhibited a broad range of thermostability (30 to 50 °C) and pH stability (5.0-8.0). Thus, the versatility and high activities of various enzymes on different lignocellulosic substrates suggests a robust and compatible system for the cost-effective breakdown of lignocellulosic biomass.

Bio-compost Impact on Soil Quality

Soil quality includes the physical, chemical and biological properties of the soil which together influence their functions and play vital role in agricultural sustainability. Improving soil quality is very important for maintaining and increasing food production from agricultural land. The effectiveness of compost with regard to beneficial effects on soil physical, chemical and biological properties, as well as constituting a nutrient source, depends on the quality of the compost. The quality criteria for compost are established in terms of nutrient content, humified and stabilised organic matter (OM), the maturity degree, the hygienisation and the presence of certain toxic compounds such as heavy metals, soluble salts and xenobiotics. The production of compost with a high nutrient content requires the control and reduction of nutrient losses during the process, whilst to ensure a high degree of OM humification enough time should be allowed for the maturation phase. Finally, a high degree of compost maturity requires the establishment of adequate maturity indices. However, it is very difficult to distinguish between the direct and the indirect effects of an compost amendment on the behavior of soil microorganisms. In soils amended with compost or other raw organic materials, even in association with nitrogenous fertilizer, microbiological activity and growth can be stimulated. However, Ros *et al.* (2006a) and Kaur *et al.* (2008) highlighted that a direct effect from microorganisms introduced with the compost is also possible. Several long-lasting experiments have demonstrated that soil biological properties, such as microbial biomass -C, basal respiration and some enzymatic activities, are significantly improved by compost treatments. This is particularly evident in the upper layers of the soil because of the added labile fraction of organic matter, which is the most degradable one (Tejada *et al.*, 2009). Since, compost is slowly decomposed in the soil, the continuous release of nutrients can sustain the microbial biomass population for longer periods of time in comparison to mineral fertilizers (Murphy *et al.*, 2007). In fact, an interesting residual effect of composts on the microbial activity has often been observed in many experimental seasons after their application, which also results in a longer availability of plant nutrients. For instance, four years after the application of compost and manure, Ginting *et al.* (2003) observed 20-40 per cent higher soil microbial biomass-C compared with the N-fertilizer treatment as residual effect.

As a thumb rule, the quantity and quality of organic material added to soils are the major factors in controlling the abundance of different microbial groups and the activity of microorganisms involved in nutrient cycling. In a rice–wheat system, farmyard manure application @ 20 t ha^{-1} showed higher organic carbon concentration compared with NPK fertilizers in the 0–15 cm soil layer, after 32 years of amendments (Kukal *et al.*, 2009). On the other hand, contrary to common belief, over 25-yea, of intensive rice-wheat cropping, a depletion of organic carbon did not occur, but rather an improved organic carbon concentration of 38 per cent was observed (Benbi and Brar 2009). In terms of sustainability, only farmyard manure fertilization maintained the total organic carbon level of 40 t C ha^{-1} in the top soil layers at the start of a 40-year experiment, while the average total organic C depletion was 23 per cent with liquid manure and mixed fertilization treatments (Nardi *et al.*, 2004). Furthermore, the presence of weakly acidic chemical functional groups on organic molecules makes organic matter an effective buffer, as supported by the findings of García-Gil *et al.* (2004). They reported a long- and short-term improvement in the soil humic acid buffering capacity in municipal solid waste compost-amended soils, derived from a residual effect of a single amendment and cumulative effects from repeated ones. These distributions favor the general soil fertility status and crop production.

It is well known that many microorganisms convert organic-N into inorganic-N forms by mineralization. A large number of reports confirmed that N mineralization from compost is very limited in the short term. However, there is a significant residual effect from the cumulative applications which becomes visible after 4–5 years, resulting in tardy higher N-availability and yields (Blackshaw *et al.*, 2005; Barbarick and Ippolito, 2007; Leroy *et al.*, 2007). Regular addition of organic material to soil for one decade, through compost or manures, enhanced both soil-C and N stocks and resulted in build-up of N, indicating a physical protection of this nutrient within macro-aggregates (Whalen and Chang 2002; Mallory and Griffin 2007; Sodhi *et al.*, 2009). In summary, the repeated application of composted materials can enhance soil organic N content, storing it for mineralization in the following cropping seasons.

Composting and Soil Carbon Sequestration

In the arena of climate change, global concern over increased atmospheric CO_2 and methane (CH_4) emissions has raised interest about the potential role of soils as a source or sink of C and in studying organic matter dynamics and related C sequestration capacity. In fact, organic matter varies in both decomposition rate and turnover time. And the soil-C pool can be a source or sink for the atmospheric pool, depending on land use and management (Van-Camp *et al.*, 2004). When organic materials, such as composted wastes, are added to soil, at least a share of their organic carbon (C) is decomposed producing CO_2, which is a good indicator of the decomposition rate, while another part is sequestered in the soil. Mandal *et al.* (2007) observed that the total quantity of soil-C sequestered over a long period was linearly related to the cumulative crop residue C inputs, and the rate of the conversion to organic carbon was higher in the presence of added organic materials. Indeed, it has been calculated that the rate of organic carbon sequestration is on average 0.3-0.5 t C ha^{-1} year^{-1} under intensive agricultural practices (Lal, 2007). As reported by Mondini and Sequi (2008), organic matter is the largest C stock of the continental biosphere, with 1550 billion tons. On the time scale of several decades, in arable soils which receive high organic materials, relatively less organic matter is stabilized either by association with the silt plus clay mineral fractions, or by its inherent biochemical recalcitrance (Sleutel *et al.*, 2006). Shindo *et al.* (2006) reported that continuous compost application in a field subjected to 25 years of double cropping could increase both the amounts of fulvic and humic acids, and the total humus content. From the organic carbon sequestered quality standpoint, Nardi *et al.* (2004) found that, over 40 years, farmyard manure fertilization improved by 116 per cent the production of humus with a high degree of polycondensation, a high-quality fraction usually linked to soil fertility. Conversely, the absence of organic fertilizer inputs determined the opposite, with a higher percentage of non-complex and light-weight humus. According to Sodhi *et al.* (2009), a ten-year application of rice straw compost, either alone or in combination with inorganic fertilizers, results in C-sequestration in macro-aggregates. In fact, with the application of 8 t compost ha^{-1}, the C concentration (1–2 mm size fraction) enhanced by 180-191 per cent, respectively, over unfertilized control. From ongoing trials for more than 10 years, there is evidence that the organic carbon sequestration rate increased more due to farmyard manure and composted farmyard manure in comparison with mineral fertilizers or other organic materials (Monaco *et al.*, 2008; Kukal *et al.*, 2009). Besides, the soil C sequestration should not be restricted to a mere quantification of C storage or CO_2 balance. All green-house gas fluxes must be computed at the plot level in $C–CO_2$ or CO_2 equivalents, incorporating as many emission sources and sinks as possible across the entire soil-plant system (Feller and Bernoux, 2008). For instance, experimental results of Ginting *et al.* (2003) highlighted that fluxes of CH_4 and nitrous oxide (N_2O) were nearly zero after four years of manure and compost applications. This evidence indicates that the residual effects had no negative influence either on soil C and N storage, or global warming. However, there is a lack of information and a strong need to further understand the greenhouse gas fluxes as related to organic material input dynamics in the soil.

Bio-compost and Biotic Stress Management

The microbial communities present in compost are considered to be one of the major driving forces for plant pathogen suppressiveness of composts. There are several reports on beneficial microbes in compost, which compete with pathogens and thus, suppress their activity. Microbial communities present in compost amended container media function as biocontrol agents against disease caused by *Pythium* and *Phytopthora* spp. (Boehm *et al.*, 1993). A significant reduction in cucumber wilt (up to 61 per cent), caused by *F. oxysporum* f. sp. *cucumerinum*; was also recorded after inoculation with *B. subtilis* SQR 9 (Cao *et al.*, 2011). *T. asperellum* strain T34, isolated from *Fusarium*-suppressive compost (Trillas and Cotxarrera, 2003), has been reported to control Fusarium wilt in tomato and carnation plants (Cotxarrera *et al.*, 2002; Sant *et al.*, 2010) and *R. solani* in cucumber plants (Trillas *et al.*, 2006). *Trichoderma* spp. T34 have also been reported to be able to reduce foliar pathogens numbers when applied to the roots of host plants (Segarra *et al.*, 2009).This maybe happened due to the fact that the bio-composting process, where raw materials are degraded and converted into humus, provides an ideal environment for the development of these microbes.

Protective strains of *F. oxysporum* occur naturally in almost all agricultural soils including different composts (Cotxarrera *et al.*, 2002) and live their life partly inside plant tissue as endophytes, without harming plant tissue (Ito *et al.*, 2005). Non-pathogenic *F. oxysporum* have been shown to control Fusarium wilt in several crops, including asparagus, banana, basil, carnation, chickpea, cucumber, cyclamen, gladiolus, melon, tomato, spinach and watermelon (Larkin and Fravel, 1999; Elmer, 2004; Forsyth *et al.*, 2006, Mehta *et al.*, 2013). In contrast to other biocontrol agents (BCAs) such as *Trichoderma* spp., protective strains of *F. oxysporum* are mostly effective against pathogenic *F. oxysporum*. The advantage of non-pathogenic *F. oxysporum* in the control of the same and closely related pathogens is that both require similar environmental conditions, thus creating a competition between them when both are present (Larkin and Fravel, 2002). No parasitism, hyphal interference or toxin production was observed between pathogenic and non-pathogenic strains of *F. oxysporum*. The reason hypothesized for this suppression is competition with the pathogen for infection site and nutrients (Freeman *et al.*, 2002). Only a few papers have reported on the efficacy of non-pathogenic strains of *F. oxysporum* against *P. ultimum* (Benhamou *et al.*, 2002), *P. capsici* (Silvar *et al.*, 2009) and *V. dahlia* (Pantelides *et al.*, 2009). Moreover, some endophytic strains of non-pathogenic *F. oxysporum* have been shown to reduce damage caused by *Meloidogyne incognita* in tomato roots (Dababat and Sikora, 2007).

In addition to fungi, some bacteria belonging to genus *Pseudomonas*, *Bacillus*, *Burkholderia*, *Lysobacter*, *Pantoea* and *Streptomyces* are present in compost and reported to act as disease suppressers and plant growth promoters (Castano *et al.*, 2011). Sporulating Gram-positive bacteria like *Bacillus* species have been used successfully in plant disease control (Kloepper *et al.*, 2004). *B. cepacia* has shown great potential to be used as an effective biocontrol agent of *Fusarium* dry rot of potatoes (Recep *et al.*, 2009). Tomato plants inoculated with *B. subtilis* showed biocontrol activity against damping off and root rot disease and gave high yields of tomato (Morsy, 2005). A possible explanation for growth promotion and pathogen resistance by *B. subtilis* is that the microbes compete with other microorganisms that would otherwise adversely affect the plant and activate the host defense system. By doing so, the plant is poised to resist potential pathogens. It also makes certain nutrients like phosphorus and nitrogen more readily available to the plant (Nagorska *et al.*, 2007). Shanmugam and Kanoujia (2011) reported that activation of the host defense system could be one mechanism for enhanced tomato growth. *B. subtilis* strains isolated from cow dung have also been reported for their antimicrobial activity against *F. oxysporum* and *Botryodiplodia theobromae* on postharvest rotting fungi of yam tubers (Ray *et al.*, 2000; Naskar *et al.*, 2003).

The genus *Pseudomonas*, ubiquitous in soil and compost environments, contains endophytic bacteria known for their plant growth promotion and antagonistic properties towards different pathogens (Gibello *et al.*, 2011). The endophytic nature of *P. fluorescens* means that the bacterium competes with pathogens, stimulates plant growth and reduces the incidence of plant disease (Kloepper and Schroth, 1978). Several fluorescent pseudomonad species such as *P. fluorescens* (Sakthivel and Gnanamanickam, 1987), *P. putida* (de Freitas and Germida, 1991), *P. chlororaphis* (Chin-A-Woeng *et al.*, 1998) and *P. aeruginosa* (Anjaiah *et al.*, 2003) have been used to suppress pathogens as well as to promote growth and yields in many crop plants. Fishal *et al.* (2010) reported the ability of the endophytic bacteria *Pseudomonas* sp. UPMP3 to control *F. oxysporum* f. sp. *cubense* race 4. The presence of beneficial microbes like *Bacillus* spp., *Enterobacter* spp., *Pseudomonas* spp., *Streptomyces* spp., *Trichoderma* spp. and *Gliocladium* spp. in compost indicates its disease suppressiveness behavior.

The disease suppressiveness property of composted bio-wastes has been well reported (Tuitert *et al.*, 1998; Lievens *et al.*, 2001). Blok *et al.* (2002) reported that disease suppressiveness of commercial biowaste composts towards *P. ultimum*, *P. cinnamomi* and *R. solani* differed considerably, ranging from slightly conducive to highly disease suppressive. Further, Hoitink *et al.* (2001) reported that specific disease suppression of composts can only be guaranteed when composts are colonized by specific antagonists during bio-composting process. Adding thermo-tolerant, phosphate-solubilising microbes including bacteria, actinomycetes, and fungi can shorten the period of maturity, improve the quality, increase the soluble phosphorus content, and enhance the populations of phosphate-solubilising and proteolytic microbes in biofertilisers (Chang and Yang 2009). Application of microbes such as *Agrobacterium, Bacillus, Enterobacter, Pseudomonas, Aspergillus, Trichoderma* and *Glomus* to the roots of plants, to soils, and in fertilisers has been shown to release soluble phosphorus, promote plant growth, and protect plants from pathogen infection (Rodriguez and Fraga, 1999; Rudresh *et al.*, 2005; Zayed and Abdel-Motaal, 2005; Biswas and Narayanasamy, 2006; Ouahmane *et al.*, 2007). In addition, cellulolytic organisms including the fungal species, *Trichoderma, Humicola, Penicillium,* and *Aspergillus* and a wide variety of Gram-positive and Gram-negative species including *Clostridium thermocellum, Streptomyces* spp., *Ruminococcus* spp., *Pseudomonas* spp., *Cellulomonas* spp., *Bacillus* spp., *Serratia, Proteus, Staphylococcus* spp., and *B. subtilis* play an important role in the degradation of cellulosic material in composts and in municipal solid wastes which are composed of 40–50 per cent cellulose (Gautam *et al.*, 2009; Gautam *et al.*, 2010). However, the application of these microbes to compost could increase the degradation process and also the efficacy of compost towards the control of different soil-borne plant pathogens.

Bio-compost and Abiotic Stress Management

Soil salinity and nutrients deficiency pose a severe problem throughout the globe and around 20 per cent of the world's cultivated land and 50 per cent of cropland are affected (Lakhdar *et al.*, 2009). The excessive salt amounts adversely affect soil physical and chemical properties, as well as the microbiological processes. Concerning affected soils, a wide array of organic soil amendments, with varying levels of processing and characterization are used for their reclamation. Compost amendments most frequently are used to provide essential nutrients (N and P and K) to rebuild soil physico-chemical properties, and re-establish microbial populations and activities (Hanay *et al.*, 2004; Lakhdar *et al.*, 2008). The amendment of saline soil with bio-compost enhances their subsequent mineralization with microflora with a concomitant increase in CO_2 release and consequently soil aeration (Muhammad *et al.*, 2007) presumably owing to their enzymatic activities stimulus. During the microbial degradation and humification of residues, microbes generally control the mineralization of organic carbon in soil. The incorporation of organic manure including bio-compost, into soils significantly stimulates soil microbial biomass and activity, due to the high quantities of readily utilizable energy sources introduced (Shen *et al.*, 1997). The enhanced soil enzyme and biological activities are believed to be direct indicators of the enhancement of soil fertility resulting from the incorporating organic manure (Lakhdar *et al.*, 2008), which helps increase nitrogen (N) and phosphorus (P) uptake by plants. In saline soil, Stehouwer and Macneal (2003) found an increase of microbial activity reported to both: inoculation of previously sterile soil and providing a reduced energy source for microbes. The additional organic material, which provides additional substrates in the salt-affected soil for the microbial population, may also relieve osmotic and pH stress on the microorganisms (Pathak and Rao, 1998). Soil basal respiration rate, a parameter used to monitor microbial activity, was also increased for eight years after municipal solid waste (MSW) compost application as compared to a control (Pascual *et al.*, 1999). Liang *et al.* (2003) reported that soil urease and alkaline phosphatase activity, and respiration rate as well as salt tolerance of plants were significantly stimulated by incorporation of organic amendment in rice-barley rotation system. Incorporation of organic materials influences enzymatic activities in the soil because the added organic fractions may contain intra- and extracellular enzymes, which stimulate microbial activity in the soil (Garcia *et al.*, 1993; Goyal *et al.*, 1993). In the same way, microbial metabolic quotient (qCO_2) is used as indicator of biological activity through estimating the efficiently of microbial biomass to utilize available carbon for biosynthesis. By using this method, Pascual *et al.* (1997) observed a sharp increase in qCO_2 in an arid soil amended with an amount of municipal solid waste sufficient to raise its organic matter by 1.5 per cent after 1 year of incubation. This may be explained by the fact that microorganisms living in a stressed environment put up defense mechanisms by increasing their respiration per unit biomass, so increasing qCO_2. Further, Pascual *et al.* (1999) also demonstrated that an eight year amendment of an arid soil with the organic fraction of a MSW had a positive effect on the activity of enzymes involved in the C, N, P cycles as well as on biomass C, constituting a suitable technique to restore soil quality. In fact, even if microbial activity is depressed by salts, biochemical mineralization by soil enzymes (amidases and deaminases) could still proceed provided the activity is not adversely affected at high salinity and alkalinity (Pathak and Rao 1998). Organic matter incorporation under high salinity or sodicity can provide a buffer to the soil solution and to soil microbiological properties (Mccormick and Wolf 1980). According to Tripathi *et al.* (2007), phosphatase activity responded more to pH than to the differences in soil salinity. Under high pH alkaline form predominates. Ozur *et al.* (2008) observed obvious increase of this enzyme activity following compost treatment. Since, higher plants are devoid of alkaline phosphatase, the alkaline phosphatase of soils seems to be derived totally from microorganisms (Juma and Tabatabai 1988). Consequently, such amendment could stimulate phoshatase activity under saline media. Microbes can produce and release large amounts of extracellular phosphatase due to their large combined biomass, high metabolic activity and short life cycles (Lakhdar *et al.*, 2009). Lauchli and Epstein (1990) showed appreciable stimulation of urease and protease enzymes and they related this to higher microbial biomass produced in response to high doses of organic amendment. Pathak and Rao (1998) found a stimulation of nitrogen mineralization under saline medium in the later stages of experiment. Indeed, with time of incubation the early adverse effect of pH on nitrites is relieved due the remedial influence resulting from organic matter decomposition (reduction of pH and ESP). As well, the application of vermicompost, sheep manure, poultry manure, pig manure, and urban waste at the level of 1 per cent to soil characterized by high exchangeable sodium percentage increased the rate of urea hydrolysis (Wali *et al.*, 2003). Accordingly, the composted olive oil mill wastes are characterized by high organic load and a substantial quantity of plant nutrients (N, P, K, Ca, Mg and Fe) that could increase both soil fertility and crop production (Montemurro *et al.*, 2009). We believed that bio-compost amendments do not completely solve the underlying cause of the salinity problem. However, the use of bio-compost on saline soils still improves soil physico-chemical properties, microbial biomass and growth of plants. Therefore, compost use can offer a short-term reprieve for farming on soils with medium to high salinity, or soils with temporary high salinity. As well, high application of bio-compost may be a very useful tool for ameliorating severely salt-affected areas through the establishment of plant cover, including deep-rooted crops.

Bioremediation through Bio-composting

Soil serves as a sink for all the pollutants present in the environment. In the era of industrial development and urbanization, the contamination of raw city effluent with toxic substances is on the rise concurrently with its increasing volume being generated on a daily basis. Among these toxic substances, heavy metals are important due to their persistence in the environment for longer periods and to their non-biodegradable nature (Sabir *et al.*, 2015). Heavy metals like iron (Fe), molybdenum (Mo), manganese (Mn), zinc (Zn), copper (Cu) and nickel (Ni) are essential for various physiological/metabolic functions in living cells while others like lead (Pb), cadmium (Cd), mercury (Hg), arsenic (As) and uranium (U) are toxic even at very low concentrations (Meharg, 2005). Several workers reported that composts and other organic by-products amendments in soil improves chemical properties like pH, increased plant nutrients availability and crop yields (Ouédraogo *et al.*, 2001; Hartl and Erhart, 2005). It could be used as an alternate for traditional manures in areas of intensive agriculture due to poor availability of traditional manures. Effect of compost and other organic amendments on availability of nickel (Ni) and other metals depends on amount and composition of organic matter, heavy metal contents, salts and degree of stabilization (Sabir *et al.*, 2008; Murillo *et al.*, 1995). Addition of sewage-sludge compost and green waste compost decreased available metals (Cd, Cu, Zn and Pb) in contaminated soil compared to non-amended soils (van Herwijnen *et al.*, 2007). Bio-solid compost and sugar beet lime enhanced the natural vegetative cover and biomass production in mine spill contaminated soil by decreasing metal concen-tration in plants, incremental nutrients due to decomposition of composts and improved soil physico-chemical properties (Madejón *et al.*, 2006). The decrease in DTPA extractable Ni concentration in soil due to application of mushroom compost and other organic amendments was reported (Karaca, 2004b). Application of green waste compost decreased the uptake of Cu, Pb and Zn in Greek Cress by 21, 54 and 16 per cent, respectively, in contaminated calcareous soils (van Herwijnen *et al.*, 2007). Kiikkilä *et al.* (2002) concluded that exchangeable Cu decreased after mulched treatment with garden soil, sewage sludge, compost and compost mixed with bark chips and woodchips. The authors opined that Cu concentration may decrease due to formation of complexes with particulate organic matter contributed by mulch onto soil at pH > 6 (Kiikkilä *et al.*, 2002). Bio-solid compost decreased arsenic (As) concentration in carrot (*Daucus carota* L.) and lettuce (*Lactuca sativa* L.) compared to the control (Cao and Ma, 2004). This could be due to adsorption of As by organic matter that is evidenced by decreased arsenic concentration in extractable fractions. In general, heavy metal uptake by crops increases in leafy plants and it is higher in cereal leaves than in grain. For example, lettuce, had higher assimilative capacity for Zn and Cd uptake than other non-leafy crops (Sukkariyah *et al.*, 2005). Mantovi *et al.* (2005) reported that bio-solid applications significantly increased the content of Zn and Cu in wheat grain and only Cu in both sugar-beet roots and maize grain. Cadmium is one of the most significant potential contaminants of food supplies on arable lands and may limit sewage sludge suitability for soil amendment, because this organic material presents a large amount of this metal (Singh and Pandeya 1998). Being relatively soluble in soils, it is readily taken up by crops and it is quite toxic to humans (Miller and Miller 2000). This behavior is particularly evident at low soil pH. Cd solubility in equilibrium extracts of $Ca(NO_3)_2$ increased, during the 41-year long experimental trial, by a factor of 20 in the sludge treatment compared to control (Bergkvist *et al.*, 2003). This was reflected in the Cd concentration of the straw fraction in barley, which was almost doubled in the sewage sludge treatment, compared with the control. The grain fraction, on the other hand, showed no significant increase in Cd concentration. Liu *et al.* (2009) reported that application of compost effectively reduced Cd toxicity to wheat by decreasing > 50 per cent Cd uptake by wheat tissue and improving wheat growth. Alleviating effect of compost could be due to high soil pH, Cd complexation with organic matter and co-precipitation with P content (Liu *et al.*, 2009). It can be concluded that there is no tangible evidence demonstrating negative impacts of heavy metals applied to soil, particularly when high-quality compost was used for a long period. Composting processes also inherently reduce metal availability compared with other organic waste stabilization methods (Smith, 2009).

Conclusion and Future Perspectives

Management of agricultural and organic waste products through bio-composting yields a nutritionally rich product, which can be used to help fight plant disease, reducing the need for the application of chemicals in addition to environment management. Successful biocontrol of plant disease requires an intricate array of interactions. Understanding these interactions at the molecular and ecological levels will make possible the rational development of bio-compost based consortia or formulations for agriculture and environment management. However, there is a lot to discover and understand the bio-composting processes and interaction with microbial biota in light of the present challenge like drought, soil fertility loss, climate change and soil contamination. Therefore, in order to secure the food security through sustainable agricultural, role of bio-compost is certainly at high priority for productivity of agro-ecosystem and thus following points need to be taken into condensation.

☆ Bio-compost utilization in agriculture should be legislated in such a way to prevent harmful effects on soil, vegetation, animals and human health, thereby encouraging the correct use of bio-solids and protect the environment.

☆ The use of immature amendments of bio-compost should be avoided as it can cause phytotoxic effects, as well as nitrogen (N) deficiency and reduction in plant yield.

☆ The biggest challenge is to refine the use of composts in the farming systems to maximise the potential benefits. Combinations of compost applications with minimum tillage may help to further extend the soil quality benefits. More information on the yield response of the different crops in any given farming system to compost applications along with economic analyses is required at priority.

☆ Combined application of bio-compost, bio-fertilizer and synthetic fertilizers are need of the day for obtaining excellent value produce. Therefore, knowledge about appropriate composting technology, handling and its storage is imperative for increasing its nutritive value.

☆ Soil management strategies for sustainable agriculture should focus not only on increasing organic matter in the soil, but also on the uptake or stocking of soil residual nutrients in such a way to prevent excess nutrient leaching into the groundwater.

References

1. Acharya A, Joshi DR, Shrestha K, Bhatta DR (2012). Isolation and screening of thermophilic cellulolytic bacteria from compost piles. *Scientific World* 10(10): 43–46.

2. Adani F, Genevini PL, Tambone F (1995). An new index of organic matter stability. *Compost Sci Util* 3(2): 25–37.

3. Amore A, Pepe O, Ventorino V, Birolo L, Giangrande C, Faraco V (2013). Industrial waste based compost as a source of novel cellulolytic strains and enzymes. *FEMS Microbiol Lett* 339: 93–101.

4. Anjaiah V, Cornelis P, Koedam N, (2003). Effect of genotype and root colonization in biological control of Fusarium wilts in pigeonpea and chickpea by *Pseudomonas aeruginosa* PNA1. *Can J Microbiol* 49: 85–91.

5. Barbarick KA, Ippolito JA (2007). Nutrient assessment of a dryland wheat agroecosystem after 12 years of biosolids applications. *Agron J* 99: 715–722.

6. Beffa T, Blanc M, Marilley L, Fischer JL, Lyon PF, Aragno M, (1996). Taxonomic and metabolic microbial diversity during composting. In: De Bertoldi, M., Sequi, P., Lemmes, B., Papi, T. (Eds.), *The Science of Composting*. Chapman and Hall, London, pp. 149–161.

7. Benbi DK (2009). Soil organic carbon sequestration in relation to organic and inorganic fertilization in rice-wheat and maize-wheat systems. *Soil Till Res* 102: 87–92.

8. Benbi DK, Brar JS (2009). A 25-year record of carbon sequestration and soil properties in intensive agriculture, *Agron Sustain Dev* 29: 257–265.

9. Benhamou N, Garand C, Goulet A, (2002). Ability of non-pathogenic *Fusarium oxysporum* Fo47 to induce resistance against *Pythium ultimum* infection in cucumber. *Appl Environ Microbiol* 68: 4044–4060.

10. Bergkvist P, Jarvis N, Berggren D, Carlgren K (2003). Long-term effects of sewage sludge applications on soil properties, cadmium availability and distribution in arable soil. *Agr Ecosys Environ* 97: 167–179.

11. Bhattacharya A and Pletschke BI (2014). Thermophilic bacilli and their enzymes in composting. In: D. K. Maheshwari (ed.), *Composting for Sustainable Agriculture*, Springer International Publishing, Switzerland, pp. 103-124. DOI 10.1007/978-3-319-08004-8_6.

12. Biswas DR, Narayanasamy G (2006). Rock phosphate enriched compost, an approach to improve low-grade Indian rock phosphate. *Bioresour Technol* 97: 2243-2251.

13. Blackshaw RE, Molnar LJ, Larney FJ (2005). Fertilizer, manure and compost effects on weed growth and competition with winter wheat in western Canada. *Crop Prot* 24: 971-980.

14. Blok WJ, Coenen GCM, Pijl AS, Termorshuizen, AJ (2002). Disease suppression and microbial communities in potting mixes amended with composted biowastes. In: Michel FJ, Rynk RF, Hoitink HAJ (Eds.), *Proceedings of the International Symposium Composting and Compost Utilization*. Columbus, Ohio, 6-8 May 2002, disc format. JG Press, Emmaus, USA.

15. Boehm M, Madden LV, Hoitink HAJ, (1993). Effect of organic matter decomposition level on bacterial species diversity and composition in relationship to *Pythium* damping-off severity. *Appl Environ Microbiol* 59: 4171-4179.

16. Boulter JI, Boland GJ, Trevors, JT (2000). Compost: a study of the development process and end product potential for suppression of turfgrass disease. *World J Microbio Biotech* 16: 115-134.

17. Cao X, Ma LQ (2004). Effects of compost and phosphate on plant arsenic accumulation from soils near pressure-treated wood. *Environ Pollut* 132: 435-442.

18. Cao Y, Zhang Z, Ling N, Yuan Y, Zheng X, Shen B, Shen Q, (2011). *Bacillus subtilis* SQR 9 can control *Fusarium* wilt in cucumber by colonizing plant roots. *Biol Fert Soils* 47: 495-506.

19. Castaldi P, Garau G, Melis P (2008). Maturity assessment of compost from municipal waste through the study of enzyme activities and water soluble fractions. *Waste Manage* 28: 534-540.

20. Castano R, Borrero C, Aviles M, (2011). Organic matter fractions by SP-MAS 13C NMR and microbial communities involved in the suppression of *Fusarium* wilt in organic growth media. *Biol Control* 58: 286-293.

21. Cayuela ML, Mondini C, Sanchez-Monedero MA, Roig A (2008). Chemical properties and hydrolytic enzyme activities for the characterization of two phase olive mill wastes composting. *Bioresour Technol* 99: 4255-4262.

22. Chang CH, Yang SS, (2009). Thermo-tolerant phosphate-solubilizing microbes for multi-functional biofertilizer preparation. *Bioresour Technol* 100: 1648-1658.

23. Chin-A-Woeng TFC, Bloemberg GV, Vander Bij AJ, Vander Drift, KMGM, Schripsema, J, Kroon B, Scheffer, RJ, Keel C, (1998). Biocontrol by phenazine-1-

carboxamide-producing Pseudomonas chlororaphis PCL1391 of tomato root rot caused by *Fusarium oxysporum* f. sp. *radicis lycopersici*. *Mol Plant Microbe Interact* 11: 1069–1077.

24. Cooperband LR, Stone AG, Fryda MR, Ravet JL (2003). Relating compost measures of stability and maturity to plant growth. *Compost Sci Util* 11: 113–124.

25. Cotxarrera L, Trillas-Gay MI, Steinberg C, Alabouvette C, (2002). Use of sewage sludge compost and *Trichoderma asperellum* isolates to suppress *Fusarium* wilt of tomato. *Soil Biol Biochem* 34: 467–476.

26. Dababat AEFA, Sikora RA, (2007). Induced resistance by the mutualistic endophyte, *Fusarium oxysporum* strain 162, toward *Meloidogyne incognita* on tomato. *Biocontrol Sci Technol* 17: 969–975.

27. Davis CL, Donkin CJ, Hinch SA, Germishuizen P, (1992). The microbiology of pine bark composting, an electron-microscope and physiological study. *Bioresour Technol* 40: 195–204.

28. de Freitas JR, Germida JJ, (1991). *Pseudomonas cepacia* and *Pseudomonas putida* as winter wheat inoculants for biocontrol of *Rhizoctonia solani*. *Can J Microbiol* 37: 780–784.

29. Dees PM, Ghiorse WC (2001). Microbial diversity in hot synthetic compost as revealed by PCR amplified rRNA sequences from cultivated isolates and extracted DNA. *FEMS Microbiol Ecol* 35: 207–216.

30. Diaz LF, de Bertoldi M, Bidlingmaier W (2007). *Compost Science and Technology*. Elsevier 380 pp.

31. Eida MF, Nagaoka T, Wasaki J, Kouno K (2012). Isolation and characterization of cellulose decomposing bacteria inhabiting sawdust and coffee residue composts. *Microbes Environ* 27(3): 226–233.

32. Elmer WH, (2004). Combining nonpathogenic strains of *Fusarium oxysporum* with sodium chloride to suppress Fusarium crown rot of asparagus in replanted fields. *Plant Pathol* 53: 751–758.

33. Epstein E, (1997). *The Science of Composting*. Technomic Publishing Company, Lancaster, PA, USA, p. 1429.

34. Feller C, Bernoux M (2008). Historical advances in the study of global terrestrial soil organic carbon sequestration, *Waste Manage* 28: 734–740.

35. Fishal, EMM, Meon S, Yun WM, (2010). Induction of tolerance to *fusarium* wilt and defense-related mechanisms in the plantlets of susceptible berangan banana pre- inoculated with *Pseudomonas* sp. (UPMP3). and *Burkholderia* sp. (UPMB3). *Agric Sci China* 9: 1140–1149.

36. Forsyth LM, Smith LJ, Aitken EAB, (2006). Identification and characterization of non-pathogenic Fusarium oxysporum capable of increasing and decreasing *Fusarium* wilt severity. *Mycol Res* 110: 929–935.

37. Freeman S, Shalev Z, Katan J, (2002). Survival in soil of *Colletotrichum acutatum* and *C. gloeosporioides* pathogenic on strawberry. *Plant Dis* 86: 965–970.

38. Garcia C, Hernandez T, Costa F, Ceccanti B, Ciardi C (1993). Changes in ATP content, enzyme activity and inorganic nitrogen species during composting of organic wastes. *Can J Soil Sci* 72: 243–253.

39. García-Gil JC, Ceppi SB, Velasco MI, Polo A, Senesi N (2004). Long-term effects of amendment with municipal solid waste compost on the elemental and acidic functional group composition and pH-buffer capacity of soil humic acids. *Geoderma* 121: 135–142.

40. Gautam SP, Bundela PS, Pandey AK, Awasthi MK, Sarsaiya S (2009). Prevalence of fungi in municipal solid waste of Jabalpur city (M.P.). *J Basic Appl Mycol* 8: 80–81.

41. Gautam SP, Bundela PS, Pandey AK, Awasthi MK, Sarsaiya S (2010). Composting of municipal solid waste of Jabalpur City. *Global Environ Res* 4: 43–46.

42. Ghazifard A, Kasra-Kermanshahi R, Far ZE (2001). Identification of thermophilic and mesophilic bacteria and fungi in Esfahan (Iran). municipal solid waste compost. *Waste Manage Res* 19: 257–261.

43. Gibello A, Vela AI, Martín M, Mengs G, Alonso PZ, Garbi C, Fernández-Garayzábal JF, (2011). *Pseudomonas composti* sp. nov., isolated from compost samples. *Int J Syst Evol Microbiol* 61: 2962–2966.

44. Ginting D, Kessavalou A, Eghball B, Doran JW (2003). Greenhouse gas emissions and soil indicators four years after manure and compost applications. *J Environ Qual* 32: 23–32.

45. Golueke CG (1972). *Composting: A Study of the Process and its Principles.* Rodale Press Inc, Emmaus, Pennsylvania USA p. 110.

46. Goyal S, Dhull SK, Kapoor KK (2005). Chemical and biological changes during composting of different organic wastes and assessment of compost maturity. *Bioresour Technol* 96: 1584–1591.

47. Goyal S, Mishra MM, Dhankar SS, Kapoor KK, Batra R (1993). Microbial biomass turnover and enzyme activities following the application of farmyard manure to field soil with and without previous long-term applications. *Biol Fertil Soils* 15: 60–64.

48. Hanay A, Buyuksonmez F, Kiziloglu FM, Canbolat MY (2004). Reclamation of saline sodic soils with gypsum and MSW compost, *Compost Sci Util* 12: 175–179.

49. Hartl W, Erhart E (2005). Crop nitrogen recovery and soil nitrogen dynamics in a 10-year field experiment with biowaste compost. *J Plant Nutr Soil Sci* 168: 781–788.

50. Haug RT (1993). *The Practical Handbook of Composting Engineering.* Lewis Publishers Boca Raton, FL.

51. He Y, Xie K, Xu P, Huang X, Gu W, Zhang F, Tang S (2013). Evolution of microbial community diversity and enzymatic activity during composting. *Res Microbiol* 164: 189–198.

52. Heckman J (2006). A history of organic farming: transitions from Sir Albert Howard's War in the Soil to USDA National Organic Program. *Renew Agric Food Syst* 21: 143–150.

53. Hellmann B, Zelles L, Palojarvi A, Bai Q (1997). Emissions of climate-relevant trace gases and succession of microbial communities during open-windrow composting. *Applied Environ Microbiol* 63: 1011–1018.

54. Hoitink HAJ, Krause MS, Han DY (2001). Spectrum and mechanisms of plant disease control with composts. In: Stofella PJ, Kahn BA (Eds.). *Compost Utilization in Horticultural Cropping Systems.* CRC Press LLC, Boca Raton, Fla., USA, pp. 263–273.

55. Hoorman JJ, Islam R, (2010). *Understanding Soil Microbes and Nutrient Recycling. FACT SHEET, Agriculture and Natural Resources* The Ohio State University.

56. Insam H, de Bertoldi M, (2007). Microbiology of the composting process. In: LF Diaz, M de Bertoldi, W Bidlingmaier, E Stentiford (ed.). *Compost Science and Technology.* Elsevier Amsterdam p. 25-49.

57. Itavaara M, Vikman M, Venelampi O (1997). Windrow composting of biodegradable packaging materials. *Compost Sci Util* 5: 84–92.

58. Ito S, Nagata A, Kai T, Takahara H, Tanaka S, (2005). Symptomless infection of tomato plants by tomatinase producing *Fusarium oxysporum* formae specials non-pathogenic on tomato plants. *Physiol Mol Plant Path* 66: 183–191.

59. Juma NG, Tabatabai MA (1988). Hydrolysis of organic phosphates by corn and soybean roots. *Plant Soil* 107: 31–38.

60. Karaca A (2004). Effect of organic wastes on the extractability of cadmium, copper, nickel, and zinc in soil. *Geoderma* 122: 297–303.

61. Kaur T, Brar BS, Dhillon NS (2008). Soil organic matter dynamics as affected by long-term use of organic and inorganic fertilizers under maize–wheat cropping system. *Nutr Cycl Agroecosys* 81: 59–69.

62. Keener HM, Dick WA, Hoitink HAJ (2000). Composting and beneficial utilization of composted by-product materials. In: Dick WA (ed). *Land application of agricultural, industrial, and municipal by-products.* Soil Science Society of America Inc., Madison, pp. 315–341.

63. Kiikkilä O, Pennanen T, Perkiömäki J, Derome J, Fritze H, (2002). Organic material as a copper immobilising agent: A microcosm study on remediation. *Basic Appl Ecol* 3: 245–253.

64. Kim YK, Bae JH, Oh BK, Lee WH, Choi JW (2002). Enhancement of proteolytic enzyme activity excreted from Bacillus stearothermophillus for a thermophilic aerobic digestion process. *Bioresour Technol* 82: 157–164.

65. Kloepper JW, Ryu CM, Zhang S (2004). Induced systemic resistance and promotion of plant growth by *Bacillus* spp. *Phytopathology* 94: 1259–1266.

66. Kloepper JW, Schroth MN (1978). Plant growth-promoting rhizobacteria in radish. In: *Proceedings of IV International Conference Plant Pathogenic Bacteria.* Gilbert-Clarey, Tours, France, pp. 879–882.

67. Kukal SS, Rehana-Rasool, Lakhdar A, ben W, Achiba N, Jedidi C, Abdelly (2008). *Effect of MSW Compost and Sewage Sludge on Soil Biologic Activities and Wheat Yield,* 9th ed., Tunisian- Japan Symposium on Society, Science and Technology.

68. Lakhdar A, Hafsi C, Rabhi M, Debez A, Montemurro F, Abdelly C, Jedidi N, Ouerghi Z (2008). Application of municipal solid waste compost reduces the negative effects of saline water in *Hordeum maritimum* L., *Bioresour Technol* 99: 7160–7167.

69. Lakhdara A, Rabhia M, Ghnayaa T, Montemurroc F, Jedidi N, Abdellya C (2009). Effectiveness of compost use in salt-affected soil. *Journal of Hazardous Materials* 171: 29–37.

70. Lal R (2007). Carbon management in agricultural soils. *Mitigation and Adaptation Strategies for Global Change* 12: 303–322.

71. Lal R (2008). Soils and sustainable agriculture. A review. *Agron. Sustain. Dev.* 28: 57–64.

72. Larkin R, Fravel DR, (2002). Effects of varying environmental conditions on biological control of *Fusarium* wilt of tomato by nonpathogenic *Fusarium* spp. *Phytopathology* 92: 1160–1166.

73. Larkin RP, Fravel DR (1999). Mechanims of action and dose-response relationships governing biological control of *Fusarium* wilt of tomato by non-pathogenic *Fusarium* spp. *Phytopathology* 89: 1152–1161.

74. Lauchli A, Epstein E (1990). Plant response to salinity and sodic conditions, in: K.K. Tanji (Ed.), Agricultural Salinity Assessment and Management, American Society of Civil Engineers, New York, *Mann. Rep. Eng. Pract.* 71: 113–137.

75. Le Goff O, Bru-Adan V, Bacheley H, Godon JJ, Wery N (2010). The microbial signature of aerosols produced during the thermophilic phase of composting. *J Appl Microbiol* 108: 325–340.

76. Leroy BLMM, Bommele L, Reheul D, Moens M, De Neve S (2007). The application of vegetable, fruit and garden waste (VFG). compost in addition to cattle slurry in a silage maize monoculture: Effects on soil fauna and yield. *Eur J Soil Biol* 43: 91–100.

77. Liang Y, Yang Y, Yang C, Shen Q, Zhou J, Yang L (2003). Soil enzymatic activity and growth of rice and barley as influenced by organic manure in an anthropogenic soil. *Geoderma* 115: 149–160.

78. Lievens B, Vaes K, Coosemans J, Ryckeboer J (2001). Systemic resistance induced in cucumber against *Pythium* root rot by source separated household waste and yard trimmings composts. *Compost Sci. Util.* 9: 221–229.

79. Liu D, Zhang R, Wu H, Xu D, Tang Z *et al.* (2011). Changes in biochemical and microbiological parameters during the period of rapid composting of daily manure with rice chaff. *Bioresour Technol* 102(19): 9040–9049.

80. Liu L, Chen H, Cai P, Liang W, Huang Q (2009). Immobilisation and phytotoxicity of Cd in contaminated soil amended with chicken manure compost. *J Hazard Mater* 163: 563–567.

81. Loos J, Abson David J, Chappell M Jahi, Hanspach Jan, Mikulcak Friederike, Tichit Muriel, and Joern Fischer (2014). Putting meaning back into "sustainable intensification". *Front Ecol Environ*; Doi: 10.1890/130157.

82. Madejón E, De Mora AP, Felipe E, Burgos P, Cabrera F(2006). Soil amendments reduce trace element solubility in a contaminated soil and allow regrowth of natural vegetation. *Environ Pollut* 139: 40–52.

83. Mallory EB, Griffin TS (2007). Impacts of soil amendment history on nitrogen availability from manure and fertilizer. *Soil Sci Soc Am J* 71: 964–973.

84. Mandal B, Majumder B, Bandyopadhyay PK, Hazra GC, Gangopadhyay A, Samantaray RN, Mishra AK, Chaudhury J, Saha MN, Kundu S (2007). The potential of cropping systems and soil amendments for carbon sequestration in soils under long-term experiments in subtropical India. *Global Change Biol* 13: 357–369.

85. Mantovi P, Baldoni G, Toderi G (2005). Reuse of liquid, dewatered, and composted sewage sludge on agricultural land: effects of long-term application on soil and crop. *Water Res* 39: 289–296.

86. Mayende L, Wilhelmi BS, Pletschke BI (2006). Cellulases (CMCases). and polyphenol oxidases from thermophilic Bacillus spp. isolated from compost. *Soil Biol Biochem* 38: 2963–2966.

87. Mccormick RW, Wolf DC (1980). Effect of sodium chloride on CO_2, evolution ammonification, and nitrification in a sassafras sandy loam. *Soil Biol Biochem* 12: 153–157.

88. Meharg AA (2005). Mechanisms of plant resistance to metal and metalloid ions and potential biotechnological applications. *Root Physiology: From Gene to Function*. Springer, p. 163–174.

89. Mehta CM, Gupta Varun, Singh Shivom, Srivastava Rashmi, Sen Elli, Martin Romantschuk, Sharma AK (2012). Role of microbiologically rich compost in reducing biotic and abiotic stresses. In: T. Satyanarayana *et al.* (eds.), *Microorganisms in Environmental Management: Microbes and Environment*, Springer, pp 113-134. DOI 10.1007/978-94-007-2229-3_5.

90. Mehta CM, Palni U, Franke -Whittle IH, Sharma AK (2013). Compost: Its role, mechanism and impact on reducing soil-borne plant diseases. *Waste Management*. doi.org/10.1016/j.wasman.2013.11.012.

91. Miller D.M., Miller W.P. (2000). Land application of wastes, in: Sumner M. (Ed.), *Handbook of Soil Science*, CRC Press, Chap. 9.

92. Miller FC (1992). Composting as a process based on the control of ecologically selective factors. In: Metting FB Jr (ed). *Soil Microbial Ecology, Applications in Agricultural and Environmental Management*. Marcel Dekker Inc., New York, pp. 515–544.

93. Monaco S, Hatch D J, Sacco D, Bertora C, Grignani C (2008). Changes in chemical and biochemical soil properties induced by 11-yr repeated additions of different organic materials in maize based forage systems. *Soil Biol Biochem* 40: 608–615.

94. Mondini C, Sequi P (2008). Implication of soil C sequestration on sustainable agriculture and environment. *Waste Manage* 28: 678–684.

95. Montemurro F, Diacono M, Vitti C, Debiase G (2009). Biodegradation of olive husk mixed with other agricultural wastes. *Bioresour Technol* 100: 2969–2974.

96. Morsy EM (2005). Role of Growth Promoting Substances Producing Microorganisms on Tomato Plant and Control of Some Root Rot Fungi. *Ph.D. Thesis*, Faculty of Agriculture Ain shams Univ., Cairo.

97. Muhammad S, Muller T, Georg Joergensen R (2007). Compost and P amendments for stimulating microorganisms and maize growth in a saline soil from Pakistan in comparison with a nonsaline soil from Germany. *J Plant Nutr Soil Sci* 170: 745–751.

98. Murillo J, Cabrera F, López R, Martín-Olmedo P (1995). Testing low-quality urban composts for agriculture: Germination and seedling performance of plants. *Agriculture Ecosystems and Environ* 54: 127–135.

99. Murphy DV, Stockdale EA, Brookes PC, Goulding KWT (2007). Impact of microorganisms on chemical transformation in soil, in: Abbott LK, Murphy DV (Eds.). *Soil biological fertility – A key to sustainable land use in agriculture*, Springer, pp. 37–59.

100. Nagorska K, Bikowski M, Obuchowskji M, (2007). Multicellular behaviour and production of a wide variety of toxic substances support usage of *Bacillus subtilis* as a powerful biocontrol agent. *Acta Biochim Pol* 54: 495–508.

101. Nakasaki K, Sasaki M, Shoda M, Kubota H (1985). Change in microbial numbers during thermophilic composting of sewage sludge with reference to CO_2 evolution rate. *Appl Environ Microbiol* 49: 37–41.

102. Nardi S, Morari F, Berti A, Tosoni M, Giardini L (2004). Soil organic matter properties after 40 years of different use of organic and mineral fertilizers. *Eur J Agron* 21: 357–367.

103. Naskar SK, Sethuraman P, Ray RC (2003). *Sprouting in Yam by Cowdung Slurry. Validation of Indigenous Technical Knowledge in Agriculture* New Delhi, India, Division of Agricultural Extension, Indian Council of Agricultural Research, pp. 197–201.

104. Ng SI, Li CW, Yeh YF, Chen PT *et al.* (2009). A novel endo-glucanse from thermophilic bacterium *Geobacillus* sp. 70PC53 with high activity and stability over a broad range of temperatures. *Extrem* 13: 425–435.

105. Nizamudeen S, Bajaj BK (2009). A novel thermo-alkalitolerant endoglucanase production using cost-effective agricultural residues as substrates by a newly isolated *Bacillus* sp. NZ. *Food Technol Biotechnol* 47(4): 435–440.

106. Ouahmane L, Thioulouse J, Hafidi M, Prin Y, Ducousso M, Galiana A, Plenchette C, Kisa M, Duponnois R (2007). Soil functional diversity and P solubilization from rock phosphate after inoculation with native or allochtonous arbuscular mycorrhizal fungi. *Forest Ecol Manage* 241: 200–208.

107. Ouedraogo E, Mando A, Zombré NP (2001). Use of compost to improve soil properties and crop productivity under low input agricultural system in West Africa. *Agric Ecosyst Environ* 84: 259–266.

108. Ozur N, Kayikcioglu HH, Okur B, Delibacak S (2008). Organic amendment based on tobacco waste compost and farmyard manure: influence on soil biological properties and Butter-Head lettuce yield. *Turk J Agric For* 32: 91–99.

109. Pantelides IS, Tjamos SE, Striglis IA, Chatzipavlidis I, Paplomatas EJ (2009). Mode of action of a pathogenic *Fusarium oxyspoprum* strain against *Verticillium dalhiae* using Real Time QPCR analysis and biomarker transformation. *Biol Control* 50: 30–36.

110. Pascual JA, Garcia C, Hernandez T (1999). Lasting microbiological and biochemical effects of the addition of municipal solidwaste to an arid soil. *Biol Fertil Soils* 30: 1–6.

111. Pascual JA, Garcia C, Hernandez T, Ayuso M (1997). Changes in the microbial activity of an arid soil amended with urban organic wastes. *Biol Fertil Soils* 24: 429–434.

112. Pathak H, Rao DLN (1998). Carbon and nitrogen mineralization from added organic matter in saline and alkali soils. *Soil Biol Biochem* 30: 695–702.

113. Raut MP, William PSPM, Bhattacharya JK, Chakrabarti T, Devotta S (2008). Microbial dynamics and enzyme activities during rapid composting of municipal solid waste-Acompost maturity analysis perspective. *Bioresour Technol* 99: 6512–6519.

114. Rawat S, Agarwal PK, Chaudhary DK, Johri BN (2005). Microbial diversity and community dynamics of mushroom compost ecosystem. In Satyanarayana T and Johri BN (eds). *Microbial diversity: Current perspectives and applications.* International Book Distributing Co, Lucknow, pp. 181–206.

115. Ray RC, Nedunzhiyan M, Balagopalan C (2000). Microorganism associated with post harvest spoilage of yams. *Ann Trop Res* 22: 31–40.

116. Rebollido R, Martinez J, Aguilera Y, Melchor K, Koerner I, Stegmann R (2008). Microbial populations during composting process of organic fraction of municipal solid waste. *Appl Ecol Environ Res* 6: 61–67.

117. Recep K, Fikrettin S, Erkol D, Cafer E (2009). Biological control of the potato dry rot caused by Fusarium species using PGPR strains. *Biol Control* 50: 194–198.

118. Rodale JI, Rodale R, Olds J, Goldman MC, Franz M, and Minnich J (1960). *The complete book of composting.* Rodale Books, Emmaus, Pa.

119. Rodriguez H, Fraga R (1999). Phosphate solubilizing bacteria and their role in plant growth promotion. *Biotechnol Adv* 17: 319–339.

120. Ros M, Garcia C, Hernandez T (2006). A full-scale study of treatment of pig slurry by composting: kinetic changes in chemical and microbial properties. *Waste Management* 26: 1108–1118.

121. Ros M, Klammer S, Knapp B, Aichberger K, Insam H (2006). Longterm effects of compost amendment of soil on functional and structural diversity and microbial activity. *Soil Use Manage* 22: 209–218.

122. Rudresh DL, Shivaprakash MK, Prasad RD (2005). Tricalcium phosphate solubilizing abilities of *Trichoderma* spp. in relation to P uptake and growth and yield parameters of chickpea (*Cicer arietinum* L.). *Can J Microbiol* 51: 217-222.

123. Ryckeboer J, Mergaert J, Coosemans J, Deprins K, Swings J (2003). Microbiological aspects of biowaste during composting in a monitored compost bin. *J Appl Microbiol* 94: 127–137.

124. Sabir M, Ghafoor A, Saifullah M Z-u-R, Murtaza G (2008). Effect of organic amendments and incubation time on extractability of Ni and other metals from contaminated soils. *Pak J Agri Sci* 45: 1.

125. Sabir M, Zia-ur-Rehman M, Hakeem KR, and Saifullah (2015). Phytoremediation of Metal-contaminated Soils Using Organic Amendments: Prospects and Challenges. In: *Soil Remediation and Plants* (Eds.). Elsevier Inc. pp. 503-523.

126. Sairam RK and Tyagi A (2004). Physiology and molecular biology of salinity stress tolerance in plants. *Curr Sci* 86: 407–421.

127. Sakthivel N, Gnanamanickam SS (1987). Evaluation of *Pseudomonas fluorescens* for suppression of sheath rot disease and for the enhancement of grain yields in rice (*Oryza sativa* L.). *Appl Environ Microbiol* 53: 2056–2059.

128. Sant D, Casanova E, Segarra G, Avilés M, Reis M, Trillas MI (2010). Effect of *Trichoderma asperellum* strain T34 on Fusarium wilt and water usage in carnation grown on compost-based growth medium. *Biol Control* 52: 291–296.

129. Sasaay AA, Lasaridi K, Stentiford E, Budd T (1997). Controlled composting of paper sludge using the aerated static pile method. *Compost Sci Util* 5: 82–96.

130. Segarra G, Van der Ent S, Trillas I, Pieterse C (2009). MYB72, a node of convergence in induced systemic resistance triggered by a fungal and a bacterial beneficial microbe. *Plant Biol* 11: 90–96.

131. Shanmugam V, Kanoujia N (2011). Biological management of vascular wilt of tomato caused by *Fusarium oxysporum* f. sp. *lycospersici* by plant growth promoting rhizobacterial mixture. *Biol Control* 57: 85–93.

132. Shen Q, Wang Y, Chen W, Shi R (1997). Changes of soil microbial biomass C and P during wheat growth after application of fertilizers. *Pedosphere* 7: 225–230.

133. Shindo H, Hirahara O, Yoshida M, Yamamoto A (2006). Effect of continuous compost application on humus composition and nitrogen fertility of soils in a field subjected to double cropping. *Biol Fert Soils* 42: 437–442.

134. Shu-bin L, Reh-chao Z, Xia L, Chu-yi C, Ai-lin Y (2012). Solid-state fermentation with okara for production of cellobiase rich cellulases preparation by a selected *Bacillus subtilis* pa5. *African J Biotechnol* 11(11): 2720–2730.

135. Silvar C, Merino F, Diaz J, (2009). Resistance in pepper plants induced by *Fusarium oxysporum* f. sp. *lycopersici* involve different defence-related genes. *Plant Biol* 11: 68–74.

136. Singh AK, Pandeya S B (1998). Modelling uptake of cadmium by plants in sludge-treated soils. *Bioresource Technol* 66: 51–58.

137. Sleutel S, De Neve S, Németh T, Tóth T, Hofman G (2006). Effect of manure and fertilizer application on the distribution of organic carbon in different soil fractions in long-term field experiments. *Eur J Agron* 25: 280–288.

138. Smith SR (2009). A critical review of the bioavailability and impacts of heavy metals in municipal solid waste composts compared to sewage sludge. *Environ Int* 35: 142–156.

139. Sodhi GPS, Beri V, Benbi DK (2009). Soil aggregation and distribution of carbon and nitrogen in different fractions under long-term application of compost in rice-wheat system. *Soil Till Res* DOI 10.1016/j.still.2008.12.005.

140. Steel H, de la Pena E, Fonderie P, Willekens K, Borgonie G, Bert W (2009). Nematode succession during composting and the potential of the nematode community as an indicator of compost maturity. *Pedobiologia* 53: 181–190.

141. Stehouwer RC, Macneal K (2003). Use of yard trimming compost for restoration of saline soil incineration ash. *Compost. Sci Util* 11: 51–60.

142. Suárez-Estrella F, Arcos-Nievas MA, López MJ, Vargas-García MC, Moreno J (2013). Biological control of plant pathogens by microorganisms isolated from agro-industrial composts. *Biological Control* 67: 509–515.

143. Sukkariyah BF, Evanylo G, Zelazny L, Chaney RL (2005). Cadmium, Copper, Nickel, and Zinc availability in a biosolids-amended piedmont soil years after application. *J Environ Qual* 34: 2255–2262.

144. Tejada M, Hernandez MT, Garcia C (2009). Soil restoration using composted plant residues: Effects on soil properties. *Soil Till Res* 102: 109–117.

145. Tester N, Davenport R (2003). Na⁺ tolerance and Na⁺ transport in higher plants. *Annals of Botany* 91: 1–25.

146. Trillas MI, Casanova E, Corxarrera L, Ordovas J, Borrero C, Aviles M (2006). Composts from agricultural waste and the *Trichoderma asperellum* strain T-34 suppress *Rhizoctonia solani* in cucumber seedlings. *Biol Control* 39: 32–38.

147. Trillas MI, Cotxarrera L, (2003). Substrates containing a *Trichoderma asperellum* strain for biological control of *Fusarium* and *Rhizoctonia*. Patent WO03/000866 A1.

148. Tripathi S, Chakraborty A, Chakrabarti K, Bandyopadhyay BK (2007). Enzyme activities and microbial biomass in coastal soils of India. *Soil Biol Biochem* 39: 2840-2848.

149. Tuitert G, Szczech M, Bollen GJ (1998). Suppression of *Rhizoctonia solani* in potting mixtures amended with compost made from organic household waste. *Phytopathology* 88: 764–773.

150. Tuteja N, Tiburcio AF, Fortes AM and Bartels D (2011). Plant Abiotic Stress. *Plant Signaling and Behavior* 6(2): 173-174.

151. van Herwijnen R, Hutchings TR, Al-Tabbaa A, Moffat AJ, Johns ML, Ouki SK (2007). Remediation of metal contaminated soil with mineral-amended composts. *Environ Pollut* 150: 347–354.

152. Van-Camp L, Bujarrabal B, Gentile A-R, Jones RJA, Montanarella L, Olazabal C, Selvaradjou S-K (2004). *Reports of the Technical Working Groups Established under the Thematic Strategy for Soil Protection.* EUR 21319 EN/3, 872 p., Office for Official Publications of the European Communities, Luxembourg.

153. Wali P, Kumar V, Singh JP (2003). Effect of soil type, exchangeable sodium percentage, water content, and organic amendments on urea hydrolysis in some tropical Indian soils. *Aust J Soil Res* 41: 1171–1176.

154. Wang CM, Shyu CL, Ho SP, Chiou SH (2008). Characterization of novel thermophilic, cellulose degrading bacterium *Paenibacillus* sp. strain B39. *Lett Appl Microbiol* 47: 46–53.

155. Wei H, Tucker PM, Baker JO, Harris M, Luo Y, Xu Q *et al.* (2012). Tracking dynamics of plant biomass composting by changes in substrate structure, microbial community and enzyme activity. *Biotechnol Biofuels* 5: 20.

156. Whalen JK, Chang C (2002). Macroaggregate characteristics in cultivated soils after 25 annual manure applications. *Soil Sci Soc Am J* 66: 1637–1647.

157. Yang D, Weng H, Wang M, Xu W, Li Y, Yang H (2010). Cloning and expression of a novel thermo-stable cellulase from newly isolated *Bacillus subtilis* strain 115. *Mol Biol Rep* 37: 1923–1929.

158. Zambare V, Zambare A, Muthukumarappan K, Christopher LP (2011). Biochemical characterization of thermophilic lignocellulose degradating and their potential biomass for bioprocessing. *Intl J Energy Environment* 2(1): 99–112.

159. Zayed G, Abdel-Motaal H, (2005). Bio-active composts from rice straw enriched with rock phosphate and their effect on the phosphorous nutrition and microbial community in rhizosphere of cowpea. *Bioresour Technol* 96: 929-935.

2016, Environmental Biotechnology: A New Approach   *Pages 179–187*
Editors: **Dr. Rajan Kumar Gupta and Dr. Satya Shila Singh**
Published by: **DAYA PUBLISHING HOUSE, NEW DELHI**

Chapter 10

Role of Mycorrhizal Fungi in Herbicide Degradation

Raghunath Satpathy

Department of Biotechnology, MITS Engineering College
Rayagada – 765 017
E-mail: rnsatpathy@gmail.com

ABSTRACT

Chemically synthesized many types of herbicides are the most widely used in the agricultural systems. They are also causes environmental pollution that threaten to the ecosystems as well as human life. After application of the herbicides, the soil remains as the major storage and maintenance reservoir for these herbicides. More or less of the herbicides can persist in the soil for a long time, thereby causing decreased crop production and induce negative effects on soil microorganism including the mycorrhiza. Many plants having symbiotic association with the fungi in their root is called as mycorrhiza. Still, some of the mycorrhizal fungi possess enormous potential for degradation of these herbicides. In comparison to free living soil fungi and bacteria, the symbiotic mode of living with the host plant, the potential understanding about the bioremediation of herbicides of various mycorrhizal systems is complicated and less understood. This chapter highlights the mycorrhiza based bioremediation of certain herbicides along with the effect and enzyme system available on the same related to degradation of these persistent compounds.

Keywords: Biodegradation, Enzyme systems, Herbicides, Mycorrhizal fungi, Symbiotic connection.

Introduction

Modern agriculture practices are mostly relying on herbicides for the control of weeds in crops, also along with intent to maximize output (Boutin and Freemark 1995). However the harmful effect of these herbicides towards the environment is a recent matter of concern (Boutin and Jobin 1998; Gasnie *et al.*, 2009; Hussein *et al.*, 1996). The potential impacts of residual herbicides in soil and water have the effect on human, animal and crop health (Altman and Campbell 1977; Igbedioh 1991; Weisenburger 1993) Not entirely human or animal toxicity, these herbicides are also responsible for the reduced seed germination, abnormal growth of certain crops (Audus 1964). These herbicides are mainly degraded by free living bacteria and white rot fungi (Mandelbaum *et al.*, 1995; Mougin *et al.*, 1994). In case of many plants a symbiotic association between certain soil fungi and plant roots exists known as mycorrhiza and are ubiquitous in the natural environments (Muthukumar *et al.*, 2004). The typical mycorrhizons which naturally biodegrade the organic pollutants as well as herbicides are known (Bending and Read 1997). The majority of the mycorrhizas are obligated symbionts because they have little or no ability for independent growth. Their role in nutrient transport in ecosystems and protection of plants against environmental and cultural stress has long been known (Smith and Smith 2011). Various different types of mycorrhizal associations have been classied from time to time by different mycologists (Pringle *et al.*, 2009). Certain species of fungi exhibit a narrow host range of plants, whereas others have a broad range. Some associations in are specic, whereas others are non-specic. There are 3 major diverse communities of mycorrhizal fungi is known such as ectomycorrhizal (ECM), vesicular–arbuscular mycorrhizas (VAM), and ericoid mycorrhizal (ERM) fungi are known in many host plant root systems (Brundrett 2004). In a soil environment, levels of colonization of these fungi depend on seasonal variations and natural processes related to soil disturbance. The symbiotic relationship allows the fungus to tide over the harsh conditions of toxic contamination such as application of herbicide (Meharg *et al.*, 1997; Singh 2006). The role of mycorrhizal fungi in herbicide to degrade a wide range of persistent herbicides has have been described in this chapter.

Symbiotic Mechanism and Herbcidie Biodegradation Metabolism of Mycorrhizal Fungi

At present, the exact molecular mechanisms of establishment of symbiotic mycorrhizas are not understood (Bonfante and Genre 2010; Harrier 2001). A series of events are involved during interaction between the fungus and root cells, ensuing in an integrated functional structure. This seems to be made by activating and deactivating of genes in both fungus and host plant (Hogekamp and Küster 2013). Certain elicitors are produced by the root cells that regulate the expression of fungal genes to establish symbiosis (Figure 10.1). Also, certain genes are activated, which are immediately responsible for the morphogeneis of the fungi (García and Ocampo 2002). The roots of host plants also secrete signal molecule(s) that inuence the growth of hyphae in VAM fungi in soil.

The molecular signal that is created from the interaction causes fungal hyphal branching pattern of as a long, scattered manner. This branching response is a mechanism favours for the host root colonization (Barker *et al.*, 1998; Harrison 2005). This signalling process is responsible for secretion of many categories of enzymes by the mycorrhizal fungi (Khan 2006). Several types of these fungi produce a variety of enzymes for the degradation of complex organic compounds (Figure 10.2). The

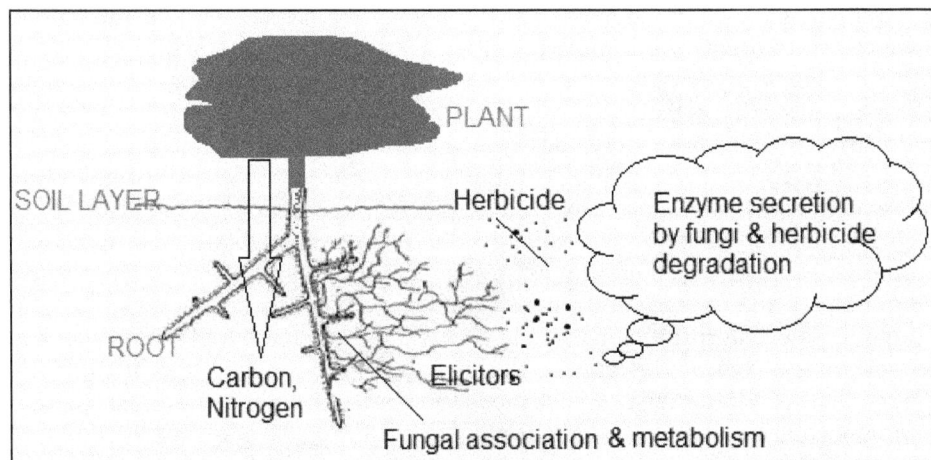

Figure 10.1: General View for Generation of Enzyme by Mycorrhizal Fungi in Soil Herbicide Degradation.

hydrolytic enzymes (proteinases and peptidases) cause the mobilization of nitrogen from organic matter in certain ericoid and ectomycorrhizal fungi. Extracellular proteinase activities are detected in these fungi when protein as a major growth substrate. The releases of these enzymes are also responsible for the role towards humication and detoxication processes in soil.

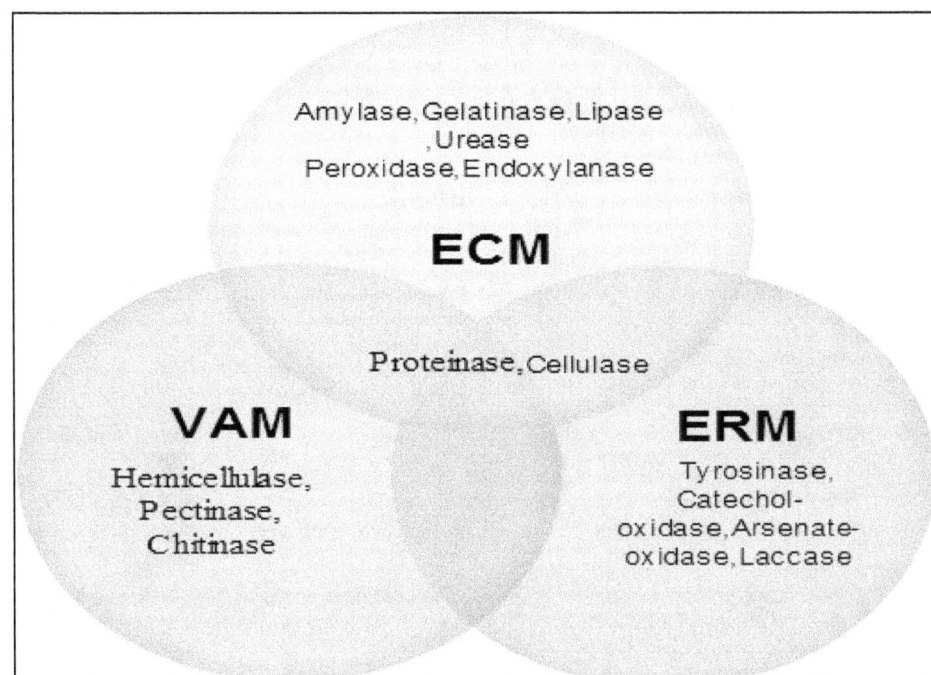

Figure 10.2: Occurrence of different Enzyme Systems Available in 3 Major Types of Mycorrizal Fungi, Proteinase and Cellulase are Common to All.

Sometimes effective mycorrhizal technology for bioremediation purpose is essential for restoring the inoculum potential in case of soils. This is achieved through *bioaugmentation* processs, by inoculating soils with suitable fungi or by using transplanted seedlings that already have the appropriate fungi in their roots (Lestan and Lamar 1996). Alternatively, the mycorrhizal populations of may be restored by the use of a crop rotation, that can stimulate and restore the fungal species in the subsequent crops (Kahiluoto and Vestberg 1999). The extreme use of herbicides in large amount may cause undesirable side effects both on biological and functional properties of biological communities through altering species composition. Several fungi display the degradation property of several harmful pesticides like 2, 4, D,glyophosate, chlorpropham,atrazine etc in the culture medium (Table 10.1). The degradation of herbicides can be observed experimentally by analyzing the intermediate product after a suitable incubation period from the time of inoculation of fungal culture. During the fungal metabolism, it has been estimated that, the nitrogen concentration is directly correlated with herbicide degradation and most cases the degradation occurred by incorporation of herbicide carbon into tissue (Entry 1999).

Table 10.1

Sl.No.	Mycorrhizal Fungal Species	Degradation of Compounds	References
1.	*Oidiodendron griseum, Hymenoscyphus ericae 1318, Gautieria crispa 4936, Gautieriaothii 6362, Radiigera atrogleba 9470, Rhizopogon vinicolor 7534, Phanaerochaete chrysosporium 1767, Sclerogaster pacificus 9011,* and *Trappea darkeri 8077*	2,4-dichlorophenoxyacetic acid (2,4-D) and 2-chloro-4-ethylamino-6-isopropyl amino-s-triazine (atrazine)	(Donnelly *et al.*, 1993)
2.	*Glomus mosseae*	Glyphosate	(Zaller *et al.*, 2014)
3.	*Hebeloma cylindrosporum, Suillus bellini* and 1 str. of *S. variegatus*	Chlorpropham	(Rouillon *et al.*, 1990)
4.	*Glomus versiforme*	2-chloro-4,6-bis-ethylamino-s-triazine(simazine), 2,6-dichlorobenzonitrile (dichlobenil), 1,1'-dimethyl-4,4'bipyridinium (paraquat)	(Hamel 1994)
5.	*Aspergillus* sp.	2,4-D	(Joshi and Gupta 2008)
6.	*Paxillus involutus*	chlorpropham	(Dix *et al.*, 1997)
7.	*Pisolithus arhizus (tinctorius)*	1-naphthalene acetic acid	(Ragonezi *et al.*, 2014)
8.	*Cenococcum graniforme. Hebeloma crustuliniforme* and *Laccaria laccata, Suillus tomentosus, Paxillus involutus*	Glyphosate, hexazinone	(Chakravarty and Chatarpaul 1990)
9.	*Glomus etunicatum*	atrazine	Huang *et al.*, 2009

Advantages and Future Aspects

Inoculation with mycorrhizal fungi offers a great opportunity to adapt plants suitable for remediation herbicides to survive. The selection of compatible host plant and fungus with suitable substrate combinations would be more effectively employed for bioremediation purpose (Khan *et al.*, 2000). But there is a lack of complete knowledge about basic biochemical factors that contribute towards a successful symbiosis that causes mycorrhizal remediation of pollutants. Therefore the mycorrhizal fungi are most often are chosen in a trial and error process based on knowledge of soil conditions and tree/fungus interactions (Chibuike 2013). Further, certain disadvantages of this method are, it is a slow process, choosing of proper fungal species as well as the type of plant specific for herbicide pollutant is required. Still many advantages exist by using the mycorrhizal fungi for the herbicide degradation process:

1. The introduction of mycorrhizal fungi can be a better choice, as they are ubiquitous in nature and (the spore) can persist for several years after their inoculation in to the soil of interest.

2. The fungi produce several enzymes responsible for degradation of wide range of pollutant.

3. It maintains a direct physiological link between root-soil –plant and increases the in transfer of different compounds including soil herbicides by enhancing the root surface area.

4. Since mycorrhizal herbicide remediation occurs through a natural process hence environmentally friendly also the process is carried out in situ, thereby eliminating the risks involved in transporting polluted soils to other locations for treatment.

5. It is assumed to be relatively cheaper and easier to accomplish compared to other methods of soil remediation (such as chemical and thermal remediation), since it does not require sophisticated technologies.

Plant and microbial biotechnology has seen tremendous developments in recent years. However, there has been little exploitation of this new technology towards mycorrhizal fungi to improve biological remediation processes. It is expected that genetically engineered microorganisms mycorrhizal fungi will show its potential for bioremediation of soil herbicides exhibiting the enhanced degrading capabilities (Figure 10.3).

In this way the mycorrhizal research would promote cost-effective, environmental friendly methods for herbicide remediation in any kind of lands.

Conclusion

Production of extracellular enzymes by the mycorrhizal fungi is now known and found out as, it is having a greater potential in soil herbicide remediation. For the effectiveness of the process various factors must be taken into the consideration *viz.* fungal mycelia, host plants, and the efficiency of the process. Experiments should be designed to decipher the herbicide degradation mechanism by mycorrhizal fungi

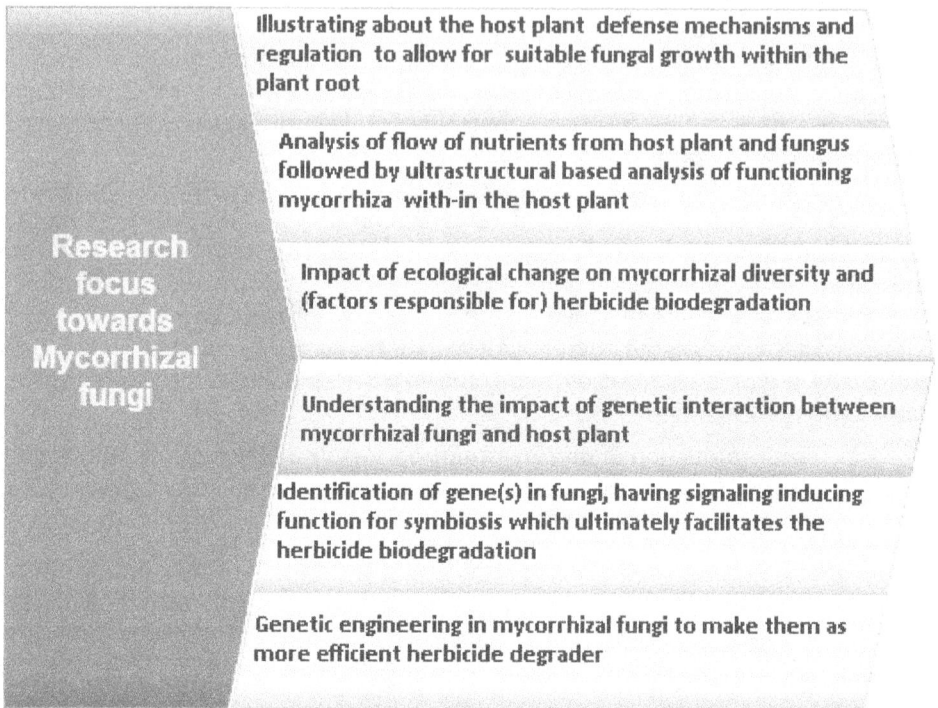

Figure 10.3: Highlights the Research Focus Towards the Mycorrhizal Fungi Based Herbicide Degradation.

during the symbiotic association with the host plant also the influence of environmental factors on degradation during the process. Little work has been performed on screening of these fungi related to degradation of persistent herbicide compounds; therefore a comprehensive screening methodology is to be developed. Several advanced molecular methods can be implemented to characterize the signalling process, genes, and related proteins necessary to maintain the symbiotic relationship. It is also possible to develop genetically engineered mycorrhizal fungi and the same can be applied to the manipulation of host plants so that such novel fungi can express herbicide degrading factors for efficient soil bioremediation.

References

1. Freemark K, Boutin C (1995). Impacts of agricultural herbicide use on terrestrial wildlife in temperate landscapes: a review with special reference to North America. *Agriculture, Ecosystems and Environment* 52(2): 67-91.

2. Hussein SY, El-Nasser MA, Ahmed SM (1996). Comparative studies on the effects of herbicide atrazine on freshwater fish *Oreochromis niloticus* and *Chrysichthyes auratus* at Assiut, Egypt. *Bulletin of Environmental Contamination and Toxicology* 57(3): 503-510.

3. Boutin C, Jobin B (1998). Intensity of agricultural practices and effects on adjacent habitats. *Ecological Applications* 8(2): 544-557.

4. Gasnier C, Dumont C, Benachour N, Clair E, Chagnon MC,Séralini GE (2009). Glyphosate-based herbicides are toxic and endocrine disruptors in human cell lines. *Toxicology* 262(3): 184-191.

5. Weisenburger DD (1993). Human health effects of agrichemical use. *Human Pathology* 24(6): 571-576.

6. Igbedioh SO (1991). Effects of agricultural pesticides on humans, animals, and higher plants in developing countries. *Archives of Environmental Health: An International Journal* 46(4): 218-224.

7. Altman J, Campbell CL (1977). Effect of herbicides on plant diseases. *Annual Review of Phytopathology* 15(1): 361-385.

8. Audus L J (1964). The physiology and biochemistry of herbicides. *The physiology and biochemistry of herbicides*.pp. Xix + 555.

9. Mandelbaum RT., Allan DL, Wackett LP (1995). Isolation and characterization of a Pseudomonas sp. that mineralizes the s-triazine herbicide atrazine. *Applied and Environmental Microbiology* 61(4): 1451-1457.

10. Mougin C, Laugero C, Asther M, Dubroca J, Frasse P, Asther M (1994). Biotransformation of the herbicide atrazine by the white rot fungus Phanerochaete chrysosporium. *Applied and environmental microbiology* 60(2): 705-708.

11. Muthukumar T, Udaiyan K., Shanmughavel P (2004). Mycorrhiza in sedges: An overview. *Mycorrhiza* 14(2): 65-77.

12. Bending GD, Read DJ (1997). Lignin and soluble phenolic degradation by ectomycorrhizal and ericoid mycorrhizal fungi. *Mycological Research* 101(11): 1348-1354.

13. Smith SE, Smith FA (2011). Roles of arbuscular mycorrhizas in plant nutrition and growth: new paradigms from cellular to ecosystem scales. *Annual review of plant biology* 62: 227-250.Doi: 10.1146/annurev-arplant-042110-103846.

14. Pringle A, Bever JD, Gardes M, Parrent JL, Rillig MC, Klironomos JN (2009). Mycorrhizal symbioses and plant invasions. *Annual Review of Ecology, Evolution, and Systematics* 40, 699-715.

15. Brundrett M (2004). Diversity and classification of mycorrhizal associations. *Biological Reviews* 79(3): 473-495.

16. Singh H (2006). Mycoremediation: fungal bioremediation. John Wiley and Sons.

17. Meharg A A, Cairney JW, Maguire N (1997). Mineralization of 2, 4-dichlorophenol by ectomycorrhizal fungi in axenic culture and in symbiosis with pine. *Chemosphere* 34(12): 2495-2504.

18. Harrier LA (2001). The arbuscular mycorrhizal symbiosis: a molecular review of the fungal dimension. *Journal of experimental botany* 52(1): 469-478.

19. Bonfante P,Genre A (2010). Mechanisms underlying beneficial plant–fungus interactions in mycorrhizal symbiosis. *Nature Communications* 1: 48.

20. Hogekamp C, Küster (2013). A roadmap of cell-type specific gene expression during sequential stages of the arbuscular mycorrhiza symbiosis. *BMC Genomics* 14(1): 306.

21. GarcíaGarrido JM, Ocampo JA (2002). Regulation of the plant defence response in arbuscular mycorrhizal symbiosis. *Journal of Experimental Botany* 53(373): 1377-1386.

22. Barker SJ, Tagu D, Delp G (1998). Regulation of root and fungal morphogenesis in mycorrhizal symbioses. *Plant physiology* 116(4): 1201-1207.

23. Harrison MJ (2005). Signaling in the arbuscular mycorrhizal symbiosis *Annual Review of Microbiology* 59: 19-42.

24. Khan AG (2006). Mycorrhizoremediation: an enhanced form of phytoremediation. *Journal of Zhejiang University Science* B 7(7): 503-514.

25. Lestan D, Lamar RT (1996). Development of fungal inocula for bioaugmentation of contaminated soils. *Applied and Environmental Microbiology* 62(6): 2045-2052.

26. Kahiluoto H, Vestberg M (1999). Impact of cropping system on mycorrhiza. *Darcof Report* (1): 305-309.

27. Entry JA (1999). Influence of nitrogen on atrazine and 2, 4 dichlorophenoxyacetic acid mineralization in blackwater and redwater forested wetland soils. *Biology and fertility of soils* 29(4): 348-353.

28. Donnelly PK, Entry JA,Crawford DL (1993). Degradation of atrazine and 2, 4-dichlorophenoxyacetic acid by mycorrhizal fungi at three nitrogen concentrations *in vitro*. *Applied and environmental microbiology* 59(8): 2642-2647.

29. Zaller, JG, Heig F,Ruess R, Grabmaier A, (2014). Glyphosate herbicide affects belowground interactions between earthworms and symbiotic mycorrhizal fungi in a model ecosystem *JA - Sci. Rep.PY*. doi: 10.1038/srep05634.

30. Rouillon R, Poulain C, Bastide J, Coste CM (1990). Degradation of the herbicide chlorpropham by some ectomycorrhizal fungi in pure culture. *Agriculture, Ecosystems and Environment* 28(1): 421-424.

31. Hamel C, Morin F, Fortin A, Granger RL,Smith DL(1994). Mycorrhizal colonization increases herbicide toxicity in apple. *Journal of the American Society for Horticultural Science* 119(6): 1255-1260.

32. Joshi N, Gupta D (2008). Soil mycofloral responses following the exposure to 2, 4-D. *Journal of Environmental Biology* 29(2): 211.

33. Dix ME, Klopfenstein NB, Zhang JW, Workman SW, Kim MS (1997). Potential use of Populus for phytoremediation of environmental pollution in riparian zones. United States Department of Agriculture Forest Service General Technical Report Rm, 206-211.

34. Ragonezi C, Teixeira D, Caldeira AT, Martins MDR, Santos-Silva C, Ganhão E,. and Zavattieri MA (2014). O-coumaric acid ester, a potential early signaling molecule in *Pinus pinea* and *Pisolithus arhizus* symbiosis established *in vitro*. *Journal of Plant Interactions* 9(1): 297-305.

35. Chakravarty P, Chatarpaul L(1990). Non target effect of herbicides: I. effect of glyphosate and hexazinone on soil microbial activity. Microbial population, and invitro growth of ectomycorrhizal fungi. *Pesticide science* 28(3): 233-241.

36. Huang H, Zhang S, Wu N, Luo L, Christie P (2009). Influence of Glomus etunicatum, Zea mays mycorrhiza on atrazine degradation, soil phosphatase and dehydrogenase activities, and soil microbial community structure. *Soil Biology and Biochemistry* 41(4): 726-734.

37. Khan AG, Kuek C, Chaudhry TM, Khoo CS, Hayes WJ (2000). Role of plants, mycorrhizae and phytochelators in heavy metal contaminated land remediation. *Chemosphere* 41(1): 197-207.

38. Chibuike GU (2013). Use of mycorrhiza in soil remediation: A review. *Scientific Research and Essays* 8(35): 679-1687.

2016, Environmental Biotechnology: A New Approach
Editors: Dr. Rajan Kumar Gupta and Dr. Satya Shila Singh
Published by: DAYA PUBLISHING HOUSE, NEW DELHI

Pages 189–203

Chapter 11

Bioprospecting Himalayan Lichens for Antimicrobial Activity: An Efficient Extraction and Microbial Assay Methodology

Himanshu Rai[1,2]*, Roshni Khare[1,2], D.K. Upreti[1]
and Rajan K. Gupta[2]

[1]Lichenology laboratory, CSIR-National Botanical research Institute,
Lucknow – 226 001, Uttar Pradesh
[2]Department of Botany, Dr. P.D.B.H. Government Post Graduate College,
Kotdwar, Uttarakhand

Introduction

Lichens, a major component of temperate-alpine habitats of Himalayan vegetation (Upreti 1998), are self-sustainable symbiotic associations formed by a fungus (mycobiont) and a green algae (phycobiont) and/or cyanobacteria (cyanobiont). Since the mycobiont constituent usually dominates the association, lichens traditionally have been classified under fungi. Lichens, in order to enable their photobionts to harvest energy from solar radiation, expose their vegetative parts at the substrate surface. The compact light-exposed vegetative bodies of lichens (the thalli) are among the most complex morphologies evolved by fungi during the past 600 million years (Yuan et al., 2005).

* Corresponding author. E-mail: himanshurai08@yahoo.com

Lichens are highly extremotolerant which allows them to live as pioneers in the alpine zone and other extreme environments, characterized by extreme regimes of temperature, moisture, wind speed and altitude. Survival under these conditions correlates with the production of a variety of compounds. Though many of these compounds are also reported in plants or in other fungi, but about 80 per cent are unique to lichens (Huneck and Yoshimura, 1996).

Lichens and their metabolites have long been used by humans. As early as 1907 Zopf published a work on the chemical, pharmacological, and technical aspects of lichen substances. In those days little was known about the chemistry of lichen substances, and it was only between 1920 and 1945 that Asahina and Shibata (1954) succeeded in elucidating the structural features of numerous compounds. The introduction of new analytical methods (*i.e.* thin layer chromatography, high-performance liquid chromatography, ultraviolet, infra red, magnetic resonance spectroscopy, mass spectrometry, X-ray crystallography) has led to the isolation of many new lichen substances, which by today, number over approximately 1050 (Stocker-Wörgötter, 2008; Molnar and Farkas, 2010).

Lichen substances basically comprise of two major groups (Stocker-Wörgötter 2008):

1. Intracellular primary compounds (proteins, amino acids, polyols, carotenoids, polysaccharides and vitamins) found in the protoplasts and cell walls. Primary compounds are products both of the algae and fungi; they are non-specific and also occur in free-living fungi, algae and higher plants. Most of them are watersoluble and can be extracted by hot water.

2. Extracellular secondary products, are produced by the fungal partner of the lichen (Elix 1996), being deposited on the surface of the fungal hyphae. Many of these secondary metabolites form crystals, and the majority can be only extracted by organic solvents. However, the carbon needed for the biosynthesis of the secondary metabolites is provided by the photosynthetic activity of the algal partner; the transfered carbohydrate depend on the type of algae, and are mainly glucose and sugar alcohols (polyols).

The distribution patterns of secondary metabolites are usually taxon specific, and therefore have been widely used in lichen taxonomy and systematics (Nylander, 1866; Cluberson, 1969; Hawksworth, 1976). Cluberson and Elix, 1989 classified lichen substances according to their biosynthetic origins and chemical structural features. Most secondary lichen metabolites are derived from the acetyl-polymalonyl pathway, while others originate from the mevalonic acid and shikimic acid pathways (Figure 11.1). Due to experimental techniques, our knowledge of the biological activities of these extracellular products has increased significantly in the last decades, and have shown to have manifold biological and pharmaceutical activity such as antimicrobial, antiviral, cytotoxic, antitumor, antiallergic, plant growth inhibitory, antiherbivore, ecological roles and enzyme inhibitory (Dülger *et al.*, 1997, 1998; Huneck 1999; Ozturk 1999; Perry *et al.*, 1999; Manojlovic *et al.*, 2002) (Figure 11.2).

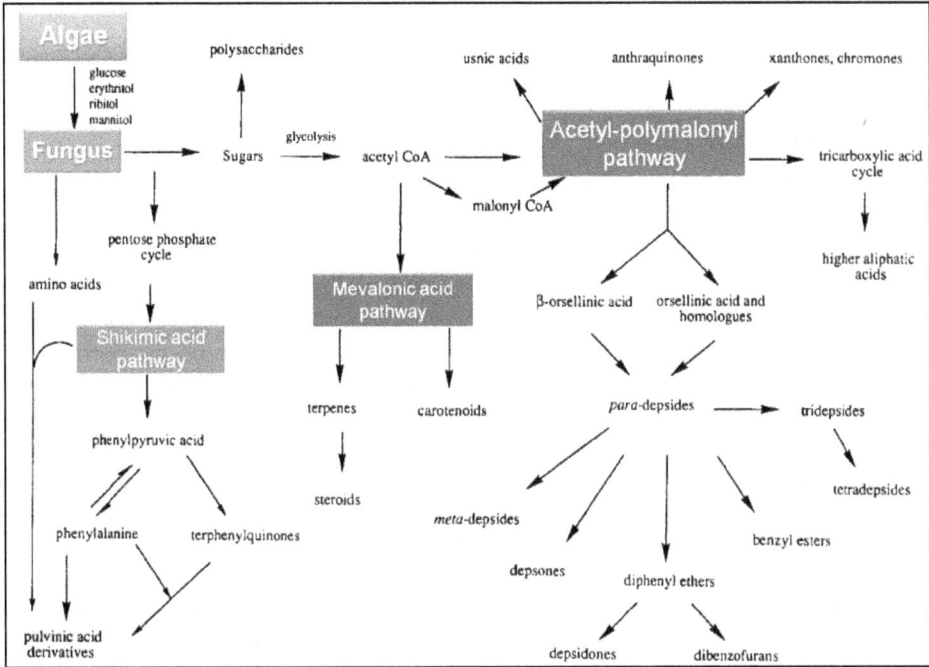

Figure 11.1: Biosynthetic Pathways of Lichen Secondary Metabolites (Modified from Elix and Stocker-Wörgötter 2008).

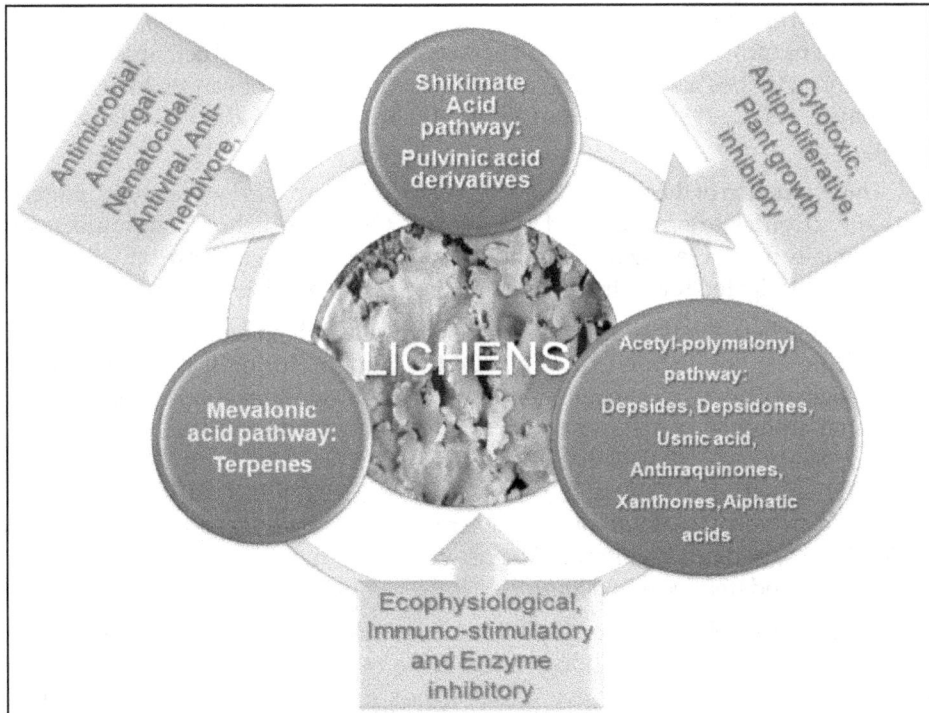

Figure 11.2: Varied Biological and Pharmaceutical Activities shown by Lichen Substances.

Lichen metabolites are known to show principally three types of antimicrobial activities *i.e.* antibacterial, antifungal and antiviral. The antibacterial properties of lichen extracts are known for long (Burkholder *et al.*, 1945) and still being assessed (Piovano *et al.*, 2002; Rankovic *et al.*, 2007; Paudel *et al.*, 2008; Schmeda-Hirschmann *et al.*, 2008; Micheletti *et al.*, 2009). Antarctic lichens were also found to be an enriched source of effective antibacterial agents (Ivanova *et al.*, 2002; Paudel *et al.*, 2008).

Usnic acid is one of the most common substances found in lichens which has been used as antibiotic (*e.g.* Binan®, Usno®), and is available as a topical antiseptic in many products (*e.g.* Gessato®shaving treatment from Italy, Camillen 60 Fudes spray and nail oil from Germany). Usnic acid and derivatives are valuable active compounds against serious pathogens such as vancomycin- resistant enteroccocci, methicillin-resistant *Staphylococcus aureus* (Elo *et al.*, 2007), mycobacteria (Ingolfsdottir *et al.*, 1998) or *Listeria monocytogenes* (Tomasi *et al.*, 2006). Lichen secondary metabolites such as protolichesterinic acid and protolichesterinic have been found to show interesting antibacterial activity (Borkowski *et al.*, 1964; Ingolfsdottir *et al.*, 1997; Ingolfsdottir *et al.*, 1998; Turk *et al.*, 2003). Lichenic substances are also known to harbour antifungal activities. Growth of *Penicillium* and *Rhizopus* were retarded by alcohol and aqueous extracts of *Parmelia molliuscula* (Burzlaff, 1950). Studies have demonstrated the inhibitory effects of aqueous lichen extracts on wood-destroying fungi (Henningsson and Lundstrom, 1970; Lundstrom and Henningsson, 1973). Wright *et al.* (1979) showed a fungistatic effect of aqueous *Nephroma arcticum* extracts on *Physarum polycephalus* (Myxomycetae). Land and Lundström (1998) observed antifungal activity with aqueous extracts from *Nephroma arcticum* on a diverse range of fungi. Tiwari *et al.*, 2011 a, b reported antifungal activity of different organic extracts of Himalayan foliose lichens on broad spectrum plant pathogenic fungi. Antiviral activities of lichens compound activities have been also reported against HIV, lung cancer, breast cancer (Hirabayashi *et al.*, 1989; Pengsuparp *et al.*, 1995; Neamati *et al.*, 1997; Mayer *et al.*, 2005).

The evaluation of these metabolites for medicinal purposes is generating interest, as their activities are varied and significant. Only few lichen compounds are commercially available whereas the thousand known compounds are only partly or not at all investigated for their medicinal potential (Stocker-Wörgötter 2008).

Indian lichen diversity accounts for 2368 species, which represents 18 per cent of world lichen population (Rai *et al.*, 2014). Though, during last fifty years the lichenological research has been mainly revolved around taxonomic studies, few studies on antibiotic and antifungal activity of Indian lichens have been initiated recently (Shahi *et al.*, 2001; Behra *et al.*, 2005; Balaji and Hariharan 2007; Sati and Joshi 2011; Tiwari *et al.*, 2011 a, b; Verma *et al.*, 2011; Shahi *et al.*, 2012). Himalayan habitats are lichen rich habitats of India where mid altitudes temperate- alpine habitat (2700-3400m), harbour maximum lichens with maximum biomass which tends to decrease beyond treeline (e" 3500m).

As, lichens are slow growing (2-0.02 mm year⁻¹) the rate of biomass accumulations is very slow in Himalayas. Though there have been attempts to cultivate lichen secondary metabolites through mass culture, the success rate is yet very slow (Behera *et al.*, 2004). Screening Himalayan lichens for their antimicrobial properties needs extraction protocol that can yield maximum amount of secondary metabolites with minimum possible dry-weight of lichen used. Here we describe an efficient protocol for lichen secondary metabolite extraction and an easy low-tech microbial assay procedure, which can be employed to faster screening of antimicrobial activity of lichens.

The screening procedure can essentially be divided into three steps: 1. Extraction of secondary metabolites from lichen thalli; 2. Isolation of secondary metabolites; 3. Antimicrobial assay using crude extract and isolated secondary metabolites. These three steps were standardized for maximum efficiency for Himalayan foliose lichens.

Secondary Metabolite Extraction form Lichen Material

The initial weight to be taken in extraction can be decided both on the availability of lichen material and sustainability of extraction protocol for future purposes. A moderate amount of 10 g lichen material was taken initially for extraction (Paudel *et al.*, 2008).

Solvent Systems

Lichen secondary metabolites are more soluble in organic solvents, which depends on the polarity of the solvents. Many solvent systems are used according to their polarity, some of them (arranged according to higher to lower polarity) used extensively are Ethyle acetate, acetone, methanol, chloroform and water(Halama and Van Haluwin 2004; Paudel *et al.*, 2008; Sharma *et al.*, 2012).

Extraction

Lichen substances were extracted using Soxhlet extractor (Soxhlet 1879; Harwood and Moody 1989) in selected solvents (Ethyle acetate, acetone, methanol, chloroform and water) equipped with a reflux condenser. Designed by Franz von Soxhlet for the extraction of a lipid from a solid material Soxhlet extractor is now not limited to the extraction of lipids only (Harwood and Moody, 1989).

Lichen material is placed inside a thimble made from thick filter paper, which is loaded into the main chamber of the Soxhlet extractor (Figure 11.3). The Soxhlet extractor is placed onto a flask containing the extraction solvent. The Soxhlet is then fitted with a reflux condenser. The solvent is heated on water bath and the vapour travelling up in distillation arm, floods into the chamber housing the thimble of solid. The condenser cools the solvent vapour, which drips back down into the chamber housing the solid material. The chamber containing the solid material slowly fills with warm solvent. Some of the desired compound will then dissolve in the warm solvent. When the Soxhlet chamber is almost full, the chamber is automatically emptied by a siphon side arm, with the solvent running back down to the distillation flask (Figure 11.3). The thimble ensures that the rapid motion of the solvent does not transport any solid material to the still pot. This cycle may is allowed to repeat for 48-72 hours. During each cycle, a portion of the non-volatile compound dissolves in the solvent. After many cycles the desired compound is concentrated in the distillation flask. The advantage of this system is that instead of many portions of warm solvent being passed through the sample, just one batch of solvent is recycled. After extraction

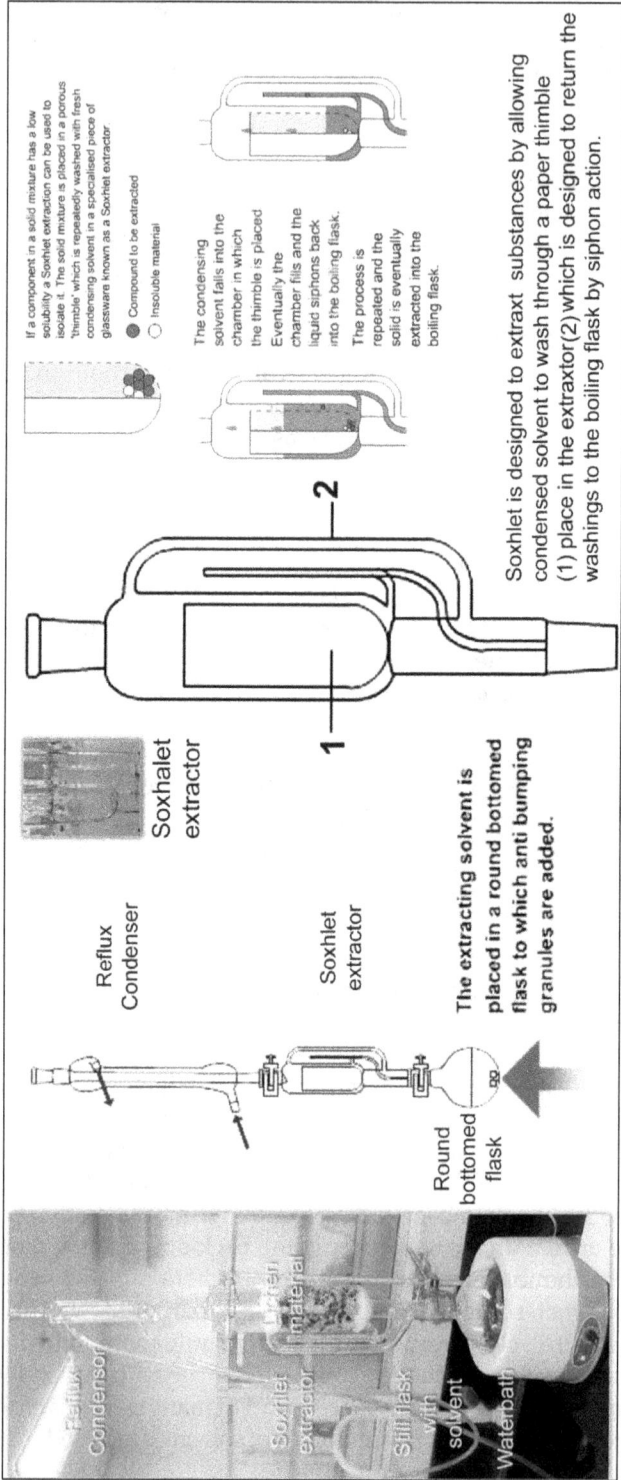

If a component in a solid mixture has a low solubility a Soxhlet extraction can be used to isolate it. The solid mixture is placed in a porous thimble which is repeatedly washed with fresh condensing solvent in a specialised piece of glassware known as a Soxhlet extractor

● Compound to be extracted
○ Insoluble material

The condensing solvent falls into the chamber in which the thimble is placed

Eventually the chamber fills and the liquid siphons back into the boiling flask.

The process is repeated and the solid is eventually extracted into the boiling flask.

Soxhlet is designed to extraxt substances by allowing condensed solvent to wash through a paper thimble (1) place in the extraxtor(2) which is designed to return the washings to the boiling flask by siphon action.

Soxhalet extractor

Reflux Condenser

Soxhlet extractor

The extracting solvent is placed in a round bottomed flask to which anti bumping granules are added.

Round bottomed flask

Figure 11.3: Extraction of Lichen Secondary Metabolites using Soxhlet Extractor Equipped with Reflux Condenser.

the solvent is removed, typically by means of a rotary evaporator, yielding the extracted compound. The non-soluble portion of the extracted solid remains in the thimble, and is usually discarded.

The extraction was carried out at specific boiling temperature of the solvents (acetone-56°C, methanol-65°C and chloroform-61.2°C).

Recovery of Secondary Metabolites

Primarily designed by Lyman C. Craig (Craig, 1950), rotary evaporator is the device used for the efficient and gentle removal of solvents from samples by evaporation. The rotary evaporator separates the condensed extracts of secondary metabolites from lichens by gentle removal of solvents under reduced pressure (vacuum) (Figure 11.4).

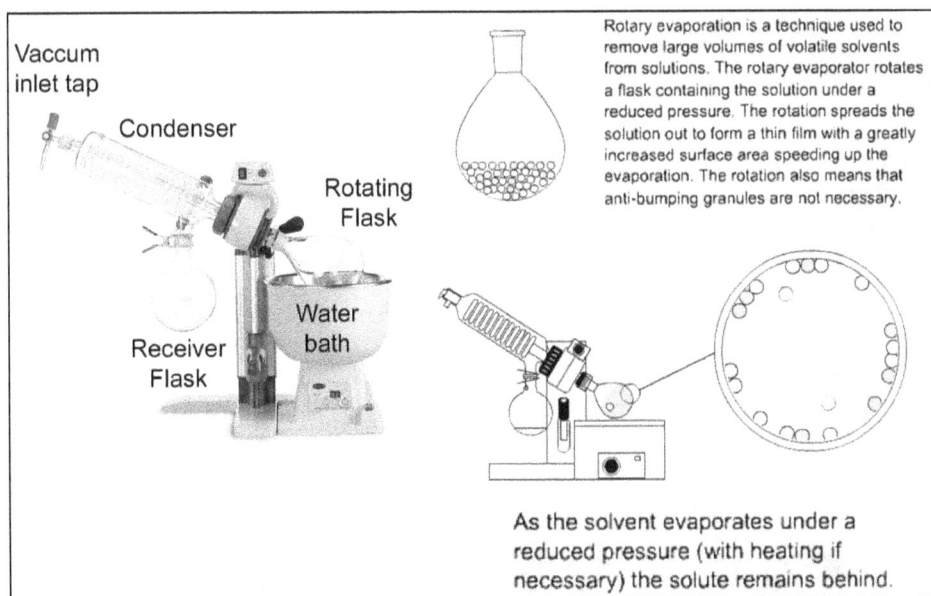

Rotary evaporation is a technique used to remove large volumes of volatile solvents from solutions. The rotary evaporator rotates a flask containing the solution under a reduced pressure. The rotation spreads the solution out to form a thin film with a greatly increased surface area speeding up the evaporation. The rotation also means that anti-bumping granules are not necessary.

As the solvent evaporates under a reduced pressure (with heating if necessary) the solute remains behind.

Figure 11.4: Recovery of Lichen Secondary Metabolites by Gentle Removal of Solvents Using Rotator Evaporator.

Antimicrobial Assay

The agar diffusion test or the Kirby-Bauer disk-diffusion method (KBDDM) is an effective and efficient assay for analyzing the susceptibility testing of bio-active compounds against microbes (Kerby *et al.*, 1957; Baur *et al.*, 1959) (Figure 11.5). Though primarily developed for measuring the effect of an antimicrobial agent against bacteria grown in culture. The KBDDM has been found to be effective even with fungi (Rankovic *et al.*, 2007, 2010, 2011; Tiwari *et al.*, 2011a, b).

The microbe (bacteria, fungi) in question is grown uniformly across a culture plate, in its specific medium a temperature and moisture conditions. Commercially available (HIMEDIA etc.) sterile susceptibility disk, impregnated with the compound to be tested, is then placed on the surface of the agar. The extracts of various solvents

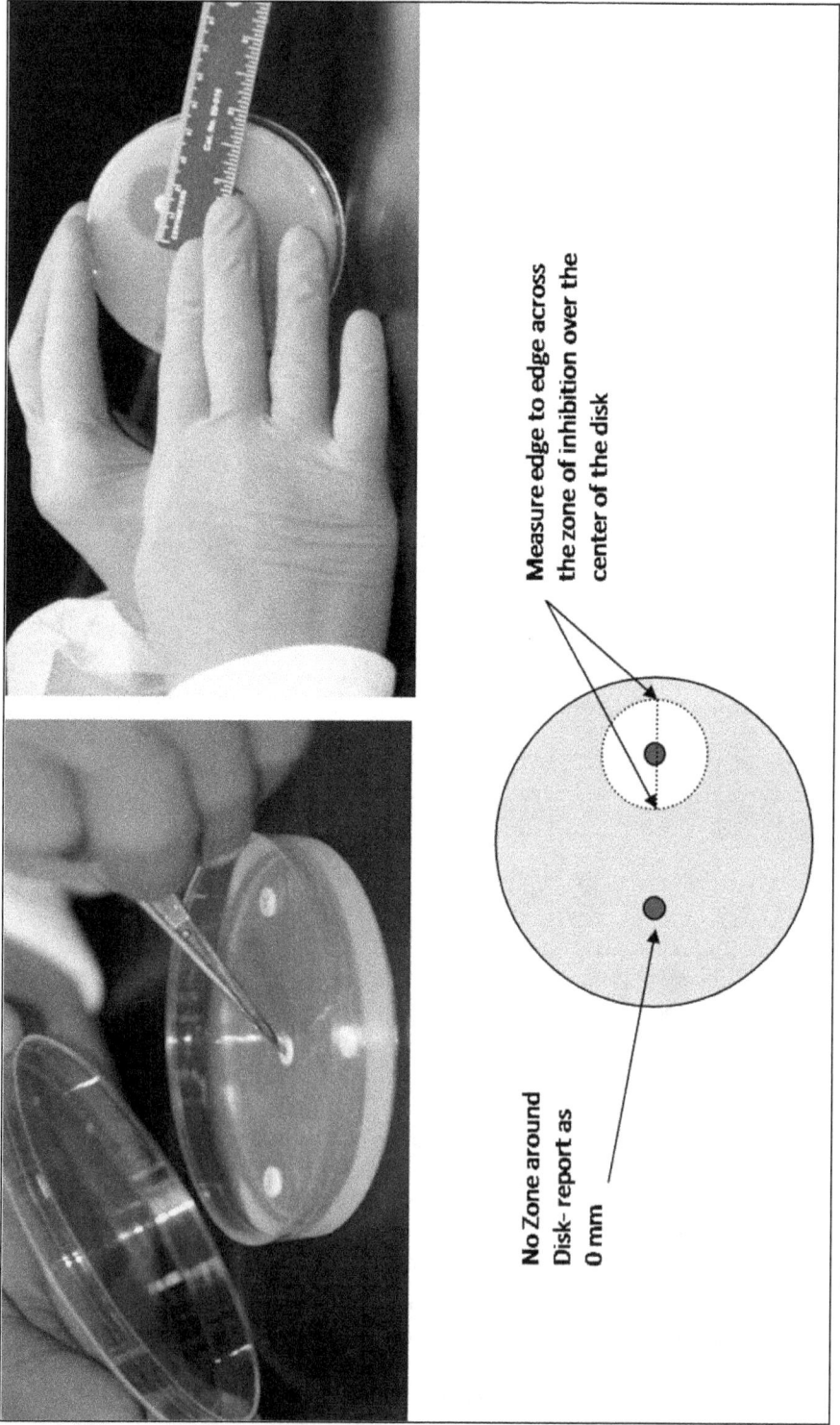

Measure edge to edge across the zone of inhibition over the center of the disk

No Zone around Disk- report as 0 mm

Figure 11.5: Disk Diffusion Method of Assessment of Antimicrobial Activity of Lichen Secondary Metabolites.

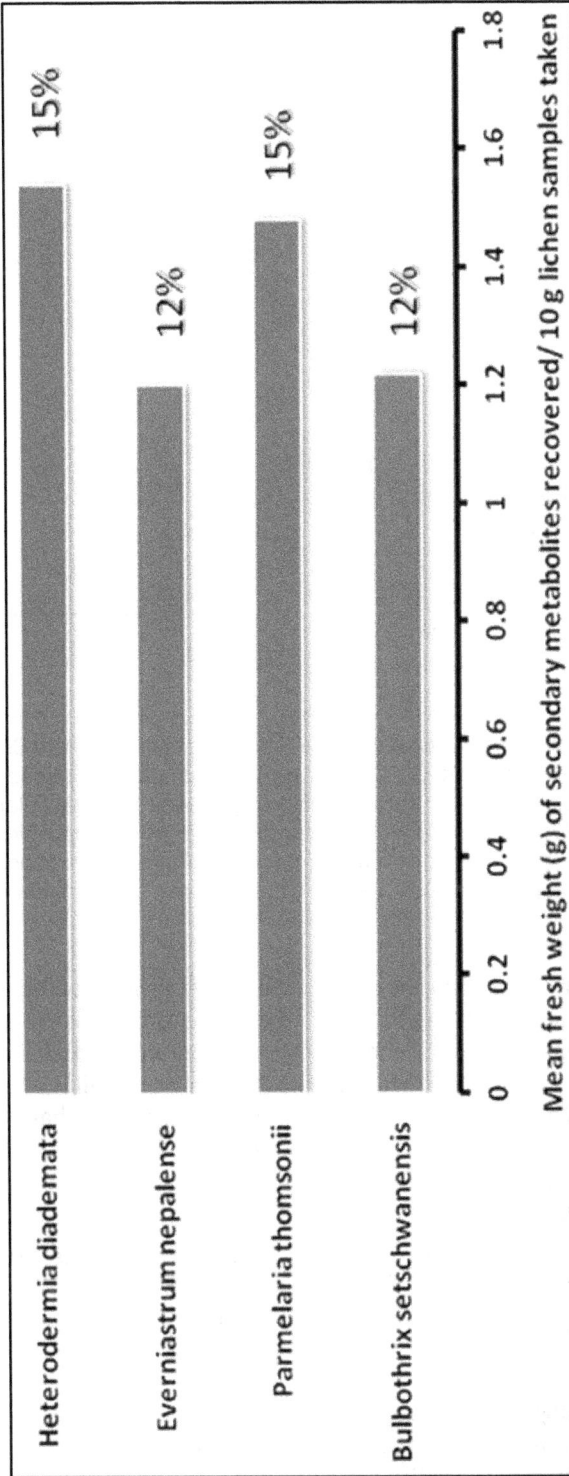

Figure 11.6: The Extraction Process Resulted in 10-15 per cent Efficiency in Yield of Secondary Metabolites (Tiwari *et al.*, 2011a).

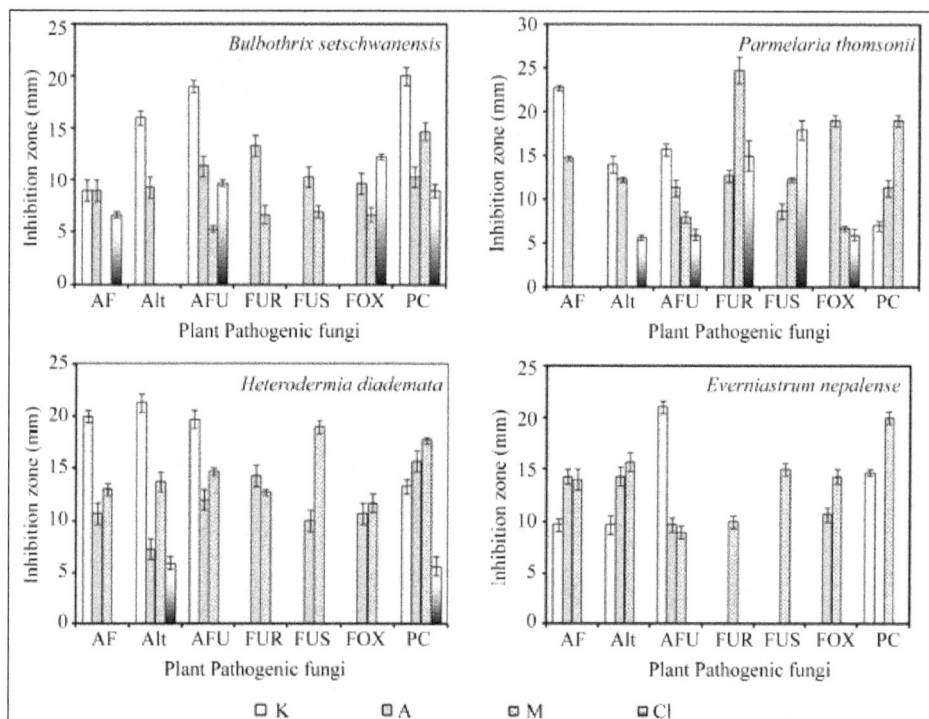

Figure 11.7: Results of Comparative Antifungal Screening of the different Solvent Extracts (A: Acetone; M: Methanol; and Cl: Chloroform) of Himalayan Foliose Lichens and Commercially Available Fungicide Ketoconazole (K); against Selected Plant Pathogenic Fungi (AF: *Aspergillus flavus*; Alt: *Alternaria alternata*; AFU: *Aspergillus fumigatus*; FUR: *Fusarium roseum*; FUS: *Fusarium soloni*; FOX: *Fusarium oxysporum*; and PC: *Penicillium citrinum*). Reported values are in Arithmetic mean ± Standard error. (Tiwari *et al.*, 2011a).

of different concentration should be used to check the highest potential. The extract diffuses from the filter paper into the agar. The concentration of the extract will be highest next to the disk, and will decrease as distance from the disk increases. If the extract is effective against microbe (bacteria, fungi) at a certain concentration, no colonies will grow where the concentration in the agar is greater than or equal to the effective concentration. This is the zone of inhibition. Thus, the size of the zone of inhibition is a measure of the extract effectiveness: the larger the clear area around the filter disk, the more effective the test species.

Commercially available antimicrobial susceptibility discs *e.g.* antifungal-Ketoconazole (Ketoconazole KT[10™] HIMEDIA®) antibacterial (Ampicillin AMP 10mcg™ HIMEDIA®) cab be used as positive control.

Validation

Studies on antifungal activities of Himalayan foliose lichens *Parmotrema tinctorum*, *Bulbothirx setschwanensis*, *Everniastrum nepalense*, *Heterodermia diademata, and Parmelaria thomsonii*, against some plant pathogenic fungi *Aspergillus niger, Aspergillus flavus*,

Aspergillus fumigatus, Alternaria alternata, Fusarium oxysporum, Fusarium solani, Fusarium roseum, Ustilago spp., *Albugo candida and Penicillium citrinum* have proven the efficiency of the proposed methodology (Tiwari *et al.*, 2011a, b). The total secondary metabolites extracted were in the range of 10-15 per cent to the total volume of lichen material taken (Tiwari *et al.*, 2011a, b). Further the study revealed good resolution of antifungal activity in comparison to positive control and found better antifungal activity against *Fusarium* spp., findings comparable to Rankovic *et al.*, 2007, 2010, 2011 (Figures 11.6 and 11.7).

References

1. Asahina, Y., Shibata, S. (1954). *Chemistry of lichen substances*. Japan Society for the Promotion of Science, Tokyo.

2. Balaji, P., Hariharan, G.N. (2007). *In vitro* antimicrobial activity of *Parmotrema praesorediosum* thallus extracts. *Research Journal of Botany* 2: 54-59.

3. Bauer, A. W., Perry, D. M., Kirby, W. M. (1959). Single-disk antibiotic-sensitivity testing of Staphylococci: An analysis of technique and results. *A.M.A. Archives of Internal Medicine* 104: 208-216.

4. Behera, B. C., Verma, N., Sonone, A., and Makhija, U. (2005). Antioxidant and antibacterial activities of lichen *Usnea ghattensis in vitro*. *Biotechnology letters*, 27(14), 991-995.

5. Behera, B.C., Verma N., Sonone A., Makhija U. (2004). Determination of antioxidative and anti-Tyrosinase potential of lichen *Usnea ghattensis in vitro*. In: Reddy M. S., Khanna Sunil (Eds). *Biotechnological Approaches for Sustainable Development*. Allied Publishers Pvt. Ltd, New Delhi, pp 94-103.

6. Borkowski, B., Wozniak, W., and Gertig, H. (1964). Bacteriostatic action of some compounds from lichen *Cetraria islandica* and of usnic acid. *Dissertationes Pharmaceuticae* 16: 189-194.

7. Burkholder, P. R., and Evans, A. W. (1945). Further studies on the antibiotic activity of lichens. *Bulletin of the Torrey Botanical Club* 72: 157-164.

8. Burzlaff, D. F. (1950). The effect of extracts from the lichen, Parmelia molliuseula, upon seed germination and upon growth rate of fungi. *Journal of the Colorado-Wyoming Academy of Science* 4: 56 (Abstr.).

9. Craig, L.C., Gregory, J.D., Hausmann, W. (1950). Versatile Laboratory Concentration Device. *Analytical Chemistry* 22: 1462-1462.

10. Cluberson, W.L. (1969). The use of chemistry in systematics of the lichens. *Taxon.* 18(2): 152-166.

11. Culberson,C. F., Elix, J. A. (1989). Lichen substances. In: Dey, P. M., Harborne, J. B. (eds.). *Methods in plant biochemistry, Vol. 1. Plant phenolics*. Academic Press, London, pp. 509–535.

12. Dülger B, Gucin F, Kara A, Aslan A (1997). Antimicrobial activity of the lichen *Usnea florida*(L.). Wigg. *Turkish Journal of Biology* 21: 103-108.

13. Dülger, B., Gucin, F., Aslan, A. (1998). Antimicrobial activity of the lichen *Cetraria islandica* (L.). Ach. *Turkish Journal of Biology* 22: 111-18.

14. Elix, J. A. (1996). Biochemistry and secondary metabolites. In: NASH T. H. III (eds.). *Lichen biology*, 1st ed. Cambridge University Press, pp. 155–180.

15. Elix, J. A.,Stocker-Wörgötter, E. (2008). Biochemistry and secondary metabolites. In: Nash, T. H. III (eds.): *Lichen biology*, 2nd ed. Cambridge University Press, pp. 104–133.

16. Elo, H., Matikainen, J., Pelttari, E. (2007). Potent activity of the lichen antibiotic (+)-usnic acid against clinical isolates of vancomycin-resistant enterococci and methicillin-resistant *Staphylococcus aureus*. *Natur Wissenschaften* 94: 465–468.

17. Halama, P.,Van Haluwin, C. (2004). Antifungal activity of lichen extracts and lichenic acids. *BioControl* 49(1), 95-107.

18. Harwood, L.M., Moody, C.J. (1989). *Experimental organic chemistry: Principles and Practice (Illustrated edition ed.)*. pp. 122–125.

19. Hawksworth, D.L. (1976). Lichen chemotaxonomy. In *Lichenology: Progress and Problems*; Proceedings of an international symposium, p. 139-184.

20. The influence of lichens, lichen extracts and usnic acid on wood destroying fungi. *Material and Organismen* 5: 19–31.

21. Henningsson, B., Lundstrom, H. (1970). The influence of lichens, lichen extracts and usnic acid on wood destroying fungi. *Material and Organismen* 5: 19–31.

22. Hirabayashi, K., Iwata, S., Ito, M., Shigeta, S., Narui, T., Mori, T., and Shibata, S. (1989). Inhibitory effect of a lichen polysaccharide sulfate, GE-3-S, on the replication of human immunodeficiency virus (HIV). *in vitro. Chemical and pharmaceutical bulletin* 37: 2410-2412.

23. Huneck, S., Yoshimura, I. (1996). *Identification of Lichen substances*. Springer, Berlin, p 493.

24. Huneck,S. (1999). The Significance of Lichens and Their Metabolites. *Natur Wissenschaften* 86: 559–570.

25. Ingólfsdóttir, K., Chung, G. A., Skúlason, V. G., Gissurarson, S. R., and Vilhelmsdóttir, M. (1998). Antimycobacterial activity of lichen metabolites *in vitro. European Journal of Pharmaceutical Sciences* 6: 141-144.

26. Ingolfsdottir, K., Hjalmarsdottir, M. A., Sigurdsson, A., Gudjonsdottir, G. A., Brynjolfsdottir, A., Steingrimsson, O. (1997). *In vitro* susceptibility of *Helicobacter pylori* to protolichesterinic acid from the lichen *Cetraria islandica*. *Antimicrobial agents and chemotherapy* 41: 215-217.

27. Ivanova, V., Aleksieva, K., Kolarova, M., Chipeva, V., Schlegel, R., Schlegel, B., Gräfe, U. (2002). Neuropogonines A, B and C, new depsidon-type metabolites from *Neuropogon* sp., an Antarctic lichen. *Pharmazie*, 57: 73-74.

28. Kirby, W.M.M., Yoshihara, G.M., Sundstedt, K., Warren, J. (1957). Clinical usefulness of a single disc method for antibiotic sensitivity testing. In: *Antibiotics Annual* 1956-1957. New York, Antibiotica, Inc. pp892.

29. Lard, C.J. and Lundstrom, H. (1998). Inhibition of fungal growth by water extracts from the lichen *Nephroma arcticum*. *The Lichenologist*, 30(3): 259-262.

30. Lundstrom, H., Henningsson, B. (1973). The effect of ten lichens on the growth of wood destroying fungi. *Material and Organismen* 8: 233–246.

31. Manojlovic, N.T., Solujic, S., Sukdolak, S. (2002). Antimicrobial activity of an extract and anthraquinones from *Caloplaca schaereri*. *Lichenologist*, 34(1): 83-85.

32. Mayer, M., O'Neill, M. A., Murray, K. E., Santos-Magalhães, N. S., Carneiro-Leão, A. M. A., Thompson, A. M., Appleyard, V. C. (2005). Usnic acid: a non-genotoxic compound with anti-cancer properties. *Anti-cancer drugs* 16: 805-809.

33. Micheletti, Ana C., Beatriz, A., Lima, D.P.D, Honda, N.K., Pessoa C.D.Ó., Moraes, M. O.D., Lotufo, L.V., Magalhães, H.I.F., Carvalho N.C.P. (2009). Chemical constituents of *Parmotrema lichexanthonicum* Eliasaro and Adler: isolation, structure modification and evaluation of antibiotic and cytotoxic activities. *Química Nova* 32: 12-20.

34. Molnár, K., Farkas, E. (2010). Current results on biological activities of lichen secondary metabolites: a review. *Zeitschrift für Naturforschung* 65C: 157–173.

35. Neamati, N., Hong, H., Mazumder, A., Wang, S., Sunder, S., Nicklaus, M.C., Milne, G.W.A., Proksa, B., Pommier, Y. (1997). Depsides and depsidones as inhibitors of HIV-I integrase: discovery of novel inhibitors through 3D database searching. *Journal of Medical Chemistry* 40: 942-951.

36. Nylander, W. (1866). Les lichens du Jardin du Luxembourg. *Bulletin Societe Botaniquede France*, 13: 364-372.

37. Ozturk, S., Guvenc, S., Arikan, N., Yilmaz, O. (1999). Effect of usnic acid on mitotic index in root tips of *Allium cepa* L. *Lagascalia* 21: 47-52.

38. Paudel, B., Bhattarai, H. D., Lee, J. S., Hong, S. G., Shin, H. W., Yim, J. H. (2008). Antibacterial potential of Antarctic lichens against human pathogenic Grampositive bacteria. *Phytotherapy Research*, 22: 1269-1271.

39. Pengsuparp, T., Cai, L., Constant, H., Fong H. H.S., Lin, L.Z., Kinghorn, A. D., Pezzuto J.M., Cordell, G.A., Ingolfsdöttir, K., Wagner, H., Hughes, S.H. (1995). Mechanistic evaluation of new plant-derived compounds that inhibit HIV-1 reverse transcriptase. *Journal of natural products* 58: 1024-1031.

40. Perry, N.B., Benn, M.H., Brennan, N.J., Burgess, E.J., Ellis, G., Gallowey, D.J., Lorimer, S.D., Tangney, R.S. (1999). Antimicrobial, Antiviral and Cytotoxic activity of New Zeland Lichens. *Lichenologist* 31: 627-636.

41. Piovano, M., Garbarino, J. A., Giannini, F. A., Correche, E. R., Feresin, G., Tapia, A., Zacchino, S., Enriz, R. D. (2002). Evaluation of antifungal and antibacterial activities of aromatic metabolites from lichens. *Boletin de la Sociedad Chilena de Química* 47: 235-240.

42. Rai, H., Khare, R. Upreti, D.K. and Nayaka (2014). Terricolous lichens of India: An introduction to field collection and taxonomic investigation. In H Rai & D.K. Upreti (Eds.), *Terricolous Lichens in India* (pp. 1-16): Springer, New York.

43. Rankovic, B., Kosani, M. (2012). Antimicrobial activities of different extracts of *Lecanora atra, Lecanora muralis, Parmelia saxatilis, Parmelia sulcata* and *Parmeliopsis ambigua. Pakistan Journal of Botany* 44: 429-433.

44. Rankovic, B., Kosani, M., and Stanojkovi, T. (2011). Antioxidant, antimicrobial and anticancer activity of the lichens Cladonia furcata, Lecanora atra and Lecanora muralis. *BMC complementary and alternative medicine*, 11: 97.

45. Rankovic, B., Misic, M., Sukdolak, S. (2007). Evaluation of antimicrobial activity of the lichens *Lasallia pustulata, Parmelia sulcata, Umbilicaria crustulosa*, and *Umbilicaria cylindrica. Microbiology* 76: 723–727.

46. Rankovic, B., Rankovic, D., Maric, D. (2010). Antioxidant and antimicrobial activity of some lichen species. *Microbiology* 79: 809-815.

47. Sati,S.C., Joshi, S. (2011). Antibacterial Activity of the Himalayan Lichen *Parmotrema nilgherrense. British Microbiology Research Journal* 1: 26-32.

48. Schmeda Hirschmann, G., Tapia, A., Lima, B., Pertino, M., Sortino, M., Zacchino, S., Rojas de Arias, A., Feresin, G.E. (2008). A new antifungal and antiprotozoal depside from the Andean lichen *Protousnea poeppigii. Phytotherapy Research* 22: 349-355.

49. Shahi, S.K., Shukla, A.C., Dikshit, A., Upreti, D.K (2001). Broad spectrum antifungal properties of the lichen *Heterodermia leucomela. Lichenologist* 33: 177-179.

50. Shahi, M.P., Shahi, S.K., Upreti, D.K. (2012). Broad spectrum antifungal activity of Indian lichen *Peltigera praetextata* extract to control fungal infections. *Current Discovery* 1: 1-6

51. Sharma, B. C., Kalikotay, S., Rai, B. (2012). Assessment of antimicrobial activity of extracts of few common lichens of Darjeeling hills. *Indian Journal of Fundamental and Applied Life Sciences* 2: 120-126.

52. Soxhlet, F. (1879). Die gewichtsanalytische Bestimmung des Milchfettes. *Polytechnisches J. (Dingler's).* 232: 461-465.

53. Stocker-Wörgötter, E. (2008). Metabolic diversity of lichen forming ascomycetous fungi: culturing, polyketide and shikimate metabolite production, and PKS genes. *Natural Product Reports* 25: 188–200.

54. Tiwari, P., Rai, H., Upreti, D. K., Trivedi, S., Shukla, P. (2011a). Antifungal activity of a common Himalayan foliose lichen *Parmotrema tinctorum* (Despr. ex Nyl.). Hale. *Nature and Science* 9: 167-171.

55. Tiwari, P., Rai, H., Upreti, D. K., Trivedi, S., Shukla, P. (2011b). Assessment of antifungal activity of some Himalayan foliose lichens against plant pathogenic

fungi. *American Journal of Plant Science* 2: 841-846.

56. Tomasi, S., Picard, S., Lainé, C., Babonneau, V., Goujeon, A., Boustie, J., Uriac, P. (2006). Solid-phase synthesis of polyfunctionalized natural products: application to usnic acid, a bioactive lichen compound. *Journal of combinatorial chemistry* 8: 11.-14.

57. Turk,O.A., Meral, Y., Merih, K., Kivanc, M., Turk, H. (2003). The antimicrobial activity of extracts of the lichen *Cetraria aculeata* and its protolichesterinic acid constituent. *Zeitschrift Fur Naturforsch* C 58: 850–854.

58. Upreti D.K. (1998). Diversity of lichens in India. In: *Perspectives in Environment* (S.K. Agarwal, J.P. Kaushik, K.K. Kaul and A.K. Jain, eds.), pp. 71–79. APH Publishing Corporation, New Delhi, India.

59. Verma, N., Behera, B. C., Parizadeh, H., and Sharma, B. (2011). Bactericidal activity of some lichen secondary compounds of Cladonia ochrochlora, Parmotrema nilgherrensis and Parmotrema sancti-angelii. *Int J Drug Dev Res*, 3, 222-232.

60. Wright, M., Tollon, Y., Lundstrom, H. (1979). Inhibitory action of *Nephroma* water extracts on the metabolism of *Physarum polycephalum*. *Planta Medica* 35: 323–330.

61. Yuan X, Xiao S, Taylor TN (2005). Lichen-like symbiosis 600 million years ago. *Science* 308: 1017–1020.

62. Zopf,W. (1907). Die Flechtenstoffe in chemischer, botanischer, pharmakologischer und technischer Beziehung. Fischer, Jena.

2016, Environmental Biotechnology: A New Approach   *Pages 205–210*
Editors: Dr. Rajan Kumar Gupta and Dr. Satya Shila Singh
Published by: DAYA PUBLISHING HOUSE, NEW DELHI

Chapter 12

Environmental Biotechnology and Human Welfare

U.K. Chaturvedi[1] and Iqbal Habib[2]

[1]*Department of Botany, Bareilly College, Bareilly – 243 005, U.P.*
[2]*Department of Botany, Govt. Degree College, Budaun – 243 601, U.P.*

ABSTRACT

Humans have been manipulating genetic material for a very long time and have been benefitted through the use of biotechnological inputs. Molecular biotechnology grew at a pace that has been unprecedented in the history of technological development. The 21st century, will belong to the biological sciences. The new biotechnology is already influencing on our agriculture, animal husbandry, forestry, pharmaceutical, bio-remediation and food and drink industry. Therefore, the benefits of biotechnology are quite significant and will help in improving the health and prosperity of human beings.

Keywords: Biotechnology, Agriculture, Industry.

Introduction

Biotechnology which consists of living systems and organisms, is being greatly utilized in the welfare processes of human beings. In the last several hundred years, biotechnology helped in agriculture, food production and in medicine. In this 21st century, biotechnology is expanding rapidly in the field of pharmaceutical therapies, immunology and in diagnostic tests. It is also in use in influencing the natural environment and at the same time commercially exploited.

Biotechnology is a new or young branch of science, which has been exploited by the scientists to improve the welfare of man. Its study encompasses a wide range of methodology for modifying living organisms according to human purposes. The role of biotechnology has lead to great advancements in the field of agriculture, medicines, plant genetics and industrial processes. For the last 20 years the use of biotechnology has been of immense importance to human beings. Therefore, scientists are trying their best to bring newer technologies every day which enable man to improve upon nature and environment.

Although, scientists have been using this branch of science for hundreds of years such as in baking of bread, making of wine and beer, in the production of cheese from milk and in the manufacture of penicillin and other antibiotics, but recently biotechnology is helping in genetic engineering, in the manufacture of new and advanced level of drugs, in the production of synthetic human insulin, in agriculture and in industrial use. The use of bio-fertilizers, replacing the excessive use of chemical fertilizers, the production of single cell protein, enhanced nitrogen fixation by Nitrogen-fixing soil bacteria such as Azotobacter vinelandii and Klebsiella, transgenic organisms, transgenic animals and plants, vaccine production and monoclonal antibodies are some of the fields where biotechnology has been of great use.

Recently, the use of biotechnology has helped in the production of transgenic plants with the genes having herbicide resistance, insect resistance, or the transgenic plants with resistance against bacteria, viruses and fungal pathogens. Studies are being conducted to produce tomato plants for hard skin to enhance self life and improved flavour, for molecular farming, DNA matching, with production of thermostable enzymes, new fuels from maize and sugarcane. Brazil is trying to reduce its oil imports by growing these plants and using plant alcohol instead of petrol.

Pharmaceutical industry is largely depended on the supply of crude plants. Most of these drug plants are normally scattered in and around forest areas in plains and in hilly tracts. The process for their collection is becoming too expensive as the transportation and collection charges are increasing alarmingly. Further, the excessive exploitation of these plants by local people leading to the extinction of some of these important medicinal plants. Therefore, it is advisable to introduce modern techniques to extract the active principles from these plants. Now, some of these plants have been brought under tissue culture, which is an exciting field of biotechnology. The significant advantages of biotechnology are in the field of physiology, phychopathological problems, growing virus free plants, in plant breeding, micro-propagation, cryopreservation, somaclonal variation, protoplast culture, gene introduction, production of artificial seeds etc. Recent developments in germplasm conservation by slow growth and cryopreservation of cells and tissues has helped in the storage of germplams.

Besides this, transgenic plants that express foreign proteins with industrial or pharmaceutical value represent an economical alternative to fermentation based industries. Therefore, the transgenic plants hold promise as low cost edible proteins and vaccines for the benefit of human beings. Another positive effect of biotechnology has come from the British scientists of Leeds and York Universities for a break through towards curing the common cold. The researches have cracked a code that governs infections by a major group of viruses, including the common cold and polio. Rhinovirus (Which causes the common cold) accounts for more infections every year than all other infectious agents put together. Professor Peter Stockley and Dr. Roman Tuma of the University of Leeds (U.K.) are able to unlock the code of this single stranded virus. These scientist have demonstrated that jamming the code can disrupt virus assembly and prevent disease.

Biotechnology is a science of life. It gives a cutting-edge demo of its latest developments. U.S. scientist Craig Venter's company became famous for running a parallel version of the Human Genome Product in 1999, says he has built a synthetic chromosome using chemicals made in a laboratory. He created a cell having a chromosome with 381 genes and 5,80,000 base pairs of genetic code, using lab- made chemicals. His team work's plan is to transfer it later into a living bacterial cell and become a new life form. The use of genetic therapies and stem cells to create new tissues has now become a common phenomenon. More over, gene targeting in particular, to target cancer causing genes or ageing and other diseases is coming to a great help in detecting them. The research also offers the best hope yet of unlocking the molecular secrets of illnesses ranging from congenital heart diseases to Alzheimer's.

The use of biotechnology is helping the molecular geneticists to achieve improvements in plant characteristics by the direct manipulation of the genes. The existing genes can be removed or modified so as to improve their nutritional value or improved production. The plants have now been modified to make them resistant to herbicides and pesticides. Commercial production of such transgenic crops will make a great contribution to food production. Further, in this context, the artificial seed technology in now rapidly developing field of plant science. The advantages of this technology include the rapid and large scale multiplication, minimum labour and low cost propagation. In addition, artificial seeds can be directly delivered to the field, thus eliminating transplantation. This technology is gradually moving towards the commercial propagation of high value crops. Leaving some technical problems, if rectified, then there would be no problem in producing the desired and designer crops. The artificial seed technology, if, commercially exploited on a large scale, will generate millions of plants in a few days and this can become a profitable multimillion dollar industry in near future. The genetic engineering techniques have great potential for increasing the quality of the grain and also the yield. The use of biotechnology to introduce a wide range of useful foreign genes that could develop new strains much more rapidly than was possible with traditional breeding methods. By using tools of biotechnology and the sophisticated techniques of genetic engineering has produced improved transgenic cereals such as wheat, rice, maize, barley, rye, oats and sorghum which are resistant to herbicides, insects, viruses and pathogens.

The marginal farmers are unable to exploit the improved agricultural technology as the do not have enough capital for various inputs including chemical nitrogen. Therefore, the process of algalization of blue green algae in the rice fields would be a cheaper source of nitrogen supply to their rice fields. According to Venkataraman (1975) the algal forms like Nostoc, Anabaena, Calothrix, Aulosira and Plectonema are found in rice fields in abundance and hence the can be exploited by poor farmers to increase their yield of rice crop. The nitrogen fixing capacity of these algae has been established in the past several years and are responsible for nitrogen recuperation under rice field conditions (Venkataraman, 1961). Many workers have also reported increase soil nitrogen as a result of algalization (Watanabe, 1967, Chopra and Dubey, 1971). Watanabe (1962) found that algal inoculation enhanced the ammonification process thereby increasing the available nitrogen. The extensive field trials during the last several years, under different agro-climatic conditions both in our country and abroad, have shown that algalization can result in a saving of upto 30 percent chemical nitrogen.

The role of biotechnology to obtain genetically modified Bt cotton crop was found satisfactory in terms of higher number of bolls, reduced number of sprays for control of boll worms and higher yields. Similarly researches are in progress to produce genetically modified and edible variety of Brassica juncea L. (Indian mustard). Now the basic question is, how safe are GM crops? So we really need them? In this context, the Bt cotton seed provider companies such as Monsanto and Proagro Seed argue that their products increase yield and bring the farmer more revenue while saving time. But the critics say that inserting genes in plants may lead to some side- effects on a plant's biochemistry, such as destroying its natural qualities and creating nutritional imbalances. It may also lead to the creation of 'superseeds' which will be resistant to herbicides. Further, there is fear that if the seeds produced by this method may become 'terminator seeds', that is, they cannot be saved and planted again. There is also likely to not get all the benefits from a second crop- "Gene jumping" can result in the transfer of the genetically induced traits to related species like cabbage, radish and cauliflower.

Genetically modified plants contain genes which may be transferred from Bacteria, Viruses and other plants and animals. In this field of science, researches are being conducted to obtain some peculiar combinations. If taken in humour or with a little pinch of salt, than some combinations have been tried such as cow genes in soyabeans, Pig virus gene in tomato to give this vegetable a hamflavour or fish genes are added to tomato to give some fishy smell. Even a human gene may be transferred to rice plant to obtain a closure cousin. Recently, American scientists have experimented on increasing the life of human beings and to feel them more energetic and young by transferring the blood of young mice into the body of an old one. And the results were astonishing. Therefore, all such and many other possibilities are in store now. So, be prepared to hear more.

Conclusion

In my opinion, the role of biotechnology should be directed especially in health care, in industry, in agriculture and in the renewable energy sectors. In the decade to come, we will witness a marriage between biotech and pharmaceutical industries. There is no doubt that biotechnology has much to offer to India and to the welfare of humanity. Indeed, biotechnology is going to be a big business in the coming years.

Acknowledgements

The authors are indebted to Prof. S.P. Khare. Principal, Govt. Degree College, Budaun (U.P.) for his encouragement. He is all the time willing to help us in promoting science. Thanks are also due to Prof. Vidyavati, former Vice Chancellor, Kakatia University, Warangal (A.P.) for her constant encouragements.

References

1. Bajaj, Y.P.S. (1997). Applied and fundamental aspects of plants cells. In: *Tissue and organ* culture (ed. Reinert, J and Bajaj YPS) pp. 467-496, Springer, Verlag, Berlin.

2. Bernard, H.V. and Helinske, D.R. (1980). In: *Genetic Engineering : Principles and Methods*, Vol. II pp. 133-168, Plenum Press, New York.

3. Bhojwani, S.S. and Razdan M.K. (1983). *Plant Tissue culture theory and practice.* Elsevier, Amsterdam.

4. Chopra, T.S. and Dubey, J.N. (1971). Changes of N content of soil, inoculated with Tolypothrix tenuis. *Plant and Soil* 35(3): 453-462.

5. Daniel A. V. (2010). *Environmental Biotechnology. A Biosystems Approach,* Academic Press, Amsterdam.

6. Doyle, JJ (1990) Isolation of plant DNA from fresh tissues. *Focus* 12: 13-15.

7. Gupta P.K. (2013-14). *Molecular Biology and Biotechnology.* Ist U.P. Unified Ed. M/s Rastogi and Co., Meerut.

8. Jagannathan, V(1987). In *Biotechnology in Agriculture,* pp. 185-191, Oxford and IBH Publication Co., New Delhi.

9. Keshav Trehan (1994). *Biotechnology.* Wiley Eastern Ltd. New Delhi.

10. Khoshoo, T.N. (1988). *Environmental concerns and strategies* II Ed. pp. 253-404. Ashish Publication House, New Delhi.

11. Kodama, K. and Yoshizawa, K. (1977). In *Economic Microbiology* (ed. A.H. Rox) pp. 423-475, Academic Press, Landon.

12. Kumar, S (2011). Biotechnological advancement in alfalfa improvement. *J. Appl. Genetics,* 52 : 11-124.

13. Mitchell, R. (1979). The analysis of India Agro Eco System pp. 180. *Interprint,* New Delhi.

14. Pant, D.D. and Singh, S.P. (1987). Energy use pattern and environmental conservation. The central Himalaya case, pp. 225, *TERI,* New Delhi.

15. Peters, P (1993). *Biotechnology: A guide to genetic engineering,* Winc. Brown Publishers, Dubuwve, P. 253.

16. Purohit S.S., Kothari P.R. And Mathur S.K. (1993). *Basic and Agricultural Biotechnology.* Agro Botanical Publishers (India).

17. Rajam M.V. (1993). Artificial Seeds and the future. *Everyman's Science,* Vol. XXVIII No. 5.

18. Rajasekaran K. (1996). Regeneration of plants from cryopreserved embryonic cell suspension and callus cultures of cotton (*Gossypium hirsutum*). *Plant Cell Reports*. 15. 859-864.

19. Roth, F.X. (1982). In: *advances in Agricultural Microbiology*. pp. 663-676. Oxford and IBH Publication Co. New Delhi.

20. Sagar V. (2005). *National Conference on Environmental Biotechnology*, Bangalore.

21. Venkataraman, G.S. (1961). Studies on nitrogen fixation by blue green algae II. Nitrogen fixation by *Cylindrospermum musicola* under various conditions. *Proc. Natl. Acad. Sci.* 31 A: 160-164.

22. Venkataraman, G.S. (1975). The role of blue-green algae in tropical rice cultivation. In: *Nitrogen fixation by free-living microorganisms* (Ed.) W.D.P. Steward, Cambridge Univ. Press, pp. 207-218.

23. Watanable, A. (1962). Effect of Nitrogen fixing blue-green alga *Tolypothrix tenuis* on the nitrogenous fertility of the paddy soil and crop yield of rice plant. *J. Cen. App. Microbial.* 8 (2): 85-91.

2016, Environmental Biotechnology: A New Approach   *Pages 211–223*
Editors: **Dr. Rajan Kumar Gupta and Dr. Satya Shila Singh**
Published by: **DAYA PUBLISHING HOUSE, NEW DELHI**

Chapter 13
Effect of Heavy Metal on Fish Gills

Shista Sharma

Department of Zoology, Panjab University, Chandigarh
E-mail: shista.sharma@gmail.com

Water is one of the valuable liquid of the nature resource available. Water covers over 70.9 per cent of earth's surface and is vital to all known forms of life. It moves continually through a cycle of evaporation, transpiration, precipitation and runoff, usually reaching the sea. Water plays an important role in world economy as it functions as a solvent for many chemical substances and facilitates industrial cooling and transportation. Only 2.5 per cent of the Earth's water is freshwater, and 98.8 per cent of that water is in ice and groundwater. Less than 0.3 per cent of all freshwater is in rivers, lakes, and the atmosphere, and an even smaller amount of the Earth's freshwater (0.003 per cent) is contained within biological bodies and manufactured products (Gleick, 1993).

Origin of life is believed to be in ocean. All over the world aquatic habitats are full of life forms ranging from small animals which form the base of aquatic food chain (plankton) to the large mammoth animals (whales). Fish are the animals which spend most of their lives in water whereas amphibians spend part of their lives in water and part on land. Aquatic animals are often of special concern to conservationists because of fragility of their environments. Among aquatic animals, fishes are important resource worldwide.

Fish are a class of important aquatic vertebrates. The combination of gills, fins and the fact that they live only in the water make fishes unique creatures in the living world. They spend all of their lives in the water habitat and are cold-blooded, with the exception of Tuna family and the Mackerel shark family means they can adjust their body temperatures according to the surrounding environment.

As is given in the literature, Devonian period is called as "Age of fishes". According to Cohen (1970), approximately 22000 fish species are present worldwide. Almost similar observations were made by Jayaram (1999). If we take a look on fish species present in India, there are 2,500 living species, out of which 939 (belonging to 326 genera) inhabit inland waters of India and adjacent countries while 1,570 are marine fish species (Talwar and Jhingran, 1991; Jayaram, 1999). The National Bureau of Fish Genetic Resources Research Advisory Committee Report (NBFGR, 2008) had estimated that the total fish diversity of India is 2,534 species, including 2,243 native species and 291 exotic ones. Among them are 765 freshwater fish species, 113 brackish water fish species and 1,365 marine fish species. The numbers of species discovered so far are the indication about the rich fish diversity all over the world waters. Though number of fish species in the marine habitat are greater in number than freshwater habitat, but the diversity per unit water area is more in the freshwater habitats due to lower level of salinity than the marine water habitat.

It is believed that culture of aquatic organisms had its beginning in Asia. Chinese are the masters of fish farming dating back to 5th century. Fish farming was common in Eastern regions of the world and South East Asia in 13th and 14th century's mainly using inherited traditional farming practices evolved through generations. The Second World War is responsible for bringing the development of aquaculture in Asian countries. The post war economic reconstruction led the Asian countries to exploit their natural resources as a result there was tremendous investment of foreign money, more efficient harvesting technology and processing of fishery products.

Fishing is an occupation that provides employment to large populations directly or indirectly. Fishing is mainly of three types based on purpose. One is commercial fishing in which fishers hunt fish for profit in wild fisheries or farm them in ponds or in cages in ocean. In subsistence fishing, fishers hunt fish for survival. Third type of fishing is recreational fishing in which fish are caught, kept as pets, raised by fish keepers, and exhibited in public aquaria. Fish have had a role in culture through ages, serving as deities, religious symbols and as the subjects of art, books and movies. Fish and also their byproducts have nutritional, ornamental, medicinal and many more benefits.

Fishery sector plays an important role in national Gross Domestic Production (GDP). Many fishes like tuna, sardines, cod, rays, salmons, herrings etc, are of commercial importance. Oils extracted from fish liver are rich source of vitamins A and D. Fish skin is also put to various uses. Shagreen is the tanned skin of shark and is used as an abrasive for polishing of wood etc. Fish meat is rich source of proteins, vitamins and minerals. Fish meal is used to feed poultry animals and cattle. Many other products obtained from fishes are of economic importance.

Aquatic habitats are directly exposed to industrial effluents, domestic waste, untreated sewage and agricultural runoff which are polluting the water bodies all over the world. The natural aquatic system have been extensively contaminated with heavy metals released from domestic, industrial and other anthropogenic activities (velez and Montare, 1998; Conacher *et.al.*, 1993). Condition is worse in the metropolitan cities which support large populations. This destruction of water bodies can sway the balance aquatic food chains resulting into death of natural flora and fauna living in fragile aquatic environments.

One important thing to keep in mind is complexity of the food chains whether it is aquatic or terrestrial food chain. Clarkson (1995) studied that humans are exposed to the hazardous chemicals through these food chains. Contaminants get entry into the food chains either from soil or from water. Terrestrial food chains are under control upto certain extent as quality checks are performed regularly at each step starting from crop production to consumption. On the other hand, due to lack of quality control, aquatic food chains are more vulnerable to accumulation of the toxicants at higher rates due to lack of inspection of aquatic food productions. Ultimately, toxicants impair the entire ecosystem and its supporting organisms.

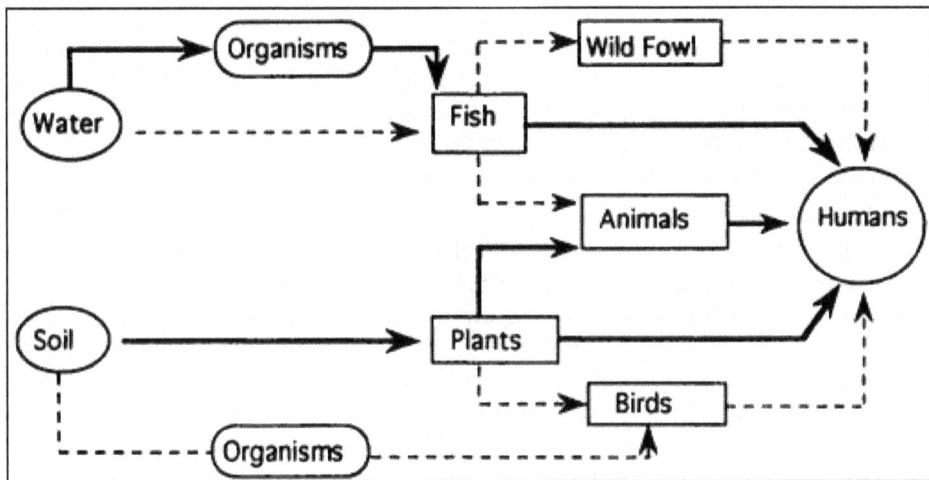

Figure 13.1: Diagrammatical Illustration of Aquatic and Terrestrial Food Chain (*Source*: Clarkson, 1995).

Nowadays, pressure on the agricultural production is high due to population explosion. During past few decades, less land and more food requirement had forced agriculture scientists to discover new ways for increasing crop production. Crops are mainly damaged by pests like group of insects, fungus and bacteria. To control the menace of these harmful agents, chemicals are used on large scale. Weedicides are also variety of chemicals used to destroy harmful plants from the agricultural fields. These pesticides and weedicides contain harmful constituents such as toxic heavy metals (zinc, arsenic, cadmium) which when come into the water bodies along with the runoff water from agricultural fields lead into suffocation of the fish and other aquatic organisms including rare or endangered ones. Dissolved oxygen level decreases, which is considered crucial for living in water. Pesticides enter into the bodies of aquatic organisms through skin by absorption, through gills by respiration and orally by drinking pesticide contaminated water.

Pesticides belong to various categories depending upon type of pest they control.

Main categories of pesticides are organophosphate pesticides, carbamate pesticides, organochlorine and pyrethroid pesticides. All of these pesticides mainly work by attacking the nervous system of pests.

Presence of heavy metals and pesticides in groundwater is found in various states of India especially in Haryana and Punjab which is great threat to fresh ground water ecosystems. Kaushik *et al.* (2010) analyzed samples of water taken from Ghaggar River whose stretch is through Punjab and Haryana and found concentrations of different pesticides like DDT, aldrin above permissible limits prescribed by European Commission Directive for drinking purpose.

Aquatic toxicology is the study of effects of environmental contaminants on aquatic organisms including effects of pesticides on health of fish or other aquatic organisms. To understand and illustrate aquatic toxicology, we take example of arsenic pesticides.

A lethal concentration is the quantity of toxicant that causes death of organism. LC_{50} is that standard toxicity dose of toxicant which kills 50 per cent of a test population of organisms within certain time period, usually 24 to 96 hrs. LC_{50} values are affected by water temperature, pH, organic contents, phosphate concentration, suspended solids and presence of other substances and toxicants as well as arsenic speciation and duration of exposure.

To understand the effect of harmful chemicals, how they metabolize in living systems and what damage they do to fish fauna, let us take general example of arsenic which is ubiquitous in natural environment. It is common element which occurs in air, water, soil and all living tissues. It ranks 20th in abundance in the earth's crust, 14th in seawater and 12th in the human body (Eisler, 1988). Most arsenic produced domestically is used in manufacture of agricultural products such as insecticides, herbicides, fungicides, algaecides, wood preservatives and growth stimulants for plants and animals. Agricultural applications provide the largest anthropogenic source of arsenic in the environment (Woolson, 1975). Inorganic arsenicals (arsenic trioxide; arsenic acid; arsenates of calcium, copper, lead, and sodium; and arsenates of sodium and potassium) have been used widely for centuries as insecticides, herbicides, algaecides and desiccants. For example Paris green (Cuprous arsenite) was successfully used in 1867 to control the Colorado potato beetle (*Leptinotarsa decemlineata*) in the eastern United States. Arsenic trioxide has been applied widely as soil sterilent. Sodium arsenite has been used to control aquatic weed, as a defoliant to kill potato vines before tuber harvest and to control crabgrass (*Digitaria sanguinalis*). Lead arsenate has been used to control insect pests of fruit trees and for many years was the only insecticide that controlled codling moth (*Carpocaspa pomonella*) in apple orchards. In recent decades, inorganic arsenicals have been replaced by organoarsenicals for herbicidal application and by carbamate and organophosphorus compounds for insect control (Woolson, 1975).

Living resources are exposed to arsenic emissions from smelters, herbicide sprays and from diet, especially from consumption of marine biota. The legal limit for As in water applied by WHO is 10μg/L. With exception of fish, most foods contain less

Table 13.1: Acute Toxicity (LC$_{50}$) of Pesticides to Fish Species (*Source*: K. Shankarmurty, B. R. Kiran and M. Venkateswarlu, 2013)

Sl.No.	Name of the Pesticides	Test Organism	Duration of Exposure	LC$_{50}$ Value	Reference
1.	Cypermethrin	*Labeo rohita*	96 hrs	4.0µl/L	Marigoudar *et al.*, 2009
2.	Methyl parathin	*Catla catla*	96 hrs	4.8ppm	Illyazhanan *et al.*, 2010
3.	Malathion	*Heteropneustes fossilis*	96 hrs	0.98ppm	Sanjoy Deka and Rita Mahanta, 2012
4.	Pyrethroid Lambela Cyhalothrin	*Danio rerio*	96 hrs	0.119µl/L	Badre Alam Ansari and Kafeel Ahmed, 2010
5.	Cypermethrin	*Colisa fasciatus*	96 hrs	0.02mg/L	Shailendra Kumar Singh *et al.*, 2010
6.	Rogor	*Puntius stigma*	96 hrs72 hrs	7.1ppm7.8ppm	Bhandare *et al.*, 2011
7.	Malathion	*Labeo rohita*	96 hrs	15mg/L	Thenmozhi *et al.*, 2011
8.	Dimethoate	*Heteropneustesfossilis*	96 hrs	2.98mg/L	Rakesh K. Pandey *et al.*, 2009
9.	Elsan	*Channa punctatus*	48 hrs	0.43ppm	Sambasiva Rao *et al.*, 2009
10.	Endosulfan	*Channa striatus*	96 hrs	0.0035ppm	Ganeshwade *et al.*, 2012
11.	Metasystox	*Nemachellus botia*	96 hrs	7.018 ppm	Nikam *et al.*, 2011
12.	Acephate	Fathead minnow	96 hrs	>1000 mg/L	Waynon Johnson and Mack Finley,1980
13.	Alaclor	Rainbow trout	96 hrs	2.4(1.8-3.1) mg/L	Waynon Johnson and Mack Finley,1980
14.	Akton	Channel catfish	96 hrs	400(295-542) µg/L	Waynon Johnson and Mack Finley,1980
15.	BHC	Gold fish	96 hrs	348(261-466) µg/L	Waynon Johnson and Mack Finley,1980
16.	Carbaryl	Lake trout	96 hrs	690(520-910) µg/L	Waynon Johnson and Mack Finley,1980
17.	Carbofuran	Yellow perch	96 hrs	147 (115-188) µg/L	Waynon Johnson and Mack Finley,1980
18.	DDT	Rainbow trout	96 hrs	8.7 (6.8-11.4) µg/L	Waynon Johnson and Mack Finley,1980
19.	Endosulfan	Channel catfish	96 hrs	1.5 (1.3-1.7) µg/L	Waynon Johnson and Mack Finley,1980

than 0.25µg/g of As. Many species of fish contain between 1 and 10µg As/gm body weight of fish. Levels of As found in bottom feeders and shellfish is 100µg As/gm body weight of fish or more than this occurring as arsenocholine and arsenobetaine (Dabeka *et al.*, 1987). In human body, total amount of As is about 0.5-15mg/kg body weight. Many arsenic compounds are excreted out unchanged in urine when man ate fish and seafood (Yamauchi and Yamamura, 1979; Freeman *et al.*, 1979; Luten *et al.*, 1982). Estimated provisional tolerable weekly intake is 0.015mg/kg body weight for human beings.

Arsenic is known more as environmental contaminant than as a nutritionally essential mineral. Low doses (less than 2µg/day) of arsenic stimulated growth and metamorphosis in tadpoles, increased viability and cocoon yield in silkworm caterpillars. Reproductive performance is adversely affected in animals like goat and pig due to deficient of Arsenic (NAS, 1977).

The adverse effects caused by As can be better, understood from its mechanism of action inside organism. Arsenic poisoning are either acute or sub-acute while cases of chronic arsenosis are rarely encountered, except in humans; early developmental stages are the most sensitive to arsenic; inorganic arsenic can traverse placental barriers, so can produce fetal deformities and even can cause its death. It is emphasized in the literature that arsenic metabolism and toxicity vary greatly between species, and that effects are significantly altered by numerous physical, chemical and biological modifiers. There is increased risk of cancers in skin, lung, liver, lymph, and hematopoietic system of humans is associated with exposure to inorganic arsenicals. These increase cancer risks mainly among smelter workers, in those engaged in production and use of arsenical pesticides (NRCC, 1978; Belton *et al.*, 1985).

Figure 13.2: Structure of Arsenic Trioxide (As_2O_3).

Fish are most important organism in the aquatic food chain, which are sensitive to heavy metals contamination. High level of heavy metals has apparent lethal as well as chronic effects on fish. Effect of pesticides and other pollutants in fish can be indicated by using tissues as biomarkers such as gills (Nowak, 1992; Johal and Dua, 1994), and scales (Johal and Dua, 1994; Sawhney and Johal, 2000) as gills and scales come in direct contact with contaminated water. Other main reason to use these biomarkers is that they show symptoms earlier than other vital organs (Richmonds and Dutta, 1989).

Gills located in the pharyngeal region as pouches on the lateral sides of the buccal cavity. Externally, they are protected by a cover called operculum. There are present four pairs of gill archs. Gill rays radiate laterally from medial base of each gill arch, and connective tissue between gill rays forms an interbranchial septum. Interbranchial septum support several rows of fleshy gill filaments called hemibranchs that run parallel on both cranial and caudal sides of gill arch.

Gill filaments are the functional unit of gills. These are long and narrow projections lateral to gill arch tapering at distal end. The surface area of filaments *i.e.* primary lamellae is further increased by folding into secondary lamellae where gas

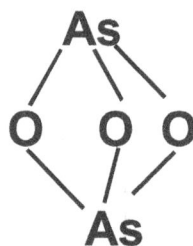

exchange takes place between water and blood (Schottle, 1931).The secondary lamellae are arranged on both sides of the filaments.

Diffusive exchange of respiratory gases or other molecules occurs through thin shelf like respiratory lamellae. Water enters pharynx from mouth, then passes over filaments and follows the inner wall of operculum until it exists via a caudal opening of operculum.

Gill epithelium covers gill filaments and lamellae provide distinct boundary between fish's external environment and extracellular fluids. At least two general types of epithelium have been described on the gill surface, one on the body of the filament and other on the lamellae (Laurent and Dunel, 1980; Laurent, 1984). Gill epithelium is composed of several distinct cell types, primarily consists of pavement cells, chloride cells (ionocytes or mitochondria rich cells) and mucous cells. Other cells such as rodlets, taste buds and merkel cells are also observed, these are rarely seen.

Gills are important for performing homeostatic functions that include respiration, osmoregulation, and acid-balance and nitrogenous waste excretion (Health, 1987). These all functions are occurring as trans-epithelial exchange between internal side of gills and external environment. When fish are exposed to environmental pollutants, these vital functions are deleteriously affected due to functional impairment of gills and ultimately can damage the death of fish (Aluzemi *et al.*, 1994, Munshi, 1993, Kumar and Tembhre, 2010).

Toxicants in the environment mainly enter into fish by means of their respiratory system (Tovell *et al.*, 1975).Fish gills are also vulnerable to pollutants in waste water because of their large surface area and external location. For this reason fish gills are considered to be the most appropriate indicators of water pollution level (Aluzemi et.al, 1996). Various fish species are frequently used as bio-indicators of heavy metal contaminations. The concentration of heavy metals in organ of fish tells about the extent upto which the aquatic environment is polluted. Heavy metal concentration in the aquatic organism depicts the past as well as the current pollution load in the environment in which the organism lives.

The large surface area and fine sieve like structure of gills make them particularly susceptible to continuous exposure to waterborne noxious agents (Lichtenfels *et al.*, 1996). Gills are good indicator of water quality (Rankin *et al.*, 1982), being model for studies of environmental impact (Bonga and Lock, 1992).

The acidic condition of aquatic environment might cause free divalent ions of many heavy metals to be absorbed by fish gills. Uptake of heavy metals and other trace elements occur through food ingestion and water via gills though accumulation of these elements may vary with the route of uptake and also with the intensity and duration of exposure (Cairdullo el al., 2008). As gills occupy a position which is in immediate contact with water, therefore the study of the impact of toxicant on the gills assumes significance. Toxicants lead into changes in morphology of the epithelial cells of gills which affect process of diffusion of gases and ultimately deteriorate overall health of fish. Temmink *et al.* (1983) suggested the direct deleterious effects on the epithelium of gills by the toxicant.

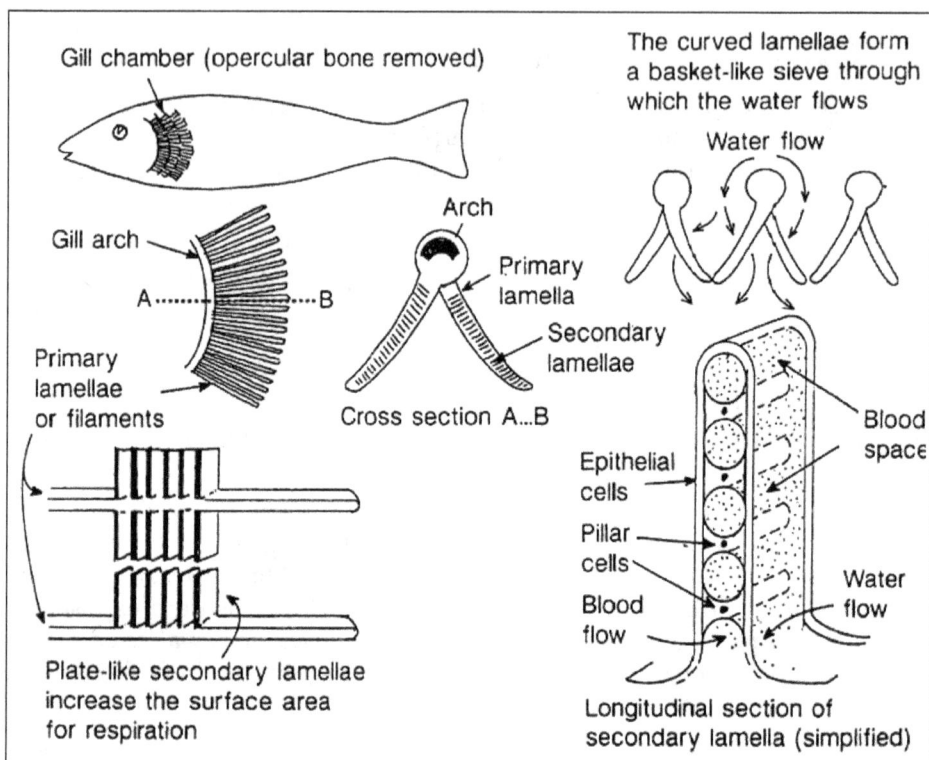

Figure 13.3: Structure of Fish Gills (from Lloyd, 1992).

On exposure to toxicant, epithelial cells of gill filament lead to its shrinkage, degeneration of microridges, inflammation in pavement cells. Sometimes gill lamellae get fused and distorted at higher toxicant concentrations. These type of alterations has also been reported on exposure of various fish species to stress conditions such as heavy metals (Oliveria-Ribeiro *et al.*, 1994; Athikesavan *et al.*, 2006), salinity (Fanta *et al.*, 1995), pesticides (Sawhney and Johal, 2000; Machado, 2003).

Hyperplasia in the epithelial cells is also reported during exposure of test fish to toxicant. Hyperplasia is a defensive mechanism as it reduces toxic agent penetration by increasing distance between blood and polluted water (Morgan and Tovell, 1973; Mallat, 1985).this condition has been reported by Prasad (1991) on the gills of *Anabas testiduneus* on exposure to crude oil. Abel (1976), Mallat (1985), Tamse *et al.* (1995); Manoj and Ragothaman (1999) observed lifting of epithelium and lamellar fusion in gills of fish which are defensive mechanisms. Bhatnagar *et al.* (1992) reported clubbed appearance of the tip of secondary lamellae. Possible reason for this was lamellar hyperplasia in which cells were derived from the primary filament and migrate to the distal end resulting into accumulation of the cells at the leading edge of the secondary lamellae. Necrosis, loss, degeneration and altered structure of microridges create problem in delivery of oxygen to gill epithelium. Loss of typical microridges pattern had also been reported in various fishes by different workers (Sawhney and Johal, 2000; Ba-Omar *et al.*, 2011).

In nutshell, toxicants cause much damage to the fish health. Lethal concentrations for various toxicants have been calculated so far in limited fish species. Their level should not exceed the legal limit. In addition to this, use of toxicants should be minimized and alternative eco-friendly methods should be implemented in agriculture and other industries in place of hazardous substances.

References

1. Abel, P. D. 1976. Toxic action of several lethal concentrations of an anionic detergent on the gills of brown trout (*Salmo trutta* L.). *J. Fish. Biol.*, **9**: 441-446.

2. Alazemi B. M., Lewis, J. W. and Andrews E. B. 1996. Gill damage in the freshwater fish *Gnathonemus ptersii* (Family: Mormyridae) exposed to selected pollutants: an ultra structural study. *Environ Technol.* **17**: 225-238.

3. Ansari, B. A. and Ahmad, M. K. 2010. Toxicity of synthetic pyrethroid Lambda-cyhalothrin and neem based pesticide Neem gold on Zebra fish *Danio rerio* (Cyprinidae). *Global Journal of Environmental Research*, **4(3)**: 151-154.

4. Athikesavan, S., Vincent, T., Ambrose, T. and Velmurugan, B. 2006. Nickel induced histopathological changes in the different tissues of freshwater fish, *Hypopthalmicthys molitrix* (Valenciennes). *J. Environ. Biol.*, **27**: 391-395.

5. Belton, J. C., Benson, N. C., Hanna, M. L. And Taylor, R.T. 1985. Growth inhibitory and cytotoxic effects of three arsenic compounds on cultured Chinese hamster ovary cells.*J. Environ. Sci. Health*, **20(10)**: 37-72.

6. Bhandare R.Y, T.S. Pathan, S.E. Shinde, P.R. More and D.L. Sonawane, 2011. Toxicity and behavioural changes in fresh water fish *Puntius stigma* exposed to pesticide (Rogor). *Am-Euras. J. Toxicol. Sci.*, **3(3)**: 149-152.

7. Bhatnagar, M. C., Banarjee, A. K. and Tyagi, M. 1992. Respiratory distress to *Clarias batrachus* (Linn.) exposed to endosulfan. A Histological Approach. *J. Environ. Biol.*, **13(3)**: 227-231.

8. Bonga, S. E. W. and Lock, R. A. C. 1991.Toxicants and osmoregulation in fish.*Netherlands J. Zoo.*, **42(2-3)**: 478-493.

9. Ciardullo, S., Aureli, F., Coni, E., Guandalini, E., Iosi, F., Raggi, A., Rufo, G. and Cubadda, F. 2008. Bioaccumulation potential of dietary arsenic, cadmium, lead, mercury and selenium in organs and tissues of rainbow trout (*Oreochromis mykiss*) as a function of fish growth. *J. Agri. Food Chem.*, **56**: 2442-2451.

10. Clarkson, T.W. 1995. Environmemtal contaminants in the food chain. *Am. J. Clin. Nutr.*, **61:** 682-686.

11. Conacher, H. B., Page, B. D. and Ryan, J. J. 1993. Industrial chemical contamination of foods [Review]. *Food Addit. Contam.* **10**(1): 129-143.

12. Cohen, D. M. 1970. How many recent fishes are there? *Proc. Calif. Acad. Sci.*, **38(17)**: 738-749.

13. Dabeka, R. W., Mckenzie, D. and Lacroix, G. M. A. 1987. Dietary intake of lead, cadmium, arsenic and fluorides by Canadian adults: A 24-hours duplicate diet study. *Food Addi. Contam.*, **4(1)**: 89-101.

14. Deka, S. and Mahanta, R. 2012. A Study on the Effect of Organophosphorus Pesticide Malathion on Hepato-Renal and Reproductive Organs of *Heteropneustes fossilis* (Bloch). *The Science Probe*, **1(1)**: 1-13.

15. Dunel-Erb, S. and Laurent, P. 1980. Ultrastructure of marine teleost gill epithelia: SEM and TEM study of the chloride cell apical membrane. *J. Morphol.*, **165**: 175-186.

16. Eisler, R. 1988. Arsenic hazards to Fish, Wildlife and Invertebrates: A Synaptic Review. Published in Biological Reports 85, *Contam.Haz. Rev.*, Report No. 12.

17. Fanta, E., Luzivotto, M. F. and Meyer, A. P. 1995. Gill structure of Antartic fishes *Notothenia* (*Gobionotothen*) *gibberifrons* and *Trematomus newnesi* (Nototheniidae) stressed by salinity changes and some behavioral consequences. *Antartic Record* (*Nankyoku shiryo*), **39**: 25-39.

18. Farkas, A., Salanki, J. and Varnka. 2000. Heavy metal concentration in Fish and lack Balaton, Lakes and Reservoirs. *Research and management*, **5(4)**: 271-279.

19. Freeman, H. C., Uthe, J. F., Fleming, R.B., Odense, P. H., Ackman, R. G., Landry, G. and Musial, C. 1979. Clearance of arsenic ingested by man from arsenic contaminated fish. *Bull. Environ. Contam.Toxicol.*, **22(1)**: 224-229.

20. Ganeshwade R.M., Dama L.B., Deshmukh D.R., Ghanbahadur A. G.and Sonawane S.R.2012. Toxicity of endosulfan on freshwater fish *Channa striatus*. *Trends in Fisheries Research*, **1(1)**: 29- 31.

21. Gleick, P.H., ed. (1993). *Water in Crisis: A Guide to the World's Freshwater Resources.* Oxford University Press. pp. 13, Table 2.1 *"Water reserves on the earth"*.

22. Gohil M.N. and Mankodi P.C., Diversity of Fish Fauna from Downstream Zone of River Mahisagar, Gujarat State, India. *Res. J. Animal, Veterinary and Fishery Sci.*,**1(3)**: 14-15 (2013).

23. Health A.G. (1987). *Water pollution and fish physiology.* CRC Press, Inc., Florida.

24. Jayaram, K. C. 1999. *The Freshwater Fishes of Indian Region* (eds. K. C. Jayaram.). Narendera Publishing House, New Delhi, India.

25. Johal, M. S. and Dua, A. 1994. Alterations in the architecture of gill surface of *Channa punctatus* produced by endosulfan treated water: SEM study. *J. Aquat. Biol. Fish.*, **1(1)**: 27-31.

26. Kotze, P., du Preez, H. H. and Van Vuren, J. H. 1999. Bioaccumulation of Copper and Zinc ion *Areochnomis* contamination and *Clarius gariepinus* from the alifants river, Mpumalanga, south Africa, Watson S. A., **25(1)**: 99-110.

27. Laurent, P. 1984. Gill internal morphology. In: Hoar, W, Randall, D. J. (eds) Fish physiology. Academic Press, New York, pp. 73–183.

28. Lichtenfels, A. J. F. C., Lorenzi-Filho, G., Guimaraes, E. T., Macchione, M. and Saldiva, P. H. N. 1996. Effect of water pollution on the gill apparatus of fish. *J. Comp. Pathol.*, **115(1)**: 47-60.

29. Lloyd R. (1992). *Pollution and Freshwater Fish.* Fishing News Books, Oxford, U. K., pp. 192.

30. Luten, J. B., Booy, G. R. and Rauchbaar, A. 1982.Occurrence of arsenic in plaice (*Pleuronectesplatessa*), nature of organoarsenic compounds present and its excretion by man. *Environ. Health Perspect.*, **45**: 165-170.

31. M. Ilavazhahan Selvi, R. T. and Jayaraj, S.S. 2010. Determination of LC50 of the Bacterial Pathogen, Pesticide and Heavy Metal for the Fingerling of Freshwater Fish *Catla catla. Global Journal of Environmental Research*, **4 (2)**: 76-82, 2010

32. Machado, M. R. 2003. Effects of the organophosphorus methyl-parathion on the branchial epithelium of the freshwater fish, *Metynnisroosevelti. Braz. Arch. Biol. Technol.*, **46**: 1-20.

33. Mallat, J. 1985. Fish gill structural changes induced by toxicants and other irritants: A 388 stastical review. *Can. J. Fish. Aquat. Sci.*, **42**: 630-648.

34. Manoj, K. and Ragothaman, G. 1999. Effect of sub-lethal concentrations of cadmium on the gills of an estuarine edible fish *Boleopthalmus dissumieri* (Curv.).*Poll. Res.*, **18(2)**: 145-148.

35. Marigoudar, S. R., Ahmed, R. N., David, M. 2009. Cypermethrin induced respiratory and behavioural responses in *Labeo rohita. Vet. Arhiv.*, **79**: 583-590.

36. Morgan, M. and Towell, P. W. A. 1973.The structure of the gill of trout, *Salmo gairdneri* (Richardson).*Z. Zellfo. Micros. Anat.*, **142**: 147-162.

37. NAS (1977). Arsenic. National Academy of Sciences, Washington DC, 332 pages.

38. N.B.F.G.R. 2008. Research Advisory Report, submitted to Research Advisory Committee, 28-29 Feb., 2008.

39. Nikam S. M., Shejule, K. B. and Patil, R. B. 2011. Study of acute toxicity of metasystox on the fresh water fish *Nemacheilus botia* from Kedrai dam in Maharashtra, India. *Biology and Medicine*, **3(4)**: 13-17.

40. Nowak, B. 1992. Histological changes in gills induced by residues of endosulfan. *Aquat.Toxicol.*, **23(1)**: 65-83.

41. Oliveria-Ribeiro, C. A., Turcati, N. M, Carvalho, C. S., Cardoso, R. and Fanta, E. 1994. Efeitotoxico dos acrosbranquiais de *Trichomycterus brasilensis* (Pisces, Siluridei). Ann. II SimpososobreMeioambiente Univ., Salgado de Oliveira, Niteroi, Brasil.

42. Pandey, Govind, Overviews on diversity of fish, Res. J. Animal, Veterinary and FisherySci.**1(8)**:12-18(2013).

43. Pandey Govind and Madhuri S., Heavy Metals Causing Toxicity in Animals and Fishes, Res. J. Animal, Veterinary and Fishery Sci., Vol. 2(2), 17-23, February (2014).

44. Pandey, R. K., Singh, R. N., Singh, S., Singh, N. N. and. Das, V. K. 2009. Acute toxicity bioassay of dimethoate on freshwater airbreathing catfish,*Heteropneustes fossilis* (Bloch). *Journal of Environmental biology*, **30(3)**: 437-440.

45. Part, P., Svanberg, O. and Kiessling, A. 1985. The Availability of Cadmium to Perfused Rainbow Trout gills in Different Water Qualities. *Water Research*, **19(2)**: 427-434.

46. Prasad, M. S. 1991. SEM study on the effects of crude oil on the gills and air breathing organs of climbing perch, Anabas testudineus. *Bull. Environ. Contam.Toxicol.*, **47**: 882-889.

47. Rankin, J. C., Stagg, R. M. and Bolis, L. 1982. *Effects of Pollutants on Gills. In*: Gills pp. 207-220 (eds: D. F. Houlihan, J. C. Rankin and T. J. Shuttleworth). Cambridge University Press, London.

48. Rao, S., Babu, K. S. and Rao, K. V. R. 2009. Toxicity of Elsan to the Indian snakehead *Channa punctatus*. (Bloch). Ind. J. Fish., **32**: 153-158.

49. Ravera, R. C. beone, G. M., Dantas, M. and Lodigiani, P. 2003. Trace Element Concentration in fresh water Mussels and Macrophytes as Related to t\Those in Their Environment, *Journal of Limnology*, **62(1)**: 61-70.

50. Richmonds, C. and Dutta, H. M. 1989. Histopathological changes induced by malathion in the gills of blue gill *Lepomis macrochirus*. *Bulletin of Environmental Contamination and Toxicology*, **43(1)**: 123-130.

51. Sawhney, A. K. and Johal, M. S. 2000. Effects of an organophosphorus insecticide, malathion, on pavement cells of the gill epithelia of *Channa punctatus* (Bloch). *Pol. Arch. Hydrobiol.*, **47(2)**: 195-203.

52. Schöttle, E. 1931. Morphologie und Physiologie der Atmung bei Wasser-, Schlamm-, und Landlebenden Gobiiformes. *Z. Wiss. Zool.*, **140**: 1–114.

53. Shankarmurthy, K., Kiran, B. R. and Venkateswarlu, M. 2013 A review on toxicity of pesticides in Fish, *International Journal of open Scientific research*, **1**: 15-36.

54. Singh, S. K., Singh, S. K., and Yadav, R. P. 2010. Toxicological and Biochemical Alterations of Cypermethrin (Synthetic Pyrethroids) Against Freshwater Teleost Fish *Colisa fasciatus* at Different Season. *World Journal of Zoology*, **5(1)**: 25-32.

55. Svobodova, Z., Celechovska, O., Kolara, J., Randak, T. and Zlabek, V. 2004. Assessment of Metal Contamination in the Upper Reaches of the Ticha Orlice River. *Czech Journal of Animal Science*, **49(4)**: 458-641.

56. Talwar, P. K. and Jhingran, A. G. 1991.*Inland Fishes of India and Adjacent Countries*, Vols.I and II. Oxford and IBH Publishing Co., New Delhi.

57. Tamse, C. T., Gacutan, R.Q. and Tamse, A. F. 1995. Changes induced in the gills of milkfish (*Chanos chanos*) fingerlings after acute exposure to nifurpirinol (Furanace: P-7138). *Bull. Environ. Contam.Toxicol.*, **54**: 591-596.

58. Temmink, J., Bowmeister, P., Jong, P. and Berg, J. V. 1983. An ultrastructural study of chromate induced hyperplasia in the gills of rainbow trout (*Salmo gairdneri*). *Aquat.Toxicol.*, **4**: 165-179.

59. Thenmozhi, C., V. Vignesh, R. Thirumurugan, S. Arun. 2011. Impacts of malathion on mortality and biochemical changes of freshwater fish *Labeo rohita. Iran. J. Environ. Health. Sci. Eng.* **8(4)**: 387-394.

60. Tovell, P.W.A., Howes, D and Newsome, C.S. (1975). Absorption, metabolism and excretion by gold fish of the anionic detergent, Sodium lauryl sulphate. *Toxicol.***6**: 17-29.

61. Velez, D. and Montoro, R.1998. Arsenic speciation in manufactured seafood products: a review. *J. Food Protect.* **61(9)**: 1240-1245.

62. Waynon W. Johnson and Mack T. Finley,1980. Handbook of acute toxicity of chemicals to fish and aquatic invertebrates. United states Department of the interior fish and wild life service/Resource Publication 137, Washington, DC.

63. Woolson, E. A. *Arsenical pesticides.* Vol. 7. Washington, DC: American Chemical Society, 1975.

64. Yamauchi, H., Kaise, T. and Yamamura, Y. 1986. Metabolism and excretion of orally administered arsenobetaine in the hamster. *Bull. Environ. Contam. Toxicol.,* **36(1)**: 350-355.

2016, Environmental Biotechnology: A New Approach   *Pages 225–232*
Editors: Dr. Rajan Kumar Gupta and Dr. Satya Shila Singh
Published by: DAYA PUBLISHING HOUSE, NEW DELHI

Chapter 14

Plant Litter Decomposition, Humus Formation and Carbon Sequestration

P. Ghosh

G.B. Pant Institute of Himalayan Environment and Development,
Kosi-Katarmal, Almora – 263643, Uttarakhand
E-mail: paroghosh@rediffmail.com

ABSTRACT

Litter decomposition is a process that leads to formation of humus layer that grow over millennia at a considerable rate and sequester carbon and nitrogen. Decomposition and photosynthesis are processes that account for a huge majority of the biological carbon processing on earth. Decomposition accounts for the transformation of nearly as much carbon as does photosynthesis however it occurs mainly on or below ground. The biochemistry of decomposition is very irregular and less understood when compared to the biochemistry of photosynthesis. It is also responsible for the formation of humic substances that contribute to soil fertility as well as the long term storage of carbon. The soil represent a major sink for carbon. Decomposition of plant litter is responsible for huge amount of carbon dioxide returned to the atmosphere. Disturbances such as fire, harvesting for forestry or agriculture and cultivation can clearly reduce soil organic matter content. Thus an understanding of factors influencing the amount of humus formed and the stability of that humus are also important in predicting global atmospheric carbon budget. Therefore the present chapter explains the role of decomposition in carbon sequestration in ecosystems and various environmental factors and their interactions affecting decomposition.

Keywords: *Carbon sequestration, Climate change, Decomposition, Factors affecting decomposition, Plant-soil interaction.*

Introduction

There are many chemical, physical and biological factors that affect decomposition at different time scales. It is not known with conviction if changes in temperature due to increase in concentration of CO_2 will lead to increase or decrease in storage of carbon in ecosystem. As one of the largest C pools in ecosystem organic matter plays an important role. In terrestrial ecosystems, turnover times of organic matter are relatively long, approximately 26 years is the estimated global average (Schlesinger, 1991). The turnover time of different soil organic carbon (SOC) pools varies from a few years for fresh litter to millennia for the most stable organic matter. Little is known as to how increase in temperature and CO_2 will affect conversion of labile C into stable C. Therefore in the present chapter we discuss the role of decomposition in carbon sequestration and then the factors and their interacting effects on decomposition.

Carbon Storage in Soil

There are five global carbon pools, of which the soil is the third largest after the oceanic and the geological pools. Globally, soils contain 2500 billion tons of carbon, and together with vegetation, hold 2.7 times more carbon than the atmosphere. Most soil carbon is in the form of soil organic matter which is mostly found at the soil surface. This organic matter is made up of dead plant and animal material at various stages of decomposition, substances synthesized by microbes and/or chemically from the breakdown products of decomposition, and living microorganisms and animals. Soils also contain inorganic carbon in the form elemental carbon and carbonate minerals, such as calcite, dolomite and gypsum.

Soils vary tremendously in the amount of organic matter they contain, with soil carbon contents varying due to a variety of factors such as vegetation type, climate, parent material, drainage, and the activity and diversity of soil biota. In general, however, the amount of carbon in soil is determined by the balance between carbon input from plant growth, in the form of dead plant litter (roots and shoots) and root exudates, and output via decomposition processes, burning and soil erosion. Hence, any factor that limits the amount of organic matter entering soil, such as a decline in plant productivity, or its breakdown in soil by the soil food web, will cause a build-up of organic matter in soil.

The largest soil carbon stores are found in extensive peat lands of boreal and tundra zone, which hold one third of the global soil carbon stock. Other important soil carbon stores are in tropical forests, savannas, and temperate grasslands and forests. Agricultural soils, which cover around 37 per cent of Earth's land surface, also contain large amounts of carbon. But, in general, agricultural activity, and especially cultivation, causes a loss of soil carbon, and as a result, cultivated soils tend to have lower soil carbon contents than unmanaged soils, such as those of native grassland and forest.

Soil Carbon and Climate Change

There is presently much concern that climate change will enhance the decomposition of soil carbon, potentially shifting soils from being sinks to sources of carbon dioxide and thereby accelerating climate change. At the same time, there is also much debate about the possibility to increase the amount of carbon sequestered in soil from the atmosphere, and hence mitigate climate change. Recent studies reveal that both the loss and gain of soil carbon are strongly regulated by plant-microbial-soil interactions. Climate change can impact on soil carbon in many ways, both direct and indirect. For direct effects, one of the most commonly discussed ideas is that global warming will accelerate rates of heterotrophic microbial activity, thereby increasing the transfer of carbon dioxide from soil to the atmosphere, thus creating a positive feedback on climate change. However, although it is well known that temperature is an important determinant of rates of organic matter decomposition, the nature of the relationship between temperature and heterotrophic respiration, and its potential to feedback to climate change are far from clear (Davidson and Janssen 2006).

Recent studies also reveal the potential for strong indirect effects of climate change on soil carbon cycling, *i.e.* responses mediated via plants. Several mechanisms occur here which can broadly be divided into two: first, as mentioned above, rising atmospheric concentrations of carbon dioxide indirectly impact on soil microbes via increased plant photosynthesis and transfer of photosynthetic carbon to soil and second, long-term climate change-induced, changes in vegetation composition alter the amount and quality of organic matter entering soil, and other soil properties, thereby affecting the belowground decomposer food web. Recent studies show that both of these mechanisms can have significant consequences for the carbon budget of terrestrial ecosystems under climate change (reviewed by Bardgett 2011).

Plant-Soil Interactions and Climate Mitigation

As mentioned in the preceeding section, a major challenge facing scientists and policy makers is to increase the amount of carbon sequestered in soil in order to mitigate climate change. A range of land management strategies based on the intervention of higher plants and soil decomposition processes have been proposed to enhance carbon pools in agricultural soils, Figure 14.1 (Woodward *et al.,* 2009). These include: the adoption of no-tillage arable agriculture, which minimizes soil disturbance and the breakdown of crop residues; the conversion of arable land to grassland, which causes a build-up of organic matter at the soil surface; and the use of cover crops in rotations. In nutrient poor situations the addition of fertilizer nitrogen has also been proposed as a way to enhance soil carbon storage due to the resulting increase in plant production and litter return to soil, and through suppressing microbial decomposition of recalcitrant organic matter. Evidence for this, however, is mixed, in that nitrogen fertilization of agricultural soils has been shown, in some situations, to enhance organic matter decomposition. Also, some of the above strategies for soil carbon storage could have trade-offs; for example, no till agriculture has been found, in some situations, to increase soil emissions of the greenhouse gas N_2O - as a result of increased denitrification in compacted soils - thereby offsetting some of the benefits of increased soil carbon storage.

Another possible way to enhance soil carbon sequestration involves the manipulation of plant-soil feedbacks, especially in grassland. For example, recent

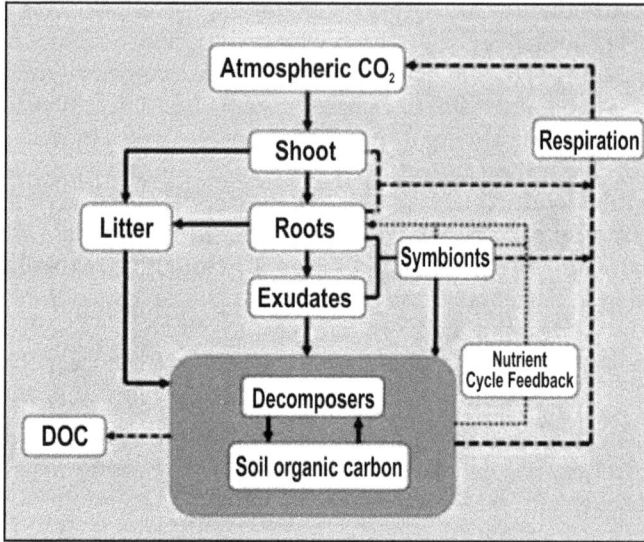

Figure 14.1: The Amount of Carbon Stored in Soil Depends on the Balance Between: (1) carbon inputs (solid lines) from plants (dead and decaying shoot and root plant tissue, and root exudates) and manures; and (2) outputs (thick dashed lines) from the respiration of roots, their symbionts (*e.g.*, nitrogen-fixing bacteria and mycorrhizal fungi), free-living soil decomposer organisms (bacteria, fungi and fauna), and from soil erosion and the leaching of dissolved organic carbon. Decomposer organisms and symbionts also influence nutrient supply to plants, thereby creating a feedback on plant productivity and hence carbon input to soil (small dashed line). Many factors determine the rate that organic carbon inputs to soil are decomposed and hence lost from soil, including the chemical composition of the organic matter, soil temperature and moisture, the abundance and activity of soil biota, and the availability of nutrients such as nitrogen. Although not shown here, soil animals (*e.g.* earthworms) can also promote soil carbon sequestration by redistributing carbon through the soil profile by channelling, mixing organic and mineral soil components, and by forming relatively stable soil aggregates and casts. DOC = dissolved organic carbon. (Taken from Woodward *et al.*, 2009. Biological approaches to global environment change mitigation and remediation. Current Biology, 19, R615-R623.).

studies show that increases in plant diversity and the introduction of certain plant species such as legumes into mixed grasslands, can reap benefits for soil carbon sequestration. The mechanisms involved in plant manipulation of soil carbon sequestration are complex and involve many types of biotic interactions between plants, their symbionts, and decomposer organisms. However, it is becoming clear that a research effort focusing on plant traits (and especially of roots) offers a potential way forward for understanding how plant-microbial-soil interactions might be manipulated to enhance soil carbon storage.

Factors Affecting Decomposition

The rates of decomposition, humification and dissolved organic carbon formation determine how much carbon is sequestered *i.e.* stored in soils and sediments and thereby protected against further decomposition. Sequestration can take place either by stimulating the conversion of labile into stable organic matter or by lowering the decomposition rate.

Chemical Composition of Organic Matter

Several chemical parameters influence the decomposition rate of soil organic matter. Among these the N content or C/N ratio is often found to be a major factor in controlling the decomposition rate. Low nitrogen content generally result in slow decomposition rate (Campbell, 1978). High lignin content retards decomposition (Spain and le Feuvre, 1987) and rate of N - mineralization (Berendse *et al.*, 1989). Initially the decomposition rate is limited by nutrient level of substrate and availability of easily degradable carbon content in the next level lignin content determines the decomposition rate.

Elevated atmospheric CO_2 concentration may decrease the N content and increase the lignin content of plant litter which in turn may affect its decomposability (Van Breemen and Van Dam, 1992).

Soil Chemistry

In acidic soils Al is known to be important in stabilizing soil organic matter (Blaser and Klemmedson, 1987). Aluminium may decrease the decomposition rate of soil organic matter due to higher chemical stability of Aluminium-organic matter complexes (Brunner and Blaser, 1989). Bases and trace elements influence decomposition when concentrations are below certain minima that vary with the element and the decomposer community. Lower base saturation and lower pH depress decomposition.

Nutrient Availability

Effects of nutrients, especially nitrogen on decomposition are best known for decomposition of litter. Studies about effects of nutrients on stabilized organic matter are fragmentary. Adding mineral nitrogen to soil does not generally speed up decomposition (Ross, 1990) however nitrogen input to ecosystems influences litter quality resulting in indirect effects of N — input on decomposition. Net N — immobilization occurs if N — immobilization by microbes exceeds gross N mineralization. If the microbial C/N increases due to low N availability decomposition will slow down if gross growth of the changing population remains N — limited.

Texture and Mineralogy

The impacts of clay minerals on soil organic carbon content is widely recognized. Sometimes interactions between organic compounds and clay particles might be of greater importance in protecting organic matter than the recalcitrant nature of certain specific compounds (Stout *et al.*, 1981). Stabilization of soil organic matter can be divided into three processes 1. adsorption 2. complexation and 3. physical protection. In soils derived from volcanic parent material, mineralogy is known to have a large impact on C accumulation/decomposition (Parfitt, 1990).

Temperature

An increase in temperature leads to an increase in decomposition rate by stimulating microbiological activity. Generally the largest increase in decomposition rate occurs in the temperature range 5 - 10°C. The optimum temperature for most decomposition processes is around 20-25°C (Swift *et al.*, 1979). If the temperature is below the optimum temperature, an increase in temperature will result in a faster decomposition of fresh litter. The decomposition rates of fresh litter are more strongly affected by temperature than more stable soil organic matter.

Moisture

Moisture can affect decomposition rates in several ways. At high moisture content gas diffusion is hampered and supply of O_2 for respiration may become limiting. Anaerobic decomposition is usually slower and results in a large amounts of poorly decomposed organic matter. Drought decreases decomposition rates by limiting microbial activity. Additionally in dry soil mobility of microbes decreases so less substrate is physically available for soil biota (Elliot and Coleman, 1988). In the top soil moisture does not appear to be a limiting factor for decomposition. Due to the high organic matter content of the top soil substrate is easily accessible to the soil fauna. In the subsoil however the organic matter content is much lower and will not be divided evenly throughout the soil matrix. An increase in moisture content could increase the mobility of the microbes causing colonization of more substrate. Increase in soil moisture disturbs aggregates. Organic matter present inside the aggregates are then available for decomposition. The high decomposition rate of the organic matter in the subsoil is most likely related to the presence of root mat which causes a substantial input of fresh root litter. In most temperate areas seasonal variation in moisture occur and according to season response of C mineralization to drying out might be reversed. In summer increased drought might limit decomposition whereas in winter better aeration might favour decomposition.

Oxygen and Carbon Dioxide Concentration

Aerobic processes take place only if sufficient O_2 is available. Usually aerobic decomposition rates decrease significantly if O_2 concentration drops below 10 per cent (Schachtschabel *et al.*, 1984). In neutral and slightly acidic soils an increase in CO_2 could cause a drop in pH due to dissociation of H_2CO_3. Usually microbial processes are sensitive to changes in pH. An increase in atmospheric CO_2 concentration is not likely to have an effect on processes taking place in the subsoil since the concentration of CO_2 is usually much higher (0.1 to 10 KPa) than the atmospheric concentration (0.03 KPa). In general CO_2 will have an indirect effect on decomposition by influencing the chemical composition of the substrate.

Soil Faunal/Soil Structure Interactions with Organic Matter Dynamics

Soil ingesting animals play a role both in aggregate formation and in litter break down (Martin and Marinissen, 1992). Faunal casts are enriched in organic matter and stable macro-aggregates are formed. Macroaggregates are stabilized by particulate organic matter acting as transient binding agents (Tisdall and Oades, 1982). Stable macro- aggregates protect organic matter against decomposition. Within the aggregates pore size is smaller than between aggregates leading to limitations in oxygen transport. The larger the aggregate size the larger the volume of aggregates material that may be oxygen limited, depending on the oxygen consumption rate of the material (Rappoldt, 1990). Another reason might be that due to small pore size the organic matter might not be accessible to soil organisms (Elliot and Coleman, 1988). In stable aggregates more clay-organic matter complexes may be present in which the organic matter is protected against decomposition (Foster, 1985).

Conclusion

Soil carbon can be stored in pools with different turnover times. In terms of C sequestration it is most desirable to fix atmospheric C in pools having long turnover times. Mechanisms based on C storage in live biomass or litter will sequester C only for a very short time and are therefore not relevant as long term C sequestration strategies. Turnover times of C pools increase if C mineralization is slowed down or if the input of carbon is increased. The main mechanism by which to sequester C in soils is to more or less permanently fix organic matter so that it is not available for decomposition. Flooding of aerated soils is likely to decrease decomposition rates but this might increase emissions of methane. A very effective mechanism by which C can be fixed in soils is by charcoal formation upon burning. Decomposition is controlled by many factors, most factors may have interactions and therefore separation of these factors is not easy. The physical, chemical and biological conditions also show tremendous spatial variability. Therefore it is needed to evaluate C sequestration strategies on a regional scale. The amount of C present in soils does not only depend on the output rates through decomposition but also the input rates through litter production.

References

1. Bardgett RD (2011). Plant–soil interactions in a changing world. *F1000 Biology Reports* 2011, 3: 16 (doi: 10.3410/B3-16).

2. Berendse F, Bobbink R, Rouwenhorst G (1989). A comparative study on nutrient cycling in wet heathland ecosystems. II. Litter decomposition and nutrient mineralization. *Oecologia* 78: 338-348.

3. Blaser P, Klemmedson JO (1987). Die Bedeutung von hohen Aluminiumgehalten für die Humusanreicherung in sauren Waldböden. Z. *Pflanzenernähr Bodenk* 150: 334-341.

4. Brunner W, Blaser P (1989). Mineralization of soil organic matter and added carbon substrate in two acidic soils with high non-exchangable aluminium. *Z Pflanzennernähr Bodenk* 152: 367-372.

5. Campbell CA (1978). Soil organic carbon, nitrogen and fertility. *In:* Schnitzer M, Kahn SU (Eds.). *Soil organic matter.* Elsevier, Amsterdam.

6. Davidson EA Janssens IA (2006). Temperature sensitivity of soil carbon decomposition and feedbacks to climate change. *Nature,* 440: 165–173.

7. Elliott ET, Coleman DC (1988). Let the soil work for us. *Ecological Bulletin* 39: 23-32.

8. Foster RC (1985). In situ localization of organic matter in soils. *Quaestions Entomologicae* 21: 609-633.

9. Martin A, Marinissen JCY (1992). Biological and physico-chemical processes in excrements of soil animals. *Geoderma* 56: 331-347.

10. Parfitt RL (1990). Allophane in New Zealand- A review. *Australian Journal of Soil Research* 28: 343-360.

11. Rappoldt C (1990). The application of diffusion models to an aggregated soil. *Soil Science* 150: 645-661.

12. Ross DJ (1990). Influence of soil mineral nitrogen content on soil respiratory activity and measurements of microbial carbon and nitrogen by fumigation-incubation procedures. *Australian Journal of Soil Research* 28: 311-321.

13. Schachtschabel P, Blume HP, Hartge KH, Schwertmann U (1984). *Lehrbuch der Bodenkunde*. Ferdinand Enke Verlag. Stuttgart.

14. Schlesinger WH (1991). *Biogeochemistry: an analysis of global change*. Academic Press, San Diego.

15. Spain AV, Le Feuvre RP (1987). Breakdown of four litters of contrasting quality in a tropical Australian rainforest. *Journal Applied Ecology* 24: 278-288.

16. Stout JD, Goh KM, Rafter TA (1981). Chemistry and turnover of naturally occurring resistant organic compounds in soil. In: Paul EA, Ladd JN (Eds.). *Soil Biochemistry*. Vol. 5. Marcel Dekker, New York.

17. Swift MJ, Heal OW, Anderson JM (1979). *Decomposition in terrestrial ecosystems*. Blackwell, Oxford.

18. Tisdall JM, Oades JM (1982). Organic matter and water-stable aggregates in soils. *Journal of Soil Science* 33: 141-163.

19. Van Breemen N, Van Dan D (1992). Studying effects of elavated CO_2 on decomposition of soil organic matter. In: Schulze ED, Mooney HA (Eds.). Design and execution of experiments on CO_2 enrichment. *Communication of European Communities*, Brussels.

20. Woodward FI, Bardgett RD, Raven JA, Hetherington AM (2009). Biological approaches to global environmental change mitigation and remediation. *Current Biology*, 19, R615-R623.

2016, Environmental Biotechnology: A New Approach   *Pages 233–244*
Editors: **Dr. Rajan Kumar Gupta and Dr. Satya Shila Singh**
Published by: **DAYA PUBLISHING HOUSE, NEW DELHI**

Chapter 15

Production of Cellulases using Agriculture Waste: Application in Biofuels Production

Neha Srivastava and Pranita Jaiswal

Central Institute of Post-Harvest Engineering and Technology (CIPHET),
Ludhiana – 141 004, Punjab
E-mail: pranitajaiswal@gmail.com, sri.neha10may@gmail.com

ABSTRACT

Lignocellulosic biomass is a potential source available for production of many value added products like biofuel. Currently slow enzymatic hydrolysis of lignocellulosic biomass is the major limiting factor. Further high cost of physical, chemical, biological pretreatment operations make the whole process less economical in comparison to presently available fossil fuels. Cellulases group of enzymes, which can be classified into many types depending on the reactions they catalyze, including endoglucanase (EG) or carboxymethyl cellulase (CMCase), exoglucanase or cellobiohydrolase (CBH), cellobiase or β-glucosidase. Auxilliary enzymes like β-xylosidase, α-L-arabinofuranosidase and feruloyl esterase if present along with xylanases help in complete conversion of hemicellulose to sugars like xylose and arabinose. The combined action of these enzymes plays an important role in the hydrolysis of cellulosic and hemicellulosic fractions. Hence it is desirable to have a consortium of all the cellulase and xylanase components for effective hydrolysis of cellulosic biomass. Present chapter covers effective production of cellulase using agriculture biomass and application of this cellulase in waste management. Limitations and possible approaches of enhancing the production of enzymes have also been discussed.

Keywords: *Bioethanol, Lignocellulosic biomass, Cellulases, Agriculture waste, Hydrolysis, Sugar, waste management.*

Introduction

Cellulases are enzymes used for the hydrolysis of cellulose into glucose. Cellulase hydrolyzes the β-1, 4-d-glucan linkages in cellulose structure and release glucose, cellobiose and cello-oligosaccharides. This is the most widely studied enzyme complex comprising of endo-glucanases (EG), cellobiohydrolases (CBH) and β-glucosidases (BGL). Endo-glucanases liberates nicks in the polymeric structure of cellulose exposing reducing and non-reducing ends whereas cellobiohydrolases produces cello-oligosaccharides and cellobiose units by acting on reducing and non-reducing ends and furthermore, β-glucosidases cleaves the cellobiose to release glucose molecules for the completion of the hydrolysis (Srivastava *et al.*, 2014). Therefore the complete cellulase system include the combined action of CBH, EG and BGL components which react synergistically to convert cellulose into glucose.

Cellulases which back large proportion in the global market of industrial enzymes show its significance as an important enzyme class in the market. At present, cellulase is the third biggest industrial enzymes in the form of dollar volume (Srivastava *et al.*, 2014) and at the same time cover approximately 20 per cent of the total enzyme market all over the world (Srivastava *et al.*, 2014). This huge demand of cellulase is due to its wide spectrum of applications in biofules production, pulp and paper, detergent, textile, food and beverages, as well as animal feed industries (Rawat *et al.*, 2014). It has been predicted earlier that the demand of cellulase will be highly driven by the commercial production of biofuel. In the field of biofules production (weather it is biohydrogen or bioethanol) cellulases are used for sugar production from cellulose. Presently, cellulases are produced using biological route such as bacterial or fungal fermentation. There are a wide range of microorganisms capable of producing cellulase such as aerobic and anaerobic bacteria, anaerobic fungi, soft rot fungi, white rot fungi (WRF) and brown rot fungi (BRF) (Rawat *et al.*, 2014). Most of the fungi are able to produce a complete cellulase system as compared to bacteria (Srivastava *et al.*, 2014). The commercial cellulase is most commonly produced from two strains of soft rot fungi (SRF), namely *Trichoderma reesei* and *Aspergillus niger*, via submerged fermentation (Rawat *et al.*, 2014). Among these two types of fungus, *T. reesei* is not capable to produce substantial amount of β-glucosidase, meanwhile endoglucanase and exoglucanase are found to be lacking in the cellulase system of *A. niger* (Rawat *et al.*, 2014). Besides that, submerged fermentation suffers from a major drawback that is associated with the low concentration of end products (Yoon *et al.*, 2014), and thus, further puriûcation is needed. The additional downstream processes required for the submerged fermentation contribute to higher cost of cellulase production. Therefore, cost effective production of cellulase is required from lignocellulosic biomass. Lignocellulosics are bioresources of enormous importance being convertible to various value added products including biofuels, chemicals (Krik *et al.*, 2002) and cheap energy sources for fermentation and improved animal feeds and nutrients. Lignocellulosic agricultural biomass has been used as substrate for the production of second generation biofuels. Lignocellulosic biomass is a complex composing of cellulose, hemicellulose and lignin and the conversion efficiency of lignocellulosic biomass into biofules depends upon the lignin content and degree of polymerization (DP) in cellulose and hemicellulose (Oberoi *et al.*, 2012). Cellulose and hemicellulose are regarded as the potential sources of sugars for second generation biofuel production and cover around two-third of the lignocellulosic biomass (Chandra *et al.*, 2010). Different pretreatment methods are applied to increase the accessibility of cellulosic substrate which helps to open the lignin sheath (Alvira *et al.*, 2010). Lignocellulose-degrading enzymes, such as cellulases and hemicellulases are used to release fermentable sugars after pretreatment of biomass. Figure 15.1 shows various applications of cellulases using lignocellulosic waste as substrate and generation of value added products.

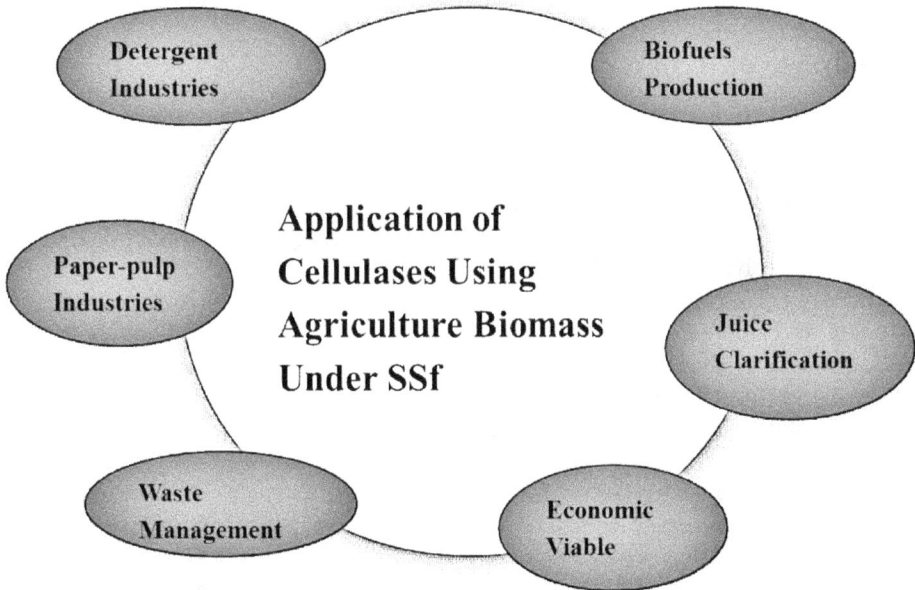

Figure 15.1: Various Application Cellulases in different Field.

Therefore, the current chapter focuses on production of cellulase enzymes using agriculture waste and its application in biofuels production.

Cellulase: Current Market Status and Challenges

Immense applications of cellulases have led to increase its demand day by day. There are numerous companies involved in the commercial production of cellulase. Presntly, "Genencor" and "Novozyme" are commonly known for commercial application for biomass conversion and have contributed in signicant manner to bring down the cost of cellulase by endorsing novel technologies as the part of their active research. Genencor has launched number of series of cellulase name Accelerase®1500, Accelerase®1000a for lignocellulosic biomass processing industries (Singhania *et al.*, 2010). Among these Accelerase®1500 is a cellulase complex obtained from genetically modified *Trichoderma reesei* strain. Genencor has also lofted Accelerase® XC which is a xylanase/cellulase enzyme complex that accommodate a broad prole of hemicellulase and cellulase activities along with enhances both xylan (C5) and glucan (C6) hydrlysis when fused with other

Accelerase® enzyme (Novozyme website, 2010). Novozymes also have a wide range of cellulase preparations such as Cellusoft®AP and Cellusoft®CR depending on the application as for different industries like bioblasting in textile mills, Carezyme® and Celluclean in detergent, Denimax® 6011 for stonewash industry (Genencor website, 2010). Amano Enzyme Inc. in Japan and MAP's India in India are also known enzyme industries actively involved in commercial production of cellulase. Although, most of the cellulase enzyme producing companies are indulged in production and marketing of cellulases but the production cost is still very high and seems to be a major bottleneck during the overall production of biofules which need to be resolved.

Mechanism of Action of Cellulase

The enzymatic hydrolysis of cellulosic polymer needs the synergistic action of three types of enzymes name: endoglucanase (1,4-β-D-glucan-4-glucanohydrolase; EC 3.2.1.4), exoglucanase (1,4-β-D-glucan cellobiohydrolase; cellobiohydrolase; EC 3.2.1.91) as well as β-glucosidase (β-glucosideglucohydrolase; cellobiase; EC 3.2.1.21) as mentioned previously in the beginning of this chapter. Endoglucanases start cellulose hydrolysis by cleaving the cellulose polymer exposing the two, reducing and non-reducing ends, whereas cellobiohydrolases act upon these producing reducing and non-reducing ends to release cello-oligosaccharides and cellobiose units, as well as β-glucosidases cleave the cellobiose to release glucose, hence completing the enzymatic hydrolysis process (Bhat M.K., 2000). Further, endoglucanases are classified into endo-acting cellulases because they cleave β-1,4-glycosidic bonds internally. They are frequently active on the amorphous region with more solubility of the cellulose crystal, enhance the concentration of chain ends and dramatically decrease degree of polymerization by attacking interior region of cellulose molecules (Schulein, 2000). Furthermore, cellobiohydrolases are classified as exo-acting cellulases because they cleave β-1,4-glycosidic bonds from chain ends position with tunnel-shaped closed active site which retains a single glucan chain and avoid it from re-joining to the cellulose crystal (Zhu *et al.*, 2009). They are generally active on the crystalline site of cellulose; shorten degree of polymerization by binding to the chain ends liberating specifically cellobiose units. Therefore, endoglucanase activity is responsible for chemical changes in solid structure cellulose that occur during the reaction of hydrolysis, but at the same time perform little role in solubilization related to exoglucanase, whereas exoglucanase activity is known to be specifically responsible for solubilization but show a minor role in observe any change in the chemical properties of residual cellulose.

Lignocellulosic Biomass as Substrate for Cellulase Production

Lignocellulosic biomass are the most abundant and promising feedstock for biofules industries. These materials as a waste products are being generated via agricultural practices specifically from different agriculture based industries (Perez *et al.*, 2002). Being the natural and renewable resources of energy, these biomasses are the main focus of modern industrial societies. A large amount of these lignocellulosic biomasses is usually disposed of by burning which contribute to environmental pollution and is not confined to developing countries only. Therefore, the huge amount of lignocellulosic biomass can potentially be converted into different high value products including cellulase production, sugar generation for bio-fuels production, value added product such as ne chemicals, and cheap energy sources for microbial fermentation and enzyme production (Asgher *et al.*, 2013; Iqbal *et al.*, 2013).

Physico-chemical Characteristics of Lignocellulosic Biomass

All lignocellulosic biomass composed of three major units name, cellulose, hemicellulose and lignin. Lignocellulosic biomass included agricultural wastes, forestry residues, grasses and woody materials which have to be very potential for bio-fuel production. Generally, majority of the agricultural lignocellulosic biomass contains around 10-25 per cent lignin, 20-30 per cent hemi- cellulose, and 40-50 per cent cellulose (Iqbal *et al.*, 2011). Cellulose is regarded as the major part of plant cell walls and have complex structure while, hemicellulose are repeated polymers of pentoses and hexoses. On the other hand, lignin forms a protective seal around the other two components *i.e.*, cellulose and hemicelluloses (Menon and Rao, 2012). The structural composition of lignocelluloses mainly depend on source from where they are obtained such as from the hardwood, softwood, or grasses. To obtain exact percentage of each proportion in structure of lignocelluloses, compositional analysis can be done by using NREL (National Renewable Energy Laboratory) protocol.

Physical Structure of Cellulose, Hemicelluloses and Lignin

Cellulose is a complex polymer having glucose and attached with linear chains up to 12,000 residues. Cellulose is mainly composed of (1,4)-D-glucopyranose units, which are bonded by b-1,4 linkages having molecular weight of around 100,000 (Himmel *et al.*, 2007). Generally, plant biomass is composed of 40-50 per cent of cellulose molecules, held together via intermolecular hydrogen bonds in native state, but they have tendency to form intra-molecular and intermolecular hydrogen bonds which provide stiffness to cellulose structure and hence it is highly insoluble and resistant to most organic solvents. In natural condition, cellulose molecules are found in the form of bundles which combined to form crystalline and amorphous regions of cellulose structure (Iqbal *et al.*, 2011).

Biotechnological Importance of Lignocellulosic Biomass for Cellulase Production

Several variety of lignocellulosic materials are available to use as potential candidate for the production of cellulase, other value added products and biofules production. From the last few years considerable attention have been given to the development of green biotechnology related to lignocellulosic materials. The randomly increasing costs of fossil fuels along with their harmful greenhouse effects are demanding to explore new pathway of alternative cheaper and eco-friendly bio-fuels resources. The issue of global warming can resolved using this agriculture waste and simultaneously producing cellulase for biofules production. Utilization of agriculture waste for cellulase as well as sugar production for generation of biofules can be proved as one of the potential alternative method (Lin and Tanaka, 2006). Moreover, the conversion of lignocellulosic waste into value added products such as bio-fuel production normally requires a multi-step processing that include (i) Cellulase production (ii) pre-treatment (mechanical, chemical, or biological etc) of cellulose (iii) enzymatic hydrolysis (iv) fermentation process (Xiao *et al.*, 2012). Figure 15.2 illustrating overall process of biofuels production with waste mangment.

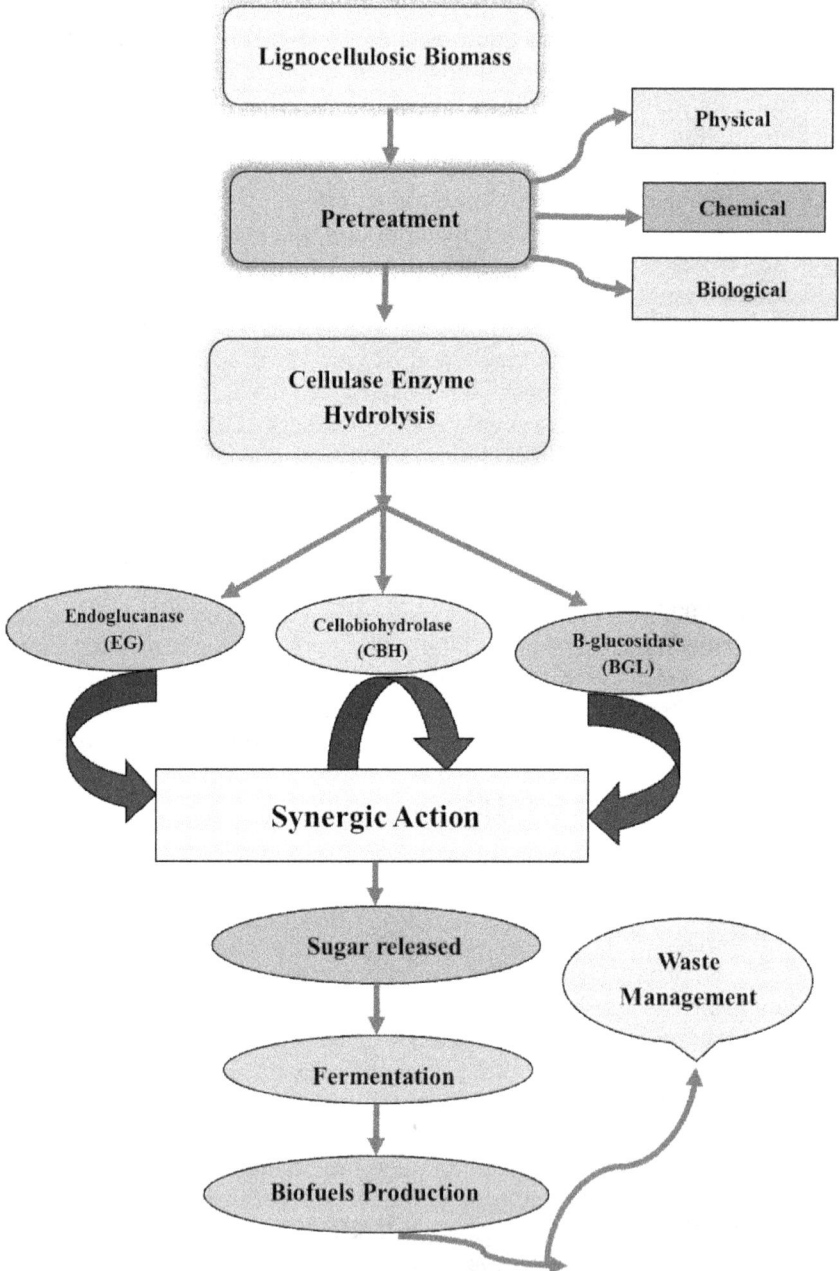

Figure 15.2: Process of Biofules Production using Agriculture Waste.

Application of Cellulases

The significant industrial importance of cellulases mainly belongs to the field of bioenergy development, textile, and detergent and cellulosic pulp paper industries. Presently, in the available industrial processes, cellulolytic enzymes are used in: (i) clarification of fruit juices and wines; (ii) detergents causing colour brightening and softening; (iii) pretreatment of biomass to improve nutritional quality of forage; and (iv) pretreatment of agro-industrial wastes (Haki and Rakshit, 2003, Bhat, 2000). The various applications of cellulase in different industries are as follow:

Cellulase for Biofuels Production

Now days, lignocellulosic waste material conversion into fermentable sugar is one of the main challenging area for biofuels production. Additionally, the lignocellulosic biomass conversion has gained more advantages when thermostable cellulases are employed during hydrolysis with low activity losses at higher temperature. Thermostable cellulases are known for retaining their half-life at their optimum incubating temperature. Some studies have shown that a long reaction time is required for complete hydrolysis when cellulases do not have this property (Barakat *et al.*, 2012). The use of thermostable enzymes like those produced from *T. aurantiacus* help in improving the hydrolysis process. In the study of Ang *et al.* (2013), optimum hydrolysis temperature of 60°C was reported by using thermostable enzyme obtained from fungus *Aspergillus fumigates* SK1. Thermostability of cellulasee can be improved in presence of nanoparticles. In one of very recent study by Srivastava *et al.* (2014), themostability of cellulase was enhanced in presence of nickel-cobaltite nanoparticle. The same authors were also reported improved production of cellulase in presence of nickel-cobaltite nanoparticle with improved sugar productivity in fungus *Aspergillus fumigates* NS (Class-*Eurotiomycetes* sp.).

Cellulase for Paper Pulp Industries

High-quality paper pulps are obtained by chemical pulping processes (kraft cooking being the most common). The target of which is to depolymerize and dissolve the lignin acting as a glue between wood bers (Fengal *et al.*, 1984). Because of its recalcitrant nature, a certain amount of residual lignin remains in pulp and, due to its oxidative alteration during cooking, it is responsible for the dark color of pulps. Cellulase enzymes are eco-friendly alternatives approach over harsh chemicals to bleach and remove dark color. The analysis of pulp lignins is one of the important aspects for development of clean industrial processes. Enzymatic isolation using cellulolytic enzymes (Horting *et al.*, 1990) represents an attractive alternative to chemical isolation of residual lignin [20], which is generally based on pulp treatment with dioxane–water–HCl (acidolysis).

Use of Cellulase in Extraction of Fruit and Vegetable Juices

The production of fruit as well as vegetable juices is important from the human health and commercial point of view. The production of fruit and vegetable juices requires methods for extraction, clarification and stabilization. In the production process of juice from fruits like apples and pears, the complete fruits are crushed to pulp mash, which, after mechanical processing (pressing, centrifuging and filtering), yields a clear fruit juice and a solid phase called pomace (Galante *et al.*, 1998b). Use of cellulase and pectinase enzymes increases both yield and performance of the process without additional capital investment. These enzymes are commonly used in two steps: (1) after crushing, to macerate the fruit pulp either to achieve partial or complete liquifaction. This increases the juice yield and reduces the processing time, and (2) just after the juice extraction, thereby pectinases are utilized for its clarification, therefore lowering the viscosity of fruit juice prior to concentration and increasing the filtration rate and stability of the final product. Therefore, the cellulase enzymes play an important role in food biotechnology.

Use of Cellulase in Laundry

The cellulase enzymes are able to modifying the structure of cellulose fibrils and therefore are added to laundry detergents to improve the colour brightness, hand feel and dirt removal from cotton and cotton blend garments. Almost all cotton or cotton blend garments become dull and fade during repeated washings. This is because of the presence of partially detached microfibrils on the surface of garments that can be removed by cellulases in order to keep the garment surface smooth. This is presently accomplished by adding a commercial cellulase preparation from *H. insolens*, active under mild alkaline conditions (pH 8.5–9.0) (Uhlig, 1998). However, the amount of cellulase added in detergent represents around 0.4 per cent of the total detergent cost, it is considered rather expensive and therefore, alternative cellulase production using lignocellulosic waste can be a potential approach for production of inexpensive cellulase.

Improvement in Cellulase Production Using Combination of Agro-Industrial Waste and Recombinant Cellulolytic Microorganisms

Modern techniques such as genetic and protein engineering are now days employed for the production of cellulase enzyme on commercial scale for improving stability of enzyme at high temperatures, extremes of pH, oxidizing agents as well as in organic solvents. Cloning and expression of genomic information present in a thermophile in a faster growing mesophilic host may provid possibilities for the production of specific thermostable enzyme (Haki and Rakshit, 2003). Thermophiles microbes have active protein structure at high temperature and therefore are capable of doing hydrolysis of lignocellulosic biomass at high temperature. Most of the thermophilic microorganisms produce low amount of cellulases at high temperatures, considering this mesophilic organism can be a potential option for expression of cellulases. For example, *Geobacillus* spp. has been reported to produce 0.0113 U/ml at 60°C (Tai *et al.*, 2004) and *Geobacillus* gene have been successfully expressed in *E. Coli* (Ng *et al.*, 2009). The recombinant xylanases with optimal activity at 70°C having same activity profiles between 20 and 90°C (Wu *et al.*, 2006) has been reported. Endoglucanases from *Thermoanaerobacter tengcongensis* MB4 (Liang *et al.*, 2011) has also been expressed in *E. coli*. Although yeasts are not natural host of cellulase production, there are increasing interests at industrial level, day by day. Some yeast such as *Pichia pastoris* has also been used for the production of recombinant cellulase production (Lindenmuth and McDonald, 2011; Sriyapai *et al.*, 2011). In a study made by Sriyapai *et al.* (2011), the authors reported that thermostable xylanses from *Actinomadura* sp. S14 became more thermostable when they express in *Pichia pastoris* was stable for 2 h at 80°C. According to Gao *et al.* (2012), at extreme condtition such as pH and temperature, post-translational modifications like protein folding, glycosylation appear when they express in mesophilic expression systems. Separation and purification of thermostable enzymes are easy when they overexpress in a heterologous host by using affinity tags, like His-Tags (Bai *et al.*, 2010), which make their production economical viable using agriculture waste from the industrial point of you.

Conclusion

The remarkable progress has been made in the production and development of cellulases using agro-industrial waste. Although cellulase enzymes are advantageous for solid state fermentation but for industrial scale processes, final selection depends on the many factors such as substrate type, energy consumption for overall process, production cost, enzyme efficiency and environmental issues of the complete process of lignocelluloses biomass conversion. Based on this, there is need of more research for development and optimization of cellulase production processes using lignocellulosic biomass and its conversion to gain an economical industrial biofuel production comparable to existing processes. This chapter may be useful to addresressing key factors related to cellulase production and utilization of agriculture waste. Reduction in environmental related problem associated with such wastes can be resolved when these wastes are adopted as substrate for cellulase and biofules production and additionally, this may also help in dropping the cost of enzyme production.

Acknowledgements

Authors thankfully acknowledge the financial assistance received from the National Fund for Basic, Strategic and Frontier Application Research in Agriculture (NFBSFARA), Indian Council of Agricultural Research (ICAR), and Government of India for conducting this study.

References

1. Ang SK, Shaza EM, Adibah Y, Suraini A, Madihah MS (2013). Production of cellulases and xylanase by *Aspergillus fumigatus* SK1using untreated oil palm trunk through solid state fermentation. *Process Biochemistry* **48**: 1293-1302.

2. Asgher M, Ahmad Z, and Iqbal H M N (2013). Alkali and enzymatic delignication of sugarcane bagasse to expose cellulose polymers for saccharication and bio-ethanol production. *Industrial Crops and Products* **44**: 488-495.

3. Bai Y, Wang J, Zhang Z, Yang P, Shi P, Luo H, Meng K, Huang H, Yao B (2010). A new xylanase from thermoacidophilic *Alicyclobacillus* sp. A4 with broad-range pH activity and pH stability. *Journal of Industrial Microbiology Biotechnology* **37**: 187–194.

4. Barakat A, Monlau F, Steyer JP, Carrere H (2012). Effect of lignin-derived and furan compounds found in lignocellulosic hydrolysates on biomethane production. *Bioresource Technology* **104:** 90–99.

5. Bhat MK (2000). Cellulases and related enzymes in biotechnology. *Biotechnology Advance* **18:** 355–83.

6. Fengel D, Wegener G. Berlin: De Gruyter (1984). *Wood: chemistry, ultrastructure, reactions.*

7. Gao L, Gao F, Wang L, Geng C, Chi L, Zhao J, Qu Y (2012). N-glycoform diversity of cellobiohydrolase I from *Penicillium decumbens* and synergism of non hydrolytic glycoform in cellulose degradation. *Journal Biochemical Engineering* **287:** 15906–15915.

8. Genencor website: dated 01/14/2010 http://www.genencor.com/wps/wcm/connect/genencor/genencor/products and services/business development/bioreneries/products/accellerase product line en.htm.

9. Haki GD, Rakshit SK. (2003). Developments in industrially important thermostable enzymes: a review. *Bioresource Technology* **89:** 17-34.

10. Himmel M E, Ding S Y, Johnson D K, Adney W S, Nimlos M R, Brady J W (2007). Biomass recalcitrance: engineering plants and enzymes for biofuel production. *Science*, 315, 804-807.

11. Hortling B, Ranua M, Sundquist J (1990). Investigation of the residual lignin in chemical pulps. Part 1. Enzymatic hydrolysis of the pulps and fractionation of the products. Nordic *Pulp Paper Research Journal*, **5:** 33–7.

12. Iqbal H M N, Ahmed I., Zia M A, and Irfan M (2011). Purication and characterization of the kinetic parameters of cellulase produced from wheat straw by *Trichoderma viride* under SSF and its detergent compatibility. *Advances in Bioscience and Biotechnology* 2(3), 149-156.

13. Iqbal H M N, Kyazze G, and Keshavarz T (2013). Advances in valorization of lignocellulosic materials by bio-technology: an overview. *BioResources*, 8(2), 3157-3176.

14. Kirk O, Borchert TV, Fuglsang CC (2002). Industrial enzyme applications. Current Openion in *Biotechnology* **13(4):** 345-51.

15. Li Wan Yoon, Teck Nam Ang, Gek Cheng Ngoh, Adeline Seak May Chua (2014). Fungal solid-state fermentation and various methods of enhancement in cellulase production. *Biomass and bioenergy* 67, 319-338.

16. Liang C, Xue Y, Fioroni M, Rodríguez-Ropero F, Zhou C, Schwaneberg U, Ma Y (2011). Cloning and characterization of a thermostable and halo-tolerant endoglucanase from *Thermoanaerobacter tengcongensis* MB4. *Applied Microbiology Biotechnology* 89, 315–326.

17. Lin Y, and Tanaka S (2006). Ethanol fermentation from biomass resources: Current state and prospects. *Applied Microbiology and Biotechnology* 69, 627-642.

18. Lindenmuth BE, McDonald KA (2011). Production and characterization of Acidothermus cellulolyticus endoglucanase in *Pichia pastoris*. *Protein Exprerimental Purification* 77: 153–158.

19. Menon V, and Rao M (2012). Trends in bioconversion of lignocellulose: biofuels, platform chemicals and biorenery concept. *Progress in Energy and Combustion Science* **38(4):** 522-550.

20. Ng IS, Li CW, Yeh YF, Chen PT, Chir JL, Ma CH, Yu SM, Ho TH, Tong CG (2009). A novel endo-glucanase from the thermophilic bacterium *Geobacillus* sp. 70PC53 with high activity and stability over a broad range of temperatures. *Extremophiles* **13:** 425–435.

21. Pandey A, Srivastava N, Sinha P (2012). Optimization of photofermentattive hydrogen production *Rhodobacter sphaeroides* NMBL-01. *Biomass and Bioenergy*, 1-6.

22. Perez J, Munoz-Dorado de la Rubia T., and Martýnez, J (2002). Biodegradation and biological treatments of cellulose, hemicellulose and lignin: an overview. *International Microbiology* **5:** 53-63.

23. Rawat R, Srivastava N, Chadha B S and Oberoi HS (2014). Generating Fermentable Sugars from Rice Straw Using Functionally Active Cellulolytic Enzymes from *Aspergillus niger* HO. *Energy and Fuel* dx.doi.org/10.1021/ef500891g Energy and Fuel | XXXX, XXX, XXX–XXX.

24. Rawat R, Srivastava N and Oberoi HS (2014). Endoglucanse characterization and its role in bioconversion of cellulosic biomass. *Recent Advances in Bioenergy Research* Vol. III.

25. Schulein M (2000). Protein engineering of cellulases. *Biochim Biophys Acta-Protein Structural Molecular Enzymology* 1543(2): 239–52.

26. Singhaniaa R R, Sukumarana R K, Patel A K, Larroche C, Pandey A (2010). Advancement and comparative proles in the production technologies using solid-state and submerged fermentation for microbial cellulases. *Enzyme and Microbial Technology* 46, 541–549 http: //www.bioenergy.novozymes.com/cellulosic-biofuel/01/14/2010.

27. Srivastava N, Rawat R, Sharma R, Oberoi H S, Srivastava M, Singh J (2014). Applied Biochemistry and Biotechnology DOI 10.1007/s12010-014-0940-0.

28. Srivastava N, Rawat R, Oberoi H S, Ramteke P W (2014). Fuel Ethanol Production from Lignocellulosic biomass. *International Journal of Green energy*, DOI: 10.1080/ 15435075.2014.890104.

29. Srivatsava N, Rawat R and Oberoi H S (2014). Application of thermostable cellulase in bioethanpl production from lignocellulosic waste. *Recent Advances in Bioenergy Research* Vol. III.

30. Sriyapai T, Somyoonsap P, Matsui K, Kawai F, Chansiri K (2011). Cloning of a thermostable xylanase from *Actinomadura* sp. S14 and its expression in

Escherichia coli and *Pichia pastoris. Journal of Biosciences Bioengineering* 111, 528–536.

31. Tai SK, Lin HP, Kuo J, Liu JK (2004). Isolation and characterization of a cellulolytic *Geobacillus thermoleovorans* T4 strain from sugar refinery wastewater. *Extremophiles* 8, 345–349.

32. Uhlig H (1998). *Industrial enzymes and their applications*, New York: John Wiley and Sons, Inc., pp. 435.

33. Wu S, Liu B, Zhang, X (2006). Characterization of a recombinant thermostable xylanase from deep-sea thermophilic *Geobacillus* sp. MT-1 in East Pacific. *Applied Microbiology Biotechnology* 72, 1210–1216.

34. Zhu JY, Pan XJ, Wang GS, Gleisner R (2009). Sulte pretreatment (SPORL). for robust enzymatic saccharication of spruce and red pine. *Bioresource Technology* 100(8): 2411–8.

2016, Environmental Biotechnology: A New Approach   Pages 245–263
Editors: Dr. Rajan Kumar Gupta and Dr. Satya Shila Singh
Published by: DAYA PUBLISHING HOUSE, NEW DELHI

Chapter 16

Impact of Light Climate on Periphyton and Autotrophic Index in Ganga River, India

Usha Pandey[1] and Jitendra Pandey[2]

[1]*Department of Botany, Faculty of Science and Technology,*
Mahatma Gandhi Kashividyapith University, Varanasi – 221 002, U.P.
E-mail: usha_pandey28@yahoo.co.in
[2]*Laboratory of Trans- Boundary Research on Ganges Basin and Climate*
Change Drivers, Environmental Science Division,
Centre of Advanced Study in Botany, Banaras Hindu University,
Varanasi – 221 005, U.P.
E-mail: jiten_pandey@rediffmail.com

ABSTRACT

Periphytic algal communities are important indicators of ecological conditions in lotic ecosystems. Among the principal drivers, hydraulic characteristics, nutrients and light intensity are major determinants of periphyton biomass accrual in river ecosystems. Effective light penetration in such water bodies are constrained by the trade- off between organic loading and benthic oxygen supply. The objective of this study was to investigate the impact of light climate, as influenced by dissolved organic carbon (DOC) and phytoplankton mediated shading effect, on periphyton biomass accrual in Ganga River during summer low flows. Periphyton chlorophyll a biomass decreased with increasing phytoplankton growth and concentration of DOC. The autotrophic index (AI) showed an opposite trend. Periphyton biomass showed negative correlation with DOC ($R^2 = 0.94$; $p < 0.001$) and phytoplankton biomass ($R^2 = 0.92$; $p < 0.001$) and positive correlation with depth of light penetration ($R^2 = 0.95$; $p < 0.001$). Chlorophyta contributed 24- 60 per cent of total standing stock with higher share at sites characterized by moderate nutrients and DOC. Cyanophyta (39- 74 per cent) contributed the larger fraction at nutrient rich sites and Bacillariophyta (2- 5 per cent) at sites with moderate nutrient concentrations. Filamentous alga Phormidium (Cyanophyceae) appeared dominant at sites characterized by high concentrations of nutrients and DOC. The study show that light limitation associated with enhanced concentration of DOC and phytoplankton mediated shading effect in River Ganga constrain the growth of benthic primary producers and alter the dominance of taxonomic divisions of periphyton which may, in long run, compromise the organic load assimilation capacity of the river and entail a shift in its trophic cascades.

Keywords: Climate change drivers, Dissolved organic carbon, Ganges River, Light climate, Periphyton.

Introduction

Changing land use pattern together with atmospheric deposition and direct and indirect release of urban- industrial wastes along the 2525 km long course of Ganga River from Gangotri in Himalaya to its confluence with Bay of Bengal add massive amount of nutrients, carbon, heavy metals and other pollutants in the river (Pandey *et al.*, 2009; Pandey *et al.*, 2013; 2014a; 2014b). Reduction of nutrients and organic loading is considered as a key measure in reducing eutrophication and associated problems in Ganga River (Pandey *et al.*, 2014a). Initiatives taken under Ganga Action Plan (GAP), which were targeted to reduce nutrient and organic loading, although delivered some improvements, the water quality of river is continuing to deteriorate. Rapidly changing water quality of the river and associated shift in its self assimilation capacity (capacity to assimilate organic load) necessitate the need for action plan towards long- term management efforts to restore the ecological status of this major river system of India.

The soils of the Indo- Gangetic plains are highly fertile. Intensive agriculture covering 73.44 per cent of total basin area renders the soil prone to erosion loss of carbon and nutrients. Nutrient enrichment driven by point- and non- point sources promotes algal growth and, as a long- term effect, causes a potential shift in ecosystem functions and trophic cascades in aquatic ecosystems. In lotic ecosystems, flow velocity and light climate of the channel also regulate algal biomass accumulation. For instance, decreased flow velocity increases the residence time of phytoplankton and thus, provide more time for these organisms to grow (Hilton *et al.*, 2006). Further, the hydrological disturbances also affect the causal relationships between resource availability (nutrient and light) and biomass accrual in surface waters (Lohman *et al.*, 1992). Accordingly, the effect of resource availability on algal growth in rivers remains generally more apparent during periods of summer low flows characterized by minimum hydrological disturbances. In tropics nutrient limitation is the primary driver of phytoplankton productivity, whereas, the biomass accrual rate of epilithic periphyton is more effectively limited by light intensity than nutrient status (Carey *et al.*, 2007; Yang *et al.*, 2008; Pandey and Pandey, 2013). More importantly, periphetic algae have access to sediment- associated nutrients and regulate their availability to phytoplankton and, phytoplankton attenuates light, limiting the periphyton growth (Vadeboncoeur *et al.*, 2003). This has relevance assessing phytoplankton- periphyton causal relationships in human impacted rivers.

Algal periphyton, the submerged micro- floral community living attached to the substrate, are important group of primary producers in rivers and streams. Periphytic growth although highly variable in rivers, help signaling eutrophication (Chetelat *et al.*, 1999), supporting higher trophic levels, maintaining river ecology and reducing the release of nutrients and greenhouse gases from the bottom sediments (Luijn *et al.*, 1995; Flury *et al.*, 2010). The wide- spread reliance of fishes and zoobenthos on carbon fixed by benthic algae cue the relative importance of different primary producers functional to aquatic food web (Bootsma *et al.*, 1996; Vadeboncoeur *et al.*, 2003). However, increased water column turbidity driven by excessive growth of phytoplankton and/or increasingly high levels of dissolved organic carbon (DOC) shift surface waters from clear to turbid states attenuating water column light penetration to constrain periphyton growth (Pandey and Pandey, 2013). Among the substances of organic origin, colored humic substances of terrestrial origin, being very optic dense, can more effectively attenuate water column light penetration (Karlsson *et al.*, 2009) and consequently, the periphyton growth. The DOC enrichment in surface waters is a derived character regulated by cross- domain causations. Anthropogenic activities in the catchment enhancing terrestrial DOC flushing (allochthonous C import) and/or pulsed phytoplankton growth (autochthonous C-pool) driven by nutrient inputs enhance surface water DOC (Evans *et al.*, 2006; Eimers *et al.*, 2008; Pandey and Pandey, 2014). Transfer of carbon from the land to the oceans through river systems is a key node in global carbon cycle (Meybeck 1988; Ramesh *et al.*, 1995; Brunet *et al.*, 2005; Baker *et al.*, 2008). Globally the river transport on average 0.8 to 1.2×10^{15}g of carbon annually (Ludwig *et al.*, 1996). About 19×10^3 km^{3}of river water (51 per cent of the world total), with over 60 per cent of world total suspended load, enters the coastal oceans in the wet tropics annually (Meybeck, 1988). Most of this input occurs from the Indo-Pacific archipelago (Ramesh *et al.*, 1995). Total and dissolved organic carbon (TOC, DOC) in the river is generated from terrestrial sources (through mobilization of natural and anthropogenic organic matter, including soil-derived organic matter), autochthonous production (within-stream generation of organic matter) and from anthropogenic sources such as sewage effluents and industrial discharge. It is further interesting to note that the DOC of terrestrial origin is extensively regulated by microbial metabolism in the soil, streams and land- water interfaces as it flows towards receiving water bodies (Jansson *et al.*, 2000; Pandey and Pandey, 2009; Pandey, 2011).

Multiple factors prevailing at different spatial and temporal scales play an important role in structuring periphytic communities in river ecosystems with local factors playing a more important role compared to broad scale landscape, climatic and geographical factors. Among the many factors, variables regulating the effective light climate (degree of shading) are more important driving biomass accrual of periphyton. Despite their ecological importance and role in maintaining ecological assimilating capacity of the river ecosystems, algal periphyton community has received little attention for this major river system of India. There is a general dearth of studies explicitly addressing the impact of effective light climate driven by phytoplankton biomass accrual and DOC concentration on periphyton growth in Ganga River. The newly established Ministry of Water Resources, River Development and Ganga

Rejuvenation by the Government of India may warrant watershed-scale research initiatives to understand how the transfer of terrestrial C to river ecosystems are influencing the ecological assimilation capacity and how these can be regulated to rejuvenate this river system. The present study was an effort to investigate algal periphyton biomass accrual and the relative contribution of taxonomic divisions regulated by phytoplankton and DOC- linked light limitation in Ganga River. An attempt was made to determine if decreased light intensity alters autotrophic index and periphyton biomass accrual rates in the river. Data on these issues can be used assessing ecological status, establishing cross- domain relationships, predicting climate change drivers and developing action plan for integrated river basin management (IRBM).

Materials and Methods

Study Area

This study was conducted during summer low flows of four consecutive years (2011 to 2014) at 10 selected sites along a 35 km stretch of Ganga River at Varanasi (25° 18'N latitude, 83° 1'E longitude and 76. 19 m above msl), India. Varanasi city is situated at the west bank of the Ganga River. The river originates in the Himalayan mountain range from the Gangotri glacier as River Bhagirathi at an altitude of about 4000 m. Along its 2525 km long course before its confluence with Bay of Bengal, the river is joined by a large number of tributaries on both banks. The Indo- Gangetic plains not only represent the agriculturally most fertile belt but also the highest population density in the country. The Ganges is one of the most utilized rivers of India including those for fisheries, irrigation, recreational and drinking water supply. However, due to increasing anthropogenic activities in the catchment, the quantity and quality of river water have massively declined over the years. The land use pattern in the watershed is a major factor controlling the regional runoff, discharge regime, erosion and sediment transport into the river. In addition, the river is receiving massive input of nutrients and pollutants from urban- industrial sources. This has strong causal relationship with downstream eutrophication. During the last two decades the region witnessed massive urban expansion. However, the expansion and development of Varanasi city does not meet the technical standards in terms of sewage treatment, collection of garbage, urban drainage and other issues. River segment and the tributaries in the urban area, therefore, receive treated, semi- treated and untreated effluents from various domestic and industrial sources. The unplanned and disorderly growth of the city results in river water quality deterioration, loss of aquatic species, shift in community composition/trophic cascades, eutrophication and many other problems (Pandey *et al.*, 2014a; 2014b; Pandey and Pandey, 2015).

Climate of the region is tropical with distinct seasonality. The year can be divided into a hot and dry summer (March to June), a humid rainy season (July to October) and, a cold winter season (November to February). Mean annual rainfall varied between 870 – 1130 mm, relative humidity between 27 and 83 per cent (summer) and 58 and 99 per cent (rainy season). More than 90 per cent of mean annual rainfall in the region occur during rainy season. During summer, day time temperature varied between 29 and 46.2 °C. During winter, night temperature some time drops below

4°C. Wind direction shifts from predominantly westerly and south- westerly in October through April to easterly and north- westerly in remaining months. The soil in the region is alluvial fluvisol associated with recurrent floods or long wetness, recent sedimentation and high natural fertility. Catchment soils differ in terms of percentage of fine soils (silt and clay), soil moisture, pH and total organic carbon (0.93 per cent to 1.67 per cent in upper 0-10 cm soil horizon).

Ten sites were established along a 35 km long transect of the Ganga River: 2 sites (Adalpura, Adp; Sultankeshwar Ghat, Stg) in the relatively less impacted agricultural and woodland catchment to act as reference site, third site (Bypass downstream, Bds) is moderately polluted with agricultural- sub urban- urban washings and 7 sites (Nagwa discharge, Ngw; Assi Ghat, Asg; Shivala Ghat, Slg; Harishchandra Ghat, Hcg; Rana Ghat, Rng; Raj Ghat, Rjg; Raj Ghat downstream, Rds) in highly polluted downstream areas along the main urban segment. Samples were collected at 15- 20m reach and to increase the validity of data comparison, equal number of complementary sites were selected opposite to the city side of the river. These sites are hereafter referred as off- side sampling locations. The main purpose of selecting these sites was to obtain a pollution gradient of the river system from relatively unpolluted sites to highly polluted urban downstream site.

Analytical Methods

Water Sampling and Analysis

The present investigation consisted of three tiers of study that include water quality variables, light climate and algal biomass (phytoplankton and periphyton). Replicates of water samples were collected from 15- 20 m reach in the river during summer (low flow periods) of four consecutive years. Replicate samples were pooled to make composite samples. The distance between replicate sampling (n = 3) was about 50 m. Water samples were collected from each site, directly below the surface (15- 25 cm depth), in acid- rinsed 5L plastic containers for analysis of total dissolved solids (TDS), dissolved oxygen (DO), dissolved organic carbon (DOC), nitrate- N (NO_3^-) and orthophosphate (PO_4^{3-}). The TDS was measured directly using a TDS meter. The dissolved oxygen (DO) was estimated following Winkler's modified method (APHA, 1998) and dissolved organic carbon (DOC) was quantified using a $KMnO_4$ digestion procedure (Michel, 1984). Water samples were mixed with acidified N/80 potassium permagnate and incubated at 37°C. Organic carbon was estimated by titrating to quantify oxygen after 4h of incubation (APHA, 1998). The efficiency of the method and validity of the analytical data was monitored by repeated calibration using different concentrations of standard potassium hydrogen phthalate solution. Nitrate-N was quantified using a brucine sulphanilic acid method (Voghe, 1971). Orthophosphate (dissolved reactive phosphorus, DRP) was quantified using Olsen's ammonium molybdate method (Mackereth, 1963). Depth of light penetration in the river was measured using Secchi disk and light at depth was calculated using mid-summer surface light intensities following Vadeboncoeur *et al.* (2003).

Algal Biomass

Chlorophyll **a** constitutes approximately 1.5 per cent of algal organic dry matter (APHA, 1998) and is considered the most authentic index of algal biomass. In this study also, the algal biomass was measured in terms of chlorophyll **a** (Chl **a**). For phytoplankton, water samples were collected in acid- rinsed 5L plastic containers from each replication points (n= 3) directly below the surface (15- 25 cm depth). From this sample, sub- samples taken for phytoplankton were preserved in Lugol's iodine and concentrated by centrifugation at 3500 rpm (APHA, 1998). Phytoplankton chlorophyll **a**, extracted from this concentrate using acetone, was determined following spectrophotometric method and values expressed as mg m^{-3}. For the determination of periphyton chlorophyll **a** biomass, 35 mm plastic slides were laid in triplicate over the experimental rock surface fixed in closed wire cage (basket sampler) of $15 \times 15 \times 15$ cm size at 2m depth for one month during summer low flow (16th April to 15th May each year of the study period). Each replicate represents a composite sample of three independent slides placed at each collection point at a distance of 2m. The material within the slide area (scrub area = 2.3 cm $\times$3.5 cm = 8 cm^2) was pulled by vacuum onto a pre-ashed (free of organic matter) glass fiber filter (Whatman GF/A) for wet and dry weights. Enough care was taken to ensure complete recovery of algal assemblages attached to the surface. Chlorophyll **a** biomass was determined in fresh material following acetone extraction procedure (Maiti, 2001) and expressed as mg m^{-2} by relating to exposed surface area. Phycocyanin (PC) pigment was extracted in 0.05 M phosphate buffer (pH 6.8) and estimated according to Bennett and Bogorad (1973). Ash free dry mass (AFDM) was used for computing the autotrophic index (AI). The AFDM was quantified by reweighing the sampled filter papers ashed at 500°C for 2 h in a muffle furnace (Bowes *et al.*, 2012) and the values thus obtained were divided by the chlorophyll **a** biomass of periphyton (APHA, 1998) to obtain autotrophic index.

Taxonomic Biomass

We used taxonomic biomass to study algal community composition along the study gradient. All the study sites were sampled (n= 3, per site) during summer low flows (16th April to 15th May each year of the study period) and thus, did not reflect the effect of seasonality. A 50 ml of subsample (scraped homogenate) was preserved in Lugol's iodine solution and identifications were made from 0.5 ml aliquot and counted in a haemocytometer using a compound microscope with phase and differential interference contrast optics. Identifications of benthic algae to genus level were made using Prescott (1973), Mizumo (1990) and Dillard (1991). Fragments, filaments, and cell densities were counted and calculated for each recorded taxa. Cell dimensions were measured and cell volumes were estimated in each sample by approximation to geometric shapes of known volume. Algal biomass was measured by converting the calculated cell volumes to biomass assuming a specific density of 1 g cm^{-3} (Chetelat *et al.*, 1999) and expressed as mg cm^{-2}. Relationships between percent taxonomic biomass and periphyton biomass were examined using a locally weighted sequential smoothing technique (LOWESS) described by Chetelat *et al.* (1999).

Statistical Analysis

Mean estimate ($\bar{x}$) of TDS, NO$_3^-$, PO$_4^{3-}$, DO, DOC and Secchi depth represents replication sites and temporal variability whereas that of Chl **a**, taxonomic biomass and AI represents replication sites only. Effects of site and time were tested using analysis of variance (ANOVA). All repeated measurements (including replication sites) were considered in ANOVA model. Values were log - transformed when needed and interaction terms were included in the analysis. Coefficients of variation (CV) were computed for expressing data variability. Correlation coefficient (R^2) and regression models were used to test linearity between variables. Relationships between percent taxonomic biomass and periphyton biomass were non- linear and were examined using a locally weighted sequential smoothing technique (LOWESS) as described by Chetelat *et al.* (1999).

Results and Discussion

Multiple factors prevailing at different spatial and temporal scales regulate water chemistry and associated shifts in algal periphyton communities in river ecosystems with local factors playing a more important role compared to broad scale landscape, climatic and geographic factors. Some of the factors, although are indirect regulator, most often found to be important in shaping distribution pattern of periphyton by regulating the influence of broad scale climatic drivers. Increased water column turbidity and presence of optically opaque substances, for instance, are one such factors regulating distribution and biomass accrual of periphyton communities through modifying effective light climate (degree of shading). In the present study, the concentrations of nutrients, DOC and TDS increased along the study gradient (Table 16.1 and Figure 16.1) and the values were lowest at Adalpura and highest at Raj Ghat site. Concentrations of TDS and nutrients were all higher at city- side indicating the direct effect of urban release. At city- side, NO_3^- concentrations varied between 227.1 and 933.1 µg L^{-1} and, PO_4^{3-} between 151.3 and 831.0 µg L^{-1}. Respective ranges at off- side locations were found to be between 217.2 and 569.1 µg L^{-1} and between 134.0 and 458.0 µg L^{-1} (Table 16.1). Dissolved oxygen however, showed an opposite trend (Table 16.1). Site- wise differences in TDS, DO and nutrients were significant ($p < 0.001$; Table 16.2). Concentration of DOC increased along the study gradient and water at Raj Ghat, compared to Adalpura, contained over 3- folds higher DOC (Figure 16.1). Concentrations of DOC towards off- side of the river were relatively lower. On the scale of the year, DOC increased over time and inter- annual variations were significant (Figure 16.1).

Secchi depth (a measure of transparency) showed a decreasing trend over time and along the study gradient. At city- side, Secchi depth declined by over 80 per cent at Raj Ghat compared to Adalpura site. Irrespective of the site and study year, Secchi disk transparency was higher at off- side locations compared to the city- side (Figure 16.1). Further, we found an increasing trend in phytoplankton Chl *a* biomass down the study gradient. This trend was common at both side of the river although the off- side locations did show relatively less accumulation of phytoplankton biomass. Contrary to phytoplankton, periphyton Chl *a* biomass declined along the study gradient spanning over 4 to 5 orders of magnitude and was lowest at Raj Ghat site (Figure 16.1) and showed dependence on factors regulating effective light climate (Table 16.3). The results show trends in periphyton Chl *a* biomass opposite to that of Secchi depth and phytoplankton biomass. Periphyton biomass varied between 22.0 and 75.2 mg m^{-2} in 2011 and between 16.0 and 65.4mg m^{-2} in 2014. The contributions

Table 16.1: Concentration of Total Dissolved Solids (TDS), Nitrate (NO_3^-), Orthophosphate (PO_4^{3-}) and Dissolved Oxygen (DO) Measured at City-Side and Off-Side in Ganga River along the Study Gradient. Values are mean (n= 24) ± 1SE.

Site	City-side				Off-side			
	TDS (mgL⁻¹)	NO_3^- (µgL⁻¹)	PO_4^{3-} (µgL⁻¹)	DO (mgL⁻¹)	TDS (mgL⁻¹)	NO_3^- (µgL⁻¹)	PO_4^{3-} (µgL⁻¹)	DO (mgL⁻¹)
Adp	470.2±27.4	227.1±20.1	151.3±11.1	9.3±0.5	457.6±26.1	217.2±13.2	134.2±9.8	9.4±0.5
Stg	472.4±29.0	267.5±21.2	175.1±11.6	8.5±0.5	462.0±28.7	222.4±13.8	138.8±9.5	8.8±0.5
Bds	752.0±49.1	396.2±26.7	271.9±15.4	7.2±0.4	492.8±28.9	236.7±15.5	148.6±10.8	7.8±0.4
Ngw	840.2±58.6	795.9±48.5	532.2±31.6	6.6±0.3	518.7±32.5	391.6±15.6	260.9±14.9	7.5±0.4
Asg	848.2±63.8	856.0±61.2	676.6±49.7	6.4±0.4	529.7±36.8	408.5±20.7	311.2±17.3	7.4±0.4
Slg	853.6±62.2	887.3±72.9	705.2±58.3	6.1±0.3	568.7±36.8	430.8±21.5	342.7±24.0	7.3±0.4
Hcg	868.3±70.6	896.8±78.8	750.9±57.2	5.8±0.3	576.8±35.9	455.4±23.5	371.8±24.7	7.2±0.4
Rng	867.37±71.9	905.7±77.3	789.1±64.6	5.1±0.3	580.8±35.8	470.7±26.8	390.6±25.7	6.9±0.3
Rjg	887.9±74.9	950.1±82.4	844.7±71.5	4.2±0.3	592.7±37.2	568.6±31.5	467.1±29.2	6.1±0.3
Rds	884.2±74.4	933.1±84.1	831.0±72.2	4.4±0.3	590.9±37.6	569.1±30.4	458.0±31.8	6.0±0.3

Table 16.2: F-ratios Resulting from Analysis of Variance (ANOVA) to Test Significant Effects of Site (s), Side (S), Year (Y) and their Interactions on Water Quality Variables and Chlorophyll a Biomass along the Study Gradient

Variable	s	S	Y	sxS	sxY	SxY	sxSxY
TDS	702.50**	180.20**	ns	77.05**	7.35*	17.67**	4.90*
NO_3^-	367.43**	314.05**	12.91**	136.32**	122.50**	156.90**	16.85**
PO_4^{3-}	483.76**	417.30**	19.83**	115.18**	123.75**	127.62**	76.98**
DO	100.33**	226.20**	ns	33.11**	38.95**	56.75**	4.87*
DOC	406.65**	365.18**	38.40**	176.59**	137.32**	176.40**	132.76**
Secchi depth	331.50**	427.00**	37.76**	107.22**	96.86**	184.35**	106.15**
Phytoplankton Chl a	770.70**	254.70**	69.54**	65.90**	105.29**	219.18**	203.67**
PeriphytonChl a	667.82**	317.15**	58.47**	79.20**	92.58**	239.50**	208.72**
AI	335.27**	196.87**	55.28**	65.70**	41.87**	77.23**	32.16**

TDS: Total dissolved solids; DO: Dissolved oxygen; DOC: Dissolved organic carbon; AI: Autotrophic index.

Values significant at: * $P < 0.05$, ** $P < 0.01$; ns: Not significant.

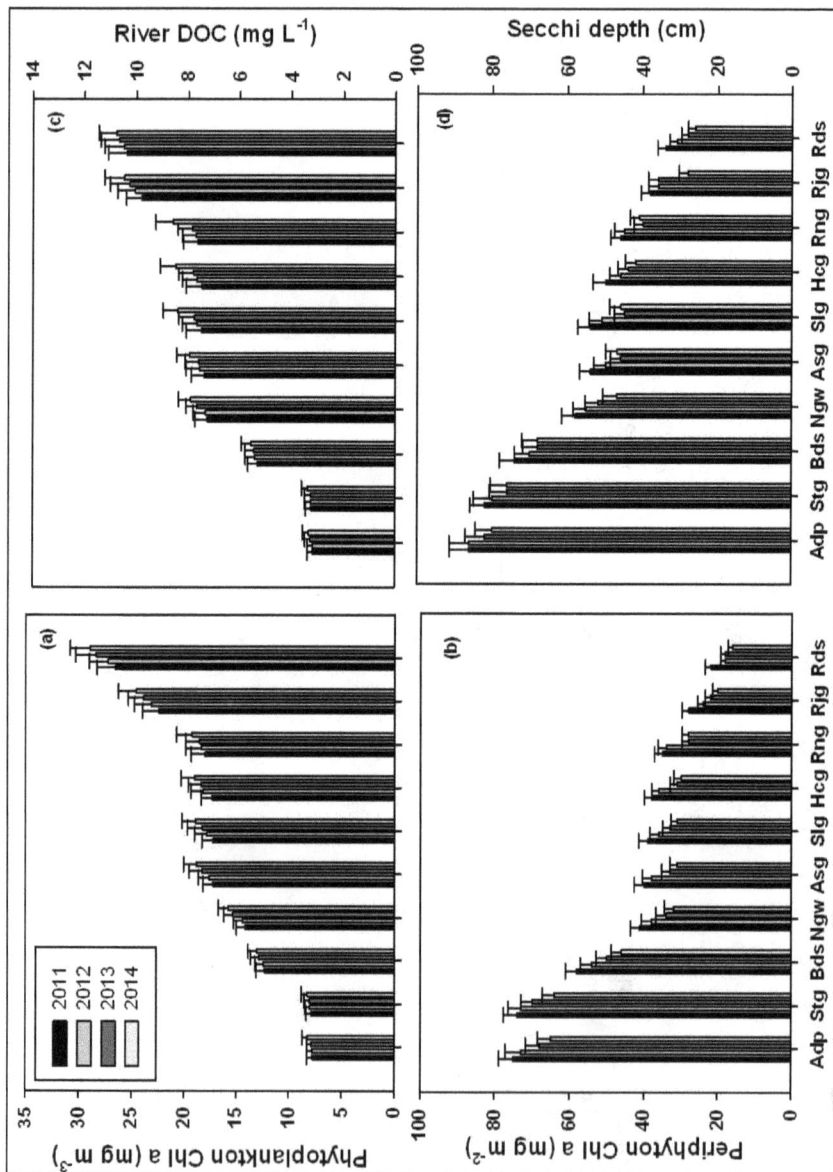

Figure 16.1: Phytoplankton Chlorophyll a Biomass (a), Periphyton Chlorophyll a Biomass (b), Dissolved Organic Carbon (c, DOC) and Sechhi Depth (d) at 10 Study Locations of Ganga River. Values are mean (n = 24 for DOC and Secchi depth and n = 12 for Chl a biomass) ± 1SE. The city- side and off- side sample values were pooled. Site codes are explained in the text.

of taxonomic divisions to overall biomass accrual varied along the study gradient. Green algae (Chlorophyta), which contributed to 24- 60 per cent of total standing stock, showed high biomass accumulation at sites characterized by moderate nutrients and DOC levels. Cyanophyta (blue green algae) contributed to 39- 74 per cent with the larger fraction at highly eutrophic sites and Bacillariophyta (diatoms), with 2- 5 per cent share, at moderately eutrophic sites. We found significant positive correlation ($R^2 = 0.95$; $p < 0.001$) between periphyton biomass and Secchi disk transparency and negative correlation between periphyton biomass and DOC ($R^2 = 0.94$; $p < 0.001$) and phytoplankton biomass ($R^2 = 0.92$; $p < 0.001$). Secchi disk transparency showed negative correlation with DOC ($R^2 = 0.94$; $p < 0.001$) and phytoplankton biomass ($R^2 = 0.93$; $p < 0.001$). The trend in autotrophic index (AI) was similar to that of the phytoplankton biomass (Figure 16.2). The AI increased from 204 at Adalpura to 405 at Raj Ghat downstream site. We used pigment ratios to further segregate the cyanophycean contribution to overall biomass accumulation. Phycocyanin to chlorophyll a (PC: Chl a) ratio varied between 2.2 and 3.2, showing an initial decrease with nutrient and DOC rise, but followed an increasing trend with further rise in nutrient and DOC. Spatial and temporal differences in these variables (DOC, Secchi depth, periphyton Chl a, phytoplankton Chl a and autotrophic index) were all, significant (ANOVA, Table 16.2).

Table 16.3: Summary of Regression Model Relating Periphyton Chl a Biomass (mg m^{-2}) to the Determinants of Light Climate in the Study River

Equation	n	R^2	p
DOC = - 0.110 Phy Chl a + 0.4980	40	0.94	<0.001
Secchi depth = 81.375 DOC - 3.6586	40	0.93	<0.001
Secchi depth = 82.760 Phy Chl a -1.8611	40	0.94	<0.001
Per Chl a = 1.12 Secchi depth + 1.0700	40	0.95	<0.001
Per Chl a = 89.011 DOC - 3.9860	40	0.94	<0.001
Per Chl a = 90.383 Phy Chl a - 2.0250	40	0.94	<0.001

DOC: Dissolved organic carbon; Phy: Phytoplankton; Per: Periphyton; Chl a: Chlorophyll a biomass; n: Number of observations.

Algal periphytons are among the most important contributors to primary productivity in rivers and streams. A shift in periphytic growth therefore, leads to a potential shift in ecosystem functions and trophic cascades (Bootsma *et al.*, 1996; Vadeboncoeur *et al.*, 2003; Hilton *et al.*, 2006). Although light and nutrient are two principal drivers of integrated (benthic + pelagic) primary production, two distinct mechanisms operate in human impacted rivers and streams. Nutrient supply promotes pelagic growth (phytoplankton production) whereas the growth of benthic algae is generally limited by light climate. Our previous studies (Pandey, 2013; Pandey and Pandey, 2013) and other available data (Karlsson *et al.*, 2009) show that both, phytoplankton and DOC reduce benthic algal production through light interception. Both, DOC of land origin (Pandey and Pandey, 2013; 2015) as well as those resulting from rapid turnover of phytoplankton (Kirchman *et al.*, 1991) constrains periphytic

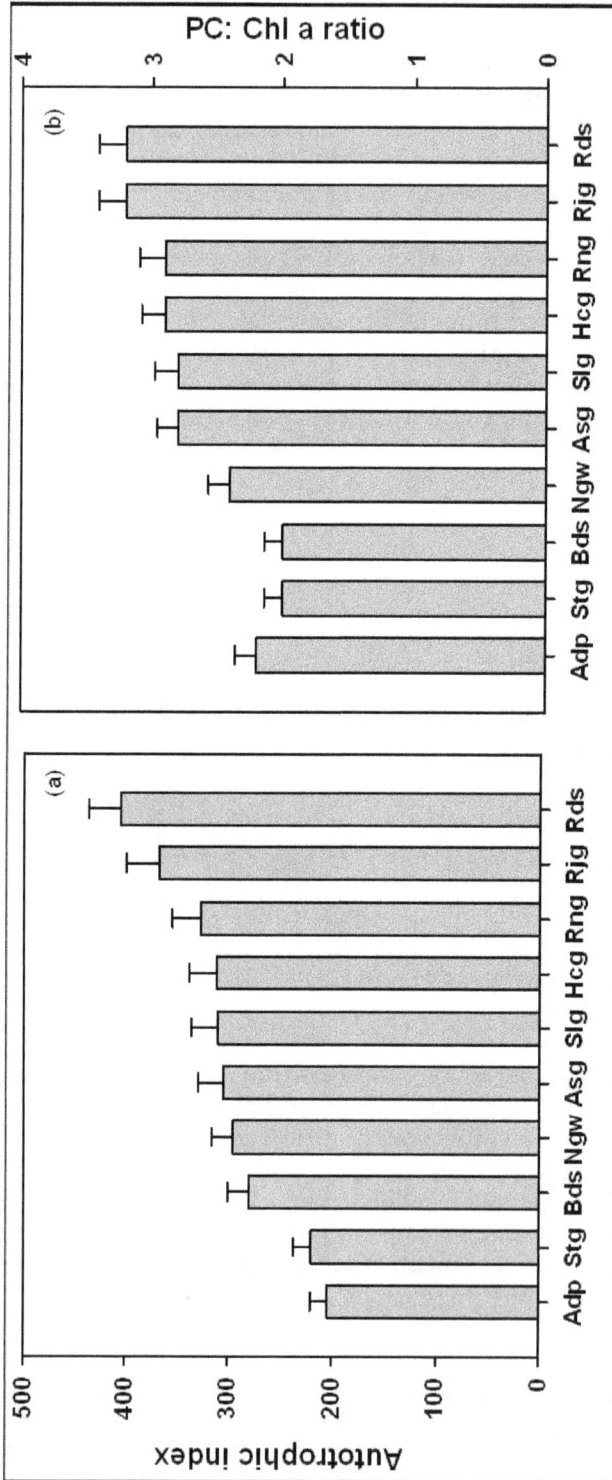

Figure 16.2: Autotrophic Index (a) and Phycocyanin (PC) to Chl **a** Ratio (b) at 10 Study Locations of the Ganga River. Values are mean (n = 12 ± 1SE). Site codes are explained in the text.

growth by reducing effective light penetration. This becomes more important in the light of the fact that anthropogenic activities have dramatically enhanced the nutrient and organic loading to this river system. Both mediate the shading effect, former through promoting phytoplankton growth and the latter through direct interception of light penetration. We found a marked increase in the concentrations of DOC over time as well as along the study gradient. Concentrations of dissolved oxygen (DO) in water showed asynchrony with DOC along the study gradient indicating the rising influence of oxygen demanding organic substances. Enhanced pelagic production (autochthonous C- pool) in response to nutrient enrichment (Pandey and Pandey, 2009) and, enhanced terrestrial input (allochthonous C- export) driven by climate, hydrology and human activities (Evans *et al.*, 2006; Monteith *et al.*, 2007; Pandey and Pandey, 2013) are considered as the major determinants of rising DOC in surface waters. We expect that the land derived DOC and nutrients in Ganga River could be massively high due to large catchment and sewage discharge from the city. More importantly, damming of the river has substantially reduced the overall stream flow enhancing further the DOC level in the river. From over two decades the Ganga River has witnessed massive change in its hydrological characteristics including a decline in water discharge by over 70 per cent from its pre- damming stage (Pekárová *et al.*, 2003). This itself is one of the major factors for substantially reducing the dilution capacity of the river and consequently increase the concentration of DOC and other analytes.

Although nutrient enrichment enhances phytoplankton development (Bergstrom *et al.*, 2008) and periphyton (Chetelat *et al.*, 1999), our results show marked asynchrony between phytoplankton and periphyton biomass accrual. With increasing N and P concentrations along the study gradient phytoplankton Chl **a** biomass increased while periphyton biomass showed a declining trend. Excess supply of organic matter and nutrients induces pronounced effect on the ecology of rivers and streams by changing the light climate and productivity (Karlsson *et al.*, 2009). In shallow reservoirs benthic algae derive sediment- associated nutrients. Light limitation under such condition can more strongly influence benthic primary production. In the present study, we hypothesized that an increase in DOC and/or phytoplankton biomass would have pronounced effect on benthic algal production by changing the effective light climate. Our data indicate that anthropogenic inputs modify the interactions of habitat boundaries (nutrient enrichment decreases light availability to benthic primary producers through increased phytoplankton- shading effect) and cross- ecosystem boundaries (land surface runoff increases allochthonous DOC level and reduce benthic production) which may, in turn, modify the river ecosystem functioning in long- run. Among the nutrients, P is most often considered as the limiting nutrient in eutrophication. Although the growth limiting concentrations of phosphorus vary greatly from river to river ranging from less than 20 μg L^{-1} soluble reactive- P (Chambers *et al.*, 2006) to greater than 100 μg L^{-1} (Bowes *et al.*, 2007), in the study segment of the river, P is unlikely to be a limiting factor (Pandey *et al.*, 2014b; Pandey and Pandey, 2015). The decrease in periphyton biomass along the study gradient appeared to be due to constrained light penetration associated with increased DOC and phytoplankton growth (Carey *et al.*, 2007). Periphyton biomass was positively correlated to depth of light penetration and the latter was negatively correlated to DOC and phytoplankton biomass. These observations suggest that DOC and phytoplankton biomass could reduce benthic algal growth through light interception. In a previous study, Vadeboncoeur *et al.* (2003) show that light attenuation associated with increased phytoplankton production along a gradient of P input in Greenland and Danish lakes caused significant loss of benthic periphyton.

We used autotrophic index (AI) to further demonstrate the relative proportion of autotrophy in the benthic biofilms. Ash free dry mass (AFDM) is used for computing the autotrophic index. The AFDM when considered in relation to chlorophyll **a** biomass of periphyton (to obtain autotrophic index) gives a correct picture of relative proportion of autotrophs and heterotrophs in the biofilm. The development of biofilms (perphyton, epilithon) both, in regard to their structure and taxonomic diversity, depends on physical and chemical factors including hydraulic turbulence, light and nutrients as major determinants. In river systems particularly, differences in flow produce different hydrological facies supporting diversity of periphyton and heterotrophs. Thus, AI can be used as an important index of periphyton biomass accrual. Our results show that both, the concentration of DOC and phytoplankton biomass accrual increased over time and along the study gradient whereas light penetration and periphyton biomass showed an opposite trend. For most of the sampling stations, the autotrophic index (AI) remained above 200 and reached over 400 at eutrophied sites indicating that the biofilm communities were relatively unbalanced between heterotrophic and autotrophic conditions (Ameziane *et al.*, 2002). A consistently rising trend in AI along the study gradient indicated that the periphytic communities have been suppressed due possibly to DOC and phytoplankton induced shading effect. The river consists of a mosaic of hydrological facies subjected to varying light climate giving rise to different periphetic communities. Although the shading effect of other water variables cannot be ruled out, DOC could be more crucial than others in causing significant reduction in periphytic growth. In particular, the terrestrially derived DOC being very optically dense reduces light penetration more effectively than the shading effect induced by phytoplankton and autochthonous DOC (Karlsson *et al.*, 2009). In the present study, the C: N ratio of DOC remained generally above 15 indicating allochthonous influences (Elser *et al.*, 2000). During rainy season high terrestrial DOC flushing could become more important contributor reducing light penetration. Our data showed that a 7 mg L^{-1} increase in DOC caused light penetration to decline by 42 per cent. Further increase in DOC showed a consistently increasing light attenuation. The trends in light interception and associated sift in periphyton biomass accrual observed in this study matches with those reported in numerous other studies. For instance, McEachern *et al.* (2000) have shown that a 9 per cent increase in DOC led Secchi disk transparency to decline by 46 per cent. Similarly, Bergstrom *et al.* (2001) observed that a 7 mg L^{-1} increase in DOC correspond to an approximately 37 per cent decrease in light climate. Luijn *et al.* (1995) have reported that even 15 to 20 per cent attenuation in Secchi depth can significantly alter the growth of benthic diatoms. Similar effects of DOC and associated shifts in light climate linked limitation of periphyton growth have been reported by Bowes *et al.* (2012) in Thames River and by Pandey and Pandey (2013) in a freshwater tropical lake of India. These studies together with the observations of the present study support the contention that light intensity is among the most important climatic variables controlling development of periphyton communities.

Along with the shifts in biomass accrual in response to changing light climate, changes in structure and diversity of taxonomic divisions can also be expected. In this study we found, corresponding to a switchover from benthic to pelagic dominance of primary productivity, a marked shift in the relative contribution of taxonomic divisions of algal periphyton. With initial increase in concentration of nutrients, Chlorophyta contributed a major share. Latter however, as the increases in nutrient levels continue, Cyanophyta represented the larger fraction in total community biomass. We used pigment ratios (phycocyanin: Chl **a** ratio) to further test these relations. Phycocyanin is a protein pigment found predominantly in members of Cyanophyta and Rhodophyta. A corresponding shift in PC: Chl **a** ratio along the gradient indicated the dominance of Cyanophyta towards nutrient rich sites. We also observed a shift in species specific dominance in algal taxa. For instance, there was a shift in dominance among blue green algal taxa from *Oscillatoria, Gloeocapsa* and *Tolypothrix* to *Phormidium* along the eutrophication gradient. Increasing dominance of *Phormidium* (Cyanophyta) along the eutrophication gradient indicate the ability of this species to exploit nutrient rich habitat replacing less adapted species. Our previous studies have shown gradual replacement of sensitive species by more opportunistic invasive species (Pandey and Pandey, 2002; 2013; Pandey, 2013). Chetelat *et al.* (1999) observed that nutrient rich sites at temperate lowland rivers in southern Ontario and western Quebec were associated with high periphyton standing crop but dominated by particular filamentous taxa. In the present study, it appears that light driven constraints offset the influence of nutrient enrichment even for exploitative algal species causing an overall loss of benthic biomass. Our results show that enhanced pelagic production and DOC enrichment affect periphyton standing crop and lead to a shift in community composition dominated by exploitative algal species that form undesired benthic mats. It is expected that, for the whole ecosystem productivity, the net reduction in benthic production may partly be offset by nutrient driven increase in pelagic production. However, since algal periphyton constitute a substantial part of the resources for higher trophic levels (Karlsson *et al.*, 2009); a shift in benthic primary production may lead to long- term changes in the river ecosystem functioning including a shift in trophic cascades. Furthermore, since DOC is an important component of aquatic ecosystems and global carbon cycle, our observations on rising DOC trends in an oceanic river like Ganges have relevance for predicting future climate change drivers.

Conclusions

The study revealed a marked change in the water quality of Ganga River as it passes along the urban segment of Varanasi indicating a strong influence of local sources. Further, the eutrophication gradient characterized in this study showed a switchover from benthic to pelagic dominance of primary productivity. Based on correlative evidences we conclude that light limitation resulting from rising concentrations of DOC and enhanced growth of phytoplankton, driven by the anthropogenic causation, are negatively influencing the benthic primary production with a shift in taxonomic divisions of periphyton. We expect that a shift in relative importance of primary production from periphyton to phytoplankton as well as in the taxonomic groups of periphyton may entail a similar shift in the diets of many invertebrates. Since benthic algae help supporting consumers of higher trophic levels, partitioning and sequestering nutrients and carbon, maintaining bottom oxygenation and consequently, reducing the bottom release of phosphorus and greenhouse gases, a shift in periphyton driven primary production may lead to long-term changes in biodiversity and food web structure as well as in waste assimilation capacity of the river. Our study provides important cue useful in understanding energetics of river ecosystem, predicting climate change drivers and designing action plan for integrated river basin management (IRBM).

Acknowledgements

We thank Head, Department of Botany, Banaras Hindu University, Dean, Faculty of Science and Technology, M. G. Kashividyapith University and Dr. A. Mishra, Department of Biochemical Engineering, Indian Institute of Technology, Banaras Hindu University for facilities.

References

1. Ameziane T, Garabetian F, Dalger D, Sauvage S, Dauta A and Capblancq J (2002). Epilithic biomass in a large gravel- bed river (The Garonne, France): a manifestation of eutrophication. *River Research and Application* 18: 343- 354.

2. APHA (1998). *Standard Methods for the Examination of Water and Wastewater*. American Public Health Association, Washington, DC.

3. Bennett A and Bogorad L (1973). Complimentary chromatic adaptation in a filamentous blue-green alga. *Journal of Cell Biology* 58: 419- 435.

4. Bergstrom AK, Jansson M, Blomqvist P and Drakare S (2001). The influence of water colour and effective light climate on mixotrophic phytoflagellates in three small Swedish dystrophic lakes. *International Association of Theoretical and Applied Limnology* 27: 1861- 1865.

5. Bergstrom AK, Jonsson M and Jansson M (2008). Phytoplankton responses to nitrogen and phosphorus enrichment in unproductive Swedish Lake along a gradient of atmospheric nitrogen deposition. *Aquatic Biology* 4: 55- 64.

6. Bootsma HA, Hecky RE, Hesslein RH and Turner GF (1996). Food partitioning along lake Malawi near shore fishes as revealed by stable isotope analysis. *Ecology* 77: 1286- 1290.

7. Bowes MJ, Smith JT, Hilton J, Sturt MM and Amitage PD (2007). Periphyton biomass response to changing phosphorus concentration in a nutrient impacted river: a new methodology for phosphorus target setting. *Canadian Journal of Fisheries and Aquatic Science* 64: 227- 238.

8. Bowes MJ, Isngs NL, Mc Call SJ, Warwick A, Barrett C, Wickham HD, Hartman SA, Armstrong LK, Scarlett PM, Roberts C, Lehmann K and Singer AC (2012).

Nutrient and light limitation of periphyton in the River Thames: Implication for catchment. *The Science of Total Environment* 434: 201- 212.

9. Carey RO, Vellidis,G, Lowrance R and Pringle CM (2007). Do nutrient limit algal periphyton in small black-water coastal plain streams. *Journal of American Water Resource Association* 5: 1183-1193.

10. Chambers PA, Culp JM, Glozier NE, Cash KJ, Wrona FJ and Noton L (2006). Northern rivers ecosystem initiatives: Nutrients, and dissolved oxygen- issues and impacts. *Environmental Monitoring and Assessment* 113: 117- 141.

11. Chetelat J, Pick FR, Morin A and Hamilton PB (1999). Periphyton biomass and community composition in rivers of different nutrient status. *Canadian Journal of Fisheries and Aquatic Science* 56: 560- 569.

12. Dillard GE (1991). *Freshwater Algae of the Southeastern United States.* J. Gramer, Berlin, Stuttgart.

13. Eimers MC, Watmough SA and Buttle JM (2008). Long- term trends in dissolved organic carbon concentration: a cautionary note. *Biogeochemistry* 87: 71- 81.

14. Elser JJ, Fafan FW, Denno RF, Dobberfuhl DR, Folarin A, Huberty A, Interlandi S, Kilham SS, Mc Cauley E, Schulz KL, Siemann EH and Sterner RW (2000). Nutritional constraints in terrestrial and freshwater food webs. *Nature* 408: 578-580.

15. Evans CD, Chapman PJ, Clark JM, Monteith DT and Cresser MS (2006). Alternative explanations for rising dissolved organic carbon export from organic soils. *Global Change Biology* 12: 2044-2053.

16. Flury S, Mc Ginnis DF and Gessner O (2010). Methane emission from a freshwater marsh in response to experimentally simulated global warming and nitrogen enrichment. *Journal of Geophysical Research* 115: 1- 9.

17. Hilton J, O'Hare M, Bowes MJ and Jones JI (2006). How green is my river? A new paradigm of eutrophication in rivers. *The Science of Total Environment* 365: 66- 83.

18. Jansson M, Bergstrom AK, Blomqvist P and Drakare S (2000). Organic carbon and phytoplankton/bacterioplankton production relationships in lakes. *Ecology* 82: 3250-3255.

19. Karlsson J, Bystrom P, Ask J, Ask P, Persson L and Jansson M (2009). Light limitation of nutrient-poor lake ecosystems. *Nature* 460: 506-509.

20. Kirchman DL, Suzuki Y, Garside C and Ducklow HW (1991). High turnover rates of dissolved organic carbon during spring phytoplankton bloom. *Nature* 352: 612- 614.

21. Luijn FV, van der Molen DT, Luttmer WJ and Boers PCM (1995). Influence of benthic diatoms on the nutrient release from sediments of shallow lakes recovering from eutrophication. *Water Science and Technology* 32: 89-97.

22. Lohman K, Jones JR and Perkins BD (1992). Effects of nutrient enrichment and flood frequency on perphyton biomass in northern Ozark streams. *Canadian Journal of Fisheries Aquatic Science* 49: 1198- 1205.

23. Mackereth, FJ H (1963). Some Methods of Water Analysis for Limnologists. *Freshwater Biology Association Scientific Publication* 21, Amleside.

24. McEachern P, Prepas EE, Gibson JJ and Dinsmore WP (2000). Forest fire induced impacts on phosphorus, nitrogen and chlorophyll **a** concentrations in boreal subarctic lakes of northern Alberta. *Canadian Journal of Fisheries and Aquatic Science* 57: 73- 81.

25. Maiti SK (2001). *Handbook of Methods in Environmental Studies (Vol.1,Water and Wastewater).* ABD Publisher, Jaipur, India.

26. Michel P (1984). *Ecological Methods for Field and Laboratory Investigation.* Tata McGraw–Hill Publication Company, New Delhi, India.

27. Mizumo T (1990). *Illustration of Freshwater Plankton of Japan.* Hoikusch Publishing Coomany, Japan.

28. Monteith DT, Stoddard JL, Evans CD, deWit HA, Forsius M, Hogasen T, Wilander A, Skjelkvale BL, Jeffries DS, Vuorenmma J, Lelle, B, Kopacek J and Vesely J (2007). Dissolved organic carbon trends resulting from changes in atmospheric deposition chemistry. *Nature* 450: 537- 541.

29. Pandey J and Pandey U (2002). Cyanobacterial flora and the physico- chemical environment of six tropical freshwater lakes of Udaipur, India. *Environmental Science* 14: 54- 62.

30. Pandey J, Shubhashish K and Pandey R (2009). Metal contamination to Ganga River (India). as Influenced by Atmospheric Deposition. *Bulletin of Environmental Contamination and Toxicology* 83: 204 – 209.

31. Pandey J and Pandey U (2009). Microbial processes at land water interface and cross-domain causal relationships as influenced by atmospheric deposition of pollutants in three freshwater lakes in India. *Lakes and Reservoirs: Research and Management* 14: 71-84.

32. Pandey J (2011). The influence of atmospheric deposition of pollutant elements on cross- domain causal relationships at three tropical freshwater lakes of India. *Lakes and Reservoirs: Research and Management* 16: 111- 119.

33. Pandey U and Pandey J (2013). Impacts of DOC trends resulting from changing climatic extremes and atmospheric deposition chemistry on periphyton biomass of a freshwater tropical lake of India. *Biogeochemistry* 112: 537- 553.

34. Pandey U (2013). The influence of DOC trends on light climate and periphyton biomass in the Ganga River, Varanasi, India. *Bulletin of Environmental Contamination and Toxicology* 90: 143- 147.

35. Pandey J, Singh AV, Singh A and Singh R (2013). Impacts of changing atmospheric deposition chemistry on nitrogen and phosphorus loading to Ganga River (India). *Bulletin of Environmental Contamination and Toxicology* 91: 184-190.

36. Pandey J, Pandey U and Singh AV (2014 a). Impact of changing atmospheric deposition chemistry on carbon and nutrient loading to Ganga River: integrating land-atmosphere-water components to uncover cross-domain carbon linkages. *Biogeochemistry* 119: 179-198.

37. Pandey J, Pandey U and Singh AV (2014 b). The skewed N : P stoichiometry resulting from changing atmospheric deposition chemistry drives the pattern of ecological nutrient limitation in the Ganges. *Current Science* 107: 956-958.

38. Pandey J and Pandey U (2014). Atmospheric deposition of nutrients shifts carbon capture and storage trends in fresh water tropical lakes in India. *Environmental Control in Biology* 52: 211-220.

39. Pandey U and Pandey J (2015). The Skewed N: P stoichiometry resulting from anthropogenic drivers regulate production of transparent exopolymer particles (TEP). in Ganga River. *Bulletin of Environmental Contamination and Toxicology*, 94: 118- 124.

40. Pekárová P, Miklánek P and Pekár J (2003). Spatial and temporal oscillation analysis of the main rivers of the world during the 19ᵗʰ- 20ᵗʰ centuries. *Journal of Hydrology* 274: 62- 79.

41. Prescott GW (1973). *Algae of the Western Great Lakes Area*. 5ᵗʰ Edition, W. C. Brown Publishers, Dubuque, Iowa.

42. Vadeboncoeur Y, Jeppesen E, Zanden MJV, Schierup HH, Christoffersen K, Lodge DM (2003). From Greenland to green lakes: Cultural eutrophication and the loss of benthic pathways in lakes. *Limnology and Oceanography* 48: 1408- 1418.

43. Voghe AL (1971). *A Text Book of Quantitative Inorganic Analysis*, 4ᵗʰ Edition, The English Language Book Society/Longman, Essex.

44. Yang Y, He Z, Lin Y, Philips EJ, Yang J, Chem, G, Stoffella PJ and Powell CA (2008). Temporal and spatial variations of nutrients in the Ten Mile Creek of South Florida, USA and effects on phytoplankton. *Jornal of Environmental Monitoring* 4: 508- 516.

2016, Environmental Biotechnology: A New Approach
Editors: Dr. Rajan Kumar Gupta and Dr. Satya Shila Singh
Published by: DAYA PUBLISHING HOUSE, NEW DELHI

Pages 265–275

Chapter 17

Phytoremediation of Environmental Pollutants: A Biotechnological Perspective

Poonam Singh[1] and A.N. Singh[2]

[1]Department of Genetics and Plant Breeding,
Banaras Hindu University, Varanasi – 221 105, U.P.
[2]Environmental Management Division,
ICFRE, FRI Campus, Dehradun – 248 006, Uttarakhand

ABSTRACT

Contaminants of inorganic and organic nature are polluting the environment and posing risk to the living organisms in one way or other. For removal of such pollutants especially from soil and water environments, the phytoremediation is emerging as a green, environment friendly and safer technology. While a great deal of research have been conducted to elucidate the physiology and biochemistry of metal hyper-accumulation in plants, there are still certain limitations that need to be overcome in order to make this technology more efficient and cost effective on a commercial scale. Biotechnology offers the opportunity to transfer hyperaccumulator phenotypes into fast-growing, high biomass plants that could be highly effective in phytoremediation. Transgenic plants produced with aid from the biotechnology can also enhance the absorption and detoxification of target pollutants from the contaminated environments.

Introduction

It is now established that the mining, manufacturing and other anthropogenic activities have contributed to extensive soil contamination over the past century, the metals being the main group of inorganic contaminants. Inorganic contaminants such as metals are commonly found in low concentration in the soil. Metals such as

Arsenic (As), Cadmium (Cd), Chromium (Cr), Mercury (Hg), Nickel (Ni), Lead (Pb), Selenium (Se), and Uranium (U) etc. are founds as contaminants and are also non-essential for plants. Likewise, there are several classes of organic pollutants: solvents (*i.e.*, trichloroethylene); explosives such as trinitrotoluene (TNT) and cyclotrimethylenetrinitramine or Research Department Explosive (RDX); polycyclic aromatic hydrocarbons (*i.e.*, naphthalene, pyrene); petroleum products including benzene, toluene, ethylbenzene, and xylene (BTEX); polychlorinated biphenyls (PCBs); and herbicides/pesticides (*i.e.*, atrazine, chlorpyrifos, 2,4-D), which are also the cause of environmental pollution.

The generic term 'Phytoremediation' consists of the Greek prefix *phyto* (plant) attached to the Latin root *remedium* (to correct or remove an evil) as has been opined by Cunningham *et al.* (1996). It can be defined as, 'the efficient use of plants to remove, detoxify or immobilise environmental contaminants in a growth matrix (soil, water or sediments) through the natural biological, chemical or physical activities and processes of the plants'. Phytoremediation is a powerful tool for cleaning up pollutants by enhancing the natural biodegradation process. This technology is an alternative or complimentary one that could be applied along with or instead of mechanical congenital cleaning methodologies which mostly require high capital input, labour and intensive energy. It is being referred now a day as safe and environment friendly green technology. The most important requirements for effective Phytoremediation are the use of fast growing, high biomass plants that are capable of uptake and accumulation of large amounts of toxic metals in their above-ground and harvestable parts. The plants which can accumulate and degrade the contaminants are known as 'hyperaccumulaters' which play a major role in Phytoremediation.

While a great deal of research have been conducted to elucidate the physiology and biochemistry of metal hyper-accumulation in plants, there are still certain limitations that need to be overcome in order to make this technology more efficient and cost effective on a commercial scale (Khan *et al.*, 2000). Biotechnology offers the opportunity to transfer hyperaccumulator phenotypes into fast-growing, high biomass plants that could be highly effective in phytoremediation. Transgenic plants can also enhance the absorption and detoxification of pollutants, thereby aiding the phytoremediation of contaminated environments (Pilon-Smits, 2005 and Doty, 2008). Application of biotechnological approaches for phytoremediation have recently been reviewed by Kawahigashi (2009), Prasad (2009), Ismail (2012) and Dhankher *et al.* (2012).

Techniques and Mechanisms of Phytoremediation

For removal of different hazardous compounds from contaminated soil and water, plant potentials have been exploited that have resulted in several technological subsets. Schwitzguebel (2000) has defined the following techniques of Phytoremediation:

1. **Phytoextraction:** The use of pollutant-accumulating plants to remove pollutants like metal organics from soil by concentrating them in harvestable plant parts.

2. **Phytotransformation:** The degradation of complex organic to simple molecules or the incorporation of these molecules into plants tissues.

3. **Phytostimulation:** Plant-assisted bioremediation or the stimulation of microbial and fungal degradation by release of exudates/enzymes into the root zone (rhizosphere).

4. **Phytovolatilization:** The use of plants to volatilize pollutants or metabolites.

5. **Phytodegradation:** Enzymatic breakdown of organic pollutants such as trichloroethylene (TCE) and herbicides, both internally and externally and through secreted plant enzymes.

6. **Phytorhizofiltration:** The use of plant roots to ab/adsorb pollutants, mainly metals, but also organic pollutants, from water and aqueous waste streams.

7. **Dendroremediation (Pump and tree):** The use of trees to evaporate water and thus to extract pollutants from the soil.

8. **Phytostabilization:** The use of plants to reduce the mobility and bioavailability of pollutants in the environment, thus preventing their migration to groundwater or their entry into the food chain.

9. **Hydraulic Control:** The control of the water and the soil field capacity by plant canopies.

Case Studies of Biotechnological Approaches for Phytoremediation of different Pollutants

As mentioned earlier, different pollutants have different fates in plant-substrate systems, so they have different rate-limiting factors for Phytoremediation that may be targeted using genetic engineering. A schematic overview of biotechnological approaches that may enhance various rate limiting steps in phytoremediation has been provided in Figure 17.1.

I. Inorganic Pollutants

Inorganic pollutants include metals/metalloids (*e.g.*, As, Cd, Cu, Hg, Mn, Se, Zn), radionuclides (*e.g.*, Cs, P, U), and plant fertilizers (*e.g.*, nitrate, phosphate). All occur in nature mainly as positively or negatively charged ions and depend on plant transporters for uptake and translocation. Inorganic can be altered (reduced/ oxidized), moved into/inside plants, or in some cases volatilized (Hg, Se), but cannot be degraded. Thus, phytoremediation methods available for inorganic include mmobilization (phytostabilization), sequestration in harvestable plant tissues (phytoextraction or rhizofiltration) and, in exceptional cases, phytovlatilization. As reviewed by Pilon-Smits (2005) and Doty (2008) biotechnological approaches that have successfully altered the capacity of plants for Phytoremediation of inorganic have focused on both tolerance and accumulation.

Arsenic

Arsenic-contaminated ground waters, apart from use for drinking, are widely used for irrigation of many crops, particularly rice (*Oryza sativa*), adding more than 1000 tons of As per year to the agricultural soils in Bangladesh alone (Ali *et al.*, 2003).

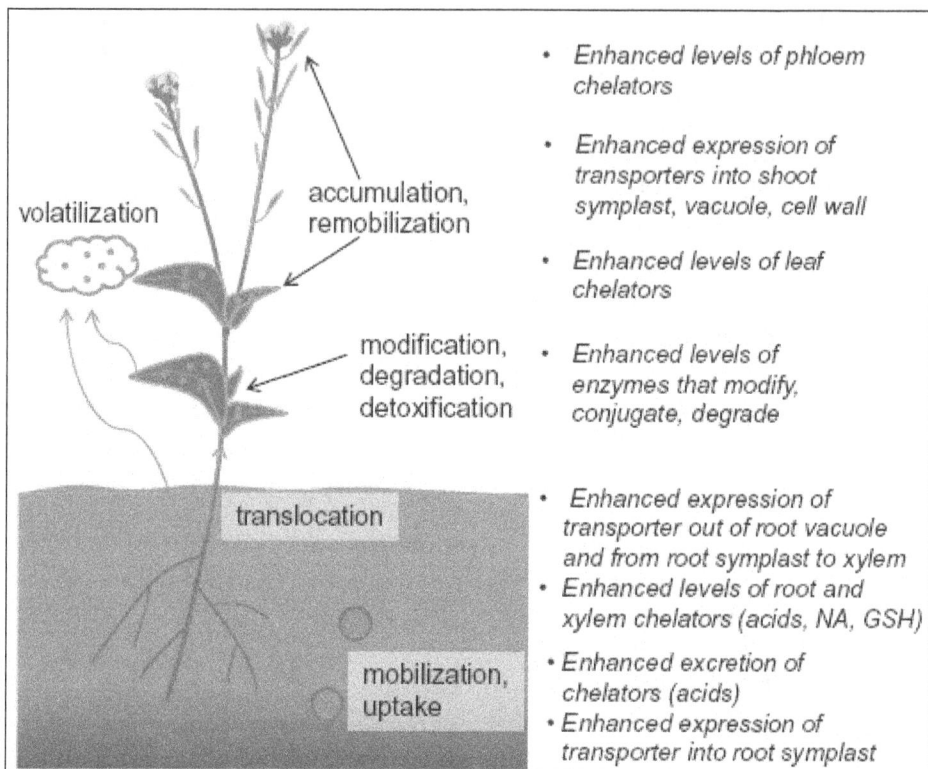

Figure 17.1: Schematic Overview of Biotechnological Approaches that may Enhance Various Rate Limiting Steps in Phytoremediation (Reproduced from Dhankher *et al.*, 2012).

Arsenic species are non-biodegradable and they remain in the surface and subsurface of agricultural soils (Juhasz *et al.*, 2003). Significantly high levels of arsenic in the edible crops grown in contaminated soils have been reported in many countries (Williams *et al.*, 2005).

The Chinese brake fern (*P. vittata*) has an exceptional ability to hyperaccumulate very high levels of As (Ma *et al.*, 2001) and thrives in tropical and subtropical places. Thus, *P. vittata* could be highly useful for phytoremediation of As in those regions. In contrast to other land plants, AsIII is the main form of accumulated As in *P. vittata*, where As is transported from rhizome to the frond region and stored as free AsIII (Zhao *et al.*, 2003). A gene, PvACR3, encoding a protein weakly homologous to the yeast ACR3 arsenite effluxer, has been shown to be localized to the vacuolar membrane in the fern gametophyte, indicating that it likely effluxes AsIII into the vacuole for sequestration (Indriolo *et al.*, 2010).

Transgenic plants with strong tolerance to As and enhanced As accumulation in the shoots were developed by co-expressing two bacterial genes (Dhankher *et al.*, 2002; Dhankher *et al.*, 2012). The *E. coli* arsenate reductase, *arsC*, gene was expressed in leaves as driven by a light-induced soybean RuBisCo small subunit 1 (*SRS1*) promoter. In addition, the *E. coli* γ-glutamylcysteine synthatase, γ-*ECS*, was expressed in both roots and shoots, driven by a strong constitutive *Actin2* promoter (Dhankher *et al.*, 2002).

Mercury

Mercury is a highly toxic pollutant and its widespread contamination in the soil and water is threatening human and environmental health (Kraemer and Chardonnens, 2001). Mercury is usually released into the environment in inorganic forms, either elemental metallic (Hg (0)) or ionic (Hg (II)) forms. Hg (II) tends to bind strongly to soil components, which reduces its availability and absorption. The world first became aware of the extreme dangers of methylmercury in the 1950s after a large, tragic incident of human Hg poisoning at Minamata Bay, Japan (Harada, 1995).

Selenium

Selenium is an essential nutrient for many organisms including humans, but is toxic at elevated levels. Selenium deficiency and toxicity are problems worldwide. There is no evidence that Se is essential for higher plants, but due to its similarity to sulfur Se is readily taken up and assimilated by plants via sulfur transporters and biochemical pathways. Plants accumulate Se in all organs including seeds, and can also volatilize Se into the atmosphere. Some species can even hyperaccumulate Se up to 1 per cent of their dry weight.

The toxicity of Se is thought to be due to non-specific incorporation of the resulting amino acids, selenocysteine (SeCys) and selenomethionine (SeMet), into proteins. To prevent this, plants may break down SeCys into relatively innocuous elemental Se (Se0) or methylate it to relatively non-toxic methyl-SeCys, which may be accumulated or further methylated to volatile dimethyldiselenide (DMDSe). SeMet can also be methylated to form volatile dimethylselenide (DMSe). Methylation of SeCys occurs primarily in Se hyperaccumulators, and is thought to be a key mechanism for their Se tolerance (Pilon-Smits and Quinn, 2010).

In a first approach to manipulate plant Se tolerance, accumulation, and/or volatilization, genes involved in sulfur/selenium assimilation and volatilization were over-expressed. *Brassica juncea* (Indian mustard) over-expressing ATP sulfurylase (APS), involved in selenate-to-selenite conversion, showed enhanced selenate reduction, judged from the finding that transgenic APS plants supplied with selenate accumulated an organic form of Se while wild-type plants accumulated selenate (Pilon- Smits *et al.*, 1999).

A second approach to manipulate plant Se metabolism focused on the prevention of SeCys incorporation into proteins. In one strategy, selenocysteine lyase (SL) was expressed in *A. thaliana* and Indian mustard, initially using a mouse SL (Pilon *et al.*, 2003). In another strategy to prevent SeCys incorporation into proteins, SeCys methyltransferase (SMT) from hyperaccumulator *A. bisulcatus* was over-expressed in *A. thaliana* or *B. juncea* (LeDuc *et al.*, 2004).

II. Organic Pollutants

In general, plants use a three-step pathway for the detoxification of organic pollutants. In the first phase, a reactive group, such as a hydroxyl, amino, or sulfhydryl group, is added to the xenobiotic. In the second phase, another compound such as a sugar moiety is conjugated via the reactive group. Finally, in the third phase the conjugated pollutant is sequestered into the vacuole or integrated into cell wall components, thus rendering the compound less toxic.

Efforts to increase the effectiveness of phytoremediation of organic pollutants involve either the over-expression of the plant genes involved in any of these steps, introduction of microbial genes known to be involved in pollutant biodegradation, or the inoculation of the plant with pollutant-degrading endophytes (reviewed in Doty, 2008; Dowling and Doty, 2009).

Solvents

Phytoremediation of solvents including TCE is effective for sites with shallow groundwater within the range of tree roots. Poplar trees are especially well suited for phytoremediation of TCE as they are deep-rooted, and a variety of herbaceous species (tobacco, *Leucaena leucocephala*, *Arabidopsis*) also have the genetic capability to degrade TCE (Doty *et al.*, 2003; Doty, 2008). However, phytoremediation of TCE is limited by the apparent low expression of the cytochrome P450 enzyme that activates TCE prior to its degradation. The metabolism of TCE in plants is often considered too slow and may lead to Phytovolatilization of the pollutant.

Strategies to improve phytoremediation of TCE include genetic engineering or endophyte-assisted phytoremediation (Doty, 2008). Therefore, the plants may be more suitable for contaminated water as simulated in the lab studies (Doty *et al.*, 2007) rather than for pump and treat systems that have a continuous source of TCE.

Explosives

At military training ranges there is a need for remediation of the nitroaromatic explosives, TNT and RDX (hexahydr-1,3,5-trinitro-1,3,5-triazine), to prevent the spread into neighboring communities. TNT causes anemia and liver damage, while RDX affects the central nervous system, causing convulsions. Extensive areas are contaminated with these pollutants, approximately 40 million acres in the United States alone (U.S. Defense Science Board Task Force, 1998; Rylott and Bruce, 2008) Nitroaromatic explosives contain an aromatic ring with attached nitro (-NO_2) groups. A difficulty in phytoremediation of TNT is the toxicity of this pollutant, and its concentration on sites can be as high as 87,000 mg (Talmage *et al.*, 1999).

Pesticides

Since pesticides can cause chronic abnormalities in humans and they generally lead to reduced environmental quality, multiple methods including incineration and land filling have been used to remove this class of pollutants; however, these physical methods are expensive and inefficient. Bioremediation using microorganisms capable of degrading the polluting pesticide and enhanced phytoremediation of pesticides using transgenic plants are emerging as more effective solutions (Hussain *et al.*, 2009).

Plant Species Used for Phytoremediation

Various plant species have been tried for removal of a variety of pollutants from inorganic and organic sources. Efforts are being made by various workers in different countries across the world. Presently, apart from using lower plant groups, herbaceous plants, microbes (fungi and bacteria), efforts are also going on for use of tree species for control of pollution. Trees are proving more effective in treating deeper contamination, as tree roots penetrate more deeply in to the ground. Major aim in phytoremediation is to find a plant species which is resistant to or tolerant to a particular pollutant to maximize its potential for phytoremediation. A summary of processes, mechanisms and related pollutants and plant species have been provided in Table 17.1. A summary of transgenic plants for phytoremediation of organic pollutants has been provided in the Table 17.2.

Table 17.1: Phytoremediation Processes, Mechanisms, and Related Pollutants *vs* Plant Species

Phytotechnology	Mechanism	Pollutants	Plants
Phytoextraction	Hyperaccumulation in harvestable parts of plants	Inorganic: Co, Cr, Ni, Pb, Zn, Au, Hg, Mo, Ag, Cd Radionuclides: Sr, Cs, Pb, U	*Brassica juncea, Thalspi caerulescens, Helianthus annus*
Rhizofilteration	Rhizosphere accumulation through sorption concentration and precipitation	Organics/Inorganics: Metals like Cd, Cu, Ni, Zn, Cr Radionuclides	*Brassica juncea, Helianthus annus,* Tobacco, Rye, Spinach and Corn
Phytovolatilization	Volatilization by leaves through transpiration	Organics/Inorganics: Chlorinated solvents. inorganics (Se, Hg, As)	*Arabidopsis thaliana* Poplars, Alfalfa, *Brassica juncea*
Phytodegradation	Pollutant eradication	Organic compounds, Chlorinated solvents, Phenols, Herbicides, Munitions	Hybrid poplars, Stonewort. Blackwillow, Algae
Phytostabilization	Complexation, sorption and precipitation	Inorganics: As, Cd, Cu, Cr, Pb, Zn, Hs	*Brassica juncea,* Hybrid poplars, Grasses

Table 17.2: List of Transgenic Plants Tried for Phytoremediation of Organic Pollutants

Compound	Gene	Plant
Atrazine	*bphC*	Tobacco, *Arabidopsis,* alfalfa
Atrazine	*hCYP1A1/A2*	Tobacco
Atrazine	*hCYP1A1/2B6/2C19*	Rice
Benzene	*rCYP2E1*	Poplar
Benzene	*hCYP2E1*	Tobacco
PCBs	*bphC*	Tobacco
PCBs	*bphA, bphE, bphF, bphG*	Tobacco
PCBs	*bphC*	Tobacco

Contd...

Table 17.2–*Contd...*

Compound	Gene	Plant
RDX	xplA	Arabidopsis
RDX	xplA, xplB	Arabidopsis
TCE	hCYP2E1	Tobacco
TCE	rCYP2E1	Atropa belledonna
TCE	rCYP2E1	Poplar
TNT	PETNr	Tobacco
TNT	nfsI	Tobacco
TNT	pnrA	Poplar
TNT	UGTs	Arabidopsis

Future Prospects

In the past decade there has been a tremendous increase in our knowledge of plant processes involved in, and limiting for phytoremediation of a wide variety of inorganic and organic pollutants. This knowledge has been obtained in part through plant biotechnology and conversely has led to plant biotechnological approaches to enhance the phytoremediation potential of plants. In some cases, as described earlier, natural plant processes involved in uptake, assimilation, or detoxification were manipulated, while in other cases entirely new processes were introduced, often by introducing bacterial genes or even entire bacterial endophytes. Results from field studies are starting to come in and tend to confirm results from initial lab and greenhouse trials. Clearly, plant biotechnological approaches have played an important role in moving the field of phytoremediation forward. For better acceptance in the remediation industry, it is important that new transgenics continue to be tested in the field. In that context it will be helpful if regulatory restrictions can be regularly reevaluated to make the use of transgenics for phytoremediation less cumbersome.

A multidisciplinary research effort that integrates the work of plant biologist, soil chemists, microbiologists and environmental engineers is essential for greater success of phytoremediation as a viable soil clean up technique. Combinational approaches such as genome shuffling are also useful for generating new genes or modifying enzymes activities to allow bioremediation. An appropriate use of transgenic plants and bacteria in the rhizosphere would provide a reliable basis for enhancing phytoremediation of contaminated environments and may overcome the current limitations such as detoxification and absorption efficiency of phytoremediation. Phytoremediation is a fast emerging field as a viable alternative to conventional remediation methods and will be most suitable for a developing country like India. However, further research is required to optimize the ecological and economical efficiencies of phytoremediation with aid from biotechnology.

Acknowledgement

Authors are thankful to all those researchers whose publications are the basis of the present review article. Figure reproduced and included in this article is duly acknowledged to the original worker.

References

1. Bizily, S. P., Kim, T., Kandasamy, M. K., and Meagher, R. B. (2003). Subcellular targeting of methylmercury lyase enhances its specific activity for organic mercury detoxification in plants. *Plant Physiology, 131,* 463–471.

2. Cunningham, S. D., Anderson, T. A., Schwat, T., and Hsu, F.C. (1996). Phytoremediation of soils contaminated with organic pollutants. *Adv. Agronomy, 56,* 55-114.

3. Dhankher, O. P., Li, Y., Rosen, B. P., Shi, J., Salt, D., and Senecoff, J. F., *et al.* (2002). Engineered tolerance and hyperaccumulation of arsenic in plants by combining arsenate reductase and γ-glutamylcysteine synthetase expression. *Nature Biotechnology, 20,* 1140–1145.

4. Dhankher, O.P., Pilon-Smits, E.A.H., Meagher, R.B., and Doty, S. (2012). Biotechnological approaches for phytoremediation. In: Altman, A.; Hasegawa, P.M., eds. *Plant biotechnology and agriculture.* pp. 309-328. Academic Press, San Diego, CA, USA.

5. Doty, S. L. (2008). Tansley Review: Enhancing phytoremediation through the use of transgenics and endophytes. *New Phytologist, 179,* 318–333.

6. Doty, S. L., James, C. A., Moore, A. L., Vajzovic, A., Singleton, G. L., and Ma, C., *et al.* (2007). Enhanced phytoremediation of volatile environmental pollutants with transgenic trees. *Proceedings of the National Academy of Sciences of the United States of America, 104,* 16816–16821.

7. Dowling, D. N., and Doty, S. L. (2009). Improving phytoremediation through biotechnology. *Current Opinion in Biotechnology, 20,* 204–206.

8. Harada, M. (1995). Minamata disease: Methylmercury poisoning in Japan caused by environmental pollution. *Critical Reviews in Toxicology, 25,* 1–24.

9. Hussain, S., Siddique, T., Arshad, M., and Saleem, M. (2009). Bioremediation and phytoremediation of pesticides: Recent advances. *Critical Reviews in Environmental Science and Technology, 39,* 843–907.

10. Indriolo, E., Na, G. N., Ellis, D., Salt, D. E., and Banks, J. A. (2010). A vacuolar arsenite transporter necessary for arsenic tolerance in the arsenic hyperaccumulating fern *Pteris vittata* is missing in flowering plants. *Plant Cell, 22,* 2045–2057.

11. Ismail, S. (2012). Phytoremediation: a green technology. *Iranian Journal of Plant Physiology, 3(1),* 567-576.

12. Juhasz, A. L., Naidu, R., Zhu, Y. G., Wang, L. S., Jiang, J. Y., and Cao, Z. H. (2003). Toxicity issues associated with geogenic arsenic in the grondwater–soil–plant–

human continuum. *Bulletin of Environmental Contamination and Toxicology, 71,* 1100–1107.

13. Kawahigashi, H. (2009). Transgenic plants for phytoremediation of herbicides. *Current Opinion in Biotechnology, 20:* 225-230.

14. Khan, A. G., Kuek, C., Chaudhry, T. M., Khoo, C.S., and Hayes, W. J. (2000). Role of lants, mycorrhizae and phytochelators in heavy metal contaminated land remediation. *Chemosphere, 41:* 197-207.

15. Kraemer, U., and Chardonnens, A. N. (2001). The use of transgenic plants in the bioremediation of soil contaminated with trace elements. *Applied Microbiology and Biotechnology, 55:* 661–672.

16. LeDuc, D. L., Tarun, A. S., Montes-Bayon, M., Meija, J., Malit, M. F., and Wu, C. P., *et al.* (2004). Overexpression of selenocysteine methyltransferase in *Arabidopsis* and Indian mustard increases selenium tolerance and accumulation. *Plant Physiology, 135:* 377–383.

17. Ma, L. Q., Komar, K. M., Tu, C., Zhang, W. H., Cai, Y., and Kennelley, E. D. (2001). A fern that hyperaccumulates arsenic: A hardy versatile fast-growing plant helps to remove arsenic from contaminated soils. *Nature, 409:* 579–579.

18. Pilon, M., Owen, J. D., Garifullina, G. F., Kurihara, T., Mihara, H., and Esaki, N., *et al.* (2003). Enhanced selenium tolerance and accumulation in transgenic *Arabidopsis thaliana* expressing a mouse selenocysteine lyase. *Plant Physiology, 131,* 1250–1257.

19. Pilon-Smith, E. (2005). Phytoremediation: *Annual Rev. Plant Biol. 56,* 15-39.

20. Pilon-Smits, E. A. H., and Quinn, C. F. (2010). Selenium Metabolism in Plants. In R. Hell and R. Mendel (Eds.), *Cell biology of metals and nutrients* (pp. 225–241). Berlin Heidelberg: Springer Press.

21. Pilon-Smits, E. A. H., Hwang, S., Lytle, C. M., Zhu, Y., Tai, J. C., and Bravo, R. C., *et al.* (1999). Overexpression of ATP sulfurylase in Indian mustard leads to increased selenate uptake, reduction, and tolerance. *Plant Physiology, 119,* 123–132.

22. Prasad, M. N. V. (2009). Emerging phytotechnologies for remediation of heavy meal contaminated/polluted soil and water. Sahyadri enews, newsletter, issue 25 (http://wgbis.ces.iisc.ernet.in/biodiversity/sahyadri_enews/newsletter/issue25/article2.htm)

23. Rylott, E. L., and Bruce, N. C. (2008). Plants disarm soil: Engineering plants for the phytoremediation of explosives. *Trends in Biotechnology, 27,* 73–81.

24. Schwitzguebel, J. P. (2000). Potential of Pytoremediation, an Emerging Green Technology. Ecosystem Service and Sustainable Watershed Management in North China, International Conference, Beijing, P.R. China, August, 23–25, 364-350.

25. Talmage, S. S., Opresko, D. M., Maxwell, C. J., Welsh, C. J., Cretella, F. M., and Reno, P. H., *et al.* (1999). Nitroaromatic munition compounds: Environmental

effects and screening values. *Reviews of Environment Contamination and Toxicology,* *161,* 1–156.

26. US Defense Science Board Task Force, *Unexploded ordnance (UXO) clearance, active* *range UXO clearance, and explosive ordnance disposal programs.* Washington, D.C.: Office of the Undersecretary of Defense for Acquisition and Technology. Ref Type: Report.

27. Williams, P. N., Price, A. H., Raab, A., Hossain, S. A., Feldmann, J., and Meharg, A. A. (2005). Variation in arsenic speciation and concentration in paddy rice related to dietary exposure. *Environmental Science and Technology, 39,* 5531–5540.

28. Zhao, F. J., Wang, J. R., Barker, J. H. A., Schat, H., Bleeker, P. M., and McGrath, S. P. (2003). The role of phytochelatinsin arsenic tolerance in the hyperaccumulator *Pteris vittata. New Phytologist, 159*: 403–410.

2016, Environmental Biotechnology: A New Approach
Editors: Dr. Rajan Kumar Gupta and Dr. Satya Shila Singh
Published by: DAYA PUBLISHING HOUSE, NEW DELHI

Pages 277–285

Chapter 18

Role of Programmed Cell Death in Plant Development and Defense

Bhupendra Mathpal[1]*, Nitin Kumar[2] and Suneel Kumar Singh[1]

[1]*Department of Biotechnology, Modern Institute of Technology, Dhalwala, Rishikesh, Uttarakhand – 249 201*
[2]*Department of Plant Physiology, College of Basic Sciences and Humanities, G.B. Pant University of Agriculture and Technology, Pantnagar, U.S. Nagar, Uttarakhand – 263 145*

ABSTRACT

The suicide of individual cells is an efficient and conserved mechanism to maintain homeostasis in multicellular organisms as a response to pathogen attack, abiotic stress, as well as in normal development. Programmed cell death (PCD) is a physiological cell death process which involves selective elimination of unwanted cells. In plants, cell death causes the deletion of the suspensor and aleurone cells, elimination of stamen primordial cells in female flowers. Cell death also causes the formation of leaf lobes, leaf perforations and differentiation of tracheary element (TE) of xylem. Senescence and cell death associated with hypersensitive response (HR) are the main forms through which PCD takes place in plants. Plants can recognize pathogen and respond to infection through programmed cell death. The first line of defense in plants is provided by the receptors, which recognize microbes or danger associated molecular patterns (MAMPs or DAMPs) and initiate immune signaling. Signals produced by hypersensitive response causes increase in concentration of salicylic acid (SA) and jasmonic acid (JA) in the distal plant parts, which leads to increased resistance against infection, called systemic acquired resistance (SAR).

Keywords: Programmed cell death, Plant development, Defense.

* *Corresponding author.* E-mail: mathpal.88@gmail.com

Introduction

Death of specific set of cells is an essential part of the growth and development of many eukaryotic organisms, including both in plants and animals. Programmed cell death is the selective and genetically regulated process which involves the elimination of certain cells to ensure the maintenance of homeostasis and overall integrity of both the plants and animals. In addition to its role in development, cell death plays a very important role in biotic and abiotic stresses. Programmed cell death (PCD) is an energy dependent physiological process that is genetically programmed with the involvement of regulatory genes, stimulatory events, signaling pathways and morphologically expressed distinct features (Ellis *et al.*, 1991). In animals PCD controls the cell number and turnover especially in the skin, gut and reproductive tract, oncogenesis and other disease incidents, defense reactions and thus, maintain overall integrity of the animals (Jacobson *et al.*, 1997; Krishnamurthy *et al.*, 2000 and Fuchs and Steller, 2011). While in plants it controls so many developmental and defense responses. Among the developmental process, the deletion of suspensor, aleurone cells, formation of leaf lobes, cell death in xylem tracheary element (TEs) and many other functions (Wyllie *et al.*, 1980; Pennell and Lamb, 1997; Roger *et al.*, 1997; Krishnamurthy *et al.*, 2000 and Williams and Dickman, 2008).

Different types of cell death occur in both plants and animals. Cells that die by process that requires energy and regulated by distinct sets of the genes are said to undergo programmed cell death and are of five types namely, apoptosis, necrosis, senescence, hypersensitive response and autophagy (Jones and Dangl, 1996; Levine *et al.*, 1996; Beers, 1997; Gunawardena *et al.*, 2001a and b; Jones, 2001; Arunika *et al.*, 2004; Collazo *et al.*, 2006; Shlezinger *et al.*, 2012 and Coll *et al.*, 2014). A number of categories of cells are recognized in both plants and animals that undergo cell death. A number of plant cells belong to this category are root cap cells, cells undergoing senescence, cells involved in abscission of plants parts, vegetative cells of embryo sac such as synergids and antipodal cells, aleurone cells of endosperm, suspensor cells of developing embryo etc (Figure 18.1). These cells lose their earlier identity, morphology and functions and are completely eliminated in animals while in plants their remnants may persist. The stamens and carpel primordial cells in male and female flowers respectively, nonfunctional megaspores in monosporic and bisporic embryo sac, nonfunctional microspore in members of cyperaceae and epicardiaceae etc. are examples of this category (Giuliani *et al.*, 2002). There are two main characteristic features of cells comes under this category. The first one is that they start functioning only after their death. The other characteristic feature of these cells is that significant changes takes place in their cell walls. The examples of cells comes under this category are xylem tracheary elements, cells of certain trichomes, thorns, spines and scelerenchyma cork cells etc. (Groover *et al.*, 1997 and Fukuda, 2000). A subcategory under this includes those cells which undergo partial death during differentiation. The phloem sieve elements belong to this subcategory.

Cells subjected to hypersensitive response (HR) due to pathogenesis and other environmental and abiotic stresses such as osmotic, oxidative (H_2O_2 and salicylic acid), temperature, salt, water, heavy metals, UV, nutrient deprivation, toxins, chemicals, etc. often undergo PCD. Here cell death is used as a defense against the

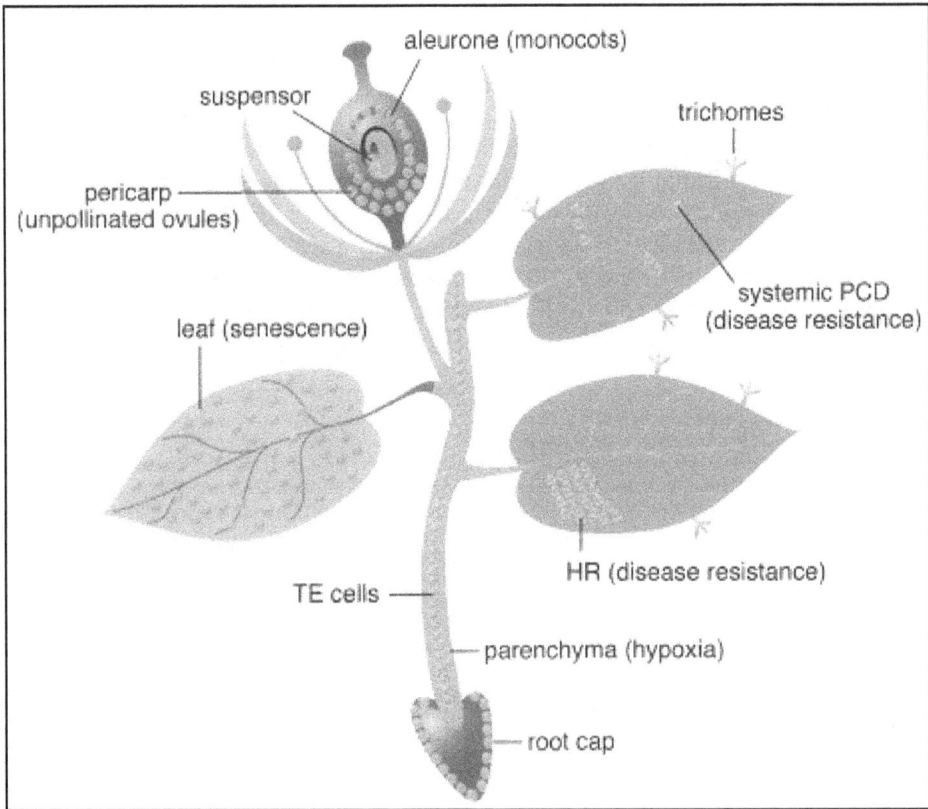

Figure 18.1: Sites of PCD in Vascular Plants. Orange Spheres Represents Internal Dead Cells.

stress or as a means of emanating signals for the other nearby cells to build up defense/immune reactions. Programmed cell death (PCD) is reported to be employed in these instances not only to kill the host cells, but also the invading pathogens. In many instances, cells experiencing lethal stress doses often react by activating their PCD mechanisms to commit suicide before they are killed. A critical survey of the available literature indicates that not all cells subjected to stress undergo PCD; there are cells which may undergo necrosis, especially if the dosage of stress is beyond a particular level (Coll *et al.*, 2011; Jones 2001; Sharon and Shlezinger, 2013 and Sharon and Finkelshtein, 2009). Cells that die to result in leaf lobe formation in lobed leaves and leaflets in compound leaves or to result in holes in the lamina of taxa like Monstera and some Croton varieties are typical examples of plant cells of this category (Simionova *et al.*, 2000 and Arunika *et al.*, 2004).

After fertilization in most angiosperms, the first mitotic division of zygote gives rise to two cells. One produces the embryonic cell and the other suspensor cell. Embryo is developed from embryonic cell. Suspensor cells help this developing embryo to take nutrition from endosperm. So, after the complete development of embryo these suspensor cells are eliminated by PCD (Figures 18.2A and 18.2B) (Pennell and Lamb

1997 and Giuliani *et al.*, 2002). In seeds of monocots aleurone cells form secretory tissue that releases hydrolases to digest the endosperm and to nourish the embryo. Aluerone cells are unnecessary for postembryonic development and die as soon as germination is complete (Figures 18.2C and 18.2D). Flower development is affected by PCD of selected cells or group of cells. In many plants that contain flowers, the developing flower initially contain primordia for both male and female organs. At early stages, male and female flowers are indistinguishable. During flower formation either the male or female parts cease growing and are eliminated via a cell death programme (Figures 18.2E and 18.2F). A cap of cells protects the root apical meristems during seed germination and seedling growth. Root cap cells are formed by initial cells in the meristems and are continually displaced by new cells. After several days

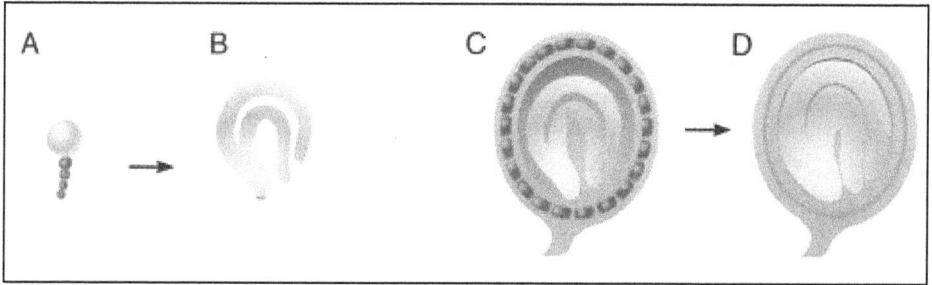

Figures 18.2: A and B: Deletion of Suspensor Cells in Embryos. **C and D**: Deletion of Aleurone Cells in Seeds.

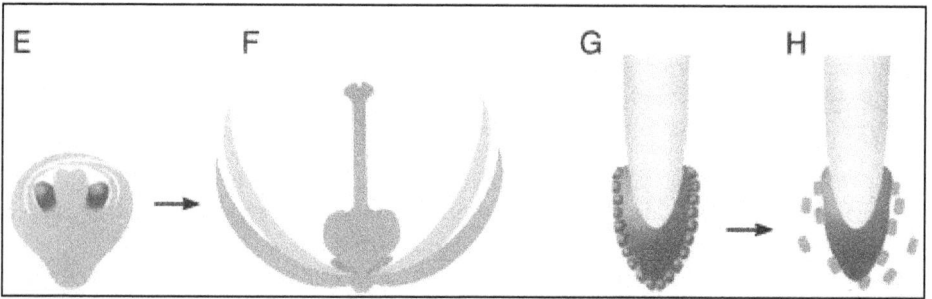

Figures 18.2: E and F: Deletion of Stamen Primordia Cells in Unisexual Flowers. **G and H**: Deletion of Root Cap Cells.

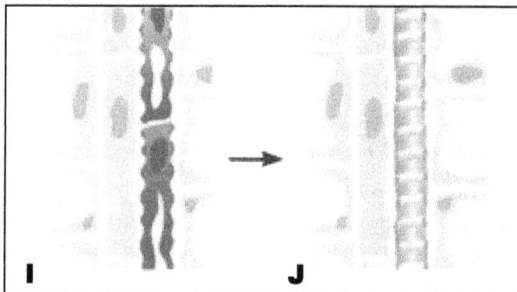

Figures 18.2: I and J: Deletion of Cells during Cell Specialization in Tracheary Elements.

the peripheral cells die (Figures 18.2G and 18.2H). This cell death also occurs in root cap when roots are grown in water. It shows that cell death is a normal part of development and not a consequence of abrasion during soil penetration (Pennell and Lamb, 1997; Beers, 1997 and Kuriyama and Fukuda 2002).

Differentiation of tracheary elements (TEs) involves cell elongation, deposition of cell wall components, including lignin and autolysis (Figures 18.2I and 18.2J). A plant *Zinnia elegans* where mesophyll cells can be cultured and induced to re-differentiating into TEs so that cell death process can be studied biochemically as well as morphologically. Treatment of differentiating *Zinnia* cells with actinomycin D or cycloheximide blocks cell death (Groover *et al.*, 1997, Pennell and Lamb, 1997 and Fukuda, 2000).

Electron microscope image of cells from an embryonic suspension culture of carrot is shown in Figure 18.3. Cultured cells of carrot plants were induced to develop into somatic embryo. In embryonic suspension culture, totipotent cells divide asymmetrically into cell pairs, one member of which stops synthesizing DNA and die whereas other members were developed into embryos. The condensation and shrinkage of cytoplasm and the small DNA fragments imply that these cells die by PCD (Pennell and Lamb, 1997). Arenchyma formation in maize roots in response to hypoxia is shown in Figure 18.4. Roots were grown under aerobic and hypoxic conditions. Under low oxygen conditions, cortical cells undergo lysigeny to form air spaces that are continuous throughout the root, thereby allowing submerged roots access to atmospheric gases obtained by above ground tissues (Buchanan *et al.*, 2000).

Plant pathogen is an organism that, to complete a part or all of its life cycle grows inside the plant and in so doing has a detrimental effect on the plant (Buchanan *et al.*, 2000). Plants are infected by many pathogens and they must continuously defend themselves against attack from bacteria, viruses, fungi, invertebrates, herbivores and even other plants. Because of their immobility they can not escape from pathogen

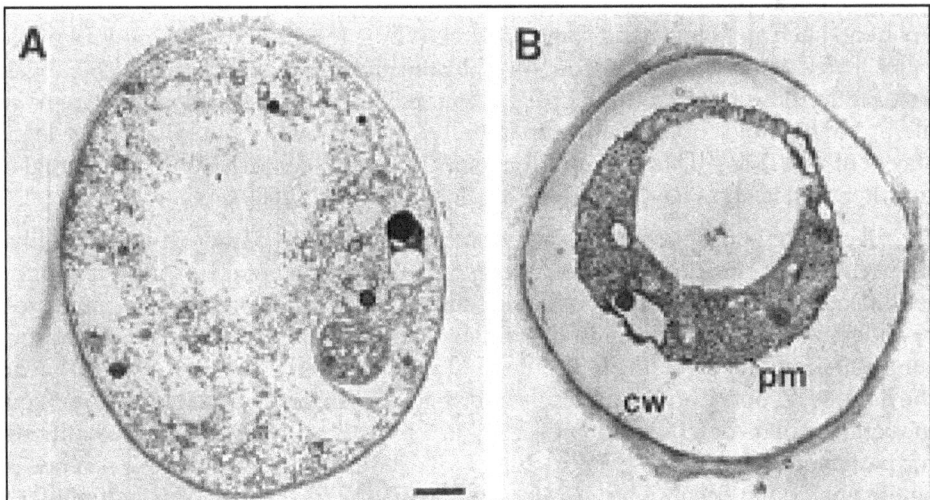

Figure 18.3: Role of Programmed Cell Death in Somatic Embryogenesis.

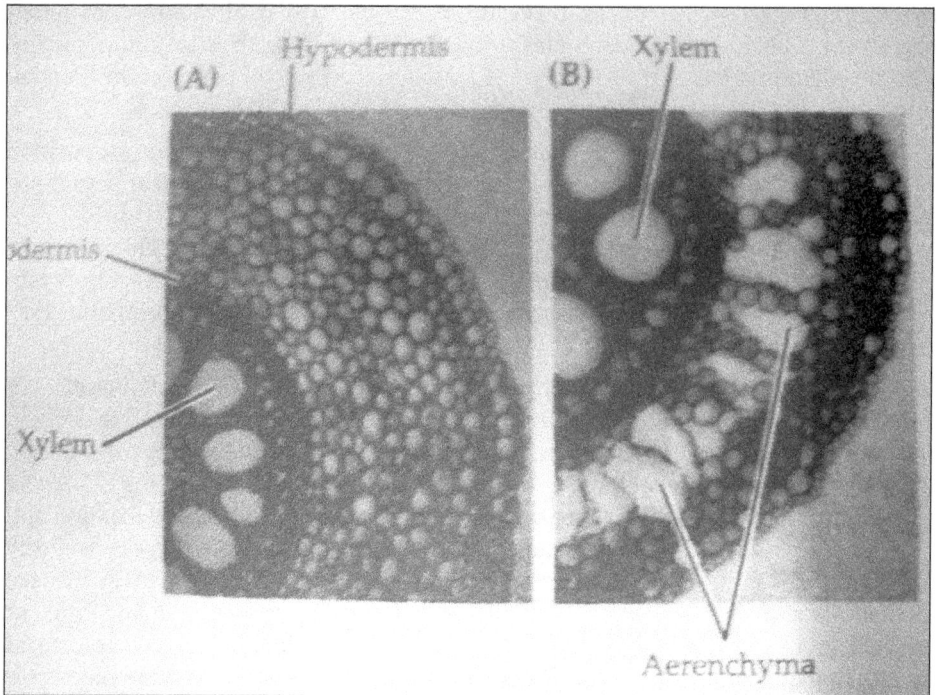

Figure 18.4 A: Roots Under Aerobic Condition; B: Roots Under Hypoxic Condition.

attack, so each plant cell possesses both preformed and an inducible defense capacity (Greenberg, 1996; Greenberg and Yao, 2004 and Sharon and Finkelshtein, 2009). All the plant parts come into direct contact with pathogens. Each pathogen has adopted a specific strategy to invade plants. Some pathogens directly penetrate surface layers by mechanical pressure or enzymatic attack while other pass through natural openings such as stomata and some pathogens invades through wounding of plant parts. Three main strategies are deployed by the pathogen to utilize the host plant as a substrate, these are necrotrophic (plant cells are killed), biotrophic (plant cell remain alive) and hemibiotrophic (pathogen initially keeps cells alive but kills them at later stages of infection). This overall process of infection, colonization and pathogen reproduction is known as pathogenesis (Buchanan *et al.*, 2000).

Plants show different types of responses against the pathogen infection like transcriptional activation of numerous defense related genes, opening of ion channels, modification of protein phosphorylation status and activation of preformed enzymes to undertake specific modification to primary and secondary metabolism. There are a range of secondary signaling molecules that are generated to ensure coordination of the defense response both temporally and spatially. This rapid and highly localized induction of plant defense responses result in the creation of unfavourable conditions for pathogen growth and reproduction. Full activation of these intense responses against pathogen occurs within 24 hrs and leads to either directly or indirectly to localized cell and tissue death. This rapid activation of defense reaction in association

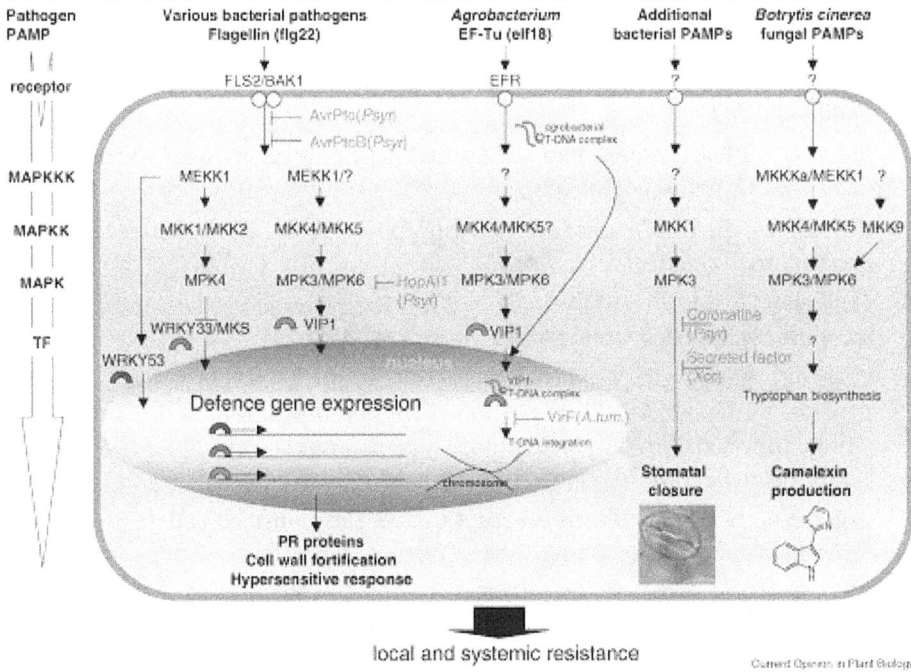

Figure 18.5: HR in the Plant Defense to Bacterial and Fungus Pathogen.

with the host cell death is called hypersensitive response (HR) (Figure 18.5). Because the dead cells contain high concentration of molecules with antimicrobial activity, they are not subsequently attacked by pathogens (Buchanan *et al.*, 2000; Greenberg and Yao, 2004 and Sharon and Finkelshtein, 2009).

Viruses, fungi and bacteria activate systematically a specific subset of defense responses in a phenomenon known as systemic acquired resistance (SAR). It involves necrosis of tissue at the initial site of pathogen invasion which triggers both a local increase in salicylic acid (SA) accumulation and initiation of phloem mobile signal. Subsequently in distal plant parts, salicylic acid concentration increased and volatile methyl salicylate (MeSA) is released. Together these, signal induce the synthesis of various pathogen related proteins in the non invaded parts of the plant (Buchanan *et al.*, 2000 and Pitzschke *et al.*, 2009).

Conclusion

Despite significant recent development and discoveries in the field of PCD, there is still a lack of information of several genes such as caspases that have been found in animals. The similarity of plant PCD with animals should need to be tested that they are conserved and are derived from a common ancestor. The application of PCD to achieve resistance against pathogens in plants should be taken into account more seriously in the future. Overall, in plants PCD is essential for development and survival but in animals it is a way to rid the organism of unwanted cells.

References

1. Arunika H. L.; Gunawardena, A. N.; Greenwood, J.S. and Dengler, N.G. 2004. Programmed Cell Death Remodels Lace Plant Leaf Shape during Development. *Plant Cell*, 16(1): 60-73.

2. Beers, E.P. 1997. Programmed cell death during plant growth and development. *Cell Death Differ.*, 4: 649-661.

3. Buchanan, B. B., Gruissem, W. and Jones, R. L. 2000. Senescence and programmed cell death. *Biochem. Mol. Biol. Plants*. 1044-1099.

4. Coll, N. S., Epple, P. and Dangl, J. L. 2011. Programmed cell death in the plant immune system. *Cell Death Differe.* 18(8): 1247-1256.

5. Coll, N. S., Smidler, A., Puigvert, M., Popa, C., Valls, M. and Dangl, J. L. 2014. The plant metacaspase AtMC1 in pathogen-triggered programmed cell death and aging: Functional linkage with autophagy. *Cell Death and Differentiation* online publication, 02 May 2014; doi: 10.1038/cdd.2014.50.

6. Collazo, C., Chacon, O. and Borras, O. 2006. Programmed cell death in plants resembles apoptosis of animals. *Biotec. Aplica.* 23: 1-10.

7. Ellis, R. E., Jacobson, D. M. and Horvitz, H. R. 1991. Genes required for the engulfment of cell corpses during programmed cell death in Caenorhabditis elegans. *Genetics*. 129(1): 79-94.

8. Fuchs, Y. and Steller, H. 2011. Programmed cell death in animal development and disease. *Cell.*147: 742-758.

9. Fukuda, H. 2000. Programmed cell death of tracheary elements as a paradigm in plants. *Plant Mol. Biol.* 44: 245-253.

10. Giuliani, C., Consonni, G., Gavazzi, G., Colombo, M., and Dolfini, S. 2002. Programmed cell death during embryogenesis in maize. *Ann. Bot.* 90: 287-292.

11. Greenberg, J. T. and Yao, N. 2004. The role and regulation of programmed cell death in plant-pathogen interactions. *Cell Microbiol.* 6: 201-211.

12. Greenberg, J.T. 1996. Programmed cell death: A way of life for plants. *Proc. Natl. Acad. Sci.* USA. 93: 12094-12097.

13. Groover, A., DeWitt, N., Heidel, A., and Jones, A. 1997. Programmed cell death of plant tracheary elements differentiating *in vitro*. *Protoplasma*. 196: 197-211.

14. Gunawardena, A.H.L.A.N., Pearce, D.M., Jackson, M.B., Hawes, C.R., and Evans, D.E. 2001a. Characterization of programmed cell death during aerenchyma formation induced by ethylene or hypoxia in roots of maize (*Zea mays* L.). *Planta*. 212: 205-214.

15. Gunawardena, A.H.L.A.N., Pearce, D.M., Jackson, M.B., Hawes, C.R., and Evans, D.E. 2001b. Rapid changes in cell wall pectic polysaccharides are closely associated with early stages of aerenchyma formation, a spatially localized form of programmed cell death in roots of maize promoted by ethylene. *Plant Cell Environ.* 24: 1369-1375.

16. Jacobson, M.D., Well, M., and Raff, M. 1997. Programmed cell death in animal development. *Cell.* 88: 347-354.

17. Jones, A.M. 2001. Programmed cell death in development and defense. *Plant Physiol.* 125: 94-97.

18. Jones, A.M., and Dangl, J.L. 1996. Logjam at the Styx: Programmed cell death in plants. *Trends Plant Sci.* 1: 114-119.

19. Krishnamurthy, K. V., Krishnaraj, R., Chozhavendan, R. and Christopher, S. F. 2000. The programme of cell death in plants and animals – A comparison. *Current Sci.* 79 (9): 1169-1181.

20. Kuriyama, H., and Fukuda, H. 2002. Developmental programmed cell death in plants. *Curr. Opin. Plant Biol.* 5: 568-573.

21. Pennell, R.I., and Lamb, C. 1997. Programmed cell death in plants. *Plant Cell* 9: 1157-1168.

22. Pitzschke, A., Schikora, A. and Hirt, H. 2009. MAPK cascade signalling networks in plant defence. *Current Opini. Plant Biol.* 12: 1–6.

23. Roger I., Pennell and Chris L. 1997. Programmed cell death in plants. *Plant Cell.* 9: 1157-1168.

24. Sharon, A. and Finkelshtein, A. 2009. Programmed cell death in fungus–plant interactions. In Esser K, Deising H, editors. The mycota. 2nd edition. Heidelberg Springer. pp. 221-236.

25. Sharon, A. and Shlezinger, N. 2013. Fungi Infecting Plants and Animals: Killers, Non-Killers, and Cell Death. Heitman J, ed. *PLoS Pathogens;* 9(8): e1003517. doi: 10.1371/journal.ppat.1003517.

26. Shlezinger, N., Goldfinger, N. and Sharon, A. 2012. Apoptotic-like programed cell death in fungi: the benefits in filamentous species. *Front Oncol.* 2: 97.

27. Simeonova, E., Sikora, A., Charzynska, M., and Mostowska, A. 2000. Aspects of programmed cell death during leaf senescence of mono- and dicotyledonous plants. *Protoplasma* 214: 93-101.

28. Williams, B. and Dickman, M. 2008. Plant programmed cell death: can't live with it; can't live without it. *Mol. Plant Pathol.* 9: 531-544.

29. Wyllie, A.H., Kerr, J.F.R., and Currie, A.R. 1980. Cell death: The significance of apoptosis. *Int. Rev. Cytol.* 68: 251306.

2016, Environmental Biotechnology: A New Approach   *Pages 287–308*
Editors: Dr. Rajan Kumar Gupta and Dr. Satya Shila Singh
Published by: DAYA PUBLISHING HOUSE, NEW DELHI

Chapter 19

Phytotechnology and its Techniques to Combat Environmental Pollution of Terrestrial and Aquatic Ecosystems

**Fouzia Ishaq[1], Santosh Kumar Singh[2]
and Amir Khan[2]***

*[1]Department of Zoology and Environmental Science,
Gurukula Kangri University, Haridwar, Uttarakhand
[2]Glocal School of Life Sciences, The Glocal Univesrity,
Mirzapur Pole, Saharanpur, Uttar Pradesh*

ABSTRACT

The scope of opportunities available to phytotechnological interventions are wide ranging, though presently, much of this potential remains unexplored. It seems likely that the increasingly prioritised initiatives to find low-cost systems to bring about effective remediation, effluent and pollution controlwill begin to reverse this situation in the near future. However, it is unavoidably true that the wider uptake of all plant-utilising bioengineering applications will depend as much on local modalities as on the actual state of the biotechnologies themselves. In this respect, phytotechnology has much in its favour. For one thing, it has the enormous benefit of virtually assured universal public acceptance, which is rare for any biotechnology. Commercially, it is a relatively low-intervention, highly 'green' and thoroughly non contentious approach to environmental management, which has a strong potentially positive contribution to corporate image, with a relatively low negative influence on the balance sheet.

---

* *Corresponding author.* E-mail: amiramu@gmail.com

The great advantage that almost all plant-based systems bring to biological engineering is the tremendous energy saving represented by their solar-powered nature. This, combined with their essentially integrated and intrinsically complex array of metabolic mechanisms makes a variety of plant species extremely useful in an environmental context, and typically without any additional need for modification. Given their inherent flexibility, acceptability, efficiency and cost effectiveness, it is difficult to imagine that phytotechnological systems will not be further developed and more widely adopted in the future.

Keywords: Phytotechnology, Environmental pollution, Terrestrial and Aquatic systems.

Introduction

Phytotechnology is the use of plants in environmental biotechnology applications. It does not represent a single unified technology or even application, but rather is a wider topic, defined solely by theeffector organisms used. Plants of one kind or another can be instrumental in the biological treatment ofa large number of substances which present many different types of environmentalchallenges. Accordingly, they may be used to remediate industrial pollution,treat effluents and wastewaters or solve problems of poor drainage or noise nuisance.

The processes of bioaccumulation, phytoextraction, phytostabilisation andrhizofiltration are collectively often referred to as phytoremediation. Although itis sometimes useful to consider them separately, in most functional respects, theyare all aspects of the same fundamental plant processes and hence there is muchmerit in viewing them as parts of a cohesive whole, rather than as distinctly differenttechnologies. Moreover, the role of phytotechnologyisnot limited solely to phytoremediationbut itis more deliberately inclusive of wider plant-based activities and uses.Despite the broad spectrum of potential action exhibited by plants in thisrespect, there are really only three basic mechanisms by which they achievethe purpose desired. In essence, all phytotechnology centres on the removal andaccumulation of unwanted substances within the plant tissues themselves, theirremoval and subsequent volatisation to atmosphere or the facilitation of in-soiltreatment. Plant-based treatments make use of natural cycles within the plant and its environment and clearly to be effective, the right plant must be chosen.Inevitably, the species selected must be appropriate for the climate and it must obviously be able to survive in contact with the contamination to be able toaccomplish its goal. It may also have a need to be able to encourage localisedmicrobial growth.

One of the major advantages of phytotechnological interventions is their almostuniversal approval from public and customer alike and a big part of the appeallies in the aesthetics. Healthy plants often with flowers make the site look more attractive and help the whole project be much more readily accepted by peoplewho live or work nearby. However, the single biggest factor in its favour is thatplant-based processes are frequently considerably cheaper than rival systems, somuch so that sometimes they are the only economically possible method. Phytoremediationis a particularly good example of this, especially when substantialareas of land are involved. The costs involved in cleaning up physically largecontamination can be enormous and for land on which the pollution is suitableand accessible for phytotreatment, the savings can be very great. Part of the reasonfor this is that planting, sowing and harvesting the relevant plants requireslittle more advanced technology or specialised equipment than is readily at thedisposal of the average farmer.The varied nature of phytotechnology makes anyattempt at formalisation inherently artificial. However, for the purposes of this discussion, this topic will be considered in two general sections, purely on thebasis of whether the applications themselves represent largely aquatic or terrestrialsystems. The reader is urged to bear in mind that this is merely a convenienceand should be accorded no particular additional importance beyond that.

1) Terrestrial Phyto-Systems (TPS)

Phytoremediation methods offer significant potential for certain applications and additionally permit much larger sites to be restored than would generally be possible using more traditional remediation technologies. A large range of species from different plant groups can be used, ranging from pteridophyte ferns to angiosperms like sunflowers, and poplar trees, which employ a number of mechanisms to remove pollutants. There are over 400 different species considered suitable for use as phytoremediators. Amongst these, some hyperaccumulate contaminants within the plant biomass itself, which can subsequently be harvested, others act as pumps or siphons, removing contaminants from the soil before venting them into the atmosphere, while others enable the biodegradation of relatively large organic molecules, like hydrocarbons derived from crude oil. However, the technology is relatively new and so still in the development phase. The first steps toward practical bioremediation using various plant-based methods really began with research in the early 1990s and a number of the resulting techniques have been used in the field with reasonable success.

In effect, phytoremediation may be defined as the direct *in situ* use of living green plants for treatment of contaminated soil, sludges or groundwater, by the removal, degradation, or containment of the pollutants present. Such techniques are generally best suited to sites on which low to moderate levels of contamination are present fairly close to the surface and in a relatively shallow band. Within these general constraints, phytoremediation can be used in the remediation of land contaminated with a variety of substances including certain metals, pesticides, solvents and various organic chemicals.

Metal Phytoremediation

The remediation of sites contaminated with metals typically makes use of the natural abilities of certain plant species to remove or stabilise these chemicals by means of bioaccumulation, phytoextraction, rhizofiltration or phytostabilisation.

Phytoextraction

The process of phytoextraction involves the uptake of metal contaminants from within the soil by the roots and their translocation into the above-ground regions of the plants involved. Certain species, termed hyperaccumulators, have an innate ability to absorb exceptionally large amounts of metals compared to most ordinary plants, typically 50–100 times as much and occasionally considerably more. The original wild forms are often found in naturally metal-rich regions of the globe where their unusual ability is an evolutionary selective advantage. Currently, the best candidates for removal by phytoextraction are copper, nickel and zinc, since these are the metals most readily taken up by the majority of the varieties of hyperaccumulator plants. In order to extend the potential applicability of this method of phytoremediation, plants which can absorb unusually high amounts of chromium and lead are also being trialled and there have been some recent early successes in attempts to find suitable phytoextractors for cadmium, nickel and even arsenic. The removal of the latter is a big challenge, since arsenic behaves quite differently from other metal pollutants, typically being found dissolved in the groundwater in the form of arsenite or arsenate, and does not readily precipitate. There have beensome advances like the application of bipolar electrolysis to oxidise arseniteinto arsenate, which reacts with ferric ions from an introduced iron anode, but generally conventional remediation techniques aim to produce insoluble forms of the metal's salts, which, though still problematic, are easier to remove. Clearly, then, a specific arsenic-tolerant plant selectively pulling the metal from the soil would be a great breakthrough. One attempt to achieve this which has shown some promise involves the Chinese ladder brake fern, *Pterisvittata*, which hasbeen found to accumulate arsenic in concentrations of 5 grams per kilogramme of dry biomass. Growing very rapidly and amassing the metal in its root and stem tissue, it is easy to harvest for contaminant removal.

Hyperaccumulation

Hyperaccumulation itself is a curious phenomenon and raises a number of fundamental questions. While the previously mentioned pteridophyte, *Pterisvittata*, tolerates tissue levels of 0.5 per cent arsenic, certain strains of naturally occurring alpine pennycress (*Thlaspicaerulescens*) can bioaccumulate around 1.5 per cent cadmium, on the same dry weight basis. This is a wholly exceptional concentration. Quite how the uptake and the subsequent accumulation isachieved are interesting enough issues in their own right. However, more intriguing is why so much should be taken up in the first place. The hyperaccumulation of copper or zinc, for which there is an underlying certain metabolic requirement can, to some extent, be viewed as the outcome of an over-efficient natural mechanism. The biological basis of the uptake of a completely nonessential metal, however, particularly in such amounts, remains open to speculation at this point. Nevertheless, with plants like *Thlaspi*showing a zinc removal rate in excess of 40 kg per hectare per year, their enormous potential value in bioremediation is very clear.

In a practical application, appropriate plants are chosen based on the type of contaminant present, the regional climate and other relevant site conditions. This may involve one or a selection of these hyperaccumulator species, dependent on circumstances. After the plants have been permitted to grow for a suitable length of time, they are harvested and the metal accumulated is permanently removed from the original site of contamination. If required, the process may be repeated with new plants until the required level of remediation has been achieved. One of the criticisms commonly levelled at many forms of environmental biotechnology is that all it does is shift a problem from one place to another. The fate of harvested hyperaccumulators serves to illustrate the point, since the biomass thus collected, which has bioaccumulated significant levels of contaminant metals, needs to be treated or disposed of itself, in some environmentally sensible fashion. Typically the options are either composting or incineration. The former must rely on co-composting additional material to dilute the effect of the metal laden hyperaccumulator biomass if the final compost is to meet permissible levels; the latter requires the ash produced to be disposed of in a hazardous waste landfill. While this course of action may seem a little un-environmental in its approach, it must be remembered that the void space required by the ash is only around a tenth of that which would have been needed to landfill the untreated soil. An alternative that has sometimes been suggested is the possibility of recycling metals taken up in this way. There are few reasons, at least in theory, as to why this should not be possible, but much of the practical reality depends on the value of the metal in question. Dried plant biomass could be taken to processingworks for recycling and for metals like gold, even very modest plantcontent could make this economically viable. By contrast, low value materials, like lead for example, would not be a feasible prospect. At the moment, nickel is probably the best studied and understood in this respect. There has been considerable interest in the potential for bio-mining the metal out of sites which have been subject to diffuse contamination, or former mines where further traditional methods are no longer practical. The manner proposed for this is essentially phytoextractionand early research seems to support the economic case for drying the harvested biomass and then recovering the nickel. Even where the actual post-mining residue has little immediate worth, the application of phytotechnological measures can still be of benefit as a straightforward clean-up. In the light of recent advances in Australia, using the ability of eucalyptus trees and certain native grasses to absorb metals from the soil, the approach is to be tested operationally for the decontamination of disused gold mines. These sites also often contain significant levels of arsenic and cyanide compounds. Managing the country's mining waste is a major expense, costing in excess of Aus$30 million per year; success in this trial could prove of great economic advantage to the industry.

The case for metals with intermediate market values is also interesting. Though applying a similar approach to zinc, for instance, might not result in a huge commercial contribution to the smelter, it would be a benefit to the metal production and at the same time, deal rationally with an otherwise unresolved disposal issue. Clearly, the metallurgists would have to be assured that it was a worthwhile exercise. The recycling question is a long way from being a workable solution, but potentially it could offer a highly preferable option to the currently prevalent landfill route.

Rhizofiltration

Rhizofiltration is the absorption into, or the adsorption or precipitation onto, plant roots of contaminants present in the soil water. The principal difference between this and the previous approach is that rhizofiltration is typically used to deal with contamination in the groundwater, rather than within the soil itself, though the distinction is not always an easy one to draw. The plants destined to be used in this way are normally brought on hydroponically and graduallyacclimatised to the specific character of the water which requires to be treated. Once this process has been completed, they are planted on the site, where they begin taking up the solution of pollutants. Harvesting takes place once the plants have become saturated with contaminants and, as with the phytoextraction, the collected biomass requires some form of final treatment. The system is less widely appreciated than the previous technology, but it does have some very important potential applications. Sunflowers were reported as being successfully used in a test at Chernobyl in the Ukraine, to remove radioactive uranium contamination from water in the wake of the nuclear power station accident.

Phytostabilisation

In many respects, phytostabilisation has close similarities with both phytoextractionand rhizofiltration in that it too makes use of the uptake and accumulation by, adsorption onto, or precipitation around, the roots of plants. On first inspection, the difference between these approaches is difficult to see, since in effect, phytostabilisationdoes employ both extractive and filtrative techniques. However, what distinguishes this particular phytoremediation strategy is that, unlike the preceding regimes, harvesting the grown plants is not a feature of the process. In this sense, it does not remove the pollutants, but immobilises them, deliberately concentrating and containing them within a living system, where they subsequently remain. The idea behind this is to accumulate soil or groundwater contaminants, locking them up within the plant biomass or within the rhizosphere, thus reducing their bio-availability and preventing their migration off site. Metals do not ultimately degrade, so it can be argued that holding them in place in this way is the best practicable environmental option for sites where the contamination is low, or for large areas of pollution, for which large-scale remediation by other means would simply not be possible.

A second benefit of this method is that on sites where elevated concentrations of metals in the soil inhibits natural plant growth, the use of species which have a high tolerance to the contaminants present enables a cover of vegetation to be re-established. This can be of particular importance for exposed sites, minimising the effects of wind erosion, wash off or soil leaching, which otherwise can significantly hasten the spread of pollutants around and beyond the affected land itself.

Organic Phytoremediation

A wide variety of organic chemicals are commonly encountered as environmental pollutants including many types of pesticides, solvents and lubricants. Probably the most ubiquitous of these across the world, for obvious reasons, are petrol and diesel oil. These hydrocarbons are not especially mobile, tend to adhere closely to the soil particles themselves and are generally localised within two metres of the surface. Accordingly, since they are effectively in direct contact with the rhizosphere, they are a good example of ideal candidates for phytoremediation. The mechanisms of action in this respect are typically phytodegradation, rhizodegradation, and phytovolatilisation.

Phytodegradation

Phytodegradation, which is sometimes known by the alternative name of phytotransformation, involves the biological breakdown of contaminants, either internally, having first been taken up by the plants, or externally, using enzymes secreted by them. Hence, the complex organic molecules of the pollutants are subject to biodegradation into simpler substances and incorporated into the plant tissues. In addition, the existence of the extracellular enzyme route has allowed this technique to be successfully applied to the remediation of chemicals as varied as chlorinated solvents, explosives and herbicides. Since this process depends on the direct uptake of contaminants from soil water and the accumulation of resultant metabolites within the plant tissues, in an environmental application, it is clearly important that the metabolites which accumulate are either nontoxic, or at least significantly less toxic than the original pollutant.

Rhizodegradation

Rhizodegradation, which is also variously described as phytostimulationor enhanced rhizospheric biodegradation, refers to the biodegradation of contaminants in the soil by edaphic microbes enhanced by the inherent characterof the rhizosphere itself. This region generally supports high microbial biomass and consequently a high level of microbiological activity, which tends to increase the speed and efficiency of the biodegradation of organic substances within the rhizosphere compared with other soil regions and microfloral communities. Part of the reason for this is the tendency for plant roots to increase the soil oxygenation in their vicinity and exude metabolites into the rhizosphere. It has been estimated that the release of sugars, amino acids and other exudates from the plant and the net root oxygen contribution can account for up to 20 per cent of plant photosynthetic activity per year, of which denitrifying bacteria, *Pseudomonas* spp., and general heterotrophs are the principal beneficiaries. In addition, mycorrhizae fungi associated with the roots also play a part in metabolising organic contaminants. This is an important aspect, since they have unique enzymatic pathways that enable the biodegradation of organic substances that could not be otherwise transformed solely by bacterial action. In principle, rhizodegradation is intrinsic remediation enhanced by entirely natural means, since enzymes which are active within 1mm of the root itself, transform the organic pollutants, in a way which, clearly, would not occur in the absence of the plant. Nevertheless, this is generally a much slower process than the previously described phytodegradation.

Phytovolatilisation

Phytovolatilisation involves the uptake of the contaminants by plants and their release into the atmosphere, typically in a modified form. This phytoremediation biotechnology generally relies on the transpiration pull of fast-growing trees, which accelerates the uptake of the pollutants in groundwater solution, which are then released through the leaves. Thus the contaminants are removed from the soil, often being transformed within the plant before being voided to the atmosphere. One attempt which has been explored experimentally uses a genetically modified variety of the

Yellow Poplar, *Liriodendron tulipifera*, which has been engineered by the introduction of mercuric reductase gene (*mer A*). This confers the ability to tolerate higher mercury concentrations and to convert the metal's ionic form to the elemental and allows the plant to withstand contaminated conditions, remove the pollutant from the soil and volatilise it. The advantages of this approach is clear, given that the current best available technologies demand extensive dredging or excavation and are heavily disruptive to the site.

The choice of a poplar species for this application is interesting, since they have been found useful in similar roles elsewhere. Trichloroethylene (TCE), an organic compound used in engineering and other industries for degreasing, is a particularly mobile pollutant, typically forming plumes which move beneath the soil's surface. In a number of studies, poplars have been shown to be able to volatilise around 90 per cent of the TCE they take up. In part this relates to their enormous hydraulic pull. Acting as large, solar-powered pumps, they draw water out of the soil, taking up contaminants with it, which then pass through the plant and out to the air. The question remains, however, as to whether there is any danger from this kind of pollutant release into the atmosphere and the essential factor in answering that must take into account the element of dilution. If the trees are pumping out mercury, for instance, then the daily output and its dispersion rate must be such that the atmospheric dilution effect makes the prospect of secondary effects, either to the environment or to human health, impossible. Careful investigation and risk analysis is every bit as important for phytoremediation as it is for other forms of bioremediation.

Using tree species to clean up contamination has begun to receive increasing interest. Phytoremediation in general tends to be limited to sites where the pollutants are located fairly close to the surface, often in conjunction with a relatively high water table. Research in Europe and the USA has shown that the deeply penetrating roots of trees allow deeper contamination to be treated. Once again, part of the reason for this is the profound effect these plants can have on the local water relations.

Hydraulic Containment

Large plants can act as living pumps, pulling large amounts of water out of the ground which can be a useful property for some environmental applications, since the drawing of water upwards through the soil into the roots and out through the plant decreases the movement of soluble contaminants downwards, deeper into the site and into the groundwater. Trees are particularly useful in this respect because of their enormous transpiration pull and large root mass. Poplars, for example, once established, have very deep tap roots and they take up large quantities of water, transpiring between 200–1100 litres daily. In situations where grassland would normally support a water table at around 1.5 metres, this action can lead to it being up to 10 times lower. The aim of applying this to a contamination scenario is to create a functional water table depression, to which pollutants will tend to be drawn and from which they may additionally be taken up for treatment. This use of the water uptake characteristics of plants to control the migration of contaminants in the soil is termed hydraulic containment, shown schematically in Figure 19.1 and a number of particular applications have been developed.

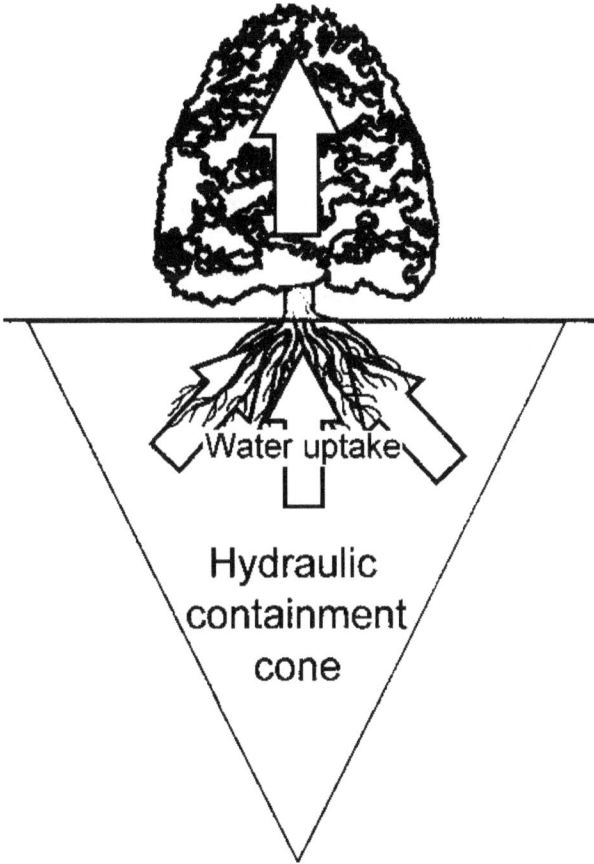

Figure 19.1: Schematic Hydraulic Containment.

Buffer strips are intended to prevent the entry of contaminants into watercourses and are typically used along the banks of rivers, when they are sometimes called by the alternative name of 'riparian corridors', or around the perimeter of affected sites to contain migrating chemicals. Various poplar and willow varieties, for example, have shown themselves particularly effective in reducing the wash-out of nitrates and phosphates making them useful as pollution control measures to avoid agricultural fertiliser residues contaminating waterways. Part of the potential of this approach is that it also allows for the simultaneous integration of other of the phytoremediating processes described into a natural treatment train, since as previously stated, all plant-based treatments are aspects of the same fundamental processes and thus part of a cohesive whole.

Another approach sometimes encountered is the production of vegetative caps, which has found favour as a means of finishing off some American landfill sites. The principle involves planting to preventing the downward percolation of rainwater into the landfill and thus minimising leachate production while at the same time reducing erosion from the surface. The method seems to be successful as a living alternative to an impermeable clay or geo-polymer barrier. The vegetative cap has also been promoted for its abilities to enhance the biological breakdown of the underlying refuse. In this respect, it may be seen as an applied form of rhizodegradation or even, arguably, of phytodegradation. How effective it is likely to be in this role, however, given the great depths involved in most landfills and the functionally anoxic conditions within them, appears uncertain. To understand the overall phytoremediation effect of hydraulic containment, it is important to realise that contaminating organics are actually taken up by the plant at lower concentration than they are found *in situ*, in part due to membrane barriers at the root hairs. In order to include this in a predictive mathematical model, the idea of a transpiration stream concentration factor (TSCF) for given contaminants has been developed, defined as $TSCF = 0.75 \exp\{-[(\log K_{ow} - 2.50)2/2.4]\}$, where K_{ow} is the octanol–water partitioncoefficients. These latter are a measure of the hydrophobia or hydrophiliaof a given organic chemical; a log K_{ow} below 1 characterises the fairly soluble,while above 3.5 indicates highly hydrophobic substances.Thus the uptake rate (U in mg/day) is given by the following equation:

$$U = (TSCF)T\,C$$

where,

   $TSCF =$ Transpiration stream concentration factor, as defined

   $T =$ Transpiration rate of vegetation, 1/day

   $C =$ Concentration in site water, mg/1

However, it must also be remembered in this context that, should the pollutants not themselves actually be taken up by the plants, then the effect of establishing a hydraulic containment regime will be to increase their soil concentration due to transpiro-evaporative concentration. Thus, the mass of affected water in the contaminant plume reduces, as does the consequent level of dilution it offers and hence, increased localised concentration can result.

The transpiration pull of plants, and particularly tree species, has also sometimes been harnessed to overcome localised water-logging, particularly on land used for agricultural or amenity purposes. To enhance the effect at the point worst affected, the planting regime may involve the establishment of close groupings, which then function as single elevated withdrawal points. The noted ability of poplars to act as solar-powered hydraulic pumps makes them of great potential benefit to this kind of phytotechnological application. Although other plant-based processes could be taking place at the same time to remediate land alongside this to clean up contaminated soils, this particular technique is not itself a type of phytoremediation. Instead, it is an example of the broader bioengineering possibilities which are offered by the appropriate use of flora species to wider environmental nuisances, which, for some sites, may be the only economic or practicable solution. This may be of particular relevance to heavy soils with poor natural inter-particulate spacing, since laying adequate artificial drainage systems can often be expensive to do in the first place and are frequently prone to collapse once installed.

Another similar example of the use of phytotechnology to overcome nuisance is the bio-bund, which consists of densely planted trees, often willows, on an engineered earthwork embankment. This system has been used successfully to reduce noise pollution from roads, railways and noisy industrial sites, the interlocking branches acting as a physical barrier to deaden the sound as well as having a secondary role in trapping wind-blown particulates. Depending on the individual site, the bio-bund can be constructed in such a way that it can also act as a buffer strip to control migrating chemical pollution, if required.

Plant Selection

It should be obvious that the major criteria for plant selection are the particular requirements for the method to be employed and the nature of the contaminants involved. For example, in the case of organic phytotransformation this means species of vegetation which are hardy and fast growing, easy to maintain, have a high transpiration pull and transform the pollutants present to nontoxic or less toxic products. In addition, for many such applications, deep rooting plants are particularly valuable. On some sites, the planting of grass varieties in conjunction with trees, often in between rows of trees to stabilise and protect the soil, may be the best route since they generate a tremendous amount of fine roots near to the surface. This particularly suits them to transforming hydrophobic contaminants such as benzene, toluene, ethyl-benzene, xylenes (collectively known as BTEX) and polycyclic aromatic hydrocarbons (PAHs). They can also be very helpful in controlling wind-blown dust, wash-off and erosion. The selection of appropriate plant species for bioengineering is not, however, limited solely to their direct ability to treat contaminants, since the enhancement of existing conditions forms as much a part of the potential applications of phytotechnology as bioremediation. For instance, legumes can be of great benefit to naturally nitrogen-deficient soils, since they have the ability, via symbiotic root nodule bacteria, to directly fix nitrogen from the atmosphere. With so much to take into consideration in plant selection, the value of a good botanist or agronomist in any interdisciplinary team is clear.

Applications

Phytotechnology has many potentially beneficial land uses, though for the most part the applications are still in the development stage. Several have been tested for the treatment of contamination, and in some cases successfully tried in the field, but generally they remain in the 'novel and innovative' category, lacking well-documented data on their performance under a variety of typical operating conditions. As a result, some researchers have voiced doubts, suggesting that the beneficial effects of plant utilisation, particularly in respect of phytoremediation, have been overstated. Some have argued that the reality may range from genuine enhancement to no effect, or even to a negative contribution under certain circumstances and that the deciding factors have more to do with the nature of the site than the plants themselves. In addition, some technologies which have been successfully used on some sites may simply serve to complicate matters on others. One such approach which achieved commercial scale use in the USA, principally for lead remediation, required the addition of chemicals to induce metal take-up. Lead normally binds strongly to the soil particles and so its release was achieved by using chelating agents like ethylene diamine tetra acetic acid (EDTA), whichwere sprayed onto the ground. With the lead rendered biologically available, it can be taken up by plants and hence removed. However, dependent on the character of the site geology, it has been suggested that this could also allow lead to percolate downwards through the soil, and perhaps ultimately into watercourses. While it may well be possible to overcome this potential problem, using accurate mathematical modelling, followed by the establishment of good hydraulic containment as an adjunct to the process, or by running it in a contained biopile, it does illustrate one of the major practical limitations of plant bioengineering. The potential benefits of phytotechnology for inexpensive, large-scale land management are clear, but the lack of quantitative field data on its efficacy, especially compared with actively managed alternative treatment options, is a serious barrier to its wider adoption. In addition, the roles of enzymes, exudates and metabolites need to be more clearly understood and the selection criteria for plant species and systems for various contamination events require better codification. Much research is underway in both public and the private sectors which should throw considerable light on these issues. Hopefully it will not be too long in the future before such meaningful comparisons can be drawn.

One area where phytoremediation may have a particular role to play, and one which might be amenable to early acceptance is as a polishing phase in combination with other clean-up technologies. As a finishing process following on from a preceding bioremediation or non-biological method first used to deal with 'hot-spots', plant-based remediation could well represent an optimal low-cost solution. The tentative beginnings of this have already been tried in small-scale trials and techniques are being suggested to treat deeply located contaminated groundwater by simply pumping to the surface and using it as the irrigant for carefully selected plant species, allowing them to biodegrade the pollutants. The lower levels of site intrusion and engineering required to achieve this would bring clear benefits to both the safety and economic aspects of the remediation operation.

2) Aquatic Phyto-Systems (APS)

Aquatic phyto-systems are principally used to process effluents of one form or another, though manufactured wetlands have been used successfully to remediate some quite surprising soil contaminants, including TNT residues. Though the latter type of application will be discussed in this section, it is probably best considered as an intergrade between the other APS described hereafter and the TPS of the previous. The major difference between conventional approaches to deal with effluents and phytotechnological methods is that the former tend to rely on a faster, more intensively managed and high energy regime, while in general, the stabilisation phase of wastewaters in aquatic systems is relatively slow. The influx and exit of effluent into and out of the created wetland must be controlled to ensure an adequate retention period to permit sufficient residence time for pollutant reduction, which is inevitably characterised by a relatively slow flow rate. However, the efficiency of removal is high, typically producing a final treated off-take of a quality which equals, or often exceeds, that of other systems. Suffice it to say that, as is typical of applications of biological processing in general, there are many common systemic considerations and constraints which will obviously affect phyto-systems, in much the same way as they did for technologies which rely on microbial action for their effect.

Many aquatic plant species have the potential to be used in treatment systems and the biological mechanisms by which they achieve some of the effects willalready be largely familiar from the preceding discussion of terrestrial systems. There are a number of ways in which APS can be categorised but perhaps the most useful relates to the natural division between algae and macrophytes, which has been adopted, accordingly, here.

3) Macrophyte Treatment Systems (MaTS)

The discharge of wastewaters into natural watercourses, ponds and wetlands is an ancient and long-established practice, though rising urbanisation led to the development of more engineered solutions, initially for domestic sewage and then later, industrial effluents, which in turn for a time lessened the importance of the earlier approach. However, there has been a resurgence of interest in simpler, more natural methods for wastewater treatment and MaTS systems, in particular, have received much attention as a result. While there has, undoubtedly, been a strong upsurge in public understanding of the potential for environmentally harmonious water cleaning *per se*, a large part of the driving force behind the newly found interest in these constructed habitats comes from biodiversity concerns. With widespread awareness of the dwindling number of natural wetlands, often a legacy of deliberate land drainage for development and agricultural purposes, the value of such manufactured replacements has become increasingly apparent. In many ways it is fitting that this should be the case, since for the majority of aquatic macrophyte systems, even those expressly intended as 'monocultures' at the gross scale, it is very largely as a result of their biodiversity that they function as they do.

These treatment systems, shown diagrammatically in Figure 19.2, are characterised by the input of effluent into a reservoir of comparatively much larger volume, either in the form of an artificial pond or an expanse of highly saturated soilheld within a containment layer, within which the macrophytes have been established. Less commonly, pre-existing natural features have been used. Although

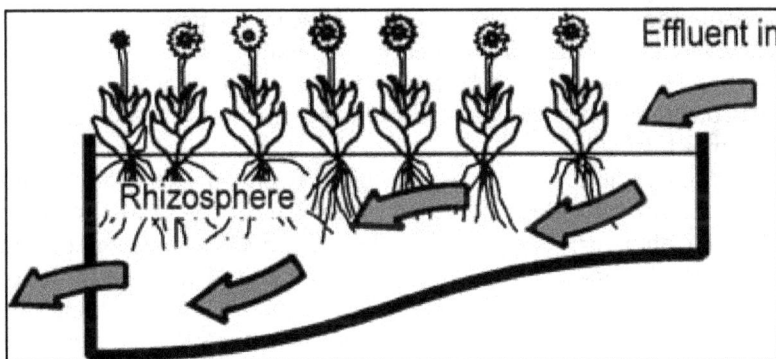

Figure 19.2: Diagrammatic Macrophyte Treatment System (MaTS).

wetlands have an innate ability to accumulate various unwanted chemicals, the concept of deliberately polluting a habitat by using it as a treatment system is one with which few feel comfortable today. A gentle hydraulic flow is established, which encourages the incoming wastewater to travel slowly through the system. The relatively long retention period that results allows adequate time for processes of settlement, contaminant uptake, biodegradation and phytotransformation to take place.

The mechanisms of pollutant removal are essentially the same, irrespective of whether the particular treatment system is a natural wetland, a constructed monoculture or polycultureand independent of whether the macrophytes in question are submerged, floating or emergent species. Both biotic and abiotic methods are involved. The main biological mechanisms are direct uptake andaccumulation, performed in much the same manner as terrestrial plants. Theremainder of the effect is brought about by chemical and physical reactions, principally at the interfaces of the water and sediment, the sediment and the root or the plant body and the water. In general, it is possible to characterise the primary processes within the MaTS as the uptake and transformation of contaminants by micro-organisms and plants and their subsequent biodegradation and biotransformation; the absorption, adsorption and ion exchange on the surfaces of plants and the sediment; the filtration and chemical precipitation of pollutants via sediment contact; the settlement of suspended solids; the chemical transformation of contaminants. It has been suggested that although settlement inevitably causes the accrual of metals, in particular, within the sediment, the plants themselves do not tend to accumulate them within their tissues. While this appears to be borne out, particularly by original studies of natural wetlands used for the discharge of mine washings this does not form any basis on which to disregard the contribution the plants make to water treatment. For one thing, planting densities in engineered systems are typically high and the species involved tend to be included solely for their desired phytoremediation properties, both circumstances seldom repeated in nature. Moreover, much of the biological pollutant abatement potential of the system exists through the synergistic activity of the entire community and, in purely direct terms, this largely means the indigenous microbes. Functionally, there are strong parallels between this and the processes of enhanced rhizosphericbiodegradation described for terrestrial applications. While exactly the same mechanisms are available within the root zone in an aquatic setting, in addition, and particularly in the case of submerged vegetation, the surface of the plants themselves becomes a large extra substrate for the attached growth of closely associated bacteria and other microbial species. The combined rhizo- and circum-phyllo- spheres support a large total microbial biomass, with a distinctly different compositional character, which exhibits a high level of bioactivity, relative to other microbial communities. As with rhizodegradation on dry land, part of the reason is the increased localised oxygenation in their vicinity and the corresponding presence of significant quantities of plant metabolic exudates, which, as was mentioned in the relevant earlier section, represents a major proportion of the yearly photosynthetic output. In this way, the main role of the macrophytes themselves clearly is more of an indirect one, bringing about local environmental enhancement and optimisation for remediative microbes, rather than being directly implicated in activities of primary biodegradation. In addition, physico-chemical mechanisms are also at work. The iron plaques which form on the plant roots trap certain metals, notably arsenic, while direct adsorption and chemical/biochemical reactions play a role in the removal of metals from the wastewater and their subsequent retention in sediments.

The ability of emergent macrophytes to transfer oxygen to their submerged portions is a well-appreciated phenomenon, which in nature enables them to cope with effective waterlogging and functional anoxia. As much as 60 per cent of the oxygen transported to these parts of the plant can pass out into the rhizosphere, creating aerobic conditions for the thriving microbial community associated with the root zone, the leaf surfaces and the surrounding substrate. This accounts for a significant increase in the dissolved oxygen levels within the water generally and, most particularly, immediately adjacent to the macrophytes themselves. The aerobic breakdown of carbon sources is facilitated by this oxygen transfer, for obvious reasons, and consequently it can be seen to have a major bearing on the rate of organic carbon biodegradation within the treatment system, since its adequate removal requires a minimum oxygen flux of one and a half times the input BOD loading. Importantly, this also makes possible the direct oxidation of hydrogen sulphide (H_2S) within the root zone and, in some cases, iron and manganese. While from the earlier investigations mentioned on plant/metal interactions their direct contribution to metal removal is small, fast growing macrophytes have a high potential uptake rate of some commonly encountered effluent components. Some kinds of water hyacinth, *Eichhornia spp.*, for example, can increase their biomass by $10 \, g/m^2/day$ under optimum conditions, which represents an enormous demand for nitrogen and carbon from their environment. The direct uptake of nitrogen from water by these floating plants gives them an effective removal potential which approaches 6000 kg per hectare per year and this, coupled with their effectiveness in degrading phenols and in reducing copper, lead, mercury, nickel and zinc levels in effluents, explains their use in bioengineered treatment systems in warm climates. Emergent macrophytes are also particularly efficient at removing and storing nitrogen in their roots, and some can do the same for phosphorus. However, the position of this latter contaminant in respect of phytotreatment in general is less straightforward. In a number of constructed wetland systems, though the overall efficacy in the reduction of BOD, and the removal of nitrogenous compounds and suspended solids has been high, the allied phosphorus components have been dealt with much less effectively. This may be of particular concern if phosphorus-rich effluents are to be routinely treated and there is a consequent risk of eutrophication resulting. It has been suggested that, while the reasons for this poor performance are not entirely understood, nor is it a universal finding for all applications of phytotreatment, it may be linked to low root zone oxygenation in slow-moving waters. If this is indeed the case, then the preceding discussion on the oxygen pump effect of many emergent macrophytes has clear implications for biosystem design.

Associated bacteria play a major part in aquatic plant treatment systems and microbial nitrification and denitrification processes are the major nitrogen-affecting mechanisms, with anaerobic denitrification, which typically takes place in the sediment, causing loss to atmosphere, while aerobic nitrification promotes and facilitates nitrogenous incorporation within the vegetation. For the effective final removal of assimilated effluent components, accessibly harvestable material is essential, and above water, standing biomass is ideal. The link between the general desire for biodiversity conservation and the acceptability of created wetlands was mentioned earlier. One of the most important advantages of these systems is their potential to create habitats not just for 'popular' species, like waterfowl, but also for many less well-known organisms, which can be instrumental in bolstering the ecological integrity of the area. This may be of particular relevance in industrial or urban districts. At the same time, they can be ascetically pleasing, enhancing the landscape while performing their function. These systems can have relatively low capital costs, but inevitably everyone must be heavily site specific, which means many aspects of the establishment financing are variable. However, the running costs are generally significantly lower than for comparable conventional treatment operations of similar capacity and efficacy. In part the reason for this is that once properly set up, a well-designed and constructed facility is almost entirely self-maintaining. However, the major contribution to low operational overheads comes from the system's low energy requirements, since gravity drives the water flow and all the remediating organisms are ultimately solar powered, either directly or indirectly,via the photosynthetic action of the resident autotroph community. Aside of cost and amenity grounds, one major positive feature is that the effluent treatment itself is as good as or better than that from conventional systems. When correctly designed, constructed, maintained and managed, plant-based treatment is a very efficient method of ameliorating wastewaters from a wide range of sources and in addition, is very tolerant of variance in organic loadings and effluent quality, which can cause problems for some of the alternative options. In addition, phyto-systems can often be very effective at odour reduction, which is often a major concern for the producers and processors of effluents rich in biodegradable substances.

Invariably, the better designed, the easier the treatment facility is to manage and in most cases, 'better' means simpler in practice, since this helps to keep the maintenance requirement to a minimum and makes maximum use of the existing topography and resources. Provision should also be made for climatic factors and most especially, for the possibility of flooding or drought. It is imperative that adequate consideration is given to the total water budget at the project planning stage. Although an obvious point, it is important to bear in mind that one of the major constraints on the use of aquatic systems is an adequate supply of water throughout the year. While ensuring this is seldom a problem for temperate lands, for some regions of the world it is a significant concern. Water budgeting is an attempt to model the total requirement, accounting for the net overall in- and outputs, together with the average steady-state volume resident within the system in operation. Thus, effluent inflow, supplementary 'clean' water and rainfall need to be balanced against off-take, evaporative and transpirational losses and the demands of the intended retention time required to treat the particular contaminant profile of a given wastewater. One apparent consideration in this process is the capacity of the facility. Determining the 'required' size for a treatment wetland is often complicated by uncertainty regarding the full range of wastewater volumes and component character likely to be encountered over the lifetime of the operation. The traditional response to this is to err on the side of caution and oversize, which, of course, has inevitable cost implications, but in addition, also affects the overall water budget. If the effluent character is known, or a sample can be obtained, its BOD can be found and it is then a relatively simple procedure to use this to calculate the necessary system size. However, this should only ever be taken as indicative. For one thing, bioengineered treatment systems typically have a lifespan of 15–20 years and the character of the effluent being treated may well change radically over this time, particularly in response to shifts in local industrial practice or profile. In addition, though BOD assessment is a useful point of reference, it is not a uniform indicator of the treatment requirements of all effluent components. For the bio-amelioration process to precede efficiently, a fairly constant water level is necessary. Although the importance of this in a drought scenario is self-evident, an unwanted influx of water can be equally damaging, disturbing the established equilibrium of the wetland and pushing contaminants through the system before they can be adequately treated. Provision both to include sufficient supplementary supplies, and exclude surface water, is an essential part of the design process. One aspect of system design which is not widely appreciated is the importance of providing a substrate with the right characteristics. A number of different materials have been used with varying degrees of success, including river sands, gravels, pulverised clinker, soils and even waste-derived composts, the final choice often being driven by issues of local availability. The main factors indetermining the suitability of any given medium are its hydraulic permeability and absorbance potential for nutrients and pollutants. In the final analysis, the substrate must be able to provide an optimum growth medium for root development while also allowing for the uniform infiltration and through-flow of wastewater. A hydraulic permeability of between 10–3 and 10–4 m/s is generally accepted as ideal, since lower infiltration tends to lead to channelling and flow reduction, both of which severely restrict the efficiency of treatment. In addition, the chemical nature of the chosen material may have an immediate bearing on system efficacy.

Soils with low inherent mineral content tend to encourage direct nutrient uptake to make good the deficiency, while highly humic soils have been shown to have the opposite effect in some studies. The difficulties sometimes encountered in relation to phosphorus removal within wetland systems have been mentioned earlier. The character of the substrate medium can have an important influence on the uptake of this mineral, since the physico-chemical mechanisms responsible for its abstraction from wastewater in an aquatic treatment system relies on the presence of aluminium or iron within the rhizosphere. Obviously, soils with high relative content of these key metals will be more effective at removing the phosphate component from effluent, while clay-rich substrates tend to be better suited to lowering heavy metal content.

Engineered reed beds are probably the most familiar of all macrophyte treatment systems, with several high profile installations in various parts of the globe having made the technology very widely accessible and well appreciated. This approach has been successfully applied to a wide variety of industrial effluents, in many different climatic conditions and has currently been enjoying considerable interest as a 'green' alternative to septic tanks for houses not joined up to mains sewerage. At its heart is the ability of reeds, often established as monocultures of individual species, or sometimes as oligo-cultures of a few, closely related forms, to force oxygen down into the rhizosphere. Many examples feature *Phragmites* or *Typha* species, which appear to be particularly good exponents of the oxygen pump, while simultaneously able to support a healthy rhizosphericmicrofloralcomplement and provide a stable root zone lattice for associated bacterial growth and physico-chemical processing of rhizo-contiguous contaminants. Isolated from the surrounding ground by an impermeable clay or polymer layer, the reed bed is almost the archetypal emergent macrophyte treatment system.

The mechanisms of action are shown in Figure 19.3 and may be categorised as surface entrapment of any solids or relatively large particulates on the growingmedium or upper root surface. The hydraulic flow draws the effluent down through the rhizosphere, where the biodegradable components come into direct contact with the root zone's indigenous micro-organisms, which are stimulated to enhanced metabolic activity by the elevated aeration and greater nutrient availability. There is a net movement of oxygen down through the plant and a corresponding take-up by the reeds of nitrates and minerals made accessible by the action of nitrifying and other bacteria. These systems are very efficient at contamination removal, typically achieving 95 per cent or better remediation of a wide variety of pollutant substances. Nevertheless, reed beds and root zone treatment techniques in general are not immune from a range of characteristic potential operational problems, which can act to limit the efficacy of the process. Thus, excessive waterlogging, surface runoff, poor or irregular substrate penetration and the development of preferential drainage channels

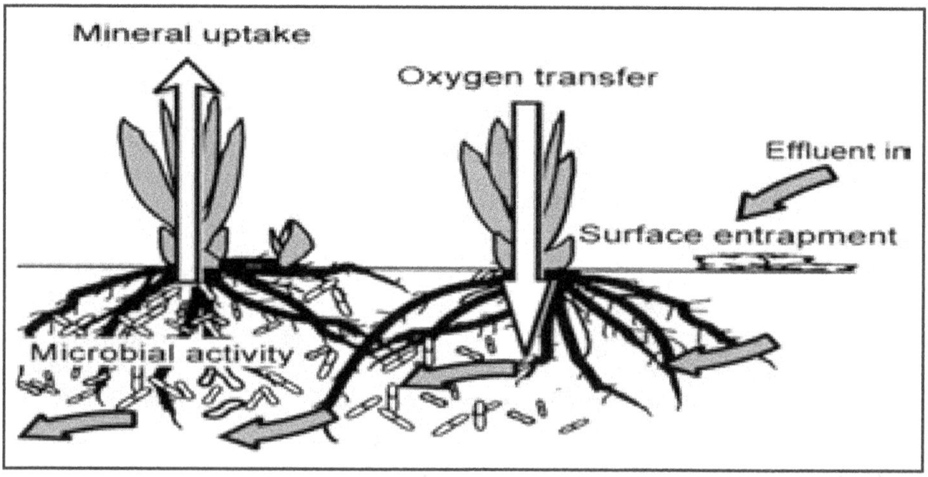

Figure 19.3: Diagrammatic Root Zone Activity.

across the beds may all contribute to a lessening of the system's performance, in varying degrees.

4) Nutrient Film Techniques (NFT)

An alternative approach to the use of aquatic macrophytes, which was tried experimentally, involved growing plants on an impermeable containment layer, in a thin film of water. In this system, the wastewater flowed directly over the root mass, thereby avoiding some of the mass transfer problems sometimes encountered by other aquatic phytotreatment regimes. Though the early work indicated that it had considerable potential for use in the biological treatment of sewage and other nutrient-rich effluents, it does not appear to have been developed further and little is known as to the conditions which govern its successful practical application. One interesting aspect which did, however, emerge was that the cultivation system could also be extended to most terrestrial plants, which may yet be of possible relevance to the future development of land-based phytotreatments.

5) Algal Treatment Systems (ATS)

Algae have principally been employed to remove nitrogen and phosphorus from wastewaters, though some organic chemicals can also be treated and a relatively new application has emerged which makes use of their efficient carbon sequestration potential.

Effluent treatment

Algal effluent treatment systems work on the basis of functional eutrophication and rely on a dynamic equilibrium between the autotrophic algae themselves and the resident heterotophic bacteria, which establishes a two-stage biodegradation/ assimilation process, as shown in Figure 19.4.

In effect this is an ecological microcosm in which organic contaminants present in the wastewater are biologically decomposed by the aerobic bacteria, which make

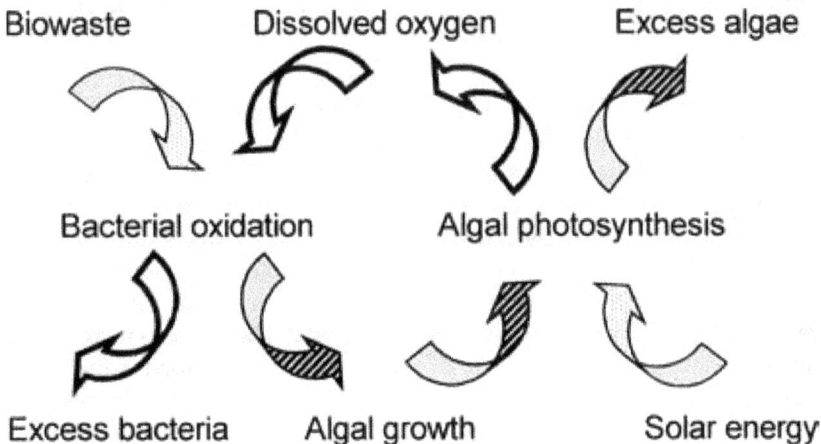

Figure 19.4: Algal and Bacterial Equilibrium.

use of oxygen providedby algal photosynthesis, while the algae grow using the nutrients produced bythis bacterial breakdown, and photosynthesise producing more oxygen.Though the process is self-sustaining, it is also self-limiting and left to preceedunchecked, will result in the well-appreciated characteristic eutrophic stagesleading to the eventual death of all component organisms, since true climacticbalance is never achieved in the presence of continuously high additional nutrientinputs. The removal of excess algal and bacterial biomass is, therefore, anessential feature, vital to maintaining the system's efficiency.Of all the engineered algal systems for effluent, the high rate algal pond(HRAP) is one of the most efficient and represents a good illustration of thisuse of phytotechnology. The system consists of a bioreactor cell in the form of a relatively shallowreservoir, typically between 0.2–0.6 metres deep, with a length to width ratio of2:1 or more, the idea being to produce a large surface area to volume ratio. Thevoid is divided with internal baffles forming walls, to create a channel throughwhich the effluent flows. A mechanically driven paddle at the end nearest tothe effluent input both aerates and drives the wastewater around the system.These ponds are not sensitive to fluctuations in daily feed, either in terms ofquantity or quality of effluent, providing that it is fundamentally of a kind suitablefor this type of treatment. Consequently, they may be fed on a continuous orintermittent basis. The main influences which affect the system's performanceare the composition of the effluent, the efficiency of mixing, the retention time,the availability and intensity of light, pond depth and temperature. The lattertwo factors are particularly interesting since they form logical constraints on thetwo groups of organisms responsible for the system's function, by affecting theautotroph's ability to photosynthesise and the heterotroph's to respire. Whilea deeper cell permits greater resident biomass, thus elevating the numbers ofmicro-organisms available to work on the effluent, beyond a certain limit, thelaw of diminishing returns applies in respect of light available to algae in thelower reaches. Warmer temperatures increase metabolic activity, at least withinreason, and the rate of straightforward chemical reactions doubles per 10 $^{\circ}$Crise, but at the same time, elevated water temperatures have a reduced oxygencarryingcapacity which affects the bacterial side of the equilibrium. As with so much of environmental biotechnology, a delicate balancing act is required.

After a suitable retention period, which again depends on the character of the effluent, the design and efficacy of the treatment pond and the level of clean-up required, the water is discharged for use or returned to watercourses. Obviously, after a number of cycles, algal and bacterial growth in a functionally eutrophic environment would, as discussed earlier in the section, begin to inhibit, and then eventually arrest, the bio-treatment process. By harvesting the algal biomass, not only are the contaminants, which to this point have been merely biologically isolated, physically removed from the system, but also a local population depression is created, triggering renewed growth and thus optimised pollutant uptake. The biomass recovered in this way has a variety of possible uses, of which composting for ultimate nutrient reclamation is without doubt the most popular, though various attempts have also been made to turn the algal crop into a number of different products, including animal feed and insulating material.

Pollution Detection

One final, and currently emerging, application of phytotechnology to the environmental context involves the possible use of plants in a variety of industrial sectors as pollution detectors. The aim is principally to provide valuable information about the toxicological components of contamination from a wide variety of sources, including the automotive, chemical and textile industries. Unlike biosensors, which tend to be designed around isolated biochemical reactions, in this approach, the plants are used as entire biological test systems. Moreover, unlike conventional chemical analytical methods which produce quantifiable, numeric measurements, the varieties used have been selected for their abilities to identify contaminants by reacting to the specific effects these substances have on the plant's vital functions. Thus, by directing the focus firmly onto the obvious and discernible biological consequences of the pollutants and then codifying this into a diagnostic tool, the assessment process is made more readily available to a wider range of those who have an interest in pollution control. The development of this technology is still in the initial stages, but it would appear to open up the way for a controllable method to determine pollutant effects. It seems likely that they will be of particular value as early detection systems in the field, since they are functional within a broad range of pH and under varied climatic conditions. An additional benefit is that they are responsive to both long-term pollution or incidental spillages and can be applied to either laboratory or on-site investigations to monitor air, soil or water, even on turbid or coloured samples, which often cause anomalous readings with spectrophotometric test methods.

References

1. Brooks, R., Chambers, M., Nicks, L. and Robonson, B., 1998.Phytomining, *Trends in Plant Science*, 3: 359–62.

2. Burken, J. and Schnoor, J., 1998. Predictive relationships for uptake of organic contaminants by hybrid poplar trees, *Environmental Science and Technology*, 32: 3379–85.

3. Chaney, R., Malik, M., Li, Y., Brown, S., Brewer, E., Angel, J. and Baker, A.,1997. Phytoremediation of soil metals, *Current Opinion in Biotechnology*, 8: 279–84.

4. Evans, G.M., 2001.*Biowaste and Biological Waste Treatment*. James and James, London.

5. Foth, H., 1990.*Fundamentals of Soil Science*, 8th edition, Wiley, New York.

6. Gareth, M. Evans. andJudith, C. Furlong., 2003. EnvironmentalBiotechnology; Theory and Application. John Wiley and Sons Ltd.pp 1-281.

7. Hardman, D.J., McEldowney, S. and Waite, S., 1994.*Pollution: Ecology andBiotreatment*, Longman, Harlow.

8. Heathcote, I., 2000.*Artificial Wetlands for Wastewater Treatment*, Prentice Hall/Pearson, New Jersey.

9. Hutchinson, G., 1975.*A Treatise of Limnology: Volume III. Limnological Botany*, Wiley, New York.

10. Murphy, M. and Butler, R., 2002. Eucalyptus could be worth weight in gold, *Chemistry and Industry*, 1: 7.

11. Otte, M., Kearns, C. and Doyle, M., 1995. Accumulation of arsenic and zinc in the rhizosphere of wetland plants, *Bulletin of Environmental ContaminationToxicology*, 55: 154–61.

12. Scragg, A., 1999. *Environmental Biotechnology*, Longman, Harlow.

2016, Environmental Biotechnology: A New Approach   *Pages 309–345*
Editors: Dr. Rajan Kumar Gupta and Dr. Satya Shila Singh
Published by: DAYA PUBLISHING HOUSE, NEW DELHI

Chapter 20

In vitro Technique for Selection of Plants Tolerant to Environmental Stresses

Nasim Akhtar*, Rajesh K. Srivastava and
Malay R. Mishra

¹Department of Biotechnology, GITAM Institute of Technology,
GITAM University, Gandhi Nagar Campus, Rushikonda,
Visakhapatnam – 530 045, Andhra Pradesh

ABSTRACT

The food grain production needs to be enhanced constantly to feed the increasing World population on a decreasing fertility of cultivable land under environmental stresses. Both abiotic and biotic stresses are the major threat to agriculture productivity. Sexual hybridization, induced mutation and genetic transformation approaches have demonstrated success in developing stress tolerance in the limited plant species. In the past two decades, cell and tissue culture based in vitro selection system has emerged as a simple, fast and economically feasible tool for developing stress-tolerant plants. The method exploits the capability to manipulate the variation for obtaining the desired characteristic of plant to the expected result. Therefore, by applying the selection agent into the media, plant tolerance to both abiotic and biotic stresses could be acquired. Chemical reagents such as NaCl (for salt tolerance), PEG and mannitol (for drought tolerance, osmotic stress tolerance), $A1Cl_3.6H_2O$ (for Al tolerance) have been incorporated in the tissue culture media as selection agents. Only the explants capable of sustaining such environments survive in the long run and are selected. In vitro selection is based on the induction of genetic variation among cells, tissues and/or organs in cultured and regenerated

---

* *Corresponding author.* E-mail: nasimakhtar01@yahoo.co.in

plants. *The selection of somaclonal variations appearing in the regenerated plants may be genetically stable and useful in crop improvement. This chapter describes a common strategy for regeneration of plants tolerant to various abiotic environmental stresses through tissue culture based in vitro selection methods. Characterization of the selected callus line, determination of callus growth rate, callus survival efficiency and frequency of regeneration in response to selection agents in the medium has been presented. Plants have evolved many physiological, biochemical and molecular defense mechanisms to survive under stress conditions by bringing certain structural and functional changes. Further, the methods for characterization of in vitro selected plants for these structural and functional changes in response to stress condition has been described for the measurements of stomatal dynamics, photosynthetic assay, leaf water loss, leaf relative water content, chlorophyll content, assays for carbohydrates, proteins, proline, ABA, glycine betain, electrolytic leakage, malondialdehyde, determination of elements such as sodium, potassium, calcium and phosphorus, detection of reactive oxygen species such as superoxide anion and hydrogen peroxide content, antioxidant enzyme activity for superoxide dismutase, gtutathione reductase, peroxidase and catalase,. Procedure for screening of salinity and drought tolerance, plantlet growth and seed germination assays has been described for greenhouse and field growing plants. The common mechanisms of tolerance and the success of in vitro selection methods for drought, salinity, oxidative, osmotic, low and high temperature, and Al toxicity stress tolerance have also been reviewed. The chapter is concluded with suggestion for the application of in vitro method to select plants tolerant to abiotic environmental stresses for improvement of various crop species in conjunction with the molecular biological approaches and functional genomics and proteomics for future research.*

Introduction

The world agriculture has seriously been challenged by increasing population (both human and animals), shrinking agricultural land (losses due to climatic changes, desertification, industrialization, over-exploitation and urbanization) and the depleting natural resources. It is acknowledge that we need to increase the food grain productivity by 50 per cent for 8.3 billion people in 2030 and must be doubled to feed 9.1 billion people in 2050 (at the rate of approximately 44 million metric tons increase per year) (http://www.fao.org/wsfs/world-summit/en/) by that time India would be the most populated country of the world with an expected population of 1.6 billion (Annonymous (2009a, b). It is projected that our food grain demand will be 294 million tons by 2020 as against the current production level of 208 million tons (Tester and Langridge, 2010). The demand for fruits, vegetables, meat and other animal products will also rise significantly. The whole world may be producing sufficient food for every one but this need to be enhanced for poor nations due to socio-economic and political issues (Annonymous (2009a, b).

Abiotic stresses arise from extremes of climate such as drought, salinity, extreme temperatures (both low and high), chemical toxicity of such elements as Na, Al and soil deficiencies of elements like P and Fe. Excessive contents of heavy metals (*e.g.* Cu, Zn, Pb, and Cd) and other inorganic ions (Na, Al, B, As and Mn) in problem soils are restricting normal cultivation of plants on one-fourth of the world soils. These abiotic stresses are serious threats to agriculture productivity. The average yield of major grain crops is expected to be reduced by more than 50 per cent primarily due to these abiotic stresses worldwide (Boyer 1982; Bray *et al.*, 2000). Drought and salinity are becoming particularly widespread in many regions, and may cause serious salinization of more than 50 per cent of all arable lands by the year 2050 (Annonymous (2012). These losses are augmented due to a series of morphological, physiological, biochemical and molecular changes that adversely affect plant growth and productivity caused by various abiotic stresses (Hasegawa *et al.*, 2000; Wang *et al.*, 2001).

Various abiotic environmental factors lead to dehydration or osmotic stress through reduced availability of water for vital cellular functions and maintenance of turgor pressure, stomata closure, reduced supply of CO_2 and slower rate of biochemical reactions during prolonged periods of dehydration, high light intensity, high and low temperatures and salinity (Ahuja *et al.*, 2010; Hasegawa *et al.*, 2000). This leads to high production of reactive oxygen species (ROS in the chloroplasts) such as super oxide (O_2^{-}), hydrogen peroxide (H_2O_2), singlet oxygen ($^1O_2$), hydroxy radical ($OH^{\cdot}$), lipidperoxide causing irreversible cellular damage and photo inhibition (Jahnke *et al.*, 1991; Apel and Hirt, 2004).

Being sessile in nature plants have evolved a number of strategies to defend the adverse effects of various biotic and abiotic environmental stresses (Ahuja *et al.*, 2010; Hasegawa *et al.*, 2000). The cellular damages due to drought, extreme temperatures, oxidative stress and salinity, are multifold, interconnected and expressed following a common cell signaling pathways (Shinozaki and Yamaguchi-Shinozaki, 2000; Knight and Knight, 2001; Zhu 2001b, 2002). It has been demonstrated that drought and/or salinity are manifested primarily as osmotic stress. This results in the disruption of homeostasis and ion distribution in the cell (Serrano *et al.*, 1999; Zhu, 2001a). Similarly, oxidative stress which is mostly accompanied by drought, high temperature or salinity stress often causes denaturation of functional and structural proteins (Smirnoff, 1998). As a consequence, several cellular defense responses such as accumulation of compatible solutes, production of stress proteins, up-regulation of anti-oxidants are activated commonly by these diverse environmental stresses (Aazami, *et al.*, 2010; Cushman and Bohnert, 2000; Ghoulam *et al.*, 2001; Vierling and Kimpel, 1992; Zhu *et al.*, 1997). Drought and salt tolerance in plants has been critically reviewed recently by Bartels and Sunkar (2005). Earlier, Bray (1997) has studied various plant molecular response to water deficit conditions. An attempt to understand the plant response to drought from gene to the whole plant level has been discussed recently (Chaves *et al.*, 2003). Kosova *et al.* (2011) have studied changes in plant proteome under stressed condition and discussed the contribution of proteomics studies to understanding plant stress response.

Water stressed selected callus line has grown better and accumulated more of Na^+, K^+ and N but lesser amount of Ca^{++} and P as compared with non-selected callus line in *Helianthus annus* (Hassan *et al.*, 2004a) and chili pepper (Santos-Diaz and Ochoa-Alejo (1994). Greater concentrations of accumulated ions obtained in selected line may have adaptive significance under water stress in order to decrease the cell water potential to keep water absorption going on and thus to prevent water loss (Morgan, 1984; Munns, 2002; Munns and James, 2003; Munns and Tester, 2008). Sodium chloride induced changes in mineral nutrients and proline accumulation in selected indica rice cultivars differing in salt tolerance have been reported by Kumar *et al.* (2008).

A significant increase in the sugar content of both selected and non-selected callus lines of *Helianthus annus* has been reported with increasing PEG concentration (Hassan *et al.*, 2004). This increase has been more in the selected line as compared with the non-selected one (except reducing sugars at low solute potential). On the other hand, PEG treatment causes a decrease in the polysaccharides content in both selected and non-selected callus lines. Similar observation has been reported by Kikuta and Richter (1988), and Kameli and Losel (1996) who found reduced level of polysaccharide and increased di- and monosaccharide concentrations under water deficit conditions. Roles of glycine betaine and proline in improving plant abiotic stress resistance has been reported by Ashraf and Foolad (2007).

Strategies for Development of Stress Tolerant Plants

Four different strategies *viz.* sexual hybridization, mutant creation, genetic transformation and *in vitro* selection have been proposed for the incorporation of high level of tolerance to abiotic stresses which is crucial for enhancing and establishing a stable yield potential of crop plants.

The conventional breeding programs are being used to integrate genes of interest through sexual hybridization from inter crossing genera and species into the crops to induce stress tolerance (Witcombe *et al.*, 2008). However, due to the multitude of abiotic stresses and their complex genetic control, the progress of breeding towards tolerance to abiotic stresses using conventional approaches have failed to provide desirable results (Jain *et al.*, 1998; Rai *et al.*, 2011).

The technique of induced mutation is applied for screening and selection of either the whole plant population in the field condition or the treatment of seeds with either a physical agent (x-ray, gamma rays, fast neutron and thermal neutron) or a chemical mutagen such as ethylene scimine (ES), diethyl sulphonate (DES), ethyl methane-sulphonate (EMS), and the azida group (Jain *et al.*, 1998). Screening of such a large seeds population or field grown plants would be very difficult, ordinarily impossible (Predieri, 2001; Jain, 2001). Moreover, the variations created by these methods are often produce chimeras, pleotrophy, genetic instability and epigenetic changes (Biswas *et al.*, 2002).

In vitro mutant creation for several traits is more easily isolated than from whole plant populations. Further, tissue culture increases the efficiency of mutagenic treatments and allows handling of large populations and rapid cloning of variants selected for disease resistance, improvement of nutritional quality, adaptation to stress conditions, *e.g.*, saline, soils, low temperature, toxic metals, resistance to herbicides and to increase the biosynthesis of plant products used for medicinal or industrial purposes (Predieri, 2001; Jain, 2001). Screening has been profitably and widely employed for the isolation of cell clones that produce higher quantities of certain biochemicals (Matkowski, 2008).

The genetic transformation of plants using molecular biological tools has become important for the development of crop plants with enhanced level of tolerance to abiotic stresses. Previous biotechnological approaches have shown that most of the abiotic stress tolerant traits are multigenic. Therefore, to improve stress tolerance several stress related genes need to be transferred. Additionally, the requirement of regeneration of transformants, expression of gene due to gene silencing, difficulties with isolation and characterization of gene of interest, susceptibility of plants to *Agrobacterium* etc. limits the wide application of genetic transformation technology for improvement of crop plants tolerant to abiotic stresses.

The tissue culture technique is the most widely applied method for improvement of plant genotype tolerance to abiotic stress such as drought, high salinity, heavy metal toxicity, acid soil, as well as biotic stress herbicides and disease through *in vitro* selection of somaclonal variants (Ahmed *et al.*, 1996; Jain *et al.*, 1998; Yusnita *et al.*, 2005). The selection agent in the culture media brings about a number of physiological, biochemical, metabolic and genetic changes in the cells of explanted tissues, callus and regenerated shoots resulting in the somaclonal variations (Biswas *et al.*, 2002). Additionally, the new variants are expected to express some other desirable traits with enhanced taste, yield and size of fruit, texture and colour of flower, through *in vitro* selection procedure (Pedrieri, 2001; Ahloowalia and Maluszynski, 2001; Witjaksono, 2003). These variation are caused either by gene amplification, the alteration of basic chromosome complement, chromosome instability, restructuring, deletions, methylation, transposition, translocation, transformation, inversion or point mutation, (Dennis, 2004; Kumar and Marthur, 2004). These plants enhance germplasm base of the plant species and serve as an excellent donor of the desirable gene(s) for crop improvements through breeding program. This selection strategy has been successfully developed for the recovery of genotypes resistant to various toxins, herbicides, high salt concentration, drought etc. (Zair *et al.*, 2003).

A Common Approach for *In vitro* Selection of Plants Tolerant to Various Abiotic Environmental Stresses

☆ In the cell selection approach, a suitable pressure is applied to permit the preferential survival/growth of variant cells. Presence of selective agents in the medium induces certain physiological, biochemical, metabolic and molecular biological changes in the explanted cells, tissues or organs allowing selecting and regenerating plants with desirable characteristics. Complete process of *in vitro* selection of plants tolerance to environmental stresses involves five steps *viz.* i. selection, ii. induction, iii. proliferation, iv. regeneration, v. evaluation. Step wise complete selection process is shown in the Figure 20.1. The experimental approaches for *in vitro* selection of plants with tolerance to various environmental stresses are described in the following protocol.

Protocol for *In vitro* Selection of Stress Tolerant Plants:

☆ An elite healthy plant of the species in trouble is selected for explants collection and development of tolerance to the said stress in question.

☆ A highly efficient callus to plant regeneration protocol is the prerequisite and needs to be developed or selected if available.

☆ The explants (preferably a leaf disk or an apical meristem/apical shoot

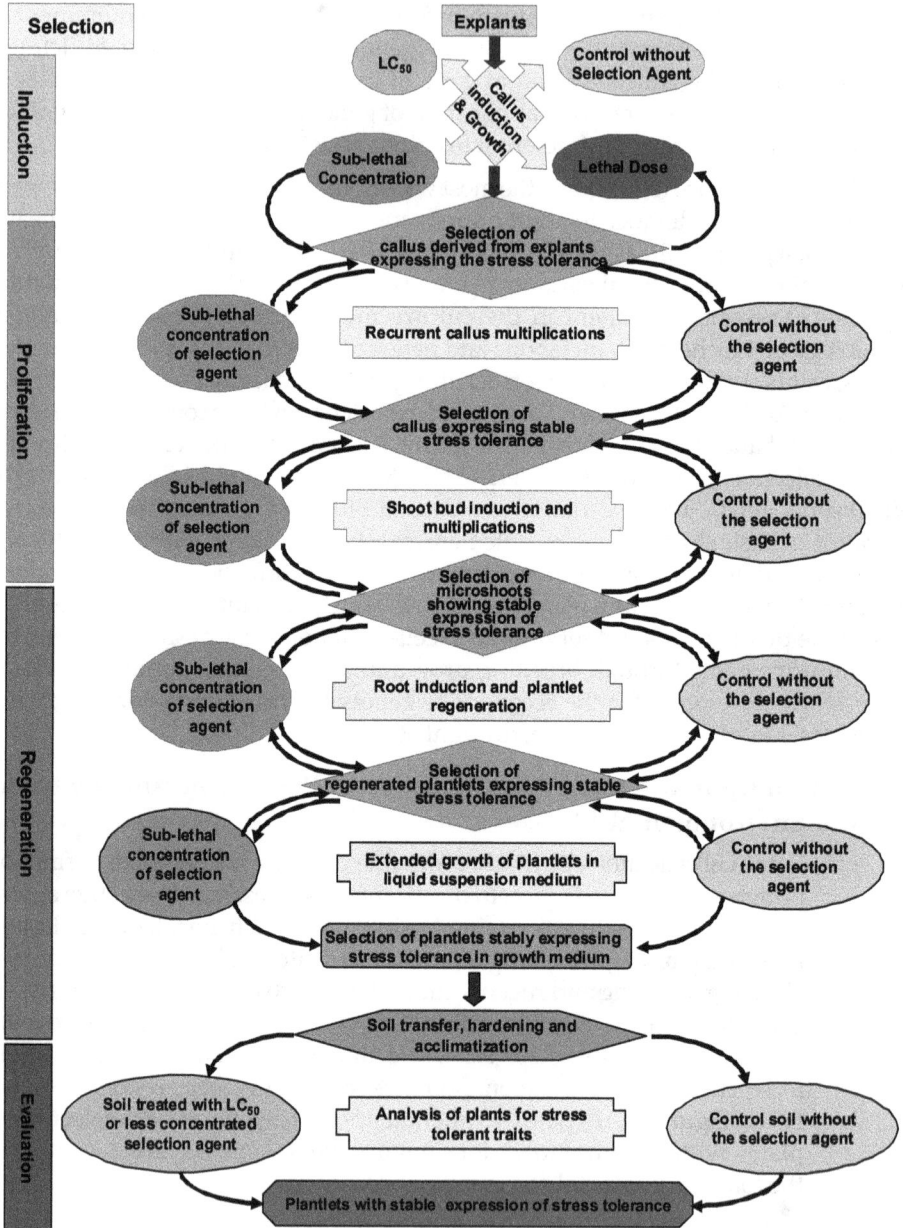

Figure 20.1: Steps Involved during *in vitro* Selection of Plants Tolerant to Environmental Stresses.

bud, an axillary bud or a nodal segment) are used for induction of callus following the protocol under the control as well as in the presence of broad range concentrations of selection agent for drought/salinity/osmotic stresses.

☆ The lethal/toxic, sub-lethal and LC_{50} concentration of the stress agent should be worked out from the growth curve on the induction and growth of callus from the explants.

☆ Callus from control, LC_{50} and sub-lethal cultures are selected for multiplication under the following treatment conditions.

Several recurrent cycle of proliferation are applied by selecting callus from sub-lethal concentrations and sub-culturing them to i. a control (without the selection agent), ii. LC_{50} concentration and iii. sub-lethal concentration of selection agent. Similarly callus from control (without the selection agent) and LC_{50} cultures are subjected to recurrent selection cycle as described above in order to select callus lines stably expressing the tolerance to particular selection agent in question.

☆ The callus lines stably expressing the tolerance to a stress agent in question from the above recurrent cycle of sub-culture are selected for plantlet regeneration.

☆ Shoot bus regeneration from the selected tolerant calli is achieved by sub-culturing on induction medium (as per the protocol) in the presence of selection agent (sub-lethal concentration), with a suitable control (without the selection agent), and the LC_{50} dose.

☆ Micro-shoots showing tolerance to stress agent are selected for clonal multiplication by sub-culturing them to i. sub-lethal concentration of selection agent, ii. control (without the selection agent), iii. in the presence of LC_{50} concentration of stress conditions. Similarly micro-shoots from control (without the selection agent) and LC_{50} cultures are subjected to recurrent selection cycle as described above regeneration of shoots stably expressing the tolerance to particular selection agent in question.

☆ Micro-shoots stably expressing tolerance to the said selection agent from the recurrent shoot multiplication cycle are subjected for root induction.

☆ Rooting and regeneration of complete plantlets are applied for micro-shoots by sub-culturing on a root induction medium containing sub-lethal concentration of selection agent with a suitable control (without the selection agent) as well as in the presence of LC_{50} concentration of selection force.

☆ The rooted plantlets expressing stable tolerance to the said selection force from the above rooting medium are grown for a short extended period of 1-2 weeks in liquid suspension medium containing the sub-lethal as well as slightly higher concentration and LC_{50} concentration of selection force in order to induce, compare, confirm and select the stress tolerant nature of regenerated plantlets.

☆ Well grown rooted plantlets are transferred to soil, hardened and acclimatized in the controlled soil (without the selection agent). After the establishment of root system and appearance of new leaflets the plants are challenged by subjecting the soil with LC_{50} or less concentration of selection force to simulate the stress condition in order to confirm the stress tolerance trait of the regenerated plants.

☆ Analysis and evaluation of various physiological, biochemical, metabolic and molecular biological parameter should be performed to confirm the nature of stress tolerance of regenerated plants as well as several seed generations.

Assays and Characterization of Stress Tolerance in callus and Regenerated Plantlets

The abiotic environmental stresses affect a number of physiological processes such as reductions of cell growth, leaf area, biomass and yield. Kumar *et al.* (2008) have studied sodium chloride induced changes in mineral nutrients and proline accumulation in indica rice cultivars differing in salt tolerance. A screening method for comparative physiological studies on salinity and water stress tolerance has been developed recently (Munns, 2002; Munns and James, 2003; Munns and Tester, 2008). As a defense mechanism higher plants generally up-regulate several antioxidant enzymes such as peroxidase (POD), catalase (CAT), superoxide dismutase (SOD), glutathione reductase (GR) etc. to scavenge reactive oxygen species (ROS) produced in response to various abiotic stresses (Mittova *et al.*, 2003a, b; Rahnama *et al.*, 2003; Ashraf and Harris 2004; Batkova *et al.*, 2008; Hossain *et al.*, 2007). Characterization of the callus lines and regenerated plantlets that survive to the imposed selection pressure under *in vitro* conditions is the critical step to obtain the genotypes more tolerant to a particular stress condition. Yang *et al.* (2010) have studied the effect of salinity on antioxidant enzymes in calli of the halophyte *Nitraria tangutorum*. There are changes in water relations, photosynthetic activity and proline accumulation in one-year-old olive trees (*Olea europaea* L. cv. Chemlali) in response to NaCl salinity (BenAhmad *et al.*, 2008). Variations in callus induction, proliferation and plantlets regeneration of two bread wheat (*Triticum aestivum* L.) genotypes under saline and heat stress conditions have been studied by Benderradji *et al.* (2012). Rai *et al.* (2011) have indicated that activation of the plant antioxidant defense system is a positively associated with salt and drought tolerance. The above-mentioned antioxidant enzymes play a necessary role in detoxification of ROS produced under stressful conditions (Rahnama *et al.*, 2003; Hu *et al.*, 2012). The enzyme SOD normally converts more toxic $O_2{}^{\cdot-}$ radicals to less toxic H_2O_2 (Scandalios, 1993) and to neutralize H_2O_2 other enzymes such as peroxidase and catalase are produced (Dionisiosese and Tobita, 1998). Hence, by measuring antioxidant activities *in vitro*, a rapid preliminary selection of tolerant genotypes is possible. An increase in the main antioxidant enzymes such as superoxide dismutase (SOD), ascorbate peroxidase (APX), catalase (CAT) and glutathione reductase (GR) have been reported by different authors (Hossain *et al.*, 2007; Lascano, 2001; Sreenivasulu *et al.*, 2000; Cherian and Reddy, 2003). Regulation of proline biosynthesis, degradation, uptake, and transport in higher plants and its implications in plant growth and abiotic stress tolerance has been reviewed and discussed recently (Kavi Kishore *et al.*, 2005).

Free radical-induced peroxidation of lipid membrane is a sign of stress-induced damage at cellular level. Consequently, measuring the level of malonyldialdehyde

(MAD), produced during peroxidation of membrane lipids has been used as an indicator of oxidative damage (Demiral and Turkan, 2005). Queiros *et al.* (2007) have found that selected callus lines of *Solanum tuberosum* subjected to NaCl showed an increase in lipid peroxidation in comparison with salt tolerant lines.

Sajid and Aftab (2014) have recurrently-selected plants have shown higher POD, CAT and SOD activities as compared to the control cultures. Quite recently, Kumar *et al.* (2008) reported that SOD activity increased in salt-treated callus cultures of *Jatropha curcas* as compared to non-treated controls. Similarly, increased SOD activity is also reported by Sreenivasulu *et al.* (2000) and Cherian and Reddy (2003). Based on the results of these investigations, Sajid and Aftab (2014) have drawn inference that the changes in growth and biochemical parameters have been rather comparable to a shifting behavior of plants from being sensitive-to-relatively-more-tolerant-ones. A relationship between salt tolerance and photosynthetic machinery performance in citrus has been studied by López-Climent *et al.* (2008).

Characterization of the Selected Callus Line

The calli are frequently observed, and the growth is characterized with assigned numerical values as: callus with no sign of browning (+++++); callus with slightly brown at the surface (++++); whole callus tissue brown with optimum growth (+++); entire callus tissue deeply brown with little growth (++); whole callus tissue deeply brown with no growth at all (+).

For the measurement of stress-tolerance, callus pieces from the two callus lines (selected and non-selected) are transferred to fresh medium added with selction agent at a concentrations finally creating a water potential of 0, -0.2, - 0.4, -0.6, -0.8, -1.0, -1.2 or -1.4 MPa. Five callus pieces per dish in three replicate dishes per treatment are used. At the end of four weeks, callus fresh weight is measured.

Callus Growth Rate (CGR)

The callus growth rate in stress and unstressed control is measured in terms of percent increase in fresh weight. Micro clumps of calli approximately of almost equal size with known weight (100 ± 10 mg) are incubated for two, three or four weeks on selection media. At the end of stress period, random samples of calli are weighted aseptically, and the mean callus growth rate is determined using following formula:

$$CGR \text{ (per cent)} = \frac{\text{Final weight} - \text{Initial Weight}}{\text{Initial Weight}} \times 100$$

Callus Survival Rate (per cent)

The calli are sub-cultured for several cycle of recurrent proliferation on callus selection medias first without any stress agents followed by re-culturing in the presence of sub-lethal concentrations of selection force (NaCl, mannitol or PEG-6000, low or high temperature etc.) for a period varying from two to three weeks in total darkness to recover putative stress (salinity, drought, water deficit or oxidative) tolerant cell lines. The well adapted calli which survived and proliferate in the continued presences of selection agent are counted and callus survival frequency is computed as:

$$\text{Callus survival (per cent)} = \frac{\text{No. of Calli survived}}{\text{Total no. of calli cultured}} \times 100$$

Frequency of Regeneration (per cent)

The callus stably expressing the tolerance to the said abiotic stress (salinity, drought, water deficit or oxidative) are transferred to shoot induction medium and several selection cycle of recurrent shoot multiplication are applied followed by rooting by re-culturing in the presence of sub-lethal concentrations of selection agent (NaCl, mannitol or PEG-6000, low or high temperature etc.) following the standard protocol to recover putative stress tolerant plants. The well adapted plantlets which survived in the continued presences of selection agent are counted and regeneration frequency is calculated as:

$$\text{Frequency of Multiple Shoot Production} = \frac{\text{Total no. of calli showing the development of more than one shoot per treatment}}{\text{Total no. calli established for regeneration per treatments}} \times 100$$

Comparison of Stomatal Dynamics

The assays for closure of stomata are conducted according to the Pei *et al*. (1997). Leaves from three control plants and three of each *in vitro* selected line are placed on the surface of a solution containing 50 µM $CaCl_2$, 10 mM KCl, 10mM MES (2-(N-morpholino) ethanesulfonic acid)- Tris, pH 6.15, and are exposed to light for 2 h. ABA is then added to the solution at concentrations of either 1 or 5 µM, respectively, and samples are incubated for a further 2 h. Following ABA treatment, 40 stomatal apertures are measured from each leaf. All assessments are carried out in triplicate.

Photosynthesis Assay

The CO_2 fixation is measured with a portable CO_2 gas analyzer before and after appropriate levels of selection agents. The net CO_2 assimilation rate is measured every hour for 12 hrs and measuring the second leaf from the tip under the following conditions: 200 µmol m^{-2}s^{-1} light intensity, 320 ppm CO_2, 28 °C and 70 per cent relative humidity. The relative net photosynthesis rate (RP) is calculated by the following formula:

Photosynthetic Rate (PR) (per cent) = $(P_n/P_0) \times 100$

where,

P_n is the average net RP on a specific day (average hourly net photosynthetic rates of that day to get P_n) and P_0 is the average net RP before treatment with selection agent.

Rate of Leaf Water Loss

Percentage of leaf water loss (LWL per cent) is evaluated according to the method of Xing *et al*. (2004). To determine the rate of water loss, leaves are harvested from 3-week-old hardened plants under the stress condition and control plants and placed on dry filter paper. Fresh weights (FW1) are measured every 10 min for 50 min to determine the rate of water loss (FWt). At every time point the rate of water loss from leaves of stress treated plants and control plants are compared. LWL was calculated by follow formula:

LWL (per cent) = {(FW1 – FWt)/FW1} × 100

Leaf Relative Water Content (RWC)

Leaf relative water content (RWC) is used as additional marker for membrane integrity. To determine the RWC leaves are harvested from 40-days hardened under the stress condition and control plants. Leaf relative water content (RWC) analysis is performed by taking fresh weight then incubating leaf sample by immersing in distilled water in dark for 24 h. At the end of treatment period, the turgid weights of saturated leaf samples are taken. The dry weights of the samples are taken by packing the leaf samples in butter paper bag and oven dried al 65 °C for 48 h. The leaf relative water content is calculated by following formula:

$$\text{Relative water content (per cent)} = \frac{\text{Fresh Weight} - \text{Dry Weight}}{(\text{Turgor Weight} - \text{Dry Weight})} \times 100$$

Chlorophyll Measurements

Chlorophyll and carotenoid contents are estimated by non-maceration method (Hiscox and Isrealstam, 1979). Fresh leaf samples (0.5 g) are incubated in 5 ml of dimethyl sulphoxide (DMSO) at 65 °C for 4 h. Absorbance is measured spectrophotometrically and recorded at 644.8, 661.6 and 470 nm. The chlorophyll content is calculated according to Arnon (1949) and carotenoids content by Llichtenthaler, (1987) as

Chlorophyll a + b = 7.05 × A661.6 + 18.09 × A644.8

Carotenoid = [1000 × A470 – 1.90 × (11.24 × A661.6 – 2.04 × A644.8) – 63.14 × (20.13 × A644.8 – 4.19 × A661.6)]/214

Measurement of Elements

Sodium, potassium and calcium are determined using emission spectrophotometry. Nitrogen is determined by Kjeldahl digestion method (Haynes, 1980). Phosphorus is determined photometrically as phosphomolybdate (John, 1970).

Determination of Na^+ and K^+

Fresh leaves are harvested from *in vitro* cultures under selection and from acclimatized plants at day 12 following stress (salt, osmotic etc.) and dried at 105 °C for 1 h, followed at 80 °C for 12 h. The dried samples are powdered in a mortar and pestle, and used for determination of Na^+ and K^+. A 100 mg samples are extracted in

1M HCl for 3 h. The solutions are analyzed for K^+ and Na^+ content by atomic absorption spectrometry (Qian *et al.*, 2000, 2001).

Measurement of Carbohydrates

Carbohydrates are measured using Naguib method (1964) and direct reducing sugars are estimated according to Nelson (1944). Total soluble sugars are estimated using anthrone reagent. Aliquot of 0.05 ml is taken in test tubes and the volume is made upto 1 ml. To this solution 4 ml of anthrone regent is added and mixture is heated in boiling water bath for 8 min followed by cooling. Optical density of green to dark green color is read at 630 nm.

Determination of Total Protein

Fresh samples of callus and leaves from *in vitro* grown plantlets as well as leaves from well acclimatized green house plants exposed to various stresses under selection pressure and control conditions are homogenized in sodium phosphate buffer (pH 6.8) and centrifuged at 1400 x g for 10 min. The total soluble proteins in the supernatant are assayed by method of Bradford (1976) or Biuret method of Racusen and Johnstone (1961).

The isozyme analysis for esterase pattern is carried out for biochemical characterization of the regenerated stress tolerant lines. Leaf tissue samples selected and non-selected shoot multiplication culture are collected and crushed in extraction buffer consisting of 15 per cent (w/v) sucrose, 5 mM tris-glycine buffer, pH 8.3. Samples are electrophoresed for 3 hr at constant current using mini vertical electrophoresis unit.

Measurement of Proline Content

To measure proline content, leaves are homogenized in 5 ml of 3 per cent sulphosalicylic acid and centrifuged at 4000 rpm for 5 min, 4 °C. Two ml of the resulting supernatants are incubated with 2 ml of ninhydrin reagent (2.5 per cent (w/v) ninhydrin, 60 per cent (v/v) glacial acetic acid, and 40 per cent 6 M phosphoric acid) and 2 ml of glacial acetic acid at 100 °C for 30 min. Reactions are terminated by snap cooling in an ice bath, followed by addition of 4 ml of toluene and mixing by vortex. The reaction mixtures are then incubated at 23 °C for 24 h. Proline concentration is calculated as to the standard curve and determined as described previously (Bates *et al.*, 1973).

Measurement of ABA Content

Callus and leaves from control and treated cultures as well as leaves from hardened plants and other seed generation challenged for said stress are collected for extraction and measurement of ABA content according to the method of Liu and Zhang (2010). The ABA are extracted with organic solvents (Liu and Zhang 2010). The HPLC (high performance liquid chromatography) performed using a C_{18} chromatography column with the mobile phase as 30 per cent of acetonitrile and 70 per cent 0.02 mol/L acetic acid buffer. Aliquots of 20 µL of the extracted solutions are used for detection at 262 nm at a column temperature of 25°C. The ABA content is determined based on the chromatographic peak area.

Estimation of Ascorbate

Ascorbic acid (As) is extracted in 5 per cent (m/v) methophosphoric acid with sand at 4 °C. The homogenate is then centrifuged at 3,000 g for 20 min at 4 °C. Ascorbate content is quantified in the supernatant as described by Sadasivam and Manickam (1992). An aliquot of 1 ml of the supernatant is mixed with 1 ml of 2 per cent Dinitrophenyl Hydrazine(DNPH) reagent. The reaction mixture is allowed to stand for 40 min at room temperature, then the absorbance is recorded at 540 nm, using ascorbic acid as standard.

Measurement of Glycine Betaine (GB)

Estimation of glycine betaine (GB) is performed in dried leaf powder. Finely powdered plant material (0.5-1.0 g) is mechanically shaken with 20 ml of deionized water for 48 h at 25 °C. The samples are then filtered and the filtrate is stored in freezer until analysis. Thawed extracts are diluted 1:1 with 2N sulfuric acid. Aliquots of 0.5 ml is taken in test tube and cooled on ice for 1 h. Cold potassium iodide-/iodine reagent (0.2 ml) is added and mixed gently by swirling and incubated at 4 °C for 16 h followed by centrifugation at 10000 x g for 15 min at 4 °C. The supernatant is carefully aspirated with 1 ml micropipette into fresh tubes and stored under cold condition until the periodite complex is separated from acid media. The periodite crystals are dissolved in 9 ml of 1,2-dichloro ethane (reagent grade) by vigorous vortex mixing. After 2.0 - 2.5 h the absorbance is measured at 365 nm with UV- visible spectrophotometer. Reference standards of GB (50 - 200 mg ml^{-1}) are prepared in 2N sulfuric acid.

Determination of ROS Production

Measurement of Electrolyte Leakage (EL) and Malondialdehyde (MDA)

Electrolyte leakage (EL) and Malondialdehyde (MDA) content are both important parameters related to plant cell damage as a result of drought stress. When the integrity and stability of cell membranes can not be maintained under drought stress, cell inclusions permeate and cell membranes produce lipid peroxidation, leading EL and MDA (the end product of lipid peroxidation) increases (Bajji *et al.*, 2001; Fu and Huang, 2001). The level of EL and MDA contents in leaves of *in vitro* selected and control plants that are dehydrated and well-watered, are compared.

Membrane Stability Index (MSI) Measured as Electrolyte Leakage (EL)

Membrane stability index (MSI) is determined by recording the electrical conductivity of leaf leachates in double distilled water at 40 and 100 °C. Leaf samples (0.1-0.5 g) are cut into discs of uniform size and taken in test tubes containing 10 ml of deionized water, shaken on a gyratory shaker at room temperature for 1 h and conductivity of the resulting solutions (C_1) is determined. Subsequently, the leaf discs are kept at 40 °C for 30 min and another set at 100 °C in boiling water bath for 15 min and their respective electric conductivity's C_{40} and C_{100} are measured by conductivity meter. The C_1 to C_2 (C_1/C_{40}, C_1/C_{100}) ratios are calculated and used as a measure of relative electrolyte leakage from the leaves (Liu *et al.*, 2006).

Measurement of Malondialdehyde (MDA)

The concentration of malondialdehyde (MDA) a product of lipid peroxidation is measured by the thiobarbituric acid (TBA) according to Wang *et al.* (2009, 2012). Selected plants from stress conditions and control variants are tested for MDA determination with three replicate. Fresh leaves (1.0 g) are detached from *in vitro* regenerating shoots at selecting agent (Mannitol, PEG, NaCl etc.), grounded in a pestle and mortar contained 5 ml 0.6 per cent TBA in 10 per cent trichloroacetic acid (TCA). The mixture is heated at 100 °C for 15 min. Samples are cooled on ice for 5 min and centrifuged at 5000 rpm for 10 min. The absorbance of the supernatant at 450, 532 and 600 nm wavelengths are recorded and MDA content is calculated on a fresh weight by (nmol MDA g^{-1} FW) = 6.45 (OD532 – OD600) – 0.56 (OD450) × 1000 (Heath and Packer *et al.*, 1968).

Detection of $O_2^{\bullet-}$ and H_2O_2

Water stress also leads to increases in reactive oxygen species, such as $O_2^{\bullet-}$ and H_2O_2, which is resulting in oxidative damage. In order to establish whether *in vitro* selected callus and plants exhibited differences in the production of reactive oxygen species is compared to control plants under stressful conditions. Histochemical staining with nitro blue tetrazolium (NBT) and diaminobenzidine (DAB) is carried out to determine the levels of superoxide anion ($O_2^{\bullet-}$) and hydrogen peroxide (H_2O_2), respectively through visual observation.

In the case of $O_2^{\bullet-}$, leaves are immersed in HEPES buffer (pH 7.5) containing 6 mM NBT for 2 h or until blue spots appeared (Kim *et al.*, 2011). In the case of H_2O_2, leaves are placed in 1 mg ml^{-1} DAB solution until brown spots became visible (Kotchoni *et al.*, 2006). In both instances, stained samples are then de-pigmented in 80 per cent (v/v) ethanol at 80°C for 2 h. Images are acquired using a digital camera. Following dehydration treatment, leaves from control plants exhibiting a deeper staining than stress treated plants, indicate that *in vitro* selected plants accumulated less $O_2^{\bullet-}$ and H_2O_2 during stress conditions.

Hydrogen peroxide is also assessed spectrophotometrically after reaction with potassium iodide (KI), according to method of Velikova and Loreto (2005). Leaf tissues (1 g) are ground and homogenized in a pestle and mortar with 10 ml 0.1 per cent trichloroacetic acid (TCA). The homogenate is centrifuged at 12000 × g for 15 min at 4 °C. Sebsequent reaction is carried out by adding 0.5 ml of the supernatant to 0.5 ml of 10 mM potassium phosphate buffer (pH 7.0) and 1 ml reagent (1 M KI in fresh double distilled water) and then the absorbance of the supernatant is read at 390 nm. The blank probe consisted of 0.1 per cent TCA in the absence of leaf extract. The content of H_2O_2 is calculated applying a standard curve prepared with identified concentrations of hydrogen peroxide.

Alternatively hydrogen peroxide content (H_2O_2) is estimated by analyzing the absorbance of titanium-hydroperoxide complex. Leaf samples (1 g) are homogenized in 10 ml of cold acetone. The homogenate is filtered through Whatman No. 1 filter paper. Reaction is performed by adding 4 ml of titanium reagent to the above extracts followed by 5 ml of concentrated ammonium solution to precipitate hydroperoxide-titanium complex. After centrifugation for 5 min at 10000 x g the supernatant is discarded and precipitate is dissolved in 1 M sulfuric acid centrifuged to remove undissolved material and absorbance is recorded at 415 nm against blank. Concentration of H_2O_2 is determined using standard curve plotted with known concentration of H_2O_2.

Antioxidant Enzymes

Fresh samples of leaf are collected and brought to the laboratory in ice bucket. Leaves are then washed with distilled water and surface moisture is wiped out. Leaf samples (0.5 g) is homogenized in ice cold 0.1 M phosphate buffer (pH 7.5) containing 0.5 mM EDTA with pre-chilled pestle and mortar centrifuged at 4 °C in for 15 min at 15000 x g. The supernatant referred to enzyme extract is used for determination of superoxide dismutase (SOD) and catalase (CAT) activity as described by Zhou *et al.* (2005).

SOD (superoxide dismutase) activity is estimated spectrophotometrically according to Maral *et al.* (1977) with modification of Sajid and Aftab, (2009) by recording the decrease in absorbance of superoxide-nitro blue tetrazolium complex by the enzyme. About 3 ml of reaction mixture containing 0.1 ml of 1.5 M sodium carbonate, 0.2 ml of 200 mM methionine, 0.1 ml of 2.25 mM Nitroblue tetrazolium, 0.1 ml of 3 mM EDTA, 1.5 ml of 100 mM potassium phosphate buffer, 1 ml distilled water and 0.05 ml of enzyme are taken in test tubes in triplicate from each enzyme sample. Three tubes without enzyme extract are taken as control. The reaction is started by adding 0.1 ml riboflavin (60 mM) and placing the tubes below a light source of two 15 W florescent lamps for 15 min. Reaction is stopped by switching off the light and covering the tubes with black cloth. Tubes without enzyme developed maximal color. A non-irradiated complete reaction mixture without color development served as blank. Absorbance is recorded at 560 nm and one unit of enzyme activity is taken as the quantity of enzyme which reduced the absorbance reading of samples to 50 per cent in comparison with tubes lacking enzymes.

CAT (catalase) activity is assayed according to Beers and Sizer (1952) with certain modification by measuring the residual H_2O_2 by titanium reagent. Reaction mixture (3 ml) with 1 ml of 6 mM H_2O_2 and 1.9 ml of 0.1 M phosphate buffer (pH 7.0) is taken in test tubes, and the reaction is initiated by adding 0.1 ml of diluted enzyme extract. The reaction is stopped after 5 min by addition of 4 ml of titanium reagent, which also forms colored complex with residual H_2O_2. Reaction mixture without enzyme is served as control and developed maximal color with titanium reagent. Aliquots are centrifuged at 10000 x g for 10 min and absorbance of supernatant is recorded at 415 nm in spectrophotometer. The residual H_2O_2 content in samples are computed with the help of standard curve. The quantitative estimation of **POD** (peroxidases) is followed by "Guaiacol-H_2O_2" method of Luck (1974) with certain modifications.

Estimation of Ascorbate Peroxidase (APX) Activity

The activity of APX is determined by measuring the decrease in absorbance at 290 nm and the amount of Ascorbate oxidized to dehydroascorbate is calculated from the extinction coefficient 2.8 (Nakano and Asada 1981). The reaction mixture consisted of 25 mM phosphate buffer (pH 7.0), 0.1 mM EDTA, 0.25 mM ascorbic acid,

1.0 mM H_2O_2 and 0.2 ml enzyme extract. The activity of APX is expressed as units per m $^{-1}$ per mg^{-1} protein. One unit is defined as micromoles of ascorbate oxidized per m $^{-1}$ per mg^{-1} protein.

GR (Gtutathione reductase) activity is assayed by recording the increase in absorbance in the presence of oxidized glutathione and DTNB (5,5-dithiobis-2-nitrbenzoic acid). The reaction mixture consists of 1 ml of 0.2 M potassium phosphate buffer (pH 7.5) containing 0.1 mM EDTA, 0.5 ml of 3 mM DTNB in 0.01 M potassium phosphate buffer (pH 7.5), 0.1 ml of 2 mM NADPH, 0.1 ml enzyme extract and distilled water to make up a final volume of 2.9 ml. Reaction is initialed by adding 0.1 ml of 2 mM GSSG (oxidized glutathione). The increase in absorbance at 412 nm is recorded at 25 °C over a period of 5 min on a spectrophotometer.

Screening of Clones for Salinity Treatments

To determine the impact of salt stress on mature plants (plants taken to seed set), selected and control plants are grown under greenhouse conditions, without salt stress for two months, and then gradually exposing plants to increasing levels of salt from 100 mM - 300 mM NaCl prior to seed production as described by Redman *et al.* (2011). Seed production is measured after 5 months of growth, of which, approximately 6 weeks of that time frame, involved exposure to the highest level of salt concentrations of 300 mM NaCl. Analysis of roots and shoot biomass (per cent) is done for selected plants in the presence and absence of salt stress, and compared to control plants. The yields of the selected and control plants is measured in terms of 100 seeds weight and compared amongst the salt stressed plants in the presence, in the absence of stress treatment and the control non-selected plants.

Screening of Clones for Drought Stress

Plantlets from growth medium are transplanted into 12 cm diameter plastic pots filled with 650 g air-dried clay soil. One week later, 10 plants are randomly selected from each selected clones and control plants. Individual plant heights are assessed and water potential (WP) measured using the pressure-chamber technique (Scholander *et al.*, 1965). Individual shoot fresh weights are assessed immediately. Shoot dry weights are measured after air drying for 2 weeks.

Plants are arranged in sets each with 24 plants from each selected clone and the control plants. About half of the plants of each type are well-watered to 100 per cent field capacity (FC) and half exposed to drought conditions at 40 per cent FC. Field capacity of the soil in pots was assessed before the treatment as reported by Tuomela (1997). Plants grown in soil with 40 per cent FC are irrigated for the first time 6 days after starting water stress treatment. Before watering, four pots from each treatment in each sets are randomly selected and weighed. The amount of water required to restore the soil to 100 per cent FC is added to the non-stressed group.

Drought stressed plants are watered to 40 per cent FC. After 2 months, plant heights are measured and then plants are cut at the soil surface. WP was measured for individual plants using 10-cm long shoot tips immediately after cutting. Individual leaf and stem fresh weights are assessed immediately and the dry weights are assessed after air drying for 2 weeks. The relative growth rates (RGR) for plant fresh and dry weights were calculated by the formula:

RGR (g g^{-1} day^{-1}) = loge W 2 - loge W 1 / T2 - T1

where,

W represent weight and T indicates harvesting time (Hunt, 1990).

Growth and Germination Assays

In order to compare the growth efficiency of controls and subjected to for various stresses (drought stress, low or high temperature or salinity) plants are allowed to grown till maturity. The seeds are allowed to set and ripen and then collected and stored at room temperature. For germination assays, approximately 100 seeds from each of the selected plants lines as well as control plants are vernalized for 4-5 d at 4°C prior to being placed in germination condition at 25 °C or following the standard protocol. The seeds are sown on either standard plant growth medium or on the same medium supplemented with selection agents to simulate the said stress. The percentage of germinated seeds is calculated based on the number of seedlings that has reached the cotyledon stage at 2 weeks. All germination assays are performed in triplicate. Average germination response on medium containing stress agent from the selected lines and control are compared to validate tolerance to the stress in question.

Mechanism of Stress Tolerance

Plants respond to their surrounding environment in a complex and integrated manner helps them to react to the specific set of stresses at a given time (Todaka *et al.*, 2012). The regulatory mechanisms underlying response to stresses allow plants to escape, avoid or tolerate them with potential biochemical, physiological and architectural modifications. Environmental stresses such as drought, salt stress, low and high temperatures are all physiologically linked result in limiting the plants access to water (Ahuja *et al.*, 2010; Wang *et al.*, 2003). The cell osmotic potential decreases due to this loss of cellular water. However, the elicitor of cell water loss differs between these stresses. For example i. the decrease of the cell water content under drought stress is due to water shortage in soil or/and in the atmosphere, ii. cold stress decreases cell water content due to the physiological drought, *i.e.*, the inability to transport the water to the living cells (leaf mesophyll) from the soil, iii. salinity stress decreases cell water content due to the decrease of external water potential, caused by the increased ion concentration (mainly Na $^+$ and Cl$^-$), turning more difficult water uptake by roots and water translocation to metabolically active cells (Ahuja *et al.*, 2010).

As a consequence of these stresses, the cell sap solute concentrations increase causing a decrease in cell osmotic potential and turgor. As a result there is sharp decrease in leaf expansion rate and finally the overall plant growth rate (Ahuja *et al.*, 2010). Additionally, low-molecular weight osmolytes such as carbohydrates sugar alcohols (mannitol and sorbitol) (Woodward and Bennett, 2005), the amino acid glycine betain (Mohamed *et al.*, 2000) and proline (Hassan *et al.*, 2004b) are synthesized that can counteract cellular dehydration and turgor loss by osmotic adjustment of cell sap (Verslues *et al.*, 1998). The synthesis of these compounds by the plant enhances tolerance to drought, cold, high temperature and salinity (Wang *et al.*, 2003). The group of genes that govern accumulation of compatible solutes, passive transport across membranes, energy-requiring water transport systems are involved in protection and stabilization of cell structures from desiccation and damage by reactive oxygen species (Ahuja *et al.*, 2010, Cheong *et al.*, 2002). The second groups of genes are the one which regulate the transduction of the stress signal and modulate gene expression. A highly complex and redundant network of four independent stress-responsive genetic regulatory pathways are known to exist in plants (Ahuja *et al.*, 2010; Cheong *et al.*, 2002; Wang *et al.*, 2003). The plant stress hormone abscisic acid (ABA) is known to play a crucial role in the regulation of the two of this gene (Rozema and Flowers, 2008).

In vitro Selection for Drought Tolerance

Applying osmotic stress by simulation of drought condition at all the stages of plant regeneration from the explants has been used as the most efficient method for selection of tolerant plants. High molecular weight polyethylene glycol (PEG) has been used in the medium *in vitro* selection for drought tolerance in several plants as non-penetrating osmotic agent lowering the water potential similar to the soil drying (Bouslama and Scapaugh, 1984). Petkova *et al.* (1995) and Lestari *et al.* (2005) found a positive correlation between the capability of germinating seed at the media containing PEG and the growth of the plants under stress condition. Dragisga *et al.* (1996) have obtained similar result and concluded that PEG can be applied as an osmotic stress on the *in vitro* selection. It is used to induce water stress in two soybean cultivars at callus and shoot regeneration stages (Sakthivelu *et al.*, 2008). It is also used for selection of drought tolerance after UV-C radiation of *in vitro* grown calli in alfalfa and potato (Eshanpour and Razavizadeh, 2005; Eshanpour *et al.*, 2007). The increasing amount of PEG added to the selective media inhibit the growth and development explants, callus formation and number of somatic embryo development among the different genotype of the soybean (Widoretno *et al.*, 2003; Husni *et al.*, 2005). *In vitro* selection for obtaining the tolerance to the drought stress has been conducted to barley (Duncan *et al.*, 1995), green grams mungbean (Gulati and Jaiwal, 1993), potato (Prakash *et al.*, 1994) and rice (Biswas *et al.*, 2002; Lestari, 2006). Drought tolerant plants of several rice varieties has been achieved by *in vitro* selection using PEG 20 per cent produced some tolerant plants after listing for their tolerance in green house (Lestari, 2006). These somaclones produces more full grains per panicle than that of the mother plants, have shown better seed germination response test, rapidly penetrate paraffin layers tests, the thicker and longer size of root and the leaves from various somaclone produce higher proline (Lestari, 2006, Lestari *et al.*, 2005).

Isolation and characterization of PEG (water stress) tolerant lines using cell suspensions of the drought-sensitive sunflower (*Helianthus annus* L. Cv. Myak) has been described recently (Hassan *et al.*, 2004a). They have found that PEG treatment causes a decrease in the polysaccharides content in both selected and non-selected callus lines of sunflower. They have speculated that the decrease in polysaccharide may be explained by its degradation into disaccharide and monosaccharide, which are increased under water stress. Similar observation has been reported by Kikuta and Richter (1988), and Kameli and Losel (1996) who have found reduced level of polysaccharide and increased di- and mono-saccharide concentrations under water deficit. It is also possible that sugars may be involved in osmotic adjustment under water deficit.

Shoot apex culture has been widely used to evaluate plant physiological responses to salinity and osmotic stress in various species, including apple (Shibli *et al.*, 1992), olive (Shibli and Al-Juboory, 2002) and tomato (Cano *et al.*, 1998).

In vitro Selection for Salinity Tolerance

Salinity is the main abiotic stress that has been addressed by *in vitro* selection, but applications to other stresses such as heat and drought have been reported (Turkan and Demiral, 2009; He *et al.*, 2009). Considerable results have been achieved for *in vitro* selection for salt tolerance (Gandonou *et al.*, 2006; Hassan *et al.*, 2008; Hossain *et al.*, 2007). The nature of correlation between high productivity and salt tolerance at whole plant level is not well established and understood (Turkan and Demiral, 2009). Investigations of *in vitro* selection for salt tolerant lines in sour orange (*Citrus aurantium* L.) are reported by Koc *et al.* (2009). *In vitro* selection of NaCl-tolerant callus line and regeneration of plantlets in a bamboo (*Dendrocalamus strictus* Nees.) has been reported by Singh *et al.* (2003). Salinity-induced changes in responses of antioxidative and proline metabolism among the two cultivars of *Vigna radiate* has been studied by Sumithra *et al.* (2006).

High salinity causes hyper-osmotic stress and ion disequilibrium in cells producing secondary effects like reduction of turgor below the yield threshold of cell wall resulting in growth cessation (Yokoi *et al.*, 2002). In sugarbeet (*Beta vulgaris* L.) salinity inhibits the growth of plants by affecting both water absorption and biochemical processes, such as nitrogen assimilation and protein biosynthesis. Fragments of shoots from tolerant clones formed new leafy shoots, roots from clones of partially tolerance only shoots, while those from non-tolerant clones did not initiate organogenesis (Flowers, 2004). The differentiated conditions for callus growth on a medium containing salt mixture rendered it possible to select in sugarbeet lines cell fragments tolerant to the salt stress. A similar result with the reduction of callus fresh weight of *Asparagus officinalis* and Globe artichoke as a result of salinity stress have been obtained by Bekheet *et al.* (2000) and El-Bahr *et al.* (2001), respectively. The relatively high variability in survival, shoot numbers, fresh weight, and chlorophyll levels in response to salt stress reflects genetic heterogeneity among the various spearmint clonal lines as compared with the unstressed controls (Al-Amier and Craker, 2007). Kashyap and Sharma (2006) have selected salt tolerant saplings of *Morus alba* (cv. Sujanpuri) raised from nodal explants with axillary buds in 0.4 per cent NaCl containing medium. A number of plant species have been selected with increased percentage of calli showing high frequency regeneration in the presence of NaCl in the medium (Aazami *et al.*, 2010; Dziadczyk *et al.*, 2003; He *et al.*, 2009; Tam, and Lang, 2003). A novel *in vitro* tissue culture approach to study salt stress responses in citrus is reported by Montoliu *et al.* (2009).

Bekheet *et al.* (2006) have found reduction of callus fresh weight of onion as a result of salinity stress. Similar results has also been reported by Bekheet *et al.* (2000) and El-Bahr *et al.* (2001) in their studies on *Asparagus officinalis* and Globe artichoke, respectively. The marked increase in protein content in callus cultures grown on saline media may be due to synthesis of new proteins (osmoprotectant protein) or inactivation of proteolytic enzymes (Bekheet *et al.*, 2000). The salt-induced proteins are also reported in potato plants by Rahnama and Ebrahimzadeh (2004). Recently, Queiros *et al.* (2007) also observed this increasing trend of soluble and insoluble proteins in potato cultures during the selection of salt-tolerant cell lines. These higher protein contents might be attributed to synthesis of stress-induced proteins (Sajid and Aftab, 2012) that may be helpful for maintaining the osmotic imbalance. Salt has two antagonistic effects on protein; firstly they tend to break electrostatic bonds and secondly increase hydrophobic interactions (Ashraf and Harris, 2004).

In vitro Selection for Oxidative stress

Several environmental stresses such as cold, drought, salinity, pathogen infection and herbicide action promote excess formation of reactive oxygen species and creates a common cross tolerance between them (Apel and Hirt, 2004). Double haploid maize plants tolerant to oxidative stress has been produced by the *in vitro* selection of microspores in the anther cultures with ROS proginators (Ambrus *et al.*, 2006, 2008). Paraquat is added to the medium and selected double haploid lines have been compared with non-selected double haploid lines, original hybrids and control lines. The oxidative stress tolerance of double haploid lines and hybrids has been determined by measurements on photosynthetic activity of PS II via Fv/Fm chlorophyll a fluorescence parameter by dtermination of chlorophyll (a+b) content and ion conductivity of leaf discs floated in different concentrations of paraquat solution (Lichtenthaler, 1987; Lichtenthaler and Rinderle, 1988). Three of the five paraquat tolerant double haploid lines possess higher cold tolerance than the control double haploids and the original hybrids during the germination period. The low temperature stress (8 °C) exposed the plant at the early autotrophic phase of development resulted in a higher cold tolerance in all the five paraquat mediated oxidative double haploid lines (Ambrus *et al.*, 2008). Relationship between salt tolerance and photosynthetic machinery performance in selected citrus species is reported by López-Climent *et al.* (2008). *In vitro* selection of salinity tolerant variants from triploid bermudagrass (*Cynodon transvaalensis* and *C. dactylon*) and their physiological responses to salt and drought stress has been reported by Lu *et al.* (2007).

In vitro Selection for Osmotic stresses

An important aspect of *in vitro* selection strategy is the improvement of plant species displaying an increased tolerance to osmotic stress (Nabors, 1990). An increase of osmotic pressure is caused by the increase of osmolytes and osmo-protectants like sugars, sugar alcohols (Elavumoottil *et al.*, 2003), quarternary amino acid derivatives (proline, glycine betaine, β-alanine betaine, praline betaine), tertiary amines (1,4,5,6-tetrahydro-2-methyl-4-carboxyl pyrimidine), and sulfonium compounds (Cherian and Reddy 2003) in the cells to enhance stress tolerance. Application of sugar alcohol, mannitol and sorbitol has been suggested as osmotic agents for selection of drought tolerance plants under *in vitro* conditions.

Sucrose, mannitol or sorbitol have been applied as osmotic stress agents for *in vitro* selection of *Chrysanthemum morifolium* (Shibli *et al.*, 1992) and of many plant species (Mohamed *et al.*, 2000, Watnabe *et al.*, 2000). Sucrose failed to elicit consistent osmotic stress symptoms and enhanced both shoots and root growth. Mannitol has an inhibitory effect on plant growth by lowering the water potential of the medium, so cultured explants are unable to take up water and nutrients from the medium. The harmful effects of the osmotic stress agent has been noticed in both shoot proliferation and rooting stages. A decrease in the dry matter accumulation of wheat embryos, rape seedlings and potato stem segments has been reported with the decrease in osmotic potential of the culture medium by mannitol used for *in vitro* selection (Lipavska and Vreugdenhil, 1996). Mohamed *et al.* (2000) have found that the drought-tolerant-selected clone of *T. minuta* have demonstrated a higher capacity to maintain membrane stability, low water potential, greater callus growth when grown on medium containing mannitol induced low water potential. However, the regenerated plants have yielded a higher biomass in green house condition.

Addition of PEG as a selective agent in the medium has also been applied for selection of osmotic stress tolerant plants. It has been possible to distinguish osmotic tolerant and sensitive alfalfa cultivars with the application of PEG (Georgieva *et al.*, 2004; Petkova *et al.*, 1995). PEG has served as an excellent selective agent for *in vitro* screening of osmotic tolerant raspberry genotype (Georgieva *et al.*, 2004) and mulberry plants (Tiwary *et al.*, 2000). There has been reduction in callus induction ability and plant regeneration efficiency with increasing levels of PEG (6000) stress, but selected plants have shown improved yield under stress condition as compared to control (Wani *et al.*, 2010).

The accumulation of proline and soluble sugars as an osmotic tolerance mechanism has been widely observed in many species (Thomas, 1997). Soluble sugars, which accumulate in the vacuole, are a major organic solute, involved in osmotic adjustment when plants are exposed to water stress. Proline may provide osmo-regulation and stabilization of proteins and membrane during stress (Thomas, 1997). In many plants during moderate water stress, there is a 10- to 100-fold increase of free proline in leaf tissue (Hanson and Hitz, 1982, Hanson *et al.*, 1994). However, no correlation has been established between the accumulation of proline and drought stress tolerance in tomato (Alian *et al.*, 2000). *In vitro* selection for drought stress in *Sorghum bicolor* have shown reduced regeneration ability (Duncan *et al.*, 1995). While, screening of regenerated plants under field conditions have shown enhanced yield under stress conditions than that of the control parents. Similar results has been reported by Errabii *et al.* (2006) in sugarcane.

In vitro Selection for Iron Deficiency

Iron deficiency is a common nutritional problem in plants and soils. There are many reports describing yield reductions caused by Fe deficiency in field crops (Hansen *et al.*, 2006) and in fruits (Rombola and Tagliavini, 2006) and the detrimental effects of Fe-chlorosis on both fruit yield and quality (Álvarez-Fernández et al., 2006). The differences in sensitivity to Fe deficiency among plant species have been reported by several researchers (Tagliavini and Rombola, 2001; Erdal et al., 2004; Álvarez-Fernández et al., 2011; Pestana et al., 2011, 2012; Zuo and Zhang, 2011). The strawberry is one of the most sensitive species to Fe deficiency (Pestana et al., 2011, 2012). Tourn et al. (2014) have evaluated the responses of several genotypes belonging to octoploid *Fragaria chiloensis* (L.) Mill. and *Fragaria virginiana* Mill. (the cultivated strawberry) against Fe treatments under *in vitro* conditions. The results of these experiments indicated that *F. chiloensis* and *F. virginiana* genotypes exhibited considerable variation under different Fe treatments *in vitro*, and the more resistant genotypes could be utilized to develop new strawberry cultivars with tolerance to low Fe concentrations.

In vitro Selection for Tolerance to Low and High Temperature

The direct effect of low temperatures on *in vitro* grown calli has not been studied extensively. Kendall et al. (1990, 1989) have obtained frost tolerant plants by cryoselection of callus in spring wheat. Adding hydroxyproline as a constituent of *in vitro* media, proline over producing variants of several spring wheat varieties are selected, which contained increased levels of free proline and proved to be more frost tolerant than the initial material (Tantau and Dorffling, 1991). Using *in vitro* techniques, it was possible to regenerate hydroxyl proline winter wheat plants with an increased proline content and a higher frost tolerance (Dorffling et al., 1993).

Liu et al. (2013) have studied cold stress is a major factor limiting the growth of turfgrass species in warm-season. Cold tolerance in warm-season turfgrass species is improved through *in vitro* selection for somaclonal variations using mature seeds of seashore paspalum (*Paspalum vaginatum*). Embryogenic calli were subjected to 2 or 6 °C treatment for 90 days for *in vitro* cold selection of somaclonal variation. Plants regenerated from calli surviving cold treatment (cold-selected) for 45 or 60 days were then exposed to low temperatures [15/10 or 5/3 °C (day/night)]. Plant variants derived from cold-selected calli exhibited significant improvement in their tolerance to low temperature of either 15/10 or 5/3 °C (day/night), as manifested by higher turf quality, leaf chlorophyll content, and membrane stability as well as lower levels of lipid peroxidation compared with the control plants.

Trolinder and Shang (1991) have used cell suspension cultures of cotton (*Gossypium hitirsutum* L. cv. Coker 312) exposed to various temperature and for different time periods in order to select cell lines resistant to high temperature stress. Cells are exposed to 45°C for 0-105 hrs by giving 3 h each including 24 h of continuous treatment at the end. After the stress treatments, the cells are plated onto embryo development medium for plantlets recovery. The embryogenic calli that are sub-cultured monthly for 6 months and tested for increased resistance to the temperature: time treatments previously determined to be lethal and to water stress as imposed by PEG. All of the selected cell lines were more resistant to both types of stress than the control cell lines.

Xiao-ying et al. (2007) have used high temperature and HYP as direct and indirect selection stress. *In vitro* selection of heat-tolerant clonal lines from tussock plantlets and callus of *Senecio cruentus* was carried out. The results showed that [1] 40°C for 16-20 h and 30 mmol·L⁻¹ HYP were optimal direct and indirect selection stress pressure, respectively; [2] Two heat- tolerant clones (Z_{1-1-1} and N_{1-2-1}) were obtained by using high temperature direct selection, HYP and high temperature selection; [3] The regenerated plants of Z_{1-1-1} were different significantly from their mother plants(Z_1) in leaf thickness, TPT/TST and stomata index. The fruit setting rate of adult plant of Z_{1-1-1} was only 1.0 per cent after artificial pollination, and the fruit setting rate of Z_1 was over 85 per cent, [4]After heat stress (40°C 24 h), the heat injury index, content of MDA and electrolyte leakage of the regenerated plants of Z_{1-1-1} and N_{1-2-1} were lower than their mother plants, but the activity of POD and SOD were higher than their mother plants (Z_1 and N_1).

In vitro Selection for Tolerance to Metal Toxicity

In vitro selection for acid soil and Al toxicity tolerance could be applied with $AlCl_3.6H_2O$ as the selection agent on the low acid media as much as pH 4 (Short *et al.*, 1987). The Al toxicity to selection media could be emerged by modifying macro nutrient MS, *i.e.* by increasing NH_4NO_3, $CaCl_2.2H_2O$ and decreasing KH_2PO_4 and the application of Fe which was not chelated by EDTA (Purnamaningsih *et al.*, 2001). This method of application has resulted several Al tolerance plants such as in rice (Van Sint Jan *et al.*, 1997), soybean (Mariska, 2003), and tobacco (Yamamoto *et al.*, 1994). Van Sint Jan (1997) showed that one of three Al tolerance was obtained from unselected callus using Al, while the others were selected with 250 and 1000 µmol from the total Al. Development of Al-toxicity tolerant lines in indica rice has been reported by Roy and Mandal (2005) through the exploitation of somaclonal variations.

Hutami *et al.* (2001) attained the somatic embryo structure of soybean which is capable of proliferation after selection at the media containing Al and low pH. It has been shown that there is a positive correlation between the mass of somatic cell tolerant to Al and low pH and their tolerance to acid field (Mariska, 2003). Stress condition caused by Al results in the decreasing capability of growth and the development of somatic cell in the tobacco (Yamamoto *et al.*, 1994), and rice (Purnamaningsih *et al.*, 2001; Edi, 2004). Plant improvement for tolerance to aluminum in acid soils has been reviewed by Samac and Tesfaye (2003). *In vitro* selection and regeneration of zinc tolerant calli from *Setaria italica* L. has been studied by Samantaray *et al.* (1999). Induction, selection and characterization of Cr and Ni-tolerant cell lines of *Echinochloa colona* (L.) Link *in vitro* has been described by Samantaray *et al.* (2001). A relationship between copper- and zinc-induced oxidative stress and proline accumulation in *Scenedesmus* sp. has been established by Tripathi and Gaur, (2004).

Conclusion

Development of plant varieties with a high level of tolerance to various abiotic stresses is the need of the present time for establishing full yield potential and to stabilize production on the increasing soil infertility, salinity, acidity and the drought prone area. Breeding for abiotic stress tolerance through sexual hybridization has been limited due to species barriers. Induction of mutation with screening and selection of population of seeds and plants as well as recessive nature of these mutants delays the developments of abiotic stress tolerance variants. Raising the transgenic genotypes has been limited because of being the most expensive, requiring high technical expertise, sensitivity of plants to *Agrobacterium*, difficulties with multigene introduction and regeneration of transformants for most of these abiotic stresses. The most successful approach is the cell and tissue culture based *in vitro* selection system which offers a convenient tool for inducing the physiological, biochemical and molecular regulation of plant stress responsive phenomena to improve growth, development and yield of food grain crops. In past two decades considerable progress has been made in the selection of stress tolerant plants by using *in vitro* techniques. The efficiency and simplicity of *in vitro* selection procedure makes it possible to evaluate large number of regenerated plants for tolerance to various abiotic stresses. However, *in vitro* selected variants should be finally field-tested to confirm the genetic stability of the selected traits under field conditions. Further, the molecular mechanisms operating during stress tolerance under the tissue culture conditions is not yet fully explored. The development of *in vitro* selection technique in conjunction with the molecular biological approaches and functional genomics need to be worked out in future for creating better opportunity to regulate stress tolerance in plants with improved food production in an environmentally sustainable manner.

Acknowledgement

Finanacial support provided for major research project by University Grant Commission, New Delhi, India is gratefully acknowledged.

References

1. Aazami, M.A., Torabi, M. and Shekari, F. (2010). Response of some tomato cultivars to sodium chloride stress under *in vitro* culture condition. *African J. Agric. Res.*; 5: 2589 – 2592.

2. Ahloowalia, B.S. and Maluszynski, M. (2001). Induced mutations-A new paradigm in plant breeding. *Euphytica*; 118: 167-173.

3. Ahmed, K.Z., Mesterhazy, A., Bartok, T. and Sagi. F. (1996). *In vitro* techniques for selecting wheat (*Triticum aestivum*.L) for *Fusarium* resistance II. Culture filtrate technique and inheritance of *Fusarium* I resistance in the somaclones. *Euphytica*; 91: 341-349.

4. Ahuja, I., de Vos, R.C.H., Bones, A.M. and Hall, R.D. (2010). Plant molecular stress responses face climate change. *Trends Plant Sci.*; 15: 664–674.

5. Al-Amier, H. and Craker, L. E. (2007). *In vitro* Selection for Stress Tolerant *Spearmint*. In: J. Janick and A. Whipkey (eds.). Issues in new crops and new uses. ASHS Press, Alexandria, VA. pp. 306-310.

6. Alian, A., Altman, A. and Heuer, B. (2000). Genotypic difference in salinity and water stress tolerance of fresh market tomato cultivars, *Plant Sci.*; 152: 59–65.

7. Álvarez-Fernández, A., Abadía, J. and Abadía, A. (2006). Iron deficiency, fruit yield and fruit quality. In: Barton, L.L., Abadia, J., editors. *Iron Nutrition in Plants and Rizospheric Microorganisms*. Dordrecht, The Netherlands: Springer, pp. 437–448.

8. Álvarez-Fernández, A., Melgar, J.C., Abadía, J., Abadía, A. (2011). Effects of moderate and severe iron deficiency chlorosis on fruit yield, appearance and composition in pear (*Pyrus communis* L.) and peach (*Prunus persica* (L.) Batsch. *Environ. Exp. Bot.;* 71: 280–286.

9. Ambrus, H., Darko, E., Szabo, L., Bakos, F., Király, Z. and Barnabás, B. (2006). *In vitro* microspore selection in maize anther culture using oxidative stress stimulatiors. *Protoplasma;* 228: 87-94.

10. Ambrus, H., Dulai, S., Király, Z., Barnabás, B. and Darko, E. (2008). Paraquat and cold tolerance in doubled haploid maize. *Acta Biologica Szegediensis;* 52(1): 147-151.

11. Annonymous (2009a). The environmental food crisis: The environment's role in averting future food crises. UNEP, 2009.

12. Annonymous (2009b). World Population Prospects: The 2008 Revision, Highlights. UNDESA. 2009.

13. Annonymous (2012). United Nations World Water Development Report 4. UNESCO, UN-Water, WWAP. March 2012.

14. Apel, K. and Hirt, H. (2004). Reactive oxygen species: metabolism, oxidative stress and signal transduction. *Ann. Rev. Plant Biol.;* 55: 373-399.

15. Arnon, D.I. (1949). Copper enzymes in isolated choloroplast: polyphenoloxidase in *Beta vulgaris*. *Plant Physiol.;* 14: 1–15.

16. Ashraf, M. and Harris. P.J.C. (2004). Potential biochemical indicators of salinity tolerance in plants. *Plant Sci.;* 166: 3-16.

17. Ashraf, M., Foolad, M.R. (2007). Roles of glycine betaine and proline in improving plant abiotic stress resistance. *Environ. Exp. Bot.;* 59: 206–216.

18. Bajji, M., Kinet, J.M., Lutts, S. (2001). The use of the electrolyte leakage method for assessing cell membrane stability as a water stress tolerance test in durum wheat. *Plant Growth Regul;* 33: 1-10.

19. Bartels, D. and Sunkar, R. (2005). Drought and salt tolerance in plants, *Crit. Rev. Plant Sci.;* 24: 23–58.

20. Bates, L.S., Waldren, R.P. and Teare, I.D. (1973). Rapid determination of free proline for water-stress studies. *Plant Soil;* 39: 205-207.

21. Batkova, P., Paspisilova, J. and Synkova, H. (2008). Production of reactive oxygen species and development of antioxidative system during *in vitro* growth and *ex vitro* transfer. *Biol Plant.;* 52: 413-422.

22. Beers, R.F. and Sizer, I.W. (1952). A spectrophotometric method for measuring the breakdown of hydrogen peroxide by catalase. *J. Biol Chem.;* 195: 133-140.

23. Bekheet, S.A., Taha, H.S., Sawires, E.S. and El-Bahr, M.K. (2000). Salt stress in tissue cultures of *Asparagus officinalis*. *Egyp. J. Hort.;* 27: 275-187.

24. Bekheet, S. A., Taha, H. S. and Solliman, M. E. (2006). Salt tolerance in tissue culture of onion (*Allium cepa* L.). *Arab J. Biotech.;* 9 (3): 467-476.

25. Ben-Ahmed, C., Ben-Rouina, B. and Boukhris, M. (2008). Changes in water relations, photosynthetic activity and proline accumulation in one-year-old olive trees (*Olea europaea* L. cv. Chemlali) in response to NaCl salinity. *Acta Physiol. Plant.;* 30: 553–560.

26. Benderradji, L., Brini, F., Kellou, K., Ykhelf, N., Djekoun, A., Masmoudi, K. and Bouzerour, H. (2012). Callus induction, proliferation, and plantlets regeneration of two bread wheat (*Triticum aestivum* L.) genotypes under saline and heat stress conditions. *ISRN Agronomy*; Article ID 367851.

27. Biswas, J., Chowdhury, B., Bhattacharya, A. and Mandal, A.B. (2002). *In vitro* screening for increased drought tolerance in rice. *In vitro Cell Dev. Biol.-Plant;* 38: 525–530.

28. Bouslama, M. and Schapaugh, W.T. (1984). Stress tolerance in soybean. I. Evaluation on three screening techniques for heat and drought tolerance. *Crop Science;* 24: 993-937.

29. Boyer, J. S. (1982). Plant productivity and environment. *Science*; 218: 443-448.

30. Bradford, M.M. (1976). A rapid and sensitive method for the quantitation of micro gram quantities of protein utilizing the principle of protein–dye binding. *Anal Biochem;* 72: 248–254. doi: 10.1016/0003-2697(76)90527-3.

31. Bray, E.A., Bailey-Serres, J. and Weretilnyk, E. (2000). Responses to abiotic stresses. In: Gruissem W., Buchannan B., Jones R. (eds) Biochemistry and molecular biology of plants. American Society of Plant Physiologists, Rockville, MD, pp 1158–1249.

32. Bray, E.A. (1997). Plant responses to water deficit, *Trends Plant Sci;* 2: 48–54.

33. Cano, E.A., Perez-Alfocea, F., Moreno, V., Caro, M. and Bolarin, M.C. (1998). Evaluation of salt tolerance in cultivated and wild tomato species through *in vitro* shoot apex culture. *Plant Cell, Tiss. Org. Cult.;* 53: 19-26.

34. Chaves, M.M., Maroco, J.P. and Pereira, J.S. (2003). Understanding plant responses to drought from genes to the whole plant. *Funct. Plant Biol.;* 30: 239–264.

35. Cheong, Y.H., Chang, H.S., Gupta, R., Wang, X., Zhu, T. and Luan, S. (2002). Transcriptional profiling reveals novel interactions between wounding, pathogen, abiotic stress, and hormonal responses in *Arabidopsis*. *Plant Physiol.;* 129: 661–677.

36. Cherian, S. and Reddy, M.P. (2003). Evaluation of NaCl tolerance in the callus cultures of *Suaeda nudiflora* Moq. *Biol Plant.;* 46: 193-198.

37. Cui, Yu, Shujun, H., Xingming, H., Wen, D., Chao, X., Chuhua, Y., Yong, L. and Bo, P. (2013). Changes in photosynthesis, chlorophyll fluorescence, and antioxidant enzymes of mulberry (*Morus* ssp.) in response to salinity and high-temperature stress. *Biologia;* 68(3): 404-413.

38. Cushman, J.C. and Bohnert, H.J. (2000). Genomic approaches to plant stress tolerance. *Curr. Opin. Plant Biol.;* 3: 117–124.

39. Demiral, T. and Turkan, I. (2005). Comparative lipid peroxidation, antioxidant defense systems and proline content in roots of two rice cultivars differing in salt tolerance. *Environ. Exp. Bot.*; 53: 247–257.

40. Dennis, E.S. (2004). Molecular analysis of the alcohol dehydrogenase (*adhI*) genes of maize. *Nuc. Acid Res.*; 12: 3983-4000.

41. Dionisiosese, M. and Tobita, S. (1998). Antioxidant response of rice seedlings to salinity stress. *Plant Sc.*; 135: 1-9.

42. Dorffling, K., Dorffling, H. and Lesselich, G. (1993). *In vitro* selection and regeneration of hydroxy proline-resistant lines of winter wheat with increased proline content and increased frost tolerance. *J. Plant Physiol.*; 142: 222–225.

43. Dragiiska, R., Djilianov, D., Denchev1, P. and Atanassov, A. (1996). *In vitro* selection for osmotic tolerance in alfalfa (*Medicago sativa* l.). *Bulg. J. Plant. Phys.*; 22(3–4): 30–39.

44. Duncan, R.R., Waskom, R.M. and Nabors, M.W. (1995). *In vitro* screening and field evaluation of tissue-culture-regenerated sorghum (*Sorghum bicolor* L. Moench) for soil stress tolerance, *Euphytica*; 85: 373–380.

45. Dziadczyk, P., Bolibok, H., Tyrka, M. and Hortynski, J.A. (2003). *In vitro* selection of strawberry (*Fragaria ananassa* Duch.) clones tolerant to salt stress. *Euphytica*; 132 (1): 49–55.

46. Edi, S. (2004). Peningkatan Ketenggangan terhadap Alumunium dan pH Rendah pada Tanaman Padi melalui Keragaman Somaklonal dan Iradiasi Sinar Gamma. [Disertasi]. Bogor: Sekolah Pascasarjana Institut Pertanian Bogor.

47. Ehsanpour, A.A. and Razavizadeh, R. (2005). Effect of UV-C on drought tolerance of alfalfa (*Medicago sativa*) callus. *Am. J. Biochem. and Biotech.*; 1(2): 107-110.

48. Ehsanpour, A.A., Madani, S. and Hoseini, M. (2007). Detection of somaclonal variation in Potato callus induced by UV-C radiation Using RAPD-PCR. Gen. *Appl. Plant Physiol.*; 33 (1-2): 3-11.

49. Elavumoottil, O.C., Martin, J.P. and Moreno, M.L. (2003). Changes in sugars, sucrose synthase activity and proteins in salinity tolerant callus and cell suspension cultures of *Brassica oleracea* L. *Biol. Plant.*; 46: 7–12.

50. El-Bahr, M. K., Okasha, Kh. A. and Bekheet, S.A. (2001). Effect of salt stress on callus cultures of Globe artichoke. *Arab Univ. J. Agric. Sci., Ain Shams Univ. Cairo*; 9: 783-792.

51. Erdal, Ý., Kepenek, K. and Kýzýlgöz, Ý. (2004). Effect of foliar iron applications at different growth stages on iron and some nutrient concentrations in strawberry cultivars. *Turk J. Agric. For.*; 28: 421–427.

52. Errabii, T., Gandonou, C.B., Essalmani, H., Abrini, J., Idomar, M. and Senhaji, N.S. (2006). Growth, proline and ion accumulation in sugarcane callus cultures under drought-induced osmotic stress and its subsequent relief. *Afri. J. Biotechnol.*; 5: 1488–1493.

53. FAO, (2006). World Agriculture: Towards 2030/2050 – Interim Report – Prospects for Food, Nutrition, Agriculture and Major Commodity Groups. FAO. 2006.

54. Flowers, T. J. (2004). Improving crop salt tolerance. *J. Exp. Bot.;* 55(396): 307-319.

55. Fu, J. and Huang, B. (2001). Involvement of antioxidants and lipid peroxidation in the adaptation of two cool-season grasses to localized drought stress. *Environ. Exp. Bot.;* 45 : 105- 114.

56. Gale, M. (2003). Applications of Molecular Biology and Genomics to Genetic Enhancement of Crop Tolerance to Abiotic Stress - A Discussion Document. Interim Science Council Secretariat Food And Agriculture Organization Of The United Nations. Information brief on Water and Agriculture in the Green Economy. UNW-DPAC, 2011.

57. Gandonou, C.B., Errabii, T., Abrini, J., Idaomar, M. and Senhaji, N.S. (2006). Selection of callus cultures of sugarcane (*Saccharum* sp.) tolerant to NaCl and their response to salt stress. *Plant Cell Tiss. Org. Cult.;* 87: 9–16.

58. Georgieva, M., Djilianov, D., Konstantinova, T. and Parvanova, D. (2004). Screening of Bulgarian raspberry cultivars and elites for osmotic tolerance *in vitro*. *Biotechnol. and Biotechnol. Eq.;* 19(2): 95-98.

59. Ghoulam, C., Ahmed, F. and Khalid, F. (2001). Effects of salt stress on growth, inorganic ions and proline accumulation in relation to osmotic adjustment in five sugar beet cultivars. *Environ. Exp. Bot.;* 47: 139–150.

60. Gulati, A and Jaiwal, P.K. (1993). Selection and characterization of mannitol-tolerant callus lines of *Vigna radiata* (L.) Wilczeck. *Plant Cell Tiss. Org. Cult.;* 34: 35-41.

61. Hansen, N.C., Hopkins, B.G., Ellsworth, J.W., Jolley, V.D., Barton, L.L. and Abadia, J. (2006). Iron Nutrition in Field Crops. Iron Nutrition in Plants and Rhizospheric Microorganisms. The Netherlands: Springer, pp. 23–59.

62. Hanson, A.D., Rathinasabapathi, B., Rivoal, J., Burnet, M., Dillon, M.O. and Gage, D.A. (1994). Osmoprotective compounds in the Plumbaginaceae: A natural experiment in metabolic engineering of stress tolerance. *Proc. Natl. Acad. Sci. USA* 91: 306–310.

63. Hanson, A.D. and Hitz, W.D. (1982). Metabolic responses of mesophytes to water deficits. *Ann. Rev. Plant Physiol.;* 33: 162–203.

64. Hasegawa, P.M., Bressan, R.A., Zhu, J.K., Bohnert, H.J. (2000). Plant cellular and molecular responses to high salinity. *Annu. Rev. Plant. Physiol. Plant. Mol. Biol.* 51, 463–499.

65. Hassan, N. S., Shaaban, L. D., Hashem, El-S. A. and Seleem, E. E. (2004a). *In vitro* Selection for Water Stress Tolerant Callus Line of *Helianthus annus* L. Cv. Myak. *Intern. J. Agr. Biol.* 1560–8530/2004/06–1–13–18

66. Hassan, N.M., Serag, M.S., El-Feky, F.M. (2004b). Changes in nitrogen content and protein profiles following *in vitro* selection of NaCl resistant mung bean and tomato. *Acta Physiol. Plant.;* 26: 165–175.

67. Hassan, N.M., Serag, M.S., El-Feky, F.M., Nemat Alla, M.M. (2008). *In vitro* selection of mung bean and tomato for improving tolerance to NaCl. *Ann. Appl. Biol.;* 152: 319–330.

68. He, S., Han, Y., Wang, Y., Zhai, H., Liu, Q. (2009). *In vitro* selection and identification of sweetpotato (*Ipomoea batatas* (L.) Lam.) plants tolerant to NaCl. *Plant Cell Tiss. Org. Cult.;* 96: 69–74.

69. Heath, R.L. and Packer, L. (1968). Photoperoxidation in isolated chloroplasts: I. Kinetics and stoichiometry of fatty acid peroxidation. *Arch. Biochem. and Biophys.;* 125: 189-198.

70. Hiscox, J.D. and Israelstam, G.F. (1979). A method for the extraction of chlorophyll from leaf tissue without maceration. *Canadian J. Bot.;* 57: 1332-1334.

71. Hossain, Z., Mandal, A.K.A., Datta, S.K. and Biswas, A.K. (2007). Development of NaCl tolerant line in *Chrysanthemum morifolium* Ramat. through shoot organogenesis of selected callus line. *J. Biotechnol.;* 129: 658–667.

72. Hu, L., Li., H., Pang, H. and Fu, J. (2012). Responses of antioxidant gene, protein and enzymes to salinity stress in two genotypes of perennial ryegrass (*Lolium perenne*) differing in salt tolerance. *J. Plant Physiol.;* 169: 146-156.

73. Husni, A., Hutami, S., Kosmiatin, M. and Mariska, D.I. (2005). Pembentukan benih somatik dewasa kedelai dan aklimatisasi serta uji terhadap indikator sifat toleransi kekeringan. *Prosiding Seminar Hasil Penelitian BB-Biogen Tahun 2004.* Bogor: BB-Biogen Badan Litbang Pertanian.

74. Hutami, S., Mariska., I., Kosmiatin, M., Rahayu, S. and Adil, D.W.H. (2001). Regenerasi massa sel embrionik tanaman kedelai setelah diseleksi dengan Al dan pH rendah. *Prosiding Seminar Hasil Penelitian Rintisan dan Bioteknologi Tanaman.* Bogor: BB-Biogen Badan Litbang Pertanian.

75. Jahnke, L.S., Hull, M.R. and Long, S.P. (1991). Chilling stress and oxygen metabolizing enzyme in *Zea diploperennis. Plant Cell Environ.;* 14: 97-104.

76. Jain, S.M. (2001). Tissue culture-derived variation in crop improvement. *Euphytica;* 118: 153–166.

77. Jain, S.M., Brar, D.S. and Ahlowalia, B.S. (Eds). (1998). Somaclonal variation and induced mutations in cropm improvement. Kluwer Academic Publisher, Dordrectht.

78. John, M.K. (1970). Coulorimetric determination of phosphorus in soil and in plant materials with ascorbic acid. *Soil. Sci.;* 109: 214–20.

79. Kameli, A. and Losel, D.M. (1996). Growth and sugar accumulation in durum wheat plants under water stress. *New Phytol.;* 132: 57–62.

80. Kashyap, S. and Sharma, S. (2006). *In vitro* selection of salt tolerant *Morus alba* and its field performance with bioinoculants. *Hort. Sci.* (Prague); 33(2): 77–86.

81. Kavi Kishore, P.B., Sangam, S., Amrutha, R.N., Sri Laxmi, P., Naidu, K.R., Rao, K.R.S.S., Rao, S., Reddy, K.J., Theriappan, P. and Sreenivasulu, N. (2005). Regulation of proline biosynthesis, degradation, uptake, and transport in higher

plants: its implications in plant growth and abiotic stress tolerance. *Curr. Sci.;* 88: 424–438.

82. Kendall, E.J. and McKersie, B.D. (1989). Free radical and freezing injury to cell membranes of winter wheat. *Physiol. Plant.;* 76: 86–94.

83. Kendall, E., Quaresshi, F.A., Kartha, K.K., Lenne, N., Caswell, N. and Chen, T.H.H. (1990). Regeneration of freezing tolerant spring wheat (*Triticum aestivum* L.) plant from cryoselected callus. *Plant Physiol.;* 94: 1756 - 1762.

84. Kikuta, S.B. and Richter, H. (1988). Rapid osmotic adjustment in detached wheat leaves. *Ann. Bot.;* 62: 167–72.

85. Kim, S.H., Woo, D.H., Kim, J.M., Lee, S.Y., Chung, W.S. and Moon, Y.H. (2011). *Arabidopsis* MKK4 mediates osmotic-stress response via its regulation of MPK3 activity. *Biochem. Bioph. Res. Co.;* 412: 150-154.

86. Knight, H. and Knight, M.R. (2001). Abiotic stress signaling pathways: Specificity and cross talk. *Trends Plant Sci.;* 6: 262-267.

87. Koc, N.K., Bas, B., Koc, M. and Kusek, M. (2009). Investigations of *in vitro* selection for salt tolerant lines in sour orange (*Citrus aurantium* L.). *Biotechnol.;* 8: 155–159.

88. Kosová, K., Vítámvás, P., Prášil, I.T. and Renaut, J. (2011). Plant proteome changes under abiotic stress - Contribution of proteomics studies to understanding plant stress response. *J. of Proteomics;* 15: 51-58.

89. Kumar, N., Pamidimarri S, D.V.N., Kaur, M., Boricha, G. and Reddy, M.P. (2008). Effect of NaCl on growth, ion accumulation, protein, proline contents and antioxidant enzymes activity in callus cultures of *Jatropha curcas. Biologia.;* 63: 378-382.

90. Kumar, P.S and Mathur. V.L. (2004). Chromosomal instability is callus culture of *Pisum sativum. Plant Cell Tiss. Org. Cult.;* 78: 267-271.

91. Kumar, V., Shriram, V., Nikam, T.D., Jawali, N. and Shitole, M.G. (2008). Sodium chloride induced changes in mineral nutrients and proline accumulation in indica rice cultivars differing in salt tolerance. *J. Plant Nutr.;* 31: 1999–2017.

92. Lascano, H.R. (2001). Antioxidant system response of different wheat cultivars under drought: field and *in vitro* studies. *Australian J. Plant Physiol.;* 28: 1095–1102.

93. Lestari, E.G. (2006). *In vitro* selection and somaclonal variation for biotic and abiotic stress tolerance. *Biodiversitas;* 7: 297-301.

94. Lestari, E.G., Guhardja, E., Harran, S. and Mariska. D.I. (2005). Uji daya tembus akar untuk seleksi somaklon toleran kekeringan pada padi varietas Gajahmugkur, Towuti dan IR64. *Penelitian Pertanian Tanaman Pangan;* 24 (2): 97-103.

95. Lichtenthaler, H.K. (1987). Chlorophylls and carotenoids: pigments of photosynthetic biomembranes. *Meth. Enzymol.;* 148: 350-382.

96. Lichtenthaler, H.K. and Rinderle, U. (1988). The role of chlorophyll fluorescence in the detection of stress conditions in plants. – *CRC Crit. Rev. anal. Chem.*; 19: S29-S85.

97. Lipavska, H. and Vreugdenhil, D. (1996). Uptake of mannitol from the medium by *in vitro* grown plants, *Plant Cell Tiss. Org. Cult.*; 45: 103–107.

98. Liu, J.H., Nada, K., Honda, C., Kitashiba, H., Wen, X.P., Pang, X.M. and Moriguchi, T. (2006). Polyamine biosynthesis of apple callus under salt stress: importance of the arginine decarboxylase pathway in stress response. *J. Exp. Bot.*; 57: 2589-2599.

99. Liu, T.X. and Zhang, Y.P. (2010). Determination of ABA content in the seedling of capsicum by HPLC. *Guangdong Agr. Sci.*; 8: 249-250.

100. Liu, J., Yang, Z., Li, W., Yu, J. and Huang, B. (2013). Improving cold tolerance through *in vitro* selection for somaclonal variations in *Seashore paspalum. J. Amer. Soc. Hort. Sci.*; 138(6): 452-460.

101. López-Climent, M.F., Arbona, V., Pérez-Clemente, R.M. and Gómez-Cadenas, A. (2008). Relationship between salt tolerance and photosynthetic machinery performance in citrus. *Environ. Exp. Bot.*; 62: 176–184.

102. Lu, S., Peng, X., Guo, Z., Zhang, G., Wang, Z., Wang, C., Pang, C., Fan, Z. and Wang, J. (2007). *In vitro* selection of salinity tolerant variants from triploid bermudagrass (*Cynodon transvaalensis* and *C. dactylon*) and their physiological responses to salt and drought stress. *Plant Cell Rep.*; 26: 1413–1420.

103. Luck, H. (1974). Methods in enzymatic analysis, 2nd edition. Bergmeyer Academic New York, pp. 885.

104. Maral, J., Puget, K. and Micheson. A.M. (1977). Comparative study of superoxide dismutase, catalase, and glutathione peroxidase levels in erythrocytes of different animals. *Biochem Biophys Res Commun.*; 77: 1525-1535.

105. Mariska, I. (2003). Peningkatan Ketahanan terhadap Alumunium pada Pertanaan Kedelai melalui Kultur *in vitro*. Laporan RUT VIII. Bidang Teknologi Pertanian. Bogor: Balai Penelitian Bioteknologi Pertanian and Kementerian Riset dan Teknologi RI-LIPI.

106. Matkowski, A. (2008). Plant *in vitro* culture for the production of antioxidants — A review. *Biotechnol. Advances*; 26: 584-560.

107. Mittova, V., Theodoulou, F.L., Kiddle, G., Gomez, L., Volokita, M., Tal, M., Foyer, C.H. and Guy, M. (2003a) Coordinate induction of glutathione biosynthesis and glutathione-metabolizing enzymes is correlated with salt tolerance in tomato. *FEBS Letter*; 554: 417-421.

108. Mittova, V., Tal, M., Volokita, M. and Guy, M. (2003b). Upregulation of the leaf mitochondrial and peroxisomal antioxidative system in response to salt-induced oxidative stress in the wild salt tolerant tomato species *Lycopersicon pennellii*. *Plant Cell Environ.*; 26: 845-856.

109. Mohamed, M.A.H., Harris, P.J.C. and Henderson, J. (2000). *In vitro* selection and characterisation of a drought tolerant clone of *Tagetes minuta*. *Plant Science,* 159: 213–222.

110. Montoliu, A.., López-Climent, M.F., Arbona, V., Pérez-Clemente, R.M. and Gómez-Cadenas, A. (2009). A novel *in vitro* tissue culture approach to study salt stress responses in citrus. *Plant Growth Reg.;* 59: 179-187.

111. Morgan, J.M. (1984). Osmoregulation and water stress in higher plants. *Annu. Rev. Plant Physiol.,* 35: 299–319.

112. Munns, R. (2002). Comparative physiology of salt and water stress, *Funct. Plant Biol.;* 25: 239–250.

113. Munns, R. and James, R.A. (2003). Screening methods for salinity tolerance: a case study with tetraploid wheat. *Plant Soil;* 253: 201–218.

114. Munns, R. and Tester, M. (2008). Mechanisms of salinity tolerance. *Annu. Rev. Plant Biol.;* 59: 651–681.

115. Nabors, M.W. (1990). Environmental stress resistance procedure and applications, In: J.D. Philip (Ed.), Plant Cell Line Selection, VCH, Weinheim, pp. 167–185.

116. Naguib, M.I. (1963). Effect of seivn on the carbohydrate and nitrogen metabolism during the germination of cotton seeds. *Ind. J. Exp. Biol.;* 2: 149–52.

117. Nakano, Y. and Asada, K. (1981). Hydrogen peroxide is scavenged by ascorbate specific peroxidase in spinach chloroplast. *Plant Cell Physiol.;* 22: 867–880.

118. Nelson, N. (1944). A photometric adaptation of the Somogy's method for determination of glucose. *J. Biol. Chem.;* 153: 375–80.

119. Pei, Z.M., Kuchitsu, K., Ward, J.M., Schwarz, M. and Schroeder, J.I. (1997). Differential abscisic acid regulation of guard cell slow anion channels in *Arabidopsis* wild-type and *abi*1 and *abi*2 mutants. *Plant cell;* 9: 409-423.

120. Pestana, M., Correia, P.J., Saavedra, T., Gama, F., Abadia, A. and de Varennes, A. (2012). Development and recovery of iron deficiency by iron resupply to roots or leaves of strawberry plants. *Plant Physiol. Bioch.;* 53: 1–5.

121. Pestana, M., Domingos, I., Gama, F., Dandlen, S., Migue, M.G., Pinto, J.C., de Varennes, A. and Correia, P.J. (2011). Strawberry recovers from iron chlorosis after foliar application of a grass-clipping extract. *Soil Sci. Plant Nutr.;* 174: 473–479.

122. Petkova, D., Nedjialkov, D. and Djilianov, D. (1995). Early screening for drought tolerance in cultivatead alfalfa. *Bulgarian J. Agricult. Sci.;*1(4): 429-432.

123. Prakash, A.H., Vajranabhaiah, S.N. and Chandrashekhara, R. (1994). Changing patterns in water relation and solukte contens during callus development under stress in sunflower (*Helianthus annum* L.) *Adv. Plant Sci.;* 7: 223-225.

124. Predieri, S. (2001). Mutation induction and tissue culture in improving fruits. *Plant Cell Tiss. Org. Cult.;* 64: 185–210.

125. Purnamaningsih, R., Mariska, I., Husni, D.A. (2001). Produksi kalus embriogenik dan regenerasinya setelah seleksi *in vitro* dengan Al dan pH rendah pada tanaman padi. *Prosiding Seminar Hasil Penelitian Rintisan dan Bioteknologi Tanaman.* BB-Biogen Badan Litbang Pertanian, Bogor 27-27 Desember.

126. Qian, Y.L., Engelke, M.C. and Foster, M.J.V. (2000). Salinity effects on Zoysiagrass cultivars and experimental lines. *Crop Sci.*; 40: 488– 482.

127. Qian, Y.L., Wilhelm, S.J. and Marcum, K.B. (2001). Comparative response of two Kentucky bluegrass cultivars to salinity stress. *Crop Sci.*; 41: 1895-1900.

128. Queiros, F., Fidalgo, F., Santos, I. and Salema, R. (2007). *In vitro* selection of salt tolerant cell lines in *Solanum tuberosum* L. *Biol. Plant.*; 51: 728–734.

129. Racusen, D. and Johnstone, D.B. (1961). Estimation of protein in cellular material. *Nature*, 191: 292-493.

130. Rahnama, H. and Ebrahimzadeh, E. (2004). The effect of NaCl on proline accumulation in potato seedlings and calli. *Acta Physiol. Plant.*; 26: 263-270.

131. Rahnama, H., Ebrahimzadeh, E. and Ghareyazie, B. (2003). Antioxidant enzymes responses to NaCl stress in calli of four potato cultivars. *Pak. J. Bot.*; 35: 579-586.

132. Rai, M.K., Kalia, R.K., Singh, R., Gangola, M.P. and Dhawan, A.K. (2011). Developing stress tolerant plants through *in vitro* selection: An overview of the recent progress. *Environ. Exp. Bot.*; 71: 89-98.

133. Redman, R.S., Kim, Y.O., Woodward, C.J.D.A., Greer, C., Espino, L., *et al.* (2011). Increased Fitness of Rice Plants to Abiotic Stress Via Habitat Adapted Symbiosis: A Strategy for Mitigating Impacts of Climate Change. *PLoS ONE* 6(7): e14823. doi: 10.1371/journal.pone.0014823

134. Rombola, A.D. and Tagliavini, M. (2006). Iron nutrition of fruit tree crops. In: Barton, L.L., Abadía, J. (eds). Iron Nutrition in Plants and Rhizospheric Microorganisms. Dordrecht, The Netherlands: Springer, pp. 61–83.

135. Roy, B., Mandal, A.B. (2005). Towards development of Al-toxicity tolerant lines in indica rice by exploiting somaclonal variation. *Euphytica*; 145: 221–227.

136. Rozema, J. and Flowers, T. (2008). Crops for a salinized world. *Science*; 322: 1478-1480.

137. Sadasivam, S. and Manickam, A. (1992). Biochemical methods for agricultural sciences. New Delhi: Wiley Eastern Limited. pp. 179–180.

138. Sajid, Z. A. and Aftab, F. (2014). Plant regeneration from *in vitro*-selected salt tolerant callus cultures of *Solanum tuberosum* L. *Pak. J. Bot.*; 46(4): 1507-1514.

139. Sajid, Z.A and Aftab, F. (2009). Amelioration of salinity tolerance in *Solanum tubersoum* L. by exogenous application of ascorbic acid. *In vitro Cell Dev. Biol-plant.*; 45: 540-549.

140. Sajid, Z.A and Aftab, F. (2012). Role of salicylic acid in amelioration of salt tolerance in potato (*Solanum tubersoum* L.) under *in vitro* condition. *Pak. J Bot.*; 44: 37-42.

141. Sakthivelu, G., Akitha Devi, M.K., Giridhar, P., Rajasekaran, T., Ravishankar, G.A., Nedev, T. and Kostturkova, G. (2008). Drought induced alteration in growth osmotic potential and *in vitro* regeneration of soybean cultivars. *Gen. Appl. Plant Physiol.* (Special issue); 34(1-2): 103-112.

142. Samac, D.A. and Tesfaye, M. (2003). Plant improvement for tolerance to aluminum in acid soils – a review. *Plant Cell Tiss. Org. Cult.;* 75: 189–207.

143. Samantaray, S., Rout, G.R. and Das, P. (1999). *In vitro* selection and regeneration of zinc tolerant calli from *Setaria italica* L. *Plant Sci.;* 143: 201–209.

144. Samantaray, S., Rout, G.R. and Das, P. (2001). Induction, selection and characterization of Cr and Ni-tolerant cell lines of *Echinochloa colona* (L.) Link *in vitro. J. Plant Physiol.;* 158: 1281–1290.

145. Santos–Diaz, M.S. and Ochoa–Alejo, N. (1994). PEG–tolerant cell clones of chili pepper: Growth, osmotic potentials and solute accumulation. *Plant Cell Tiss. Org. Cult.;* 37: 1–8.

146. Scandalios, J.G. (1993). Oxygen stresses and superoxide dismutases. *Plant Physiol.;* 101: 7-12.

147. Scholander, P. F., Hammel, H. T., Bradstreet, E. D. and Hemmingsen, E. A. (1965). Sap pressure in vascular plants: Negative hydrostatic pressure can be measured in plants. *Science;* 148: 339–3416.

148. Serrano, R., Mulet, J.M., Rios, G., Marquez, J.A., de Larrinoa, I.F., Leube, M.P., Mendizabal, I., Pascual-Ahuir, A., Proft, M., Ros, R. and Montesinos, C. (1999). A glimpse of the mechanisms of ion homeostasis during salt stress. *J. Exp. Bot.;* 50: 1023–1036.

149. Serrano, R. and Rodriguez-Navarro, A. (2001). Ion homeostasis during salt stress in plants. *Curr. Opin. Cell Biol.;* 13: 399–404.

150. Shibli, R.A. and Al-Juboory, K. (2002). Comparative response of 'Nabali' olive microshoot, callus and suspension cell cultures to salinity and water deficit. *J. Plant Nut.;* 25: 61-74.

151. Shibli, R.A., Smith, M.A.L. and Spomer, L.A. (1992). Osmotic adjustment and growth response of three *Chrysanthemum morifolium* Ramat. cultivars to osmotic stress induced *in vitro*, *Plant Nutr.;* 15: 1373–1381.

152. Shinozaki, K. and Yamaguchi-Shinozaki, K. (2000). Moelcular response to dehydration and low temperature: Differences and cross talk between two stress signaling pathways. *Curr. Opin. Plant Biol.;* 3: 217-223.

153. Short, K., Warburton, C.I. and Roberts, A.V. (1987). *In vitro* hardening of cultures cauliflower and *Chrysanthemum* plantles to humidity. *Acta Hort.;* 2(120): 319-324.

154. Singh, M., Jaiswal, U., Jaiswal, V.S. (2003). *In vitro* selection of NaCl-tolerant callus line and regeneration of plantlets in a bamboo (*Dendrocalamus strictus* Nees.). *In vitro Cell Dev. Biol.-Plant;* 39: 229–233.

155. Smirnoff, N. (1998). Plants resistance to environmental stress. *Curr. Opin. Biotechnol.*; 9: 214-219.

156. Sreenivasulu, N., Grimm, B., Wobus, U. and Weschke, W. (2000). Differential response of antioxidant compounds to salinity stress in salt-tolerant and salt-sensitive seedlings of fox-tail millet (*Setaria italica*). *Physiol. Plant.*; 109: 435-442.

157. Sumithra, K., Jutur, P.P., Carmel, B.D., Reddy, A.R. (2006). Salinity-induced changes in two cultivars of *Vigna radiata*: responses of antioxidative and proline metabolism. *Plant Growth Regul.*; 50: 11–22.

158. Tagliavini, M. and Rombola, A.D. (2001). Iron deficiency and chlorosis in orchard and vineyard ecosystems. *Eur. J. Agron.*; 15: 71–92.

159. Tam, D. M. and Lang, N. T. (2003). *In vitro* selection for salt tolerance in rice. *Omonrice*; 11: 68-73.

160. Tantau, H. and Dorffling, K. (1991). *In vitro* selection and regeneration of hydroxy - proline resistant lines of wheat (*Triticum aestivum* L.): accumulation of proline, decrease in osmotic potential and increase in frost tolerance. *Physiol.*; 82: 243 - 248.

161. Tester, M. and Langridge, P. (2010). Breeding technologies to increase crop production in a changing world. *Science*; 327: 818–822.

162. Tewary, P.K., Sharma, A., Raghunath, M.K., Sarkar, A. (2000). *In vitro* response of promising mulberry (*Morus* sp.) genotypes for tolerance to salt and osmotic stresses. *Plant Growth Regul.*; 30: 17–21.

163. Thomas, H. (1997). Drought resistance in plants. In: S.D. Amarjit, K.B. Ranjit (Eds.), Mechanisms of Environmental Stress Resistance in Plants, Harwood, pp. 1–42.

164. Todaka, D., Nakashima, K., Shinozaki, K., Yamaguchi-Shinozaki, K. (2012). Toward understanding transcriptional regulatory networks in abiotic stress responses and tolerance in rice. *Rice* 2012, 5: 6 http://www.thericejournal.com/content/5/1/6.

165. Torun, A.A., Kaçar, Y.A., Biçen, B., Erdem, N. and Serçe, S. (2014). *In vitro* screening of octoploid *Fragaria chiloensis* and *Fragaria virginiana* genotypes against iron deficiency. *Turk J. Agric. Forestry*; 38: 169-179.

166. Tripathi, B.N. and Gaur, J.P. (2004). Relationship between copper- and zinc-induced oxidative stress and proline accumulation in *Scenedesmus* sp. *Planta*; 219: 397–404.

167. Trolinder, N.L. and Shang, X. (1991). *In vitro* selection and regeneration of cotton resistant to high temperature stress. *Plant Cell Rep.*; 10(9): 448-52. doi: 10.1007/BF00233812.

168. Tuomela, K. (1997). Leaf water relations in six provenances of *Eucalyptus microtheca*: a greenhouse experiment. *Forest Ecol. Manage.*; 92: 1–10.

169. Turkan, I. and Demiral, T. (2009). Recent developments in understanding salinity tolerance. *Environ. Exp. Bot.;* 67: 2–9.

170. Van Sint Jan, V., de Mecedo, C.C., Kinet, J.M. and Bouharmont, J. (1997). Selection of Al-resistant plant from al sensitive rice cultivar, using somaclonal variation, *in vitro*, and hydroponic cultures. *Euphytica;* 97: 03-310.

171. Velikova, V. and Loreto, F. (2005). On the relationship between isoprene emission and thermotolerance in *Phragmites australis* leaves exposed to high temperatures and during the recovery from heat stress. *Plant Cell and Environ.;* 28: 00–00.

172. Vierling, E. and Kimpel, J.A. (1992). Plant responses to environmental stress. *Curr. Opin. Biotech.;* 3: 164–170.

173. Wang, W.X., Vinocur, B., Shoseyov, O. and Altman, A. (2001). Biotechnology of plant osmotic stress tolerance: physiological and molecular considerations. *Acta Hort.;* 560: 285–292.

174. Wang, W.B., Kim, Y.H., Lee, H.S., Kim, K.Y., Deng, X.P. and Kwak, S.S. (2009). Analysis of antioxidant enzyme activity during germination of alfalfa under salt and drought stresses. *Plant Physiol. Biochem.;* 47: 570-577.

175. Wang, S., Liang, D., Li, C., Hao, Y., Ma, F. and Shu, H. (2012). Influence of drought stress on the cellular ultrastructure and antioxidant system in leaves of drought-tolerants and drought sensitive apple rootstock. *Plnat Physiol. Biochem.;* 51: 81-89.

176. Wang, W., Vinocur, B. and Altman, A. (2003). Plant responses to drought, salinity and extreme temperatures: towards genetic engineering for stress tolerance. *Planta;* 218: 1–14.

177. Wani, S.H., Sofi, P.A., Gosal, S.S. and Singh, N.B. (2010). *In vitro* screening of rice (*Oryza sativa* L) callus for drought tolerance. *Commun. Biometry and Crop Sci.;* 5(2): 108–115.

178. Watanabe, S., Kojima, K., Ide, Y. and Sasaki, S. (2000). Effects of saline and osmotic stress on proline and sugar accumulation in *Populus euphratica in vitro*. *Plant Cell Tiss. Org. Cult.;* 63: 199-206.

179. Widoretno, W., Megia, R. and Sudarsono, dan (2003). Reaksi embrio somatik kedelai terhadap poliethilene glicol dan penggunaannya untuk seleksi *in vitro* terhadap cekaman kekeringan. *Hayati;* 10(4): 134-139.

180. Winicov, I. (1996). Characterization of rice (*Oryza sativa* L.) plants regenerated from salt-tolerant cell lines. *Plant Sci.;* 13: 105–111.

181. Witcombe, J.R., Hollington, P.A., Howarth, C.J., Reader, S. and Steele, K.A. (2008). Breeding for abiotic stresses for sustainable agriculture. *Philos. Trans. R. Soc. B.;* 363: 703–716.

182. Witjaksono. (2003). Peran bioteknologi dalam pemuliaan tanaman buah tropika. *Seminar Nasional Peran Bioteknologi dalam Pengembangan Buah Tropika.* Kementerian Riset dan Teknologi RI and Pusat Kajian Buah Buahan Tropika, IPB. Bogor, 9 Mei 2003.

183. Woodward, A.J. and Bennett, I.J. (2005). The effect of salt stress and abscisic acid on proline production, chlorophyll content and growth of *in vitro* propagated shoots of *Eucalyptus camaldulensis. Plant Cell Tiss. Org. Cult.*; 82: 189–200.

184. Xiao-ying, Y.U., Xiang-yang, L.U., Da, L.I., Hong-zhi, Z., Jue, Y.A.O. and Cui-lin, C. (2007). A eliminary study on *in vitro* selection of heat-tolerant clonal lines in *Senecio × hybridus. Acta Hort. Sinica*; 34(2): 513-516. http: //www.ahs.ac.cn// EN/Y2007/V34/I2/513.

185. Xing, H., Tan, L., An, L., Zhao, Z., Wang, S. and Zhang, C. (2004). Evidence for the involvement of nitric oxide and reactive oxygen species in osmotic stress tolerance of wheat seedlings: Inverse correlation between leaf abscisic acid accumulation and leaf water loss. *Plant Growth Reg.*; 42: 61–68.

186. Yamamoto, Y., Sanas., R., Chiech, Y., Ono, K., Monisu, K., and Matoomoto, H. (1994). Quantitative estimation of aluminium toxycity in cultured tobaccon cells. Corelation between alumunium uptake and growth inhibition. *Plant Cell Physiol.*; 35(4): 575-583.

187. Yang,Y.L., Shi, R.X., Wei, X.L., Fan Q. and An, L.Z. (2010). Effect of salinity on antioxidant enzymes in calli of the halophyte *Nitraria tangutorum* Bobr. *Plant Cell Tiss. Org. Cult.*; 102: 387–395.

188. Yokoi, S., Ray, A.B. and Paul, M.H. (2002). Salt tolerance of plants,. In: JIRCAS Working Report. Center for Environmental Stress Physiology, Purdue University, West Lafayette, USA. pp. 25-33.

189. Zair, I., Chlyah, A., Sabounji, K., Tittahsen, M., Chlyah, H. (2003). Salt tolerance improvement in some wheat cultivars after application of *in vitro* selection pressure. *Plant Cell Tiss. Org. Cult.*; 73: 237–244.

190. Zhou, B., Guo, Z. and Lin, L. (2005). Effects of abscisic acid on antioxidant systems of *Stylosanthes guianensis* (Aublet) Sw. under chilling stress. *Crop Sci.*; 45: 599–605.

191. Zhu, J.K. (2001a). Cell signaling under salt, water and cold stresses. *Curr. Opin. Plant Biol.*; 4: 401-408.

192. Zhu, J.K. (2001b). Plant salt tolerance. *Trends Plant Sci.*; 6: 66-71.

193. Zhu, J.K. (2002). Salt and drought stress signal transduction in plants. *Annu. Rev. Plant Biol.*; 53: 247–273.

194. Zhu, J.K., Hasegawa, P.M. and Bressan, R.A. (1997). Molecular aspects of osmotic stress in plants. *Crit. Rev. Plant Sci.*; 16: 253–277.

195. Zuo, Y. and Zhang, F. (2011). Soil and crop management strategies to prevent iron deficiency in crops. *Plant Soil*; 339: 83–95.

2016, Environmental Biotechnology: A New Approach   *Pages 347–359*
Editors: Dr. Rajan Kumar Gupta and Dr. Satya Shila Singh
Published by: DAYA PUBLISHING HOUSE, NEW DELHI

Chapter 21

Cordyceps sinensis: A Magical Medicinal Mushroom

Santosh Kumar Singh* and Amir Khan

Assistant Professor, School of Life Sciences,
The Glocal University, Saharanpur, Uttar Pradesh

ABSTRACT

Cordyceps sinensis is commonly called as Kidajadi and Yarshagumba in local villages of hilly areas of Uttarakhand. It is well known for its nucleoside derivatives especially Cordycepin. Besides it contains many metabolites that are widely used in various medicinal formulations. It is also called as 'Himalayan Viyagra' due to its potential of enhancing sexual power. Cordyceps sinensis lives as a mushroom and a parasite too in lepidopteran larvae. Its random exploitation from its natural habitats have led it into rare and endangered category of plants. Many researchers are attempting to develop somatic hybrids of Cordyceps sinensis to utilize its maximum potential and its conservation into its native habitats. The authors have presented idea about its sustainable use and benefit of its biodiversity conservation in the interest of state and country.

Keywords: Cordycepin, Nucleosides, Somatic hybrids, Ascospores, Lepidoptera larvae, Yarshagumba.

India is a Hot Spot of Biodiversity

India is widely known for its biodiversity and use of medicinal plants in various therapeutic applications from ancient years. The oldest record of the use of plants as medicines is mentioned in the 'Rigveda'(4,500-1600 B.C.) which contains many *'sholakas'* and hymns written in the praise of plants. The Charka Samhita by Agnivesa;

---

* *Corresponding author.* E-mail: tosanu_raj@rediffmail.com; researchinbotany@gmail.com

Charaka (1000-800 B.C.) and, Susruta (800-700 B.C.) describes Himalayas as the best habitat of medicinal Plants. Many Indian states including Uttarakhand has a list of thousands plant species under both lower and higher plant category. Many medicinal plants are under consideration for the evaluation of their medicinal properties and are widely used in pharmaceutical sector. Due to this, a vast variety of plants and other resources are diminishing day by day. Some have become extinct and some are at the edge of extinction. Researchers have given new remedy for this problem in form of cell and tissue culture. This technique not only help in the conservation of Biodiversity of rare plants but also provide the way to produce secondary metabolites and other useful compounds in the laboratory, without disturbing the natural form of Biodiversity.

Biodiversity in Uttarakhand

Uttaranchal State came into existence as the 27th State of Republic of India on November, 9, 2000, which was carved out from the erstwhile Uttar Pradesh. It lies between 28° 53′ 24″ and 31° 24′ 50″ N latitudes and between 77° 34′ 27″ and 81° 02′ 22″ E longitudes. It covers an area of 53, 483 sq km, of which 51, 000 sq km comes under the Himalayan region. Among the mountain peaks are Nanda Devi (7816 m, the highest peak of the State), Gauri Parvat, Kamet, Trishul, Chaukhamba, Daunagiri, Panchchuli and Nandakot. Gaumukh, Pandari and Milam are the major glaciers in the State. The State has a rich repository of herbal and medicinal plants. Many of these herbs and plants are of high repute in *Ayurveda* and other systems of natural medicine. The distribution of medicinal plants by their habit in Uttarakhand is as follows:

- ☆ Herbs – 32 per cent
- ☆ Shrubs – 20 per cent
- ☆ Trees – 33 per cent
- ☆ Climbers – 12 per cent
- ☆ Others – 3 per cent

The Uttarakhand State is blessed with thousands of species. About 320 species have been identified having medicinal value. The forest department has reported about 175 species being commercially exploited and traded. Some valuable medicinal plant species are:

Adhatoda vasica, Withania somnifera, Strychnos mux-vonica, Asparagus adscendens, Swertia chirata, Ephadra geraradiana, Ocimum basilicum, Catharanthus roseus, Barleria prionitis, Bacopa monierri etc. *Cordyceps*, a rare medicinal mushroom is also one of them reported usually at higher altitudes. It is estimated that the State is well positioned to generate revenue of about Rs. 1, 000 crores annually through medicinal herbs trade.

Cordyceps: A Medicinal Mushroom

Cordyceps is a rare and exotic medicinal mushroom, known from centuries in India and China. Most people in the West have come to know this rare herbal medicine in only the last twenty years or so. During that time, modern scientific investigation into its seemingly miraculous range of healing powers has proven what traditional practitioners have noted for centuries. Commonly the fungus is called as Yarsagumba or 'Kida-Jadi' by local practitioners in lower and upper Himalaya regions.

Name and General Description

Cordyceps sinensis is a medicinal herb (Ascomycetes fungus) closely related to the mushrooms. While not actually a mushroom in the taxonomic sense, it has been regarded as, and called, a medicinal mushroom throughout history. The name *Cordyceps* comes from the Latin words: *cord* and *ceps*, meaning "club" and "head", respectively. The Latin conjugation accurately describes the appearance of the club fungus, *Cordyceps sinensis*, whose stroma or fruitbody extend from the mummified carcasses of insect larvae, usually caterpillar larva of the Himalayan Bat Moth, *Hepialis armoricanus* (Chen Li, 1993).

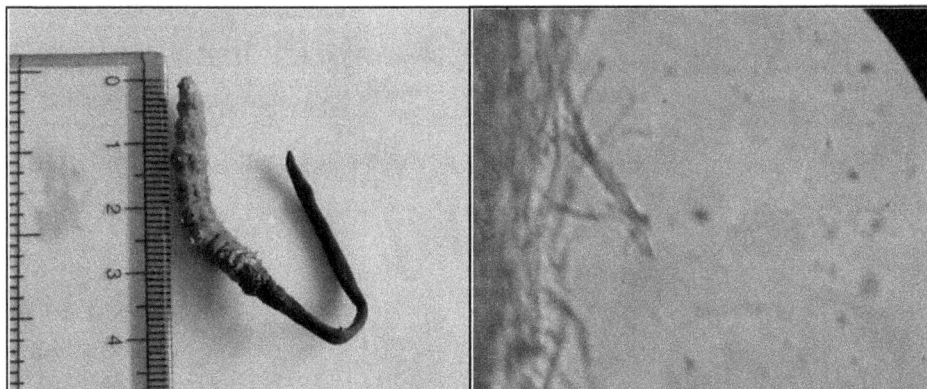

Figure 21.1. *Cordyceps sinensis* Stick and Fungal Mycelia and Bristles Emerging from the Larval Body (Cross section).

In historical and general usage the term '*Cordyceps*' usually refers specifically to the specific species *Cordyceps sinensis*, but there are also many other closely related species that come under the general term of *Cordyceps* (Gi Ho Sung *et al.*, 2007). While *Cordyceps sinensis* may be the species of *Cordyceps* that is most well known throughout the world, there are many other species in the genus *Cordyceps* in which modern science has found valuable medicinal properties in as well.

The ascocarp or fruitbody of the *Cordyceps sinensis* mushroom originates at its base on an insect larval host (usually the larva of the Himalayan bat moth, *Hepialis armoricanus*, although occasionally other insect hosts besides the bat moth are encountered.) and ends at the club-like cap, including the stipe and stroma. The fruit body is dark brown to black; and the 'root' of the organism, the larval body pervaded by the mushroom's mycelium, appears yellowish to brown in color.

Microscopic Characterization of the Fungus

Transverse section of stroma: Perithecia elliptical and embedded at the surface of the fertile portion of stroma. Central part was observed to be full with hyphae with

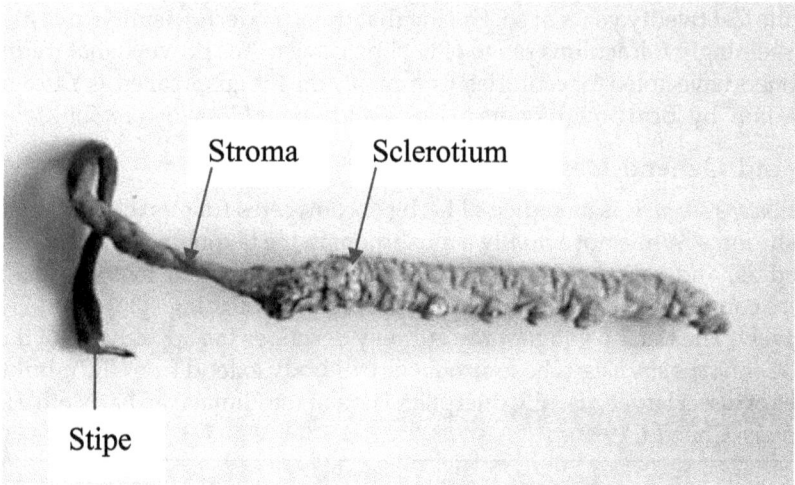

Figure 21.2: The Stroma and Sclerotium of a *Cordyceps sinensis* along with the Larval Body with 8 Pairs of Feet on Abdomen (4 pairs in the centre).

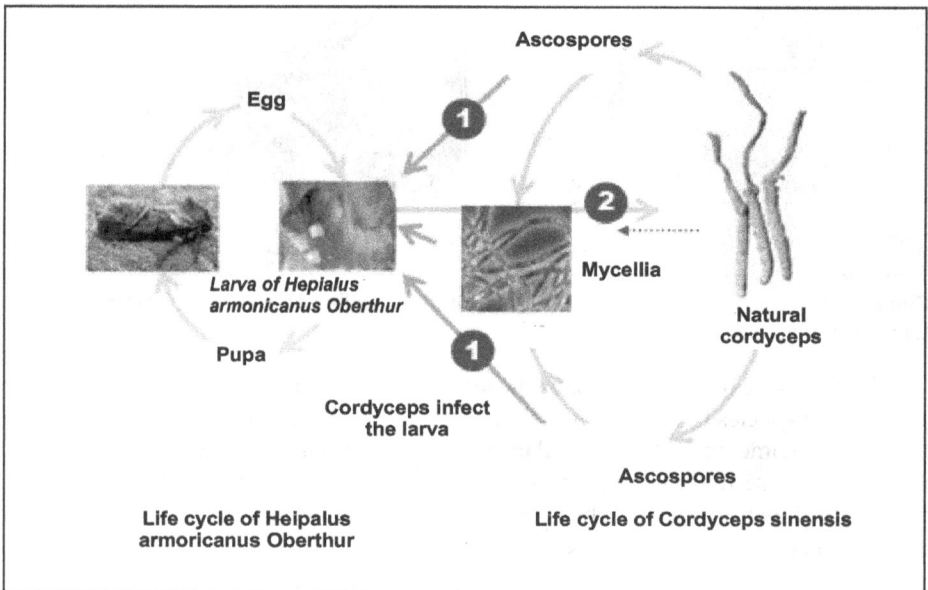

Figure 21.3. Life Cycle of *Cordyceps sinensis.*

clefts present between the hyphae. In the transverse section of larval body it was observed to be 23–85 mm thick with bristles 20–40 mm in length on the surface. Larva bod filled with whitish hyphae, with L-shaped stripes in the center. Surface view of the larval body showed an undistinguished yellowish-brown, pitchy stipes with abundant hypahe and bristles (Guo, 1986).

The immature larvae, which forms the host upon which the *Cordyceps* grows, usually lives about 6 inches below ground, is ca. 10-15 mm long and has a weight of ca. 0.05 g. The infesting spores of the *Cordyceps*, which are thought by some mycologists to be the infectious agent for the insect, are ca. 5-10 um long. As the fungus approaches maturity, it will have consumed greater than 99 per cent of the infested organism, effectively mummifying the host. As the stroma matures, it will swell and develop perihelia. The average weight of an individual stroma is only ca. 0.06 g. optimal conditions permitting; the spores are eventually discharged and taken by the wind or fall within a few centimeters of their origin (Kiho and Ukai, 1995).

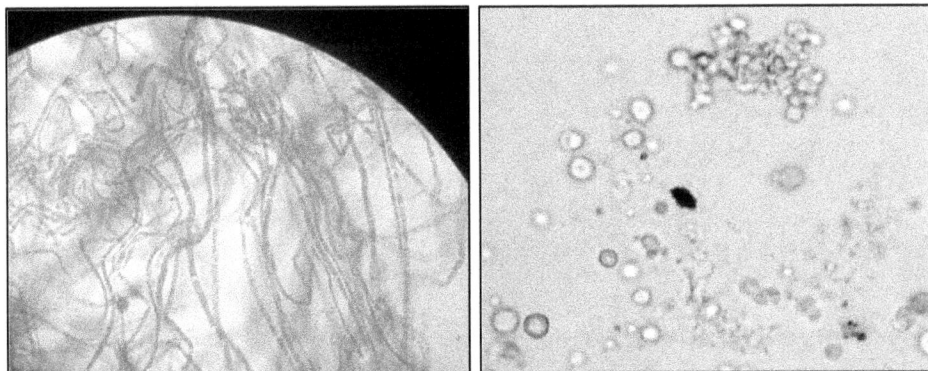

Figure 21.4. Hyphae and Ascospores of *Cordyceps sinensis* [Photograph was taken under Optical microscope (40 x resolutions) with the help of Nikon Cool pix camera (7x).

Medicinal Potential of the Fungus

Cordyceps sinensis has been known and used for many centuries in Traditional Indian and Chinese Medicine. In nature, it is found only at high altitudes on the Himalayan Plateau, and is thus difficult to find and harvest. Because of the difficulties involved in harvesting this exotic medicinal, *Cordyceps* has always been one of the most expensive medicines known. Despite its cost and rarity, the unprecedented litany of medicinal uses for *Cordyceps* has made it a highly valued staple of the medical traditions. Advance technological tools might help to obtain the full potential of this rare plant. Protoplast culture is one of them (Chamberlain, 1996). *Cordyceps* is unique source of pharmaceutically significant secondary metabolites which include unique nitrogenous bases, alkaloids, glycosides, flavonoids, volatile oils, tannins, resins etc. Demands of these secondary metabolites are not readily accomplished because of their low yield in intact plants, environmental, geographical and/or governmental restrictions. Chemical syntheses of these metabolites are either extremely difficult or economically infeasible because of their highly complex structures and stereo specific chemical nature. Plant cell culture is an attractive alternative, but to date this has had only limited commercial success. Precursor feeding has been a successful approach for enhanced production of secondary metabolites from plant cells grown *in vitro*. Elicitor-induced enhanced production of secondary metabolites is subject of attention to various researchers (Yoojeong *et al.*, 2002). This has become the new technique for the commercial production of metabolites to save money and time.

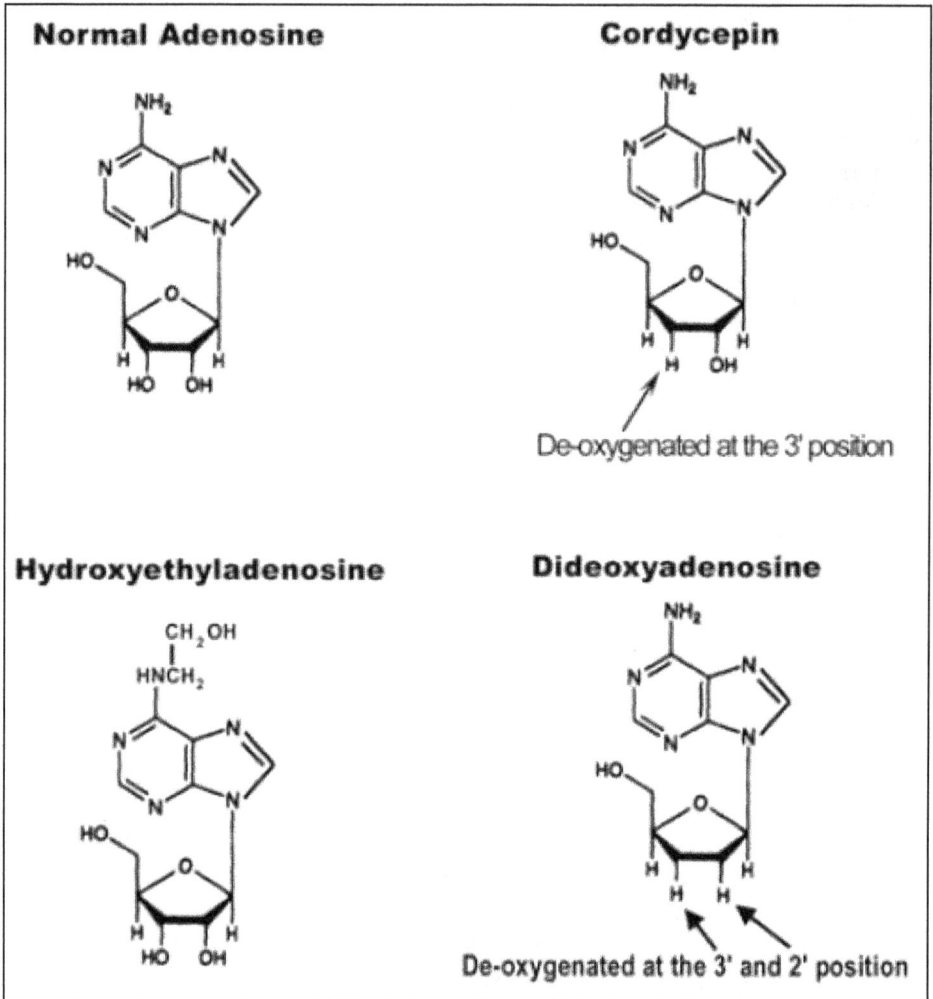

Figure 21.4: Some of the Unique Nucleosides found in *Cordyceps* sps.

Culture Conditions and Optimization of Fungal Growth in Laboratories

Fungal sticks could be harvested from Chamoli and likewise other districts in high mountains in the month of June to September, 2012. Wash collected sample specimens in running water with gentle brushing, sterilize in 0.1 per cent mercuric chloride for 10 min (surface sterilization), then wash with sterile water. Thoroughly cleaned *C. sinensis* specimens should be kept frozen during transportation and storage prior to further processing.

The pure cultures could be prepared on potato dextrose agar (PDA) plates. Use broad range antibacterial antibiotics to avoid unnecessary bacterial growth. Serial dilution method can be applied to get pure cultures of *Cordyceps sinensis*. Transfer the starter cultures to nutrient broth containing 10 g/L peptone and 30 g/L glucose at

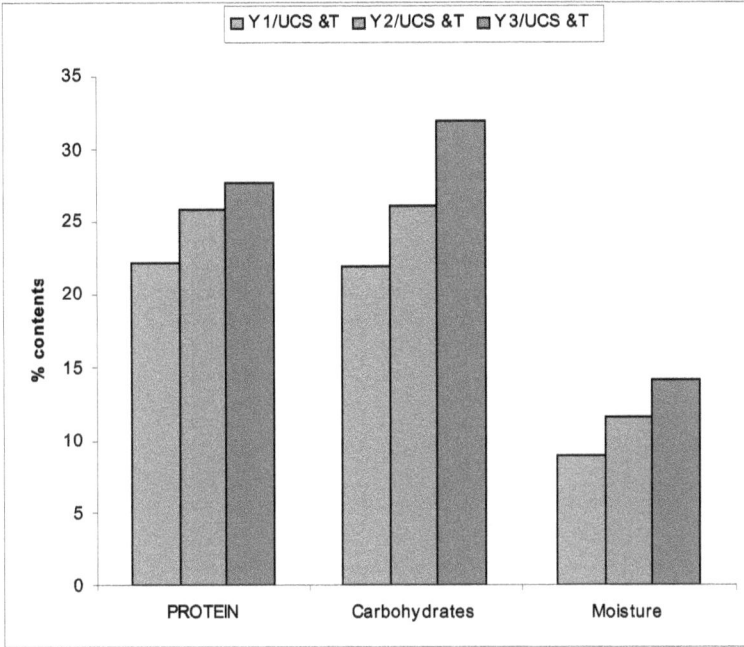

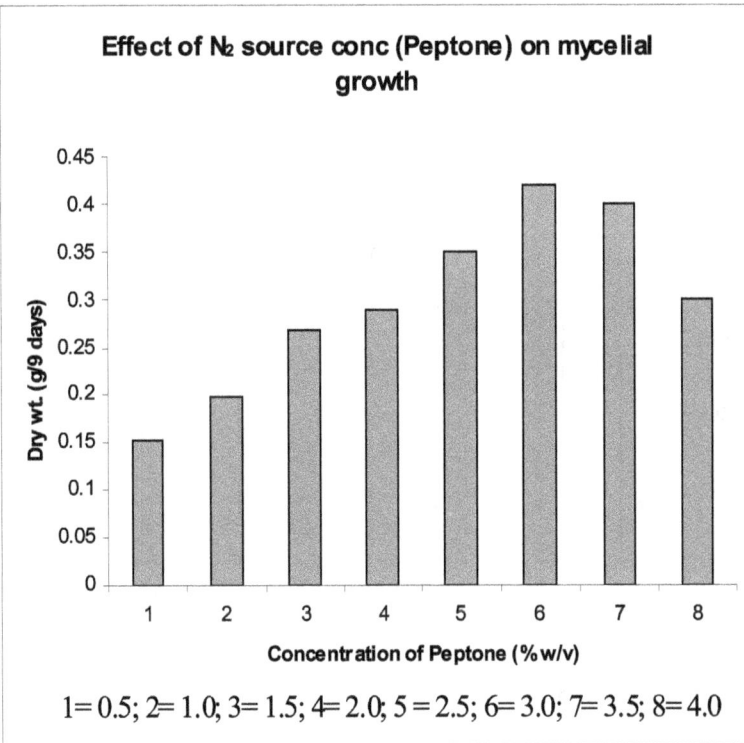

Figure 21.5a-b: Optimization Studies of Growth Conditions in *Cordyceps sinensis*.

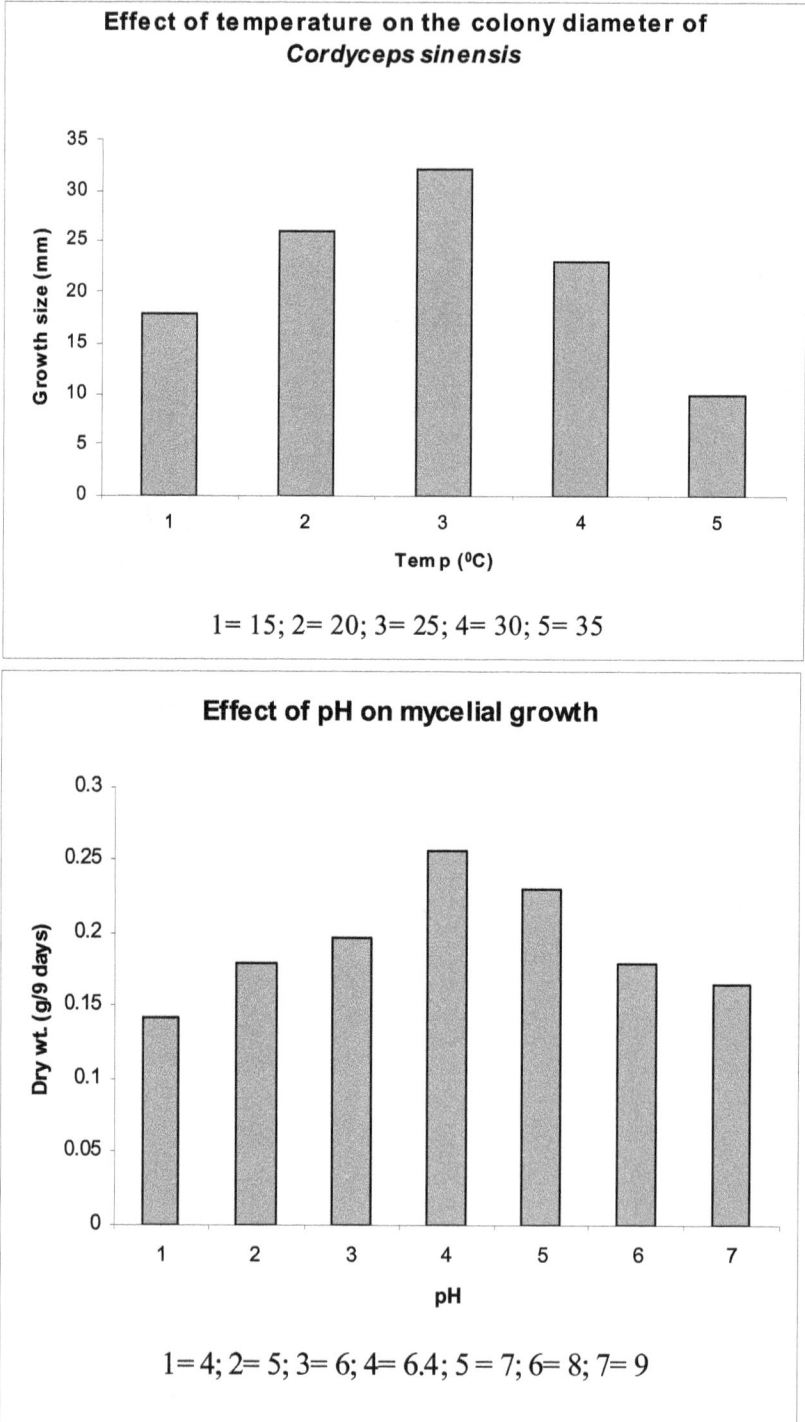

Figure 21.5c-d: Optimization Studies of Growth Conditions in *Cordyceps sinensis*.

pH 6.0, on a rotary shaker at 120 rpm for 96 h at 28°C. Collect the mycelia and store at –80°C. These mycelia can be used for chemical profiling or its suspension culture could be prepared to enhance metabolite production with the help of elicitors.

Somatic Hybridization of Fungal Species and Conservation of Natural Habitats

Protoplast culture technique is widely used to make somatic hybrids and cybrids in various plant species. Since *Cordyceps sinensis* is an exotic fungus and its overexploitation has created a highly dangerous condition. Researchers have proposed the generation of its hybrids with interrelated genera and its use in suspension culture. It can minimize the native utilization and will promote the diversity conservation with maximum benefit. Following steps can be used for optimum and sustainable uses of the endangered plant species:

1. Standardization and establishment of cell cultures of *Cordyceps sinensis in vitro* and establishment of new methodology for its rapid propagation. It can be achieved by nutritional balance and adding supplements in suspension cultures.

2. Protoplast isolation from *Cordyceps sinensis* and related taxa. Protoplast is a necked cell (cell without cell wall). Culture isolates grown in fresh suspension media overnight on rotary shaker (300-500 rpm) could be used for the isolation of fungal protoplasts. Enzyme mixture used for protoplast isolation is usually a macerozyme (crude and pure both) in a single step method (Holliday, *et al.*, 2004). Cell protoplast washing medium (CPW) should contain Chlorox beach, Mannitol solution (500mM/litre)-to maintain the omolarity of the solution and $CaCl_2$ (30 mM)-to strengthen the cell membrane. Protoplast fusion experiments should be done with Polyethylene Glycol (5, 10, 40 and 50 mM concentrations). Analysis of heterokaryons based on morphological and biochemical features can be

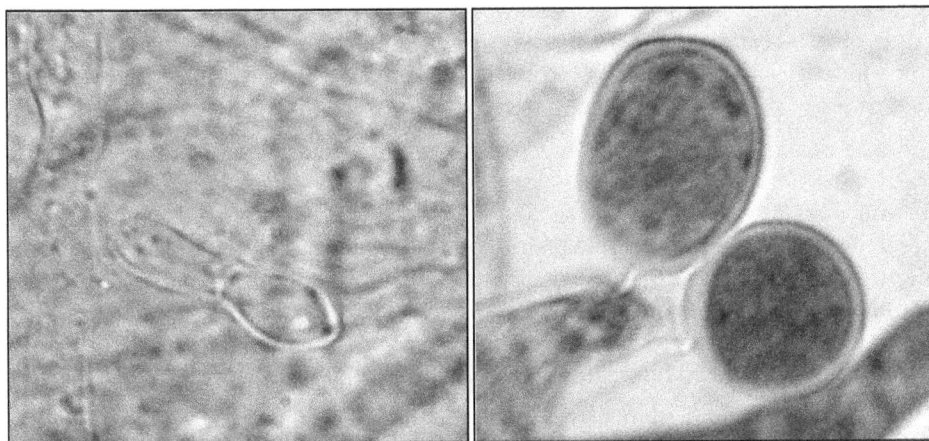

Figure 21.6: Fusion of Protoplasts of *Cordyceps sinensis* and its Purified Heterokaryons (Hybrids).

used to isolate the synkaryons (somatic hybrids).

3. Then somatic hybrids can be used for the establishment of bioreactors in which elicitor-induced production of metabolites could be practiced and standardized.

4. Extraction, purification and identification of particular metabolites, for its commercial uses.

5. Development of formulations based on the study and transfer of technology to farmers for field trials (From lab to field).

6. The study of socioeconomic impacts of the results and its commercial applications. It will not only promote the involvement of local peoples in the conservation programs but it will also help in enhancing the economic applicability to make the strategy beneficial on the Cost Benefit scale.

Scope of the Fungus and Trends in Related Research

Yarsagumba- the gold rush (*Cordyceps sinesis*) is a summer plant and winter insect. Before the rainy season begins, spores of the *Cordyceps* mushroom settle on the heads of caterpillars' that lives underground. The fungus gets so much into the body of the caterpillars' that it grows out through its head and drains all the energy from the insect and ultimately it dies. The popularity of yarsagumba is simply overwhelming. For the last couple of years, the trade of yarsagumba is increasing and it has been regarded as an expensive life saving tonic. And not only that, it is also believed to be a cure for sexual impotency – a Himalayan Herbal Viagra.

Since *Cordyceps* is an amazing substance and have passed the litmus test of scientific proof, much of research is going on to use its potentials (Halpern, 1999). The fungus not only contains an abundance of nutrients, but it also has active ingredients such as polysaccharides.

Not usually considered an edible mushroom due to its small size and rarity as well as its tough texture. Traditionally, *Cordyceps* has been consumed with a variety of meats in the form of a medicinal soup, with the type of meat used based upon the target medical condition. In the medical usage of today *Cordyceps* is often taken with some form of vitamin C, which has been found to aid the body in its digestion and absorption of the medicinal components of the mushroom.

The cap-less mushroom is eventually found and has been used in traditional medicine ever since, to treat kidney, lung, and heart ailments, male and female sexual dysfunction, fatigue, cancer and serious injury, to relieve pain, and the symptoms of tuberculosis and hemorrhoids, to restore general health and appetite, and to promote longevity. More potent than Ginseng and worth four times its weight in silver in ancient times. Due to its rarity, legend, and efficacy against a variety of health-related conditions, *Cordyceps* has held, and continues to hold, a highly esteemed position in the vast ranks of World herbal remedies (Chen *et al.*, 2001).

Indian medicinal plants and herbs in International context are paid high attention now days (Singh *et al.*, 1983). India is very rich in the Biodiversity and the scenario of Uttarakhand proves that it is one of the most important state for medicinal and herbals Research (Suresh Pal, 2008). The number of medicinal plants in India, both indigenous and exotic, has been variously put at between 3,000 to 3,500 species. Sixteen medicinal plants of exotic origin, introduced in India years back, are now considered as a part of India's medicinal plants resources. Out of these the plants providing largely or regularly used raw materials by Indian Drug and Pharmaceutical Industry are about 335, including those whose materials are imported from other countries (Subrat Sharma, 2004).

Importance and significance of the Research proposal oriented towards *Cordyceps*, mainly targets two points. One is the therapeutic application of the fungus and second is its easy availability in the market for the utilization of its maximum potentials. Since the mushroom is very rare and though many new techniques have been developed for its artificial cultures, yet it is rare and is far away from the reach of common peoples. This is the reason that its market value ranges in Black market from 10 lakhs to 25 lakhs per kg. Many local workers take risk to collect the herb from high altitudes and put their life in danger.

Cordyceps has been used to treat a wide range of conditions, including respiration, pulmonary diseases, renal, liver, cardiovascular diseases, hyposexuality, and hperlipidemia. It is also regularly used in all types of immune disorders, and as an adjunct in cancer therapy. Medicinal practitioners also recommend regular use of *Cordyceps* to strengthen resistance to infections, treat colds and flues and generally improve the homeostasis of the patient. *Cordyceps* has traditionally been, and is still most often used for kidney and lung problems, or health issues thought to stem from the lung or kidney meridian. For example, it is used to ease a wide range of respiratory ailments such as to reduce cough and phlegm, shortness of breath, bronchial discomfort and asthma. Modern science has confirmed the efficacy of *Cordyceps* for most, if not all, of the traditional uses (Guowei, 2001; Creadon and Dam, 1996; Nakamura *et al.*, 2003; Bok *et al.*, 1999). Its secondary effects on immune function help the body to more efficiently manage its immune resources while undergoing the stresses of the attack by cancer, (Shin *et al.*, 2003) allowing it to recognize, eradicate, and prevent abnormalities and disease, both at the local and the systemic level. (Koh *et al.*, 2002). Its administration demonstrated substantial antitumor activities in mice with remarkable results against murine sarcoma 180 when administered in doses of 10 mg/kg/day. This nearly complete tumor inhibition certainly heralds some positive potential in the development of new anticancer drugs.

Summary

The *Cordyceps* is a rare and highly valuable plant species. Many anticipated products are being isolated and it's utility to increase the socio economic relevance and biodiversity conservation must be paid attention to. Modern trends in related research should be used for the development of standard propagation technique to achieve a large number of individual fungal sticks in a short period of time and less expenses. The formulations to treat various diseases can be developed and the nutritional applications might be beneficial as food supplements. Its socioeconomic benefits can be drawn as under:

(A) Standardization of easy, less time and cost consuming protocols for the regeneration of Yarsagumba sticks and its cell cultures might help the upgradation of the economy of the local peoples as well as the State.

(B) Easy availability of the cell metabolites will encourage the medicinal industries to invest money that will ultimately result in more employments in the state.

(C) Yarsagumba growing in their native places will be conserved and protected from being exploited. It might help in making the state, a rich source and destination of this valuable herb.

Thus we can conclude with the statement that nature has gifted us a lot. We have to promote the sustainable use of this treasure so that we may leave this valuable gift to future generations too.

Acknowledgements

The authors wish to thank Uttarakhand State Council of Science and Technology to grant the Research project and facilities for the research work with *Cordyceps sinensis*.

References

1. Bok, J.W., Lermer, L., Chilton, J., Klingeman, H.G., Towers, G.H., 1999. Antitumor sterols from the mycelia of *Cordyceps sinensis*. *Phytochemistry*, 51(7): 891-898.

2. Chamberlain, M., 1996. Ethnomycological experiences in South West China. *Mycologist*, 10(4): 173-176.

3. Chen, K., Li, C., 1993. Recent advances in studies on traditional Chinese anti-aging. *Materia Medica. J Tradit Chin Med.*, 13(3): 223-226.

4. Chen, Y.Q., Wang, N., Qu, L., Li, T., Zhang, W., 2001. Determination of the anamorph of *Cordyceps sinensis* inferred from the analysis of the ribosomal DNA internal transcribed spacers and 5.85 rDNA. *Biochemical Systematics and Ecology*, 29: 597-607.

5. Creadon, M., Dam, J., 1996. Drink up. Time (August 19): 55. A biologically active compound from cultures mycelia of *Cordyceps* and *Isaria* species. *Phytochemistry*, 22: 2509-2512.

6. Guo, Y.Z., 1986. Medicinal chemistry, pharmacology and clinical applications of fermented mycelia of *Cordyceps sinensis* and Jin Shu Bao capsule. *Journal of Modern Diagnostics and Therapeutics*, 1: 60-65.

7. Guowei, Dai, Tiantong Bao, Changfu Xu, Raymond Cooper, and Jia, Xi, Zhu, 2001. CordyMax™ Cs-4 Improves Steady-State Bioenergy Status in Mouse Liver. *The Journal of alternative and complementary medicine*, 7: 231–240.

8. Holliday, J., Cleaver, P., Loomis-Powers, M., Patel, D., 2004. Analysis of Quality and Techniques for Hybridization of Medicinal Fungus *Cordyceps sinensis*. *International Journal of Medicinal Mushrooms*, 6: 147–160.

9. Koh, J.H., Yu, K.W., Suh, H.J., Choi, Y.M., Ahn, T.S., 2002. Activation of macrophages and the intestinal immune system by an orally administered decoction from cultured mycelia of *Cordyceps sinensis*. *Biosci Biotechnol Biochem*, 66(2): 407-411.

10. Kiho, T., Ukai, S., 1995. Tochukaso (Semitake and others), *Cordyceps* species. *Food Rev Int*, 11: 231-234.

11. Singh, U., Wadhwani, A.M., Johri, B.M., 1983. Dictionary of economic plants of India, *IARI*, New Delhi, pp. 100.

12. Nakamura, K., Yamaguchi, Y., Kagota, S., Shinozuka, K., Kunitomo, M., 1999. Activation of *in vivo* Kupffer cell function by oral administration of *Cordyceps sinensis* in rats. *Jpn J Pharmacol* 79(4): 505-508.

13. Suresh, P., S. Vimala D., Uree, Choudhary, N., 2008. Resources and priorities for Plant Biotechnology research in India. *Current Science*, 95: 1400-1402.

14. Shin, K.H., Lim, S.S., Lee, S., Lee, Y.S., Jung, S.H., Cho, S.Y., 2003. Anti-tumour and immuno-stimulating activities of the fruiting bodies of *Paecilomyces japonica*, a new type of *Cordyceps* spp. *Phytother Res*. 17(7): 830.

15. Subrat S., 2004. Trade of *Cordyceps sinensis* from high altitudes of the Indian Himalaya: Conservation and biotechnological priorities. *Current Science*, 86(12): 1614-1619.

16. Gi, Ho, Sung, Nigel, L., Hywel, Jones, Jae Mo Sung, J. Jennifer L., Bhushan, S., Joseph, W., 2007. Phylogenetic classification of *Cordyceps* and the clavicipitaceous fungi. *Studies in Mycology*, 57: 5–59.

17. Yoojeong, kim, Barbara, E., Wyslouzil, and Pamela, J.W., 2002. Secondary metabolism of hairy root cultures in bioreactors, *In vitro cell. Dev. Biol. Plant*, 38: 1–10.

2016, Environmental Biotechnology: A New Approach
Editors: Dr. Rajan Kumar Gupta and Dr. Satya Shila Singh
Published by: DAYA PUBLISHING HOUSE, NEW DELHI

Pages 361–396

Chapter 22

Biosurfactants from Micro-organisms for Broad-spectrum Environmental Biotechnological Applications

Rajesh K. Srivastava* and Nasim Aktar

*Department of Biotechnology, GITAM Institute of Technology,
GITAM University, Gandhi Nagar Campus, Rushikonda,
Visakhapatnam – 530 045, Andhra Pradesh*

ABSTRACT

Biosurfactants are more effective, selective, environmentally friendly and stable than many synthetic surfactants. Biosurfactants are surface active compound, released by many species of microorganisms. They are biodegradable and non-toxic in nature. Biosurfactants are produced by diverse group of microorganism, found in different locations on earths. Biosurfactants are secondary metabolites or membrane components and it can enhance nutrient transport across membranes, act in various host-microbe interactions, and provide biocidal and fungicidal protection to the producing organism. Microorganisms such as some of bacteria and yeasts can make use of a wide range of organic compounds as a source of carbon and energy for their growth because they excrete ionic surfactants. Insoluble form of carbon source such as hydrocarbon ($CxHy$) is taken via making their diffusion into the cell with help of biosurfactants synthesis in growth medium. And biosurfactants have significant property to reduce the surface or interfacial tension between oil and brine up to less than 0.01 m N/ m. Important applications of biosurfactant are found in the fields of enhanced oil recovery, environmental bioremediation, food processing and pharmaceuticals.

* *Corresponding author:* E-mail: rajeshksrivastava73@yahoo.co.in

Introduction

Last few years, biosurfactants have created more interest as potential candidates because of having interesting properties and also due to the utilization in the degradation of pesticides in soil and water environment. Metal ions can also be removed from soil surfaces by the biosurfactant micelles. Biosurfactant-producing micro-organisms are naturally present at hydrocarbon-impacted sites. Most common biosurfactants are glycolipids (*i.e.* carbohydrates attached to a long-chain aliphatic acid), lipopeptides, lipoproteins, and hetero-polysaccharides with more complex structures. Microbial surfactants, surface-active biomolecules are produced by a variety of microorganisms with gaining importance in the fields of enhanced oil recovery, environmental bioremediation, food processing and pharmaceuticals. The potential commercial applications of bioemulsifiers include bioremediation of oil-polluted soil and water, enhanced oil recovery, replacement of chlorinated solvents used in cleaning-up oil-contaminated pipes, vessels and machinery, use in the detergent industry, formulations of herbicides and pesticides and formation of stable oil-in-water emulsions for the food and cosmetic industries. Mukherjee *et al.* (2006) have reported the commercial production of microbial surfactants as surface-active biomolecules via produced by a variety of microorganisms. Large-scale production of these molecules has some limitations such as low yields in production processes and high recovery and purification costs. Mukherjee *et al.* (2006) have described the some practical approaches to make the biosurfactant production process economically. Mukherjee *et al.* (2006) have discussed the attractive use of cheaper raw materials, optimized and efficient bioprocesses and overproducing mutant and recombinant strains for obtaining maximum productivity. Mukherjee *et al.* (2006) have talked about hyper-producing recombinant strains in the optimally controlled environment of a bioreactor for leading towards the successful commercial production of these valuable and versatile biomolecules in near future (Mukherjee *et al.*, 2006). Ramnani *et al.* (2005) have reported the concomitant production of protease and biosurfactant via optimizing the agro-products (*i.e.* cornstarch and soy flour as carbon and nitrogen sources respectively) by use of response surface methodology (Ramnani *et al.*, 2005). In this chapter, author has tried to put more recent information on chemical nature and properties of surfactant or biosurfactant, biosurfactant production by diverse groups of microbes and biosurfactant application in biotechnological and bioremediation process.

Chemical Surfactant

Fiechter (1992) has reported about chemically synthesized surface-active compounds with wide application in the pharmaceutical, cosmetic, petroleum and food industries. Surfactants are usually organic compounds, ability to lower the surface tension (or interfacial tension) between two liquids or between a liquid and a solid. Surfactants may act as detergents, wetting agents, emulsifiers, foaming agents, and dispersants (Fiechter, 1992).

Surfactants can diffuse in water and adsorb at interfaces between air and water or at the interface between oil and water, in the case where water is mixed with oil. In the bulk aqueous phase, surfactants form aggregates, such as micelles, where the

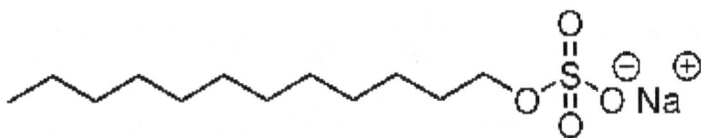

Sodium dodecyl sulfate (SDS)

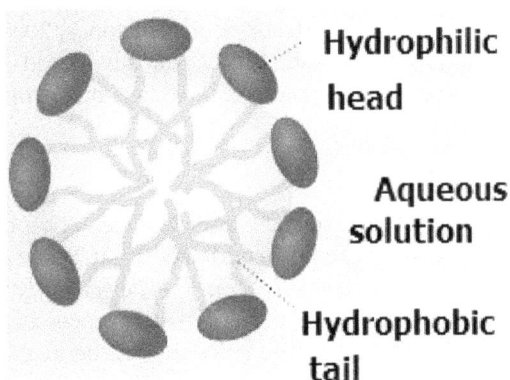

Figure 22.1: Chemical Structure of Sodium Dodecyl Sulphate ($C_{12}H_{25}NaO_4S$) as Chemical Surfactant and Form the Micelle in Aqueous Solution Due to having the Hydrophilic Head and Hydrophobic Tail (Mukerjee and Mysels, 1971; Marrakchi and Maibach, 2006).

hydrophobic tails form the core of the aggregate and the hydrophilic heads are in contact with the surrounding liquid (shown in Figure 22.1). The dynamics of adsorption of surfactants is of great importance for practical applications such as foaming, emulsifying or coating processes, where bubbles or drops are rapidly generated and need to be stabilized. The dynamics of adsorption depends on the diffusion coefficient of the surfactants (Zoller, 1994). Elasticity and viscosity of the surfactant layers plays a very important role in foam or emulsion stability. Sodium stearate is a most common component of most soap comprises about 50 per cent of commercial surfactants. Anionic surfactants contain anionic functional groups at their head, such as sulfate, sulfonate, phosphate and carboxylates. Sodium dodecyl sulphate and sodium laureth sulphate are the examples of anionic surfactant. Zwitterionic (amphoteric) surfactants have both cationic and anionic centers attached to the same molecule. The cationic part could be primary, secondary, or tertiary amines or quaternary ammonium cations (Castillo *et al.*, 1999). The anionic part can be more variable and include sulfonates or others. Sludge application, wastewater irrigation, and remediation processes can have different concentration of anionic surfactants in soils. Relatively high concentrations of surfactants together with multi-metals could cause an environmental risk (Renner, 1997).

There is little serious risk to the environment from commonly used anionic surfactants and cationic surfactants are known to be much more toxic. Linear alkyl benzene sulphonates (LAS) and the alkyl phenol ethoxylates (APE) are widespread use of detergent components. These pass into the sewage treatment plants where they are partially aerobically degraded and partially adsorbed to sewage sludge that is applied to land. Application of sewage sludge to soil can result in surfactant levels generally in a range 0 to 3 mg kg^{-1}, in the aerobic soil environment, a surfactant can undergo further degradation so that the risk to the biota in soil is very small, with margins of safety that are often at least 100. In the case of APE, their breakdown products, principally nonyl and octyl phenols adsorb readily to suspended solids and are known to exhibit oestrogen-like properties, possibly linked to a decreasing male sperm count and carcinogenic effects (Scott and Jones, 2000). Surfactant has advantages of biodegradability and beside it; biosurfactants could eventually replace their chemically synthesized counterparts. Till today, biosurfactants have been economically uncompetitive.

Biosurfactants

Biosurfactants are microbial products and surface-active compounds. Few species of bacteria, yeast or fungi are known to produce it. Some examples such as *Bacillus licheniformis* strain JF-2, *Bacillus subtilis*, or *Pseudomonas fluorescens* are known for biosurfactants production or used in bioremediation processes. It needs to gain a greater understanding of the physiology, genetics and biochemistry of biosurfactant-producing strains which can improve the process technology to reduce production costs.

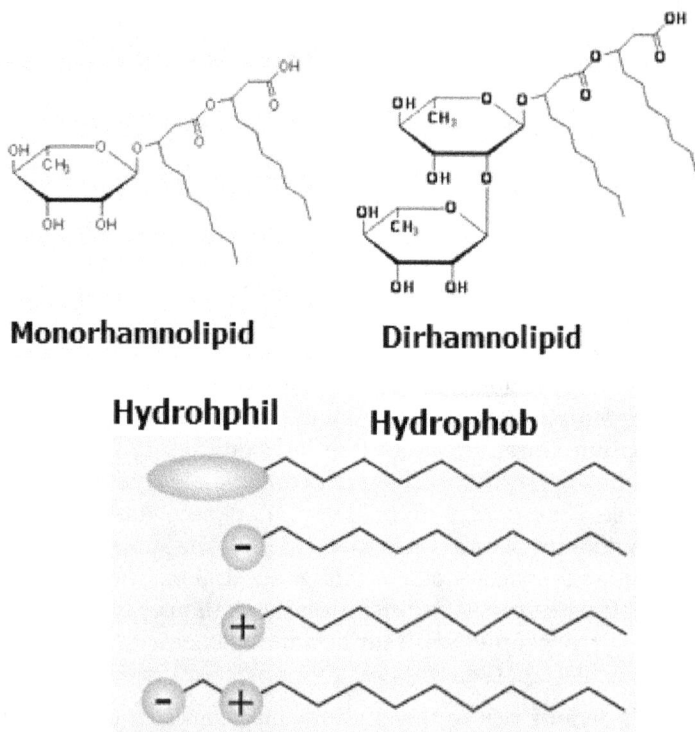

Monorhamnolipid   **Dirhamnolipid**

Hydrohphil   **Hydrophob**

Figure 22.2: Chemical Structures of Mono-Rhamnolipid and Dirhamnolipid with Hydrophilic and Hydrophobic Moieties (Maier and Chavez, 2000; Moussa *et al.*, 2014).

Structure of Biosurfactant

Biosurfactants contain both hydrophilic and hydrophobic moieties, causing them to aggregate at interfaces between fluids with different polarities such as water and hydrocarbons. The hydrophobic moiety could be on long-chain fatty acids or fatty acid derivatives, whereas the hydrophilic moiety can be a carbohydrate, phosphate, amino acid, or cyclic peptide. Biosurfactants could partition at the interfaces of liquid/liquid, gas/liquid or solid/liquid, occur in nature as an amphiphilic compounds.

Properties

Biosurfactants have better biocompatibility, lower toxicity, higher biodegradability, higher stability, extreme stability in extreme temperature and pH when compared with properties of chemical surfactants. Biosurfactants have low toxicity, eco-friendly nature and the wide range of potential industrial applications in bioremediation, health care, food processing and oil industries makes them a highly useful group of chemical compounds. Biosurfactants could be used in emulsification, foaming, detergency and dispersing. It has reported from last few decades that the microbially produced surfactants are used in bioremediation and enhanced oil recovery processes. A biosurfactant can exhibit the antimicrobial activity, surface tension reduction ability, the hydrocarbon removal from environment and its efficacy in metal removal. Biosurfactants are capable of lowering the surface tension and the interfacial tension of the growth medium and it is discussed by Cameotra *et al.* (2010). Biosurfactants possess different chemical structures-lipopeptides, glycolipids, neutral lipids and fatty acids. They are nontoxic biomolecules with biodegradable.

Biosurfactants also exhibit strong emulsification of hydrophobic compounds and form stable emulsions. The low water solubility of these hydrophobic compounds limits their availability to microorganisms, which is a potential problem for bioremediation of contaminated sites. Microbially produced surfactants enhance the bioavailability of these hydrophobic compounds for bioremediation. Biosurfactant-enhanced solubility of pollutants has potential applications in bioremediation. The biosurfactants is useful in a variety of industrial processes with vital importance to the microbes in adhesion, emulsification, bioavailability, desorption and defense strategy (Cameotra *et al.*, 2010).

Type of Biosurfactants

Biosurfactants are surface-active compounds, usually extracellular, produced by bacteria, yeast or fungi. Biosurfactants are rhamnolipids (*i.e.* produced by different *Pseudomonas* sp.), sophorolipids (*i.e.* produced by several *Torulopsis* sp.), and ornithinlipids (*i.e.* synthesized by *Pseudomonas rubescens, Gluconobacter cerinus*, and *Thiobacillus ferroxidans*). Rosenberg and Ron (1999) have reported the microbial ability to synthesize a wide variety of high- and low-molecular-mass bioemulsifiers. The low-molecular-mass bioemulsifiers are generally glycolipids, such as trehalose lipids, sophorolipids and rhamnolipids, or lipopeptides, such as surfactin, gramicidin S and polymyxin. The high-molecular-mass bioemulsifiers are amphipathic polysaccharides, proteins, lipopolysaccharides, lipoproteins or complex mixtures of these biopolymers (Rosenberg and Ron 1999).

The low-molecular-mass bioemulsifiers lower surface and interfacial tensions, whereas the higher-molecular-mass bioemulsifiers are more effective at stabilizing oil-in-water emulsions (Table 22.1). Three natural roles for bioemulsifiers have been proposed: (i) increasing the surface area of hydrophobic water-insoluble growth substrates; (ii) increasing the bioavailability of hydrophobic substrates by increasing their apparent solubility or desorbing them from surfaces; (iii) regulating the attachment and detachment of microorganisms to and from surfaces. Bioemulsifiers become prominent in industrial and environmental applications.

Table 22.1: Micro-organisms with different Classes of Biosurfactants

Surfactant Class	Micro-organism	Reference
Glycolipids (low-mass)		
Rhamnolipids	*Pseudomonas aeruginosa*	Robert *et al.*, 1989
Trehalose lipids	*Rhodococcus erithropolis*	Shulga *et al.*, 1990
	Arthobacter sp.	Suzuki *et al.*, 1974
Sophorolipids	*Candida bombicola, C. apicola*	Gobbert *et al.*, 1984
Mannosylerythritol lipids	*C. antartica*	Kitamoto *et al.*, 1993
Lipopeptides (low-mass)		
Surfactin/iturin/fengycin	*Bacillus subtilis*	Arima *et al.*, 1968
Viscosin	*P. fluorescens*	Neu *et al.*, 1990
Lichenysin	*B. licheniformis*	Horowitz *et al.*, 1990
Serrawettin	*Serratia marcescens*	Chan *et al.*, 2013
Phospholipids (low-mass)		
	Acinetobacter sp.	Suzuki *et al.*, 1974
		Itoh *et al.*, 1974
	Corynebacterium lepus	Cooper *et al.*, 1979
Polymeric surfactants (high-mass)		
Emulsan	*Acinetobacter calcoaceticus*	Chamnrokh *et al.*, 2008
Alasan	*A. radioresistens*	Barkay *et al.*, 1999
Liposan	*C. lipolytica*	Rufino *et al.*, 2014
Lipomanan	*C. tropicalis*	Kapadia and Yagnik, 2013
Surface-active antibiotics (high-mass)		
Gramicidin	*Brevibacterium brevis*	Peypoux *et al.*, 1999
Polymixin	*B. polymyxa*	Karanth *et al.*, 1999
Antibiotic TA	*Myxococcus xanthus*	Xiao *et al.*, 2012
Fatty acids or neutral lipids		
	Corynomicolic acids	
	Corynebacterium insidibasseosum	Akit *et al.*, 1981
Particulate biosurfactants		
	Vesicles whole microbial cells (high-mass)	
	Acinetobacter calcoaceticus	Kappeli and Finnerty, 1979
	Cyanobacteria	Karanth *et al.*, 1999

Screening Method for Biosurfactant and its Producing Microbes

The three methods such as drop collapse, oil spreading and blood agar lysis are found to detect biosurfactant production (Youssef *et al.*, 2004). All three methods are compared for their ease of use and reliability in relation to the ability of the cultures to reduce surface tension. Those methods are used to test for biosurfactant production in 205 environmental strains with different phylogenetic affiliations. Blood agar plates could be used to screen and isolate the biosurfactant-producing bacteria. Full-strength plate-count agar is used for the maintenance the isolates. The use of the drop collapse method (*i.e.* primary method) is found to detect biosurfactant producers. 16 per cent of the strains are found to lye the blood agar with negative for biosurfactant production. 38 per cent of the strains did not lye blood agar with positive for biosurfactant production. In drop collapse method, drops of cell suspensions are placed on an oil-coated surface. And drop contained biosurfactants is observed to collapse where as non-surfactant-containing drops are remained stable. Bacteria are able to form clearing zones on mineral salt agar plates sprayed with solutions of polycyclic aromatic hydrocarbon (PAHs) (*i.e.* containing the naphthalene and phenanthrene) utilized as sole substrates (Deziel *et al.*, 1996).

In oil spreading method, it is reported that the very strong, negative, linear correlation between the diameter of clear zone is obtained with the oil spreading and surface tension (r_s=–0.959). And a weaker negative correlation between drop collapse method and surface tension (r_s=–0.82) is found (Youssef *et al.*, 2004; Burch *et al.*, 2010). Axisymmetric drop shape analysis by profile (ADSAP) could help to identify biosurfactant-producing microbes or quantify their production capabilities via simultaneously determining the contact angle and liquid surface tension from the profile of a droplet resting on a solid surface (Van der Vegt *et al.*, 1991). Anionogenic bacterial peptidolipid biosurfactants is used to determine the ability of anionic surfactants to form a coloured complex with the cationic indicator methylene and this method of anionic surfactants determination is described by Shulga *et al.* (1993). A surfactant producing microorganism is screened which is isolated from a polymer dump site. It is characterized as *Bacillus subtilis* YB7. This microorganism is capable of producing 0.5 g/l of crude extract (dry weight) in the mineral salt medium (Bodour *et al.*, 2003).

Biosurfactant-producing microorganisms are found in most soils even by using a relatively limited screening assay. Distribution is dependent on soil conditions with gram-positive biosurfactant-producing isolates tending to be from heavy metal-contaminated or uncontaminated soils and gram-negative isolates tending to be from hydrocarbon-contaminated or cocontaminated soils (Bodour *et al.*, 2003). The biosurfactant produced by *Paenibacillus alvei* has ability to lower the surface tension of media to 35 mN/m. Thin layer chromatography (TLC) and FT-IR has been used to determine the compositional analysis of the produced biosurfactant as lipopeptide derivative (Najafi *et al.*, 2011). The parameters for better growth of the bacterial strain are optimized and production of surfactant is studied. The crude biosurfactant is extracted and the emulsification potency is assessed using different vegetable edible oils. Rhamnolipid is detected from the extracted biosurfactant and is confirmed by Infrared spectroscopy. Strain has showed high surfactant activity over the Gingelly oil, required mesophilic temperature and pH-7 for its better growth. The surfactant showed comparatively high emulsification index over Gingelly oil at the rate of 71 per cent with typical pattern of stretches for CH_2, CH_3 and C-O groups (Chander et al., 2012).

The FTIR and LC/MS analysis of the crude extract of surfactant has showed the presence of similar functional groups and six isoforms respectively as the commercially available Sigma surfactin. However, the proportions of the peaks are found to vary from sigma surfactin. Molecular mass of the major peak in the crude extract is 1023 $(M+H)^+$ which is equivalent to C14 surfactin analog whereas it is 1035 (C15 surfactin) in case of sigma surfactin. The crude extract reduced the surface tension of water from 72 to 30 mN/m at the CMC of 40 µM. This particular strain produced unique type of cycliclipopeptide analogs (Chander et al., 2012). The amount of surfactin or other biosurfactant could be determined by Water PICO-TAG amino acid analysis system. Surfactin can be isolated by adding concentrated HCL to the broth after removal of cells by centrifugation. Then precipitated crude surfactin is extracted by 3 times with equal amount of dichloromethane (Chander et al., 2012). Haghighat and Akhavan sepahy (2009) have found only two strains (i.e. P1, L2) to have the ability of high salt tolerance (8 per cent) and their successful production of biosurfactant and surface tension reduction capacity (value to given in Table 22.2). The product of L2 and P1 is mainly lipopeptide in nature with emulsify index E24 of 90 per cent -91 per cent. These strains are identified by morphological and biochemical technique using the taxonomic scheme of Bergey's manual of determinative bacteriology and 16s r-RNA gene. Those strains are found to the Bacillus subtilis and Bacillus licheniformis (Haghighat and Akhavan sepahy, 2009).

Chemical structures of the purified biosurfactants produced by Rhodotorula muciliginosa and Candida rugosa are identified as diacetate acidic sophorolipid and mono acetate lactonic sophorolipid using FTIR and GC-MS (Chandran and Das, 2011). Sari et al. (2014) have screened the novel biosurfactant-producing yeasts belonging to the genus Pseudozyma. It is characterized by colony pigmentation and cell morphology. They have found their ability to produce biosurfactant by cultivating on liquid mineral salt medium using soybean oil as a substrate (Sari et al., 2014).

Molecular Study of Genes Involved in Biosynthesis of Biosurfactant

Simpson et al. (2011) have designed the degenerate primers which can detect the presence of the surfactin/lichenysin (srfA3/licA3) gene involved in lipopeptide biosurfactant production in members of Bacillus subtilis/licheniformis group where as rhlR gene is involved in regulation of rhamnolipid production in pseudomonads. Polymerase chain reaction amplification, cloning and gene sequencing have confirmed the presence of the srfA3/licA3 genes in brines collected microbes from all nine wells. The presence of B. subtilis/licheniformis strains is confirmed by sequencing two other genes such as gyrA (gyrase A) and the 16S r-RNA gene. rhlR and 16S r-RNA gene is not found to pseudomonads which are detected in any of the brines. Intrinsic levels of surface-active compounds in brines are found very less, but biosurfactant production is found to stimulate by nutrient addition. Supplementation with a known biosurfactant-producing Bacillus strain together with nutrients has increased the biosurfactant production. The genetic potential to produce lipopeptide biosurfactants (*e.g.*, srfA3/licA3 gene) is prevalent and nutrient addition is stimulated the biosurfactant production in brines from diverse reservoirs. It is found as biostimulation approach for biosurfactant-mediated oil recovery (Simpson *et al.*, 2011). There are rapid and reliable methods which are available for screening and selection of biosurfactant-producing microorganisms and also evaluation of their activity. Genes involved in rhamnolipid synthesis (rhlAB) and regulation (rhlI and rhlR) in *Pseudomonas aeruginosa* are characterized and expression of rhlAB in heterologous hosts is reported. There are few characterized genes such as sfp, srfA, and comA for surfactin production in *Bacillus* species.

Molecular Biology Techniques for Identification of Biosurfactant Producer Isolates

Bodour *et al.* (2003) have grouped the 45 isolates by using repetitive extragenic palindromic (REP)-PCR analysis, which has yielded 16 unique, isolates (Bodour *et al.*, 2003). The re-identification of the halotolerant biosurfactant-producing *Bacillus licheniformis* strain JF-2 as *Bacillus mojavensis* strain JF-2 is confirmed by analysis of DNA-DNA similarity, gene sequence data and phenotypic characteristics and it is reported by Folmsbee *et al.* (2006). The restriction patterns with AluI, analysis of gyrA and 16S r-RNA gene sequences have grouped *Bacillus* strain JF-2 with *B. mojavensis*. DNA-DNA similarity of JF-2 strain is found more similar to *B. mojavensis* (75 per cent) than to *B. licheniformis* (11 per cent). Both strain JF-2 and *B. mojavensis* is required DNA for anaerobic growth where as *B. licheniformis* did not. *B. mojavensis* and strain JF-2 can not grow anaerobically in thioglycollate medium or aerobically with propionate while *B. licheniformis* can be grown under these conditions (Folmsbee *et al.*, 2006). Phylogenetic relationships are determined by comparing the 16S r-RNA gene sequence of each unique isolate with known sequences, revealing one new biosurfactant-producing microbe, a *Flavobacterium* sp. Sequencing results has indicated only 10 unique isolates (in comparison to the REP analysis, which indicated 16 unique isolates). Surface tension reduction results have demonstrated that isolates are similar according to sequence analysis but unique according to REP analysis in fact produced different surfactant mixtures under identical growth conditions. These results suggest that the 16S r-RNA gene database commonly used for determining phylogenetic relationships may miss diversity in microbial products (*e.g.* biosurfactants and antibiotics) that are made by closely related isolates. The potential of an indigenous bacterial strain for the production of biosurfactant is reported by Najafi *et al.* (2011) which are isolated from an Iranian oil field. Further this bacterium is characterized as *Paenibacillus alvei* by biochemical tests and 16S- ribotyping (Najafi *et al.*, 2011).

UV Radiated Bacterial Mutant Strain

Biosurfactants are produced by various bacteria and fungi with reduction of surface and interfacial tension. Enhanced productions of surfactin (*i.e.* a lipopeptide in nature) are found by using *Bacillus subtilis* ATCC 21332 as strain selection or manipulation of environmental or genetic or nutritional factor. *Bacillus subtilis* prototroph strain ATCC 21332 (potential to produce only 328 mg surfactin) are grown up to logarithmic phase and 3000 cells are plated on nutrient agar plates and plates are then UV irradiated for 35 seconds with 10- 20 per cent survival rate colonies. The mutant strain developed from UV irradiation experiment is *Bacillus subtilis* ATCC 53813 with having enhanced potential up to 1124 mg concentration of surfactin (Mulligan *et al.*, 1989; Symmank *et al.*, 2002).

Production of Biosurfactant

Many microorganisms can produce different type of biosurfactants. Fermentative production of biosurfactants is primarily dependent on the microbial strain, source of carbon and nitrogen, pH, temperature, and concentration of oxygen and metal ions. Addition of water-immiscible substrates to media and nitrogen and iron limitations in the media causes an overproduction of some biosurfactants. The use of water-soluble substrates and agro-industrial wastes for production, development of continuous recovery processes and production through biotransformation is also reported. Many bacterial and yeast species can synthesize the large amounts of fatty acids and phospholipids surfactants during their growth on n-alkanes. Phosphatidylethanolamine is synthesized by *Rhodococcus erythropolis* grown on n-alkane and it can lower the interfacial tension between hexadecane and water to less than 1 mN/m. And it has a critical micelle concentration (CMC) of 30 mg/l (Kuyukina *et al.*, 2005). Vesicles of *Acinetobacter* species are found to have diameter of 20–50 nm and a buoyant density of 1.158 cubic g/cm. Vesicles is consisted of protein, phospholipids and lipopolysaccharide (Menezes *et al.*, 2005).

Biosurfactants could either remain attached to the cell surface or be secreted into the culture medium as extra-cellular compounds. It is mostly produced by aerobic and very less by anaerobic mode fermentation (biosurfactant biosynthesis pathway shown in Figure 22.3). A biosurfactant could show any structure or chemical nature such as glycolipids, mycolic acid, polysaccharide–lipid composite, lipoprotein/lipopeptide, phospholipid or the microbial cell surface itself. As it is known that most oil reservoirs are anaerobic and aerobic biosurfactant-producing organisms could not be well suited for in situ microbial enhanced oil recovery processes. Type and quantity of the microbial surfactants produced is mainly dependent on the producer organism and also factors like nitrogen, carbon, temperature, aeration and trace elements. Biosurfactant-producing bacteria are found to constitute a significant proportion (up to 35 per cent) of aerobic heterotrophs. There are many reports on biosurfactant production by diverse types of micro-organism in different locations. It may be occurred in different species of bacteria, yeast or fungi. Commercial production of microbial surfactants has some limitations such as low yields in production processes and high recovery and purification costs (Mukherjee *et al.*, 2006).

Menezes *et al.* (2005) have characterized biosurfactant producing microbial populations, isolated from a Long Beach soil, California (USA) and a Hong Kong soil (China) contaminated with diesel oil. Menezes *et al.*, have isolated total 33 hydrocarbon-utilizing microorganisms from those soils. Twelve isolates and three defined consortia are found for biosurfactant production and emulsification activity (Menezes *et al.*, 2005). Krieger *et al.* (2010) have reported that the use of biosurfactants in current period is extremely limited due to their high cost in relation to chemical α-D-glucose-6-phosphate

Pgm*
phosphoglucomutase
5.4.2.2

D-glucose-1-phosphate

dTTP

RmlA
glucose-1-phosphate
thymidylyltransferase
2.7.7.24

PP$_i$

dTDP-D-glucose

RmlB
dTDP-glucose 4,6-
dehydratase
4.2.1.46

H$_2$O

4,6-Dideoxy-4-oxo-dTDP-D-glucose

RmlC
dTDP-4-dehydrorham-
nose 3,5-epimerase
5.1.3.13

dTDP-4-oxo-L-rhamnose

NADPH

RmlD
dTDP-4-dehydro-
rhamnose reductase
1.1.1.133

NADP$^+$

dTDP-L-rhamnose

dTDP

RhlC
rhamnosyl-
transferase II
2.4.1.-

di-
rhamnolipid

RhlB
rhamnosyl-
transferase I
2.4.1.-

dTDP

mono-
rhamnolipid

Acetyl-CoA

HCO$_3^-$ ATP

ACP

ACCC
acetyl-CoA
carboxylase
6.4.1.2

Malonyl-CoA
ACP

ADP+P$_i$

FabH
3-oxoacyl:ACP
synthase III
2.3.1.180

CoA

CoA

FabD
ACP S-malonyl-
transferase
2.3.1.39

Acyl-ACP

Malonyl-ACP

FabB/F
3-oxoacyl-ACP
synthases I / II
2.3.1.41

CO$_2$+ACP

β-ketoacyl-ACP

NADPH+H+

FabG
3-oxoacyl-ACP
reductase
1.1.1.100

NADP$^+$

β-hydroxyacyl-ACP

RhlA
3-hydroxyacyl-ACP:3-hydroxyacyl-
ACP O-3-hydroxyacyltransferase
2.4.1.-

H$_2$O   CoA

ACP

PhaG
beta-hydroxyacyl-
ACP:CoA transacylase
2.4.1.-

2 ACP

β-D-(β-D-
hydroxyalkanoyloxy)alkanoic
acid (HAA)

β-hydroxyacyl-
CoA

PhaC1/2
poly(3-hydroxyalkanoic acid)
synthases 1 / 2
2.3.1.-

Polyhydroxyalkanoates
(PHA)

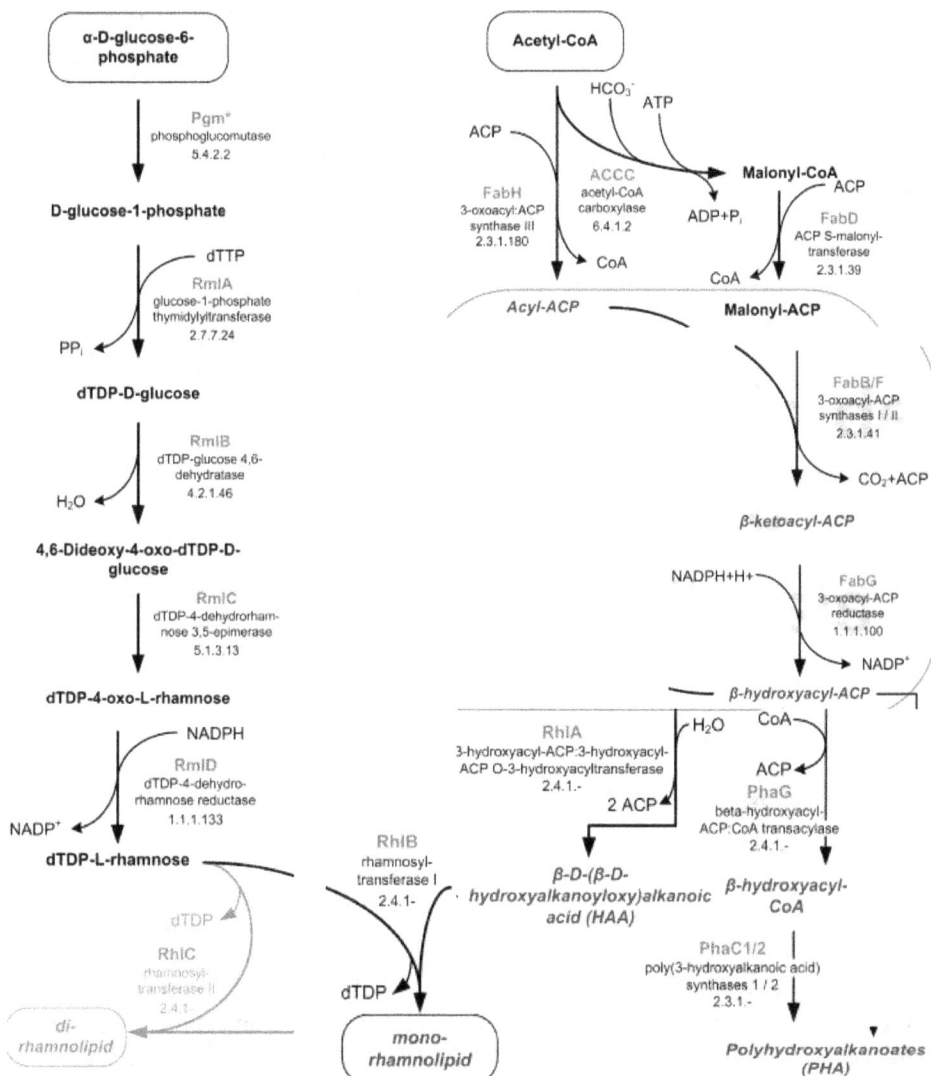

Figure 22.3: Metabolic Pathways for Rhamnolipid Production in *Pseudomonas putida*. Substrates and different intermediates are shown with different enzymes involves in rhamnolipid biosynthesis (Wittgens *et al.*, 2011)

surfactants. Biosurfactants have attracted more attention due to their low toxicity, biodegradability and ecological acceptability.

Biosurfactants could not use in industry because of their low yields and relatively high production and recovery costs. The use of mutants and recombinant hyper producing microorganisms can be used along with cheaper raw materials. And optimal growth and production conditions and more efficient recovery processes can be make economically feasible for the production of biosurfactant. High-level production of biosurfactants can done by the development of appropriate combinations of hyper-producing microbial strains, optimized cheaper production media and optimized process conditions, which could lead to economical and commercial level of biosurfactant production (Joshi and Desai 2010).

Utilization of renewable agro substrates for biosurfactant production is found more suitable carbon source due to cost effective and industrial alternative substrates (Makkar *et al.*, 2011). Ron and Rosenberg (2001) have discussed the natural role of biosurfactants and they have reported that microorganisms can produce a variety of surface-active agents (or surfactants) with low-molecular-weight molecules (*i.e.* lower surface and interfacial tensions efficiently) and high-molecular-weight polymers (*i.e.* bind tightly to surfaces) (Ron and Rosenberg, 2001). Biosurfactants are produced by a wide variety of microorganisms having different chemical structures and surface properties which is mentioned below in different sections.

Submerged Cultivation

Anaerobic Micro-organism

A 7-days-old culture of *Desulfovibrio desulfuricans* has the property of reduction of surface tension of 50 m N/m which is anaerobic in nature to produce the biosurfactant (Hitzman, 1988). *Bacillus licheniformis* JF-2 can produce a biosurfactant under strictly anaerobic conditions which can significantly lower the surface tension of the medium (*i.e.* <30 m N/m). And there are other reports that similar reductions in surface tension are occurred by different type of biosurfactants which are produced by various clostridial species (Javaheri *et al.*, 1985).

Aerobic Micro-organisms

In Bacterial Species

Production of biosurfactant from different species of bacteria is reported which are isolated from petrochemical wastes and marine water. And these strain of bacteria are Gram (+ve) *Bacillus* (SJ101), Gram (+ve) coccus (SJ301), Gram (+ve) *Bacillus* (JC101), small Gram (–ve) coccus (JB201), Gram (–ve) coccus (PW03) and Gram (+ve) *Bacillus* (PW05). There are some characterized genes such as sfp, srfA, and comA for surfactin production in *Bacillus* spp. Biosurfactant productions are found in submerged cultures fermentation.

Bacillus Strains/Species

Biosurfactant-producing *Bacillus* species is found to present in produced brines from Oklahoma oil reservoirs which has a wide range of salinities and it is reported by Simpson *et al.* (2011). Simpson *et al.*, have used the nine wells producing from six different reservoirs with salinities ranging from 2.1 per cent to 15.9 per cent. Wells are found with presence of surface-active compounds and biosurfactant-producing microbes (Simpson *et al.*, 2011). The production of biosurfactant from *Bacillus subtilis* MUV4 is found by use of shake-flask culture (200 rpm) at 30°C with modified Mckeen medium which is consisted of glucose (carbon source), mono-sodium glutamate and yeast extract (nitrogen sources). And it is reported by Suwansukho *et al.* (2008). The supernatant or fermentation broth of *B. subtilis* MUV4 is found to contain the biosurfactant which has reduced the surface tension of the medium after 48 h of cultivation (value shown in Table 22.2). *B. subtilis* MUV4 has produced the 0.30 g/L of crude biosurfactant after 96 h of cultivation in the fermentor with agitation rate of 200 rpm without aeration and uncontrolled pH condition. And the final yield of crude biosurfactant from *B. subtilis* MUV4 is reported up to 0.652 g/L after precipitating the supernatant with 6N HCl (Suwansukho *et al.*, 2008). A thermophilic *Bacillus subtilis* strain is found to produce the biosurfactant on sucrose and at high temperature and it is reported by Makkar *et al.* (1997). This biosurfactant has property to reduce the surface tension of the medium from 68 dynes cm^{-1} to 28 dynes cm^{-1}. Thermophilic *Bacillus subtilis* has the capacity to produce the biosurfactant at high NaCl concentrations and a wide range of pH. The biosurfactant produced by thermophilic *Bacillus subtilis* strain has found to retain its surface-active properties after heating at 100°C for 2h and at different pH values (*i.e.* in the range of 4.5–10.5). A maximum amount of biosurfactant is produced when urea or nitrate ions are supplied as nitrogen source. The use of the biosurfactant is reported at high temperatures, acidic, alkaline and saline environments. By biosurfactant action, 62 per cent of oil in a sand pack column could be recovered, indicating its potential application in microbiologically enhanced oil recovery (Makkar *et al.*, 1997).

A *Bacillus subtilis* strain is isolated from contaminated soil from a refinery and it is reported by Fonseca *et al.* (2007). *Bacillus subtilis* strain is screened for biosurfactant production in crystal sugar (sucrose) with different nitrogen sources (*i.e.* NaNO$_3$, (NH$_4$)$_2$SO$_4$, urea, and residual brewery yeast). The highest reduction in surface tension is reported in a 48-h old fermentation containing crystal sugar and ammonium nitrate. Optimization of carbon/nitrogen ratio (3, 9 and 15) and agitation rate (50, 150 and 250 rpm) for biosurfactant production is reported using complete factorial design and response surface analysis. The condition of C/N ratio (*i.e.* the value 3) and agitation rate (*i.e.* the value 250) rpm is found optimum to allow the maximum increase in surface activity of biosurfactant. Preliminary characterization of the bioproduct has reported it as lipopeptide with some isomers differing from a commercial surfactin (Fonseca *et al.*, 2007).

The commercial medium (Pharmamedia) is used for the production of surfactin by *Bacillus subtilis* MZ-7 and it is reported by Al-Ajlani *et al.* (2007). Al-Ajlani *et al.* (2007) have compared the different media (defined, semi-defined, and complex media) for the production of surfactin after fixing the least influential variables in standardized fermentation conditions. Carbohydrate and nitrogen supplements are also added to improve the production in Pharmamedia. Al-Ajlani *et al.* (2007) have found that optimized and brain heart infusion media are best for production of surfactin (280 mg/l) and a relatively more production found comparable with Pharmamedia (220 mg/l). And supplementing Pharmamedia with Fe$^+$ (4.0 mM) and sucrose (2 g/l) could lead to a maximum production of about (300 mg/l). Cottonseed-derived medium has found a suitable substrate for the production of bioactive substances including surfactin in both medical and biotechnological fields. The medium provided are found the higher product accumulations with considerably lower cost and it has good potential for large scale industrial applications (Al-Ajlani *et al.*, 2007).

Table 22.2: Different Types of *Bacillus* Species or Strains Produced Biosurfactants with different Surface Tension Reduction Properties with Specific Media or Process Conditions.

Bacillus Species/ Strain	Surface Tension Reduction (mN/m) at Period	Media Components/ Process Condition	Biomass or Biosurfactant Yield/Concentration	Reference
Bacillus subtilis MUV4	53.05 to 33.5 at 48h	0.3 yeast extract 1 per cent glucose 1 per cent monosodium glutamate	Y (p/s): 0.072 Y(x/s): 0.713 Y (p/x):101 µ: $0.14h^{-1}$	Suwansukho *et al.* (2008)
Thermophilic *Bacillus subtilis* strain	68 to 28	2 per cent sucrose at 45ºC 4 per cent NaCl pH:4-10.5	—	Makkar *et al.*, 1997
Bacillus subtilis strains CCTCC M201162	72 to 22.8	Beer wastewater: 40 g/l, diammonium phosphate $(NH_4H_2PO_4)$: 4 g/l,NaCl: 5 g/l.	Production biosurfactant: 1.26 g/l	Fonseca *et al.*, 2007; Liu *et al.*, 2014
Bacillus subtilis (P1) and *Bacillus licheniformis* (L2)	69 to 30	Sodium nitrate	Lipopeptide with E24: 90 per cent - 91 per cent	Haghighat and Akhavan-sepahy, 2009
Bacillus subtilis MZ-7		Pharmamedia with Fe^+ (4.0 mM) and sucrose (2 g/l)	Surfactin production: 280-300 mg/l	Al-Ajlani *et al.*, 2007
Bacillus subtilis 20B	Up to 29.5	Salts up to 7 per cent Temperature: 55ºC Different carbon: glucose, sucrose starch, date molasses, cane molasses	Interfacial tension: 3.79 mN/m —	Joshi *et al.*, 2008
B. subtilis ATCC 21332 *B. subtilis* LB5a	Up to 25.9 Up to 26.6	Substrate is manipueira (*i.e.* cassava flour-processing effluent)	Crude biosurfactant concentration: 2.2 g/l3.0 g/l	Nitschke and Pastore, 2004

The optimization of inocula conditions is reported for enhanced biosurfactant production by *Bacillus subtilis* SPB1 in submerged culture by using Box–Behnken Design and it is discussed by Mnif *et al.* (2013). Inoculum age and cell density can considerably influence the production yield and cost of the fermentation process. Mnif *et al.* (2013) have used the two-stage inocula to enhance metabolite production. Optimization studies are done in order to define the best inocula conditions supporting a maximum biosurfactant production by *Bacillus subtilis* SPB1. Lipopeptide production is effectively enhanced up to almost 3.4 g/l as estimated by gravimetrically. The new defined parameters are as follows; a first inoculum age of 23 h followed by a second inoculum age and size of 4 h and 0.01 per cent respectively (Mnif *et al.*, 2013).

The *Bacillus subtilis* 20B, a biosurfactant producing a strain is found to capable of growth at higher temperature and higher salts concentration and it is reported by Joshi *et al.* (2008). *Bacillus subtilis* 20B can utilize the different sugars, alcohols, hydrocarbons and oil as a carbon source with preference for sugars. It can produce the biosurfactant which can reduce the surface tension and interfacial tension (values are shown in Table 22.2). *Bacillus subtilis* 20B produced biosurfactant can give stable emulsion with crude oil and n-hexadecane with its activity stable at high temperature, a wide range of pH and salt concentrations for five days. The application of this organism as bio-control agent and use of biosurfactant in microbial enhanced oil recovery (MEOR) is found (Joshi *et al.*, 2008).

The fermentative production of biosurfactants by *Bacillus subtilis* strain B30 is found based on enhanced oil recovery using core-flood. The biosurfactants have found stable over wide range of pH, salinity and temperatures. The crude biosurfactant preparation has enhanced the light oil recovery and heavy oil recovery by 17–26 per cent and 31 per cent respectively in core-flood. Different carbon sources (such as glucose, sucrose, starch, date molasses, cane molasses) are tested to determine the optimal biosurfactant production. Produced biosurfactants could reduce the surface tension and interfacial tension up to 26.63 N/m and 3.79 mN/m respectively in less than 12 h of fermentation containing the both glucose or date molasses based media. A crude biosurfactant concentration of 0.3–0.5 g/l and critical micelle dilution (CMD) values of 1:8 are reported. Potential of the *Bacillus subtilis* strain (B30) has found in development of ex situ microbial enhanced oil recovery processes using glucose or date molasses based minimal media (Al-Wahaibia *et al.*, 2014).

The microbial enhanced oil recovery (MEOR) is found by using of microorganisms present in soil near the petroleum industry which has reported by Haghighat and Akhavan sepahy (2009). Eighty strains of *Bacillus* are isolated from contaminated area close to storage and distribution centre of oil products in Tehran refinery. The ability of isolates in biosurfactant production is analysed by haemolysis test, emulsification test and measurement of surface tension. *Bacillus* species have been widely used as model organisms during MEOR. An important characteristic of *Bacillus* species is their ability to produce spores. Spores are essential for MEOR because of their small size compared to vegetative cells and their ability to withstand the harsh environmental condition. Haghighat and Akhavan sepahy (2009) have studied the effect of different pH, salinity concentration, temperatures and variety of carbon and nitrogen sources on biosurfactant production. The optimum pH and temperature for *Bacillus subtilis* and *Bacillus licheniformis* is found to 6.8 and 37°C respectively for biosurfactant production. The best source of carbon is crude oil and best source of nitrogen is sodium nitrate. These strains are suitable for Ex-situ MEOR (Haghighat and Akhavansepahy, 2009).

A substrate (cassava flour-processing effluent *i.e.* manipueira) is found for surfactant production by two *Bacillus subtilis* strains such as ATCC 21332 and LB5a and it is reported by Nitschke and Pastore (2004). Manipueira has the difference in concentration of surfactant due to having different protease activity. Water is extracted from the cassava roots by squeezing (*i.e.* manipueira). Surfactant produced by *B. subtilis* ATCC 21332 has reduced the surface tension of the medium with a higher crude biosurfactant concentration. Surfactant produced by wild-type strain, *B. subtilis* LB5a has more reduced the surface tension of the medium with little more crude biosurfactant concentration compared to produced by *B. subtilis* ATCC 21332 (Nitschke and Pastore, 2004).

A study on increase of production of biosurfactant is done by using the inexpensive medium materials and the optimized fermentation conditions which is reported by Liu *et al.* (2014). In this regards, five biosurfactant producer strains are isolated from oilfield wastewater. The most efficient strain S2 is identified through 16S r-DNA sequence analysis with exhibit of highest similarities to *Bacillus subtilis* strains CCTCC M201162. The emulsification index (E24) is optimized using a box-behnken experimental design and response surface methodology (RSM). The RSM has found the optimal condition such as pH; 7.2, temperature; 42.1°C, inoculum concentration, 5.2 per cent (v/v) and rotate speed; 163 rpm. Under above mentioned optimal conditions, the optimal E24 of the biosurfactant produced by *Bacillus subtilis* strains CCTCC M201162 is 81.20 per cent and the production of crude biosurfactant is increased from 0.72 g/l to 1.26 g/l and also the surface tension of fermentation broth is found to reduce up to 22.8 mN/m. A quadratic response model is constructed through RSM designs, leading to a 75 per cent increase of biosurfactant production by *Bacillus subtilis* strains CCTCC M201162 (Liu *et al.*, 2014).

Youssef *et al.* (2007) have discussed the biosurfactants (*i.e.* lipopeptide in nature) production in situ at concentrations needed to mobilize oil in an oil reservoir by using the two *Bacillus* strains (a mixture of *Bacillus* strain RS-1 and *Bacillus subtilis* subsp. *spizizenii* NRRL B-23049) and nutrients (glucose, sodium nitrate, and trace metals). Youssef *et al.* (2007) have found the two wells with just nutrients, and one well with only formation water. In situ metabolism and biosurfactant production, the average concentration of lipopeptide biosurfactant in the produced fluids of the inoculated wells is found about 90 mg/l. And this concentration is approximately nine times more to the minimum concentration required to mobilize the entrapped oil from sandstone cores. Carbon dioxide, acetate, lactate, ethanol, and 2, 3-butanediol are detected in the produced fluids of the inoculated wells. Only CO_2 and ethanol are detected in the produced fluids of the nutrient-treated wells. Youssef *et al.*, have reported the essential data such as growth rates ($0.06h^{-1}$), carbon balances (107 per cent), biosurfactant production rates ($0.02h^{-1}$), and biosurfactant yields (0.015 mol biosurfactant/mol glucose) for modelling microbial oil recovery processes in situ (Youssef *et al.*, 2007).

Bodour *et al.* (2003) have discussed about distribution of biosurfactant-producing bacteria in the environment such as in undisturbed and contaminated sites by determining the common culturable biosurfactant producing bacteria. A series of 20 contaminated (*i.e.* with metals and/or hydrocarbons) and undisturbed soils is collected by Bodour *et al.* (2003) and plated it on Reasoner´s 2A agar or R(2)A agar. R(2)A agar is consisted of Proteose peptone (0.5g/l), Casamino acids (0.5g/l),Yeast extract (0.5g/l),Dextrose (0.5g/l), Soluble starch (0.5g/l), Dipotassium phosphate (0.3 g/l), Magnesium sulfate 7H2O (0.05 g/l), Sodium pyruvate (0.3 g/l), Agar (15 g/l) and final pH 7. The Forty-five of the isolates from 1,305 colonies are found to positive for biosurfactant production in mineral salts medium containing 2 per cent glucose. Chander *et al.* (2012) have worked on production and characterization of biosurfactant from *Bacillus subtilis* MTCC441 and its evaluation to use as bioemulsifier for food bio-preservative. Biosurfactant activity is studied against different vegetable edible oils (Chander *et al.*, 2012). The combination of central composite rotatable design (CCRD) and response surface methodology (RSM) is used to optimize biosurfactant production for variations of four impressive parameters such as pH, temperature, glucose and salinity concentrations. A maximum reduction in surface tension is reported under the optimal conditions of 13.03 g/l glucose concentration, 34.76 °C, 51.39 g/l total salt concentration and medium pH of 6.89 (Chander *et al.*, 2012).

Pseudomonas Species/Strains

The capacity of polycyclic aromatic hydrocarbon (PAH)-utilizing bacteria to produce biosurfactants is found via isolating twenty-three bacteria from a soil contaminated with petroleum wastes and it is discussed by Deziel *et al.* (1996). Biosurfactant production is found from 10 strains out of 23 strains which have shown the surface tension lowering property and emulsifying activities. Bacteria are found to grow in an iron-limited salt medium, supplemented with high concentrations of dextrose or mannitol as well as with naphthalene or phenanthrene. Presence of glycolipid biosurfactant in cultures of *Pseudomonas aeruginosa* 19SJ is found. Cultures of *Pseudomonas aeruginosa* 19SJ on naphthalene has the maximal productivity of biosurfactants than cultures grown on mannitol. There are small amounts of biosurfactants and naphthalene degradation intermediates at the onset of the cultivation. Production of biosurfactants is accompanied by an increase in the aqueous concentration of naphthalene or phenanthrene with promotion in the solubility of its substrate (Deziel *et al.*, 1996). There are pseudomonas strains with its ability to biosynthesize rhamnose-containing surfactants (*i.e.* rhamnolipids) which is discussed by Nitschke *et al.* (2005). Rhamnolipids compounds are reported with respect to chemical structure, properties, biosynthesis, and physiological role. Nitschke *et al.* (2005) have focussed on rhamnolipids production by using of low-cost substrates (*i.e.* wastes from food industries) as alternative carbon sources. The use of inexpensive raw materials such as agro-industrial wastes or oils and lipid-rich wastes is an attractive strategy to reduce the production costs associated with biosurfactant production. This can also contribute to the reduction of environmental impact, generated by the discard of residues, and the treatment costs. Carbohydrate-rich substrates generated low rhamnolipid levels (Nitschke *et al.*, 2005).

Pseudomonas aeruginosa PACL strain is isolated from oil-contaminated soil. And oil-contaminated soil is taken from a lagoon. *Pseudomonas aeruginosa PACL* is used to investigate the efficiency and magnitude of biosurfactant production by using the different waste frying soybean oils via submerged fermentation in stirred tank reactors of 6 and 10 litre capacities. It is reported by De Lima *et al.* (2009). A complete factorial experimental design is used with the goal of optimizing the aeration rate (0.5, 1.0 and 1.5 vvm) and agitation speed (300, 550 and 800 rpm). Aeration is identified as the primary variable affecting the process with a maximum rhamnose concentration occurring at an aeration rate of 0.5 vvm. A maximum rhamnose concentration of 3.3 g/l, an emulsification index of 100 per cent, and a minimum surface tension of 26.0 dynes/cm are found at optimum levels. Under these conditions, the biosurfactant production is compared with the use of a mixture of waste frying soybean oil (WFSO) to non-used soybean oil (NUSO), or waste soybean oils used to fry specific foods as a carbon source. NUSO has produced the highest level of rhamnolipids and the waste soybean oils also resulted in biosurfactant production of 75–90 per cent of the maximum value. Under ideal conditions the kinetic behavior and the modelling of the rhamnose production, nutrient consumption, and cellular growth are studied. The resulting model has predicted data points to correspond well to the empirical information (De Lima *et al.*, 2009).

Lactococcus Strains/Species

The potential use of alternative fermentative medium for biosurfactant production is reported by *Lactococcus lactis* 53 and *Streptococcus thermophilus* A and it is reported by Rodrigues *et al.* (2006). Rodrigues *et al.* (2006) have discussed the response of the experiments pertaining to glucose, lactose or sucrose consumption, cell growth and biosurfactant production. Synthetic media MRS and M17 broth are used as control experiments. Synthetic media in fermentations can be replaced by cheaper alternative media such as cheese whey and molasses with effectively high yields and productivities of biosurfactant. There are 1.2–1.5 times more production of biosurfactant (*i.e.* per gram cell dry weight) and 60–80 per cent costs reduction medium preparation for both strains. M17 Agar is used for isolating and enumerating lactic streptococci in yogurt, cheese starters and other dairy products. M17 Broth contains peptones and meat derivatives as sources of carbon, nitrogen, vitamins and minerals. Yeast extract supplies B-complex vitamins which stimulate bacterial growth. Di-sodium-β-glycerophosphate buffers the medium as acid is produced from fermentation of lactose. Ascorbic acid stimulates growth of lactic streptococci. Magnesium sulphate provides essential ions for growth. M17 or MRS broth is consisted of common composition of different concentration of whey/molasses, yeast extract and peptone. But M17 broth contains sodium glycerophosphate also. Supplemented cheese whey and molasses media can be used as a relatively inexpensive and economical alternative to synthetic media for biosurfactant production by probiotic bacteria and it could be attractive alternative for biosurfactants formation (Rodrigues *et al.*, 2006).

Rodriguesa *et al.* (2006) have done the screening of biosurfactant-producing ability of four *Lactobacillus* strains and all the tested strains are used for biosurfactant production in the first 4 h (three strains are shown in Table 22.3). The *Lactobacillus* strains have showed the zones of clearing in the blood agar (*i.e.* the diameter of <1 cm). The minimum surface tension value of the fermentation broth is found for *Lactobacillus pentosus* CECT-4023. It has reduced the surface tension up to 10.5 mN/m by comparing with the control (*i.e.* 39.5 mN/m). Time courses of glucose, biomass and biosurfactant are modelled by using MRS broth for lactobacilli strains. Using whey as production medium, the values estimated by the modeling of biosurfactant for *L. pentosus* CECT-4023 are P_{max} = 1.4 g of biosurfactant/l and r_p/X = 0.093 g/(lh). *L. pentosus* CECT-4023 is strong biosurfactant producer strain and cheese whey is found as an alternative medium for biosurfactant production (Rodriguesa *et al.*, 2006).

Table 22.3: Maximum Biosurfactant Concentration and Productivity by Using MRS Broth (Man, Rogosa and Sharpe medium) *i.e.* Culture Medium for different Lactobacilli Strains (Rodriguesa *et al.*, 2006)

Sl.No.	Micro-organism	Maximum Production Concentration (P_{max} value) (g/l)	Productivity (r_pX value) g/(lh)	Reference
1.	*Lactobacillus casei* CECT-5275	1.6	0.091	Rodriguesa *et al.*, 2006
2.	*Lactobacillus rhamnosus* CECT-288	1.7	0.217	Rodriguesa *et al.*, 2006
3.	*Lactobacillus coryniformis* subsp. *torquens* for CECT-25600	1.8	0.090	Rodriguesa *et al.*, 2006

Biosurfactant such as mycolates and corynomycolates are synthesized by *Rhodococcus* sp., *Corynebacteria* sp., *Mycobacteria* sp., and *Nocardia* sp. Trehalose lipids are obtained from *Rhodococcus erythropolis* and *Arthrobacter* sp. which can be reduced the surface tension and interfacial tension in culture broth. Biosurfactant production using petrol is found as a good carbon source because of better production of biosurfactants. *Clostridium pasteurianum* has ability to produce an extracellular neutral lipid which could lower the surface tension of water up to 55 m N/m (Cooper *et al.*, 1979).

Consortium of L1 (*Acinetobacter junii*), L2 (*Actinomyces* species) and L3 (*Pseudomonas* species) isolates from a Long Beach soil have shown the highest reduction property of surface tension (*i.e.* 41.4 mN/m to 46.5 mN/m). The highest emulsifying capacity (*i.e.* 74 per cent) is evaluated by the E24 emulsification index in the culture of isolate L5. Based on surface tension and the E24 index results analysis, isolates *Bacillus cereus* (F1), *Bacillus sphaericus* (F2), *B. fusiformis* (F3 and F4), *Acinetobacter junii* (L1), *Actinomyces* sp (L2) and *Pseudomonas* sp (L3) and *Bacillus pumilus* (L4) are identified. Cluster analyses of 16S r-RNA gene sequencing have revealed the 70 per cent similarity amongst hydrocarbon-degrading bacterial community present in both soils such as Long Beach soil, California (USA) and a Hong Kong soil (China) contaminated with diesel oil. F1, F2, F3, F4 and L4 from Firmicutes order and L1 and L3 from Proteobacteria order and one isolate (L2) from *Actinomyces* species are reported. Menezes *et al.* (2005) have used the Simpson's index (1-D) and the Shannon-Weaver index (H) for more diversity of HDM in the Hong Kong soil, while evenness (E) and the equitability (J) data have indicated to not a dominant population. These bacterial isolates can be applied in the bioremediation of soils contaminated with petroleum hydrocarbons (Menezes *et al.*, 2005).

Unicellular Fungal/Yeast Species/Strain

Pathogenic nature of some producers restricts the wide application of biosurfactant compounds. The production of biosurfactants from yeasts has been reported during the last decade (Amaral *et al.*, 2010). The industrial importance of yeasts and their potential to biosurfactant production has been discussed with its characteristics and the production costs. Nonionic or lipopolysaccharides surfactants could change the cell wall of some of micro-organism and when it is produced in it. *Candida lipolytica* and *C. tropicalis* can produce cell wall-bound lipopolysaccharides surfactant when they grow on n-alkanes. Sophorolipids are found as a mixture of free acid form and macrolactones and it is produced by *Torulopsis bombicola*. Liposan is an extracellular water-soluble emulsifier synthesized by *Candida lipolytica* and is composed of 83 per cent carbohydrate and 17 per cent protein. Formation a micro-emulsion plays a very important role in alkane uptake by microbial cells and it is possible due to extracellular membrane vesicles partition hydrocarbons. Amaral *et al.* (2010) have reported about the biological surfactant production which has grown the significantly importance due to having advantages over synthetic compounds such as biodegradability, low toxicity, diversity of applications and functionality under extreme conditions (Amaral *et al.*, 2010).

Campos-Takaki *et al.* (2010) have discussed the some yeast as preferred sources for biosurfactants production. It is mainly due to their generally recognized as safe (GRAS) status for environmental and health safety reasons. The production of biosurfactants is occurred by some yeast cultures utilizing renewable resources like fatty wastes from household and vegetable oil refineries as major substrates. The application of response surface methodology and artificial neural network techniques are used for the optimization of biosurfactant production by yeasts. Fontes *et al.* (2010) have reported the improvement in biosurfactant production by using *Yarrowia lipolytica* IMUFRJ 50682 by a factorial design method. Full factorial design is used to investigate the effects of nitrogen sources (urea, ammonium sulfate, yeast extract, and peptone) on maximum variation of surface tension (ST) and emulsification index (EI). They have reported the most favorable medium for biosurfactant production which is composed of glucose (4 per cent w/v) and glycerol (2 per cent w/v) (Fontes *et al.*, 2010).

The yeast species such as *Rhodotorula muciliginosa* and *Candida rugosa* is isolated from hydrocarbon contaminated sites. They are found to capable of producing biosurfactants in the presence of 2 per cent (v/v) diesel as sole source of carbon and energy. When diesel oil is emulsified with biosurfactant, the surface charge of the diesel is modified via more adsorption of diesel on yeast cell surface (Chandran and Das, 2011).

Solid-State Cultivation

Solid-state cultivation can represent an alternative technology for biosurfactant production with use of inexpensive substrates. Solid-state cultivation can avoid the problem of foaming in submerged cultivation processes. It has little attention for biosurfactant production with its own challenges, such as the selection of a bioreactor type with adequate size, heat removal, substrates with appropriate physico-chemical properties and methods for monitoring of the cultivation process and recovering the biosurfactants from the fermented solid. Solid-state cultivation can be used for large-scale production of biosurfactants.

Actinomycete Species/Strains

Potential biosurfactant producers and economic production processes can be achieved using industrial waste and it is reported by Kiran *et al.* (2010). The surface active compound produced by strain *Nocardiopsis lucentensis* MSA04 is characterized as glycolipid with a hydrophobic non-polar hydrocarbon chain (nonanoic acid methyl ester) and hydrophilic sugar, 3-acetyl 2,5 dimethyl furan. It is done by developing solid-state culture (SSC) of a marine actinobacterium (*i.e.* MSA04 strain) for biosurfactant production which is isolated from the marine sponge *Dendrilla nigra* (Kiran *et al.*, 2009; Kiran *et al.*, 2010). Wheat bran has increased the biosurfactant production significantly (25 per cent) followed by oil seed cake and industrial waste (*i.e.* tannery pre-treated sludge, treated molasses (distillery waste) and pre-treated molasses). Enhanced biosurfactant production is reported under SSC conditions using kerosene as carbon source, beef extract as nitrogen source and wheat bran as solid substrate. The maximum production of biosurfactant by MSA04 is occurred at a C/N ratio of 0.5 (Morikawa *et al.*, 1993; Kiran *et al.*, 2010).

The kerosene and beef extract interactively increase the production and a stable production is found with the influence of both factors independently. A significant interactive influence of secondary control factors (*i.e.* copper sulfate and inoculum size) is validated by response surface methods (Kiran *et al.*, 2010). In other report, the biosurfactant production from marine actinobacterium *Brevibacterium aureum* MSA13 is found by optimizing the fermentation condition or experiments in solid state culture (SSC) and it is reported by Kiran *et al.* (2010). They have used the industrial and agro-industrial solid waste residues as substrates. The biosurfactant production by MSA13 under optimization condition is found to increase threefold over the original isolate under SSC conditions with pre-treated molasses as substrate and olive oil, acrylamide, $FeCl_3$ and inoculums size as critical control factors (Kiran *et al.*, 2009; Kiran *et al.*, 2010). The strain *B. aureum* MSA13 has produced a new lipopeptide biosurfactant with a hydrophobic moiety of octadecanoic acid methyl ester and a peptide part as a short sequence of four amino acids including pro-leu-gly-gly. This biosurfactant can be used for the microbially enhanced oil recovery processes in the marine environments (Kiran *et al.*, 2009; Kiran *et al.*, 2010; Kiran *et al.*, 2014).

Multi-cellular Fungal Species/Strain

Colla *et al.* (2010) have worked on the simultaneous production of lipases and biosurfactants by submerged (SmgB) and solid-state bioprocesses (SSB) by using the oil substrates. Colla *et al.* (2010) have isolated the *Aspergillus* species from a soil contaminated by diesel oil via metabolizing it. Colla *et al.* (2010) have reported the relationship between the production of lipases and biosurfactants. SSB has the highest production of lipases, with lipolytic activities of 25.22 U, while SmgB has 4.52 U (Colla *et al.*, 2010). There is no production of biosurfactants in the SSB where as in the SmgB, correlation coefficients are obtained between lipolytic activity and oil in water (*i.e.* 91 per cent) and water in oil and emulsifying activities (*i.e.* 87 per cent). A correlation of 84 per cent is obtained between lipolytic activity and reduction of surface tension up to 28 mN/m in the culture medium with indication of biosurfactants production (Colla *et al.*, 2010).

Martins *et al.* (2006) have reported the solid state biosurfactant production in a fixed-bed column bioreactor by using the *Aspergillus fumigatus* and *Phialemonium* species and they have used two media such as rice husks alone (simple support) and rice husks plus defatted rice bran (complex support) which are enriched with either soy oil or diesel oil (Martins *et al.*, 2006). Marine endosymbiotic fungi *Aspergillus ustus* (MSF3) produces high yield of biosurfactant (*i.e.* glycolipoprotein in nature) which is isolated from the marine sponge *Fasciospongia cavernosa*, peninsular coast source of India. Kiran *et al.* (2009) have reported the maximum production of biosurfactant by *Aspergillus ustus* (MSF3) under optimized bioprocess conditions (*i.e.* pH-7.0 and temperature-20°C, salt concentration 3 per cent, glucose and yeast extract as carbon source and nitrogen sources) from Sabouraud dextrose broth (Kiran *et al.*, 2009). The biosurfactant yield is analysed by response surface methodology based of carbon and nitrogen ratio.

Hua *et al.* (2003) have discussed the influence of the biosurfactant BS-UC on surface properties of microbial cells and biodegradation of petroleum hydrocarbons. The biosurfactant BS-UC (a new kind of surfactant) is found to produce by *Candida antarctica* from n-undecane (as the substrate). And biosurfactant BS-UC name is given by Hua *et al.* (2003). It is reported that the addition of BS-UC can affect the biodegradation rate of a variety of n-alkanes substrates via influencing positively the emulsification process. BS-UC can also change the hydrophobicity and zeta potential of the cell surface to help in microbial cell attach to the hydrophobic substrate easily. Both zeta potential of the cell and the porous media can improve the retention of the cells in the media. It could be promising choice for use in bioremediation of petroleum contamination (Kiran *et al.*, 2009). Biosurfactants (BS) are of recent interest due to their unique properties including structural diversity, higher biodegradability, lower toxicity and exhibit stable activity at extreme conditions including pH, temperature and ionic strength. They are found to have applications in medical and environmental fields. Some of the BS are potential antimicrobial agents and are suitable alternatives to synthetic antibiotics (Hua *et al.*, 2003).

Methods for Biosurfactant Concentration/Recovery

The biosurfactant concentrations (g/l) are determined for each strain using a calibration curve from surface tension (mN/m) and concentration (g/l) which could have negative value (*i.e.* 8.65) for surface tension and positive value (*i.e.* 76.984) for concentration with its correlation (r^2) = 0.9729. The calibration curve is calculated for a commercial biosurfactant produced by several *Bacilli* (surfactin lowers the surface tension of water up to 27 mN/m at 5.0×10^{-4} M) using different concentrations of biosurfactant solution below the critical micelle concentration with known surface tension (Morikawa *et al.*, 1993; Rodrigues *et al.*, 2006). The surface activity of biosurfactants produced by the bacterial strains can be determined by measuring the surface tension of the broth samples by the Ring method using a KRUSS Tensiometer equipped with a 1.9 cm De Nouy platinum ring at room temperature (Kim *et al.*, 2003).

Oil spreading technique has found better predicted biosurfactant production than the drop collapse method. The oil spreading technique is used to determine the biosurfactant concentration as a quick and easy protocol to screen and quantify biosurfactant production. Blood agar lysis method is not a reliable method to detect biosurfactant production due to poor correlation to surface tension (r_s=–0.15) demonstration activity (Youssef *et al.*, 2004).

The crude biosurfactant produced by *B. subtilis* MUV4 is dissolved in methanol and dried by vacuum evaporator (crude methanol). The supernatant contains the crude biosurfactant and the crude methanol which has retained the biosurfactant activity over the pH range of 1-6, 7-10 and 4-10 and the emulsion stability at 24 h (E24) are 66.67 per cent, 33.33 per cent and 33.33 per cent respectively. With sand pack column technique, crude methanol could be enhanced the recovery of crude oil and kerosene oil by 41.85 per cent and 75.00 per cent, respectively. Barcelos *et al.* (2014) have analyzed the concentration of biosurfactants, use by of Spray Drying which are produced by *Bacillus subtilis* LBBMA RI4914 and isolated from a heavy oil reservoir. Kaolinite and maltodextrin 10DE or 20DE are used as drying adjuvants. Surface activity of the biosurfactant is analyzed by preparing dilution of crude biosurfactant and surface activity curves of biosurfactant, crude biosurfactant plus adjuvants and the dried products after their reconstitution in water. The shelf life of the dried products can be also evaluated. Spray drying is effective method in the recovery and concentration of biosurfactant, while keeping its surface activity.

Surfactin production can be confirmed using the RP-HPLC and MALDI-TOF-MS technique. Drying adjuvants could help to obtain a solid product with the desired characteristics without interfere with tenso-active properties of the biosurfactant molecules. The dehydrated product could be maintained its surfactant properties during storage at room temperature and during the evaluation period (*i.e.* 120 days) without any loss of activity (Barcelos *et al.*, 2014). Removal of solvent from surfactin is done by using the evaporation under pressure. Ramnani *et al.* (2005) have reported the concomitant production of protease and biosurfactant via optimizing the cornstarch and soy flour as carbon and nitrogen sources respectively by response surface methodology. An overall 2.3-fold increase for both protease (2954 U/ml) and biosurfactant (41 per cent) is reported. They have used the ultrafilteration (*i.e.* 100 kDa membrane) to allow a cost-effective purification of both protease and biosurfactant. The surfactant traps the protease on the membrane and detains both in the retentate. These products are found to have a shelf life of 1 year in the dry form via mixing with sodium sulphate in the retentate. Thus, concomitant production of protease and surfactant and their easy formulation can make them attractive detergent additives (Ramnani *et al.*, 2005). The supernatant and the crude biosurfactant have showed surface tension activity at 4°C, room temperature (30°C) and 50°C after incubation for 5 h. However, only crude methanol still retained surface tension activity after 100°C for 5 h. The surface tension activity of the supernatant and the crude biosurfactant is stable in 3-10 per cent (w/v) NaCl while crude methanol showed stability in 3-20 per cent (w/v) NaCl. However, all samples lost emulsion stability when NaCl concentration is higher than 5 per cent (w/v) (Suwansukho *et al.*, 2008).

Application of Biosurfactant

Potential advantages of biosurfactant are found with over their synthetic counterparts in many fields spanning environmental, food, biomedical, petrochemical and other industrial applications and it is discussed by Makkar *et al.* (2011). Biosurfactants are found as an amphiphilic molecule, showing both hydrophilic and hydrophobic moieties which partition preferentially at the interfaces such as liquid/liquid, gas/liquid or solid/liquid interfaces. Biosurfactant characteristics are found to enable the emulsifying, foaming, detergency and dispersing properties. Biosurfactant has low toxicity and environmental friendly nature and the wide range of potential industrial applications in bioremediation, health care, oil and food processing industries makes them a highly sought after group of chemical compounds (Makkar *et al.*, 2011). Mukherjee *et al.* (2006) have reported the important applications of biosurfactant in the fields of enhanced oil recovery, environmental bioremediation, food processing and pharmaceuticals. It also exhibits its unique properties such as higher biodegradability, lower toxicity, and effectiveness at extremes of temperature, pH and salinity. Desai and Banat (1997) have reported the commercialization of biosurfactants in the cosmetic, food, health care, pulp- and paper-processing, coal, ceramic, and metal industries. Bodour *et al.* (2003) has reported the biosurfactants which could exhibit a variety of potential applications in field of bioremediation of organic and metal-contaminated sites via enhanced transport of bacteria. Biosurfactants can be used in enhanced oil recovery, as cosmetic additives, and in biological control (Bodour *et al.*, 2003). We can find the many applications of biosurfactant in different fields which are reported by different researcher.

In Environmental Process

Bioremediation of Hydrocarbon Mixtures Contaminated Soil

Perfumo *et al.* (2010) have reported the possibilities and challenges for biosurfactants use in petroleum industry. In this regards, Perfumo *et al.* (2010) have discussed the biosurfactants as a group of microbial molecules with identification of their unique capabilities to interact with hydrocarbons. Emulsification and de-emulsification, dispersion, foaming, wetting and coating are some of the numerous surface activities of biosurfactants with achievement of application within systems such as immiscible liquid/liquid (*e.g.*, oil/water), solid/liquid (*e.g.*, rock/oil and rock/water) and gas/liquid (Perfumo *et al.*, 2010). The possibilities of exploiting these bioproducts in oil-related sciences are vast and made petroleum industry with their largest possible market at present. The role of biosurfactants in enhancing oil recovery from reservoirs is certainly the best known with their effectively application in many other fields from transportation of crude oil in pipeline to the clean-up of oil storage tanks and even manufacturing of fine petrochemicals. Biosurfactants can be properly used, comparable to traditional chemical analogues with performances and offer advantages to environment protection/conservation (Perfumo *et al.*, 2010).

Pseudomonas aeruginosa UG2 can produce biosurfactant in the soil which can help in enhanced degradation of hydrocarbon mixture of hexadecane, tetradecane, pristine with exception of 2-methylnaphthalene in two months of incubation. Youssef *et al.* (2007) have reported the biosurfactant-mediated oil recovery, entrapped in oil reservoirs. It could be an economic approach for recovery of significant amounts of oil. In hydrocarbon degradation application study, the crude biosurfactant is added to the culture medium containing 0.3 per cent crude oil as carbon source and the microorganism consortium from oil contaminated soil. The result showed that the saturated hydrocarbon is reduced by 96.63 per cent after 7 day cultivation, while the control without crude biosurfactant showed only 19.96 per cent reduction. Crude oil is used in both for domestic as well as industrial purposes because of very good source of energy. The productions of crude oil have been economically developed by many nations because of the increasing demand in different fields such as industries, automobiles, households, etc (Youssef *et al.*, 2007).

Crude oil is also one of the major pollutants, causes serious problems in the environment. It forms oil sludge. The heavy hydrocarbons in the oil sludge penetrate from the top soil into the sub-soil slowly, presenting a direct risk of contamination to subsoil and groundwater. And light hydrocarbons in the oil sludge vaporize, leaving behind a layer of oil containing dust of soil which blows upwards to pollute the air. It needed to control or prevent oil sludge formation to minimize the environmental harmness. Eco-friendly technologies could be used to clean the environment such as degradation by microorganisms. Treatment of oil pollution is needed by biosurfactant which is produced by bacterial colonies because of bacteria can metabolize various classes of hydrocarbons compound. Biosurfactant is utilised in bioremediation and is also harnessed relentlessly for biotechnologiocal purposes. Most promising applications are found in cleaning of oil-contaminated tankers, oil spill management, transportation of heavy crude oil, enhanced oil recovery, recovery of crude oil from sludge, and bioremediation of sites contaminated with hydrocarbons, heavy metals, and other pollutants. They have discussed about many microorganisms, especially bacteria which could produce biosurfactants on water-immiscible substrate (Desai and Banat, 1997).

Kuyukina *et al.* (2005) have discussed the microbially produced biosurfactants which enhance the crude oil desorption and mobilization in model soil column systems. The ability of biosurfactants from *Rhodococcus ruber* is found 1.4-2.3 times greater to remove the oil from the soil core than a synthetic surfactant (*i.e.* Tween-60). Biosurfactant-enhanced oil mobilization is found to temperature-related process with slower at 15°C than at 22-28°C (Kuyukina *et al.*, 2005). Kuyukina *et al.* (2005) have used the mathematical modelling by using a one-dimensional filtration model to simulate the process of oil penetration through a soil column in the presence of biosurfactants. A strong positive correlation (R^2=0.99) is reported between surfactant penetration through oil-contaminated soil and oil removal activity (Kuyukina *et al.*, 2005).

Biosurfactant are found to less adsorb to soil components than synthetic surfactant due to rapidly penetrating through the soil column with effectively remove of 65-82 per cent of crude oil. Crude oil (*i.e.* a lower proportion of high-molecular-weight paraffins and asphaltene with non-biodegradable compounds nature) is removed by biosurfactant compared to initial oil composition. Oil mobilized by biosurfactants could be easily biodegraded by soil bacteria. *Rhodococcus* biosurfactants can be used for in situ remediation of oil-contaminated soils (Kuyukina *et al.*, 2005). Analogous remediation of metal-contaminated soils is more complex because microbial cells or large exopolymers do not move freely through the soil. The use of microbially produced surfactants (biosurfactants), is found an alternative with potential for remediation of metal-contaminated soils. The distinct advantage of biosurfactants over whole cells or exopolymers is their small size, generally biosurfactant molecular weights are less than 1500. Hommel (1990) has reported the microbial growth on water-insoluble carbon sources (*i.e.* hydrocarbons) via accompanying the metabolic and structural alterations of the cell. The appearance of surface-active compounds (*i.e.* biosurfactants) in the culture medium or attached to the cell boundaries is a prerequisite for initial interactions of hydrocarbons with the microbial cell. The production of biosurfactants is normally connected with growth limitation in the late logarithmic and the stationary growth phase due to specific enzymes induction or derepression (Hommel 1990).

Singh and Cameotra (2006) have discussed the main commercial use of biosurfactants in pollution remediation due to their ability to stabilize emulsions. Biosurfactants enhance the solubility and availability of hydrophobic pollutants by increasing their potential for biodegradation. Ron and Rosenberg (2002) have discussed the oil pollution as an environmental problem. And they have reported the hydrocarbon-degrading microorganisms which could be adapted to grow and thrive in oil-containing environments. These microorganisms have an important role in the biological treatment of this pollution. One of the limiting factors in this process is the bioavailability of many fractions of the oil. The hydrocarbon-degrading microorganisms produce biosurfactants of diverse chemical nature and molecular size. These surface-active materials increase the surface area of hydrophobic water-insoluble substrates and increase their bioavailability, thereby enhancing the growth of bacteria and the rate of bioremediation. Addition of purified biosurfactants to microbial cultures could be caused inhibitory or stimulatory effects on growth. Biosurfactants are found as prerequisites of hydrocarbon uptake and also as secondary metabolic products. Joshi and Desai have discussed the major applications of (bio) surfactants which are found in environmental bioremediation field. Most synthetic organic compounds are present in contaminated soils and they are only weakly soluble or completely insoluble in water. So they exist in the subsurface as separate liquid phase, called non-aqueous phase liquids (NAPL), which poses as threat to environment (Joshi and Desai, 2010). Several factors limit the use of surfactants in environmental remediation, mainly persistence of surfactants or their metabolites which potentially pose an environmental concern. Biosurfactants could provide a more cost-effective approach for sub-surface remediation when used alone or in combination with synthetic surfactants (Joshi and Desai, 2010).

Bioremediation of Soil Contaminated with Toxic/Heavy Metal

Miller (1995) has reported the bioremediation of metal-contaminated waste streams. Normally, whole cells or microbial exopolymers are used to concentrate and/or precipitate metals in the waste stream to aid in metal removal. Biosurfactants have a wide variety of chemical structures with different metal selectivities with their metal removal efficiencies. Complexation capacity of a rhamnolipid biosurfactant is found very good, produced by *Pseudomonas aeruginosa* ATCC 9027 (Miller 1995). Juwarkar *et al.* (2007) have conducted column experiments to evaluate the potential of environmentally compatible rhamnolipid biosurfactant, produced by *Pseudomonas aeruginosa* strain BS2 to remove heavy metals (Cd and Pb) from artificially contaminated soil. Di-rhamnolipid is found to remove the leachable or available fraction of Cd and Pb with bind to metals by comparing to tap water (*i.e.* remove the mobile fraction only). Washing of contaminated soil with tap water could remove approximately 2.7 per cent of Cd and 9.8 per cent of Pb in contaminated soil via in freely available or weakly bound forms (Juwarkar *et al.*, 2007). Juwarkar *et al.* (2007) have reported that washing with rhamnolipid can be removed the 92 per cent of Cd and 88 per cent of Pb after 36 h of leaching. Di-rhamnolipid can selectively favour the mobilization of metals in the order of Cd >Pb.

Biosurfactant specificity can be observed towards specific metal which help in preferential elution of specific contaminant using di-rhamnolipid. pH of the leachates, collected from heavy metal contaminated soil column is further observed treated with di-rhamnolipid solution with low (6.60-6.78) as compared to leachates from heavy metal contaminated soil column treated with tap water (pH 6.90-7.25). There are high dissolution of metal varieties from the contaminated soil and effective leaching of metals with treatment with biosurfactant. The microbial population of the contaminated soil is found to increase after removal of metals by biosurfactant due to decrease of toxicity of metals to soil microflora. This is effective and nondestructive method for bioremediation of cadmium and lead contaminated soil (Juwarkar *et al.*, 2007).

Nitschke and Costa (2007) have reviewed the chemical surfactants compounds essentially due to their low toxicity and biodegradable nature the increasing with environmental concern. The biosurfactants can be used for bioremediation of heavy metal contaminated soil via their ability to form complexes with metals. The anionic biosurfactants could form complex with metals in a non-ionic form and metal-biosurfactant complexes are desorbed from the soil matrix to the soil solution due to the lowering of the interfacial tension. Cationic biosurfactants could replace the same charged metal ions by competition for some but not all negatively charged surfaces (ion exchange) (de Lima *et al.*, 2009). At present as environmental concern, biosurfactants are predominantly used in remediation of pollutants with showing potential applications in many sectors of food industry.

In Biotechnological Process

Joshi and Desai (2010) have reported the surfactants and biosurfactants molecules which are amphipathic in nature with both hydrophilic and hydrophobic moieties. It partitions preferentially at the interface between fluid phases that have different degrees of polarity and hydrogen bonding which confers excellent detergency, emulsifying, foaming and dispersing traits, making them most versatile process chemicals (Joshi and Desai, 2010). Associated with emulsion forming and stabilization, anti-adhesive and antimicrobial activities, biosurfactants could be explored in food processing and formulation. Potential applications of microbial surfactants in food area and the use of agro-industrial wastes as alternative substrates for their production are also reported in many literatures (Nitschke and Costa, 2007). Ron and Rosenberg (2001) have reported the natural role of biosurfactants. Different groups of biosurfactants have different natural roles in the growth of the producing microorganisms. Several bioemulsifiers have antibacterial or antifungal activities. Other bioemulsifiers enhance the growth of bacteria on hydrophobic water-insoluble substrates by increasing their bioavailability, presumably by increasing their surface area, desorbing them from surfaces and increasing their apparent solubility (Ron and Rosenberg, 2001).

Several biosurfactants have another useful property such as strong antibacterial, antifungal and antiviral activity. Biosurfactants can have medically relevant uses such as their role as anti-adhesive agents to pathogens, making them useful for treating many diseases and as therapeutic and probiotic agents in the field of biomedical sciences. Sachdev *et al.* (2013) have the reported the application of biosurfactants in agriculture. They have discussed the use of green compounds to achieve the sustainable agriculture is the present necessity. Agricultural productivity could fulfil the meet of growing demands of human population for all countries. Biosurfactants is produced by bacteria, yeasts, and fungi can serve as green surfactants to avoid the enormous use of harsh surfactants in agricultural soil and agrochemical industries. Biosurfactants are less toxic and eco-friendly and several types of biosurfactants have extensive applications in pharmaceutical, cosmetics, and food industries (Sachdev *et al.*, 2013).

Sachdev *et al.,* have reported that the biosurfactants can be synthesized by environmental isolates (*i.e.* many rhizosphere and plant associated microbes). And it also has promising role in the agricultural industry via playing vital role in motility, signaling, and biofilm formation or plant-microbe interaction. In agriculture, biosurfactants can be used for plant pathogen elimination and for increasing the bioavailability of nutrient for beneficial plant associated microbes. Biosurfactants can widely be applied for improving the agricultural soil quality by soil remediation via minimizing the expenditure of million dollar pesticide industries. There is need to explore novel biosurfactant from uncultured microbes in soil biosphere by using advanced methodologies like functional metagenomics (Sachdev *et al.*, 2013).

Ron and Rosenberg have reported the bioemulsifiers which could also play an important role in regulating the attachment-detachment of microorganisms to and from surfaces. Emulsifiers are involved in bacterial pathogenesis, quorum sensing and biofilm formation. A high-molecular-weight bioemulsifier can coat the bacterial surface via transferring horizontally to other bacteria and it can change their surface properties and interactions with the environment. Each emulsifier probably can provide the advantages in a particular ecological niche due to different chemical structures and surface properties (Ron and Rosenberg, 2001).

Future Prospectus

In this chapter, authors have focussed on biosurfactant properties and its characteristics for its applications in many environmental remediation processes and biotechnological processes. Diverse nature of biosurfactant compounds could be found which are produced by many varieties of micro-organisms. Several types of biosurfactants have extensive applications in pharmaceutical, cosmetics, and food industries. Diverse nature of biosurfactant could enhance the advantages in a particular ecological niche due to different chemical structures and surface properties. It has played an important role in regulating the attachment-detachment of microorganisms to and from surfaces. Biosurfactants can improve the agricultural soil quality by soil remediation via minimizing the expenditure of million dollar pesticide industries. In the agricultural industry, it could play a vital role in motility, signaling and biofilm formation or plant-microbe interaction.

Conclusion

Biosurfactants are produced by a wide variety of microorganisms with different chemical structures and surface properties. Biosurfactants have become more popular due to their low toxicity, biodegradability and ecological acceptability. Biosurfactants has not used in many industries due to their low yields and relatively high production and recovery costs. Many species of bacteria, yeast and fungi have been found to produce the different nature of biosurfactant which shown more applications in environmental remediation and biotechnological processes. Surfactants have property to diffuse in water and adsorb at interfaces between air and water or at the interface between oil and water, in the case of water it mixed with oil. Hyper-producing recombinant strains in the optimally controlled environment of a bioreactor could lead towards the successful commercial production of these valuable and versatile biomolecules in near future. Biosurfactant can reduce the surface or interfacial tension between oil and brine up to 0.01 mN/m. The bioremediation of oil-polluted soil and water, enhanced oil recovery, replacement of chlorinated solvents used in cleaning-up oil-contaminated pipes, vessels and machinery, use in the detergent industry, formulations of herbicides and pesticides and formation of stable oil-in-water emulsions for the food and cosmetic industries are commercial applications of biosurfactant.

References

1. Akit, J., Cooper, DI., Manninen, KI., Zajic, JE. Investigation of potential biosurfactant production among phytopathogenic Corynebacteria and related soil microbes. *Curr. Microbiol.*, 1981, **6**: 145-50.

2. Al-Ajlani, MM., Sheikh, MA., Ahmad, Z., Hasnain, S. Production of surfactin from *Bacillus subtilis* MZ-7 grown on pharmamedia commercial medium. *Microbial Cell Factories*, 2007, **6**: 17.

3. Al-Wahaibi, Y., Joshi, S., Al-Bahry, S., Elshafie, A., Al-Bemani, A., Shibulal, B. Biosurfactant production by *Bacillus subtilis* B30 and its application in enhancing oil recovery. *Colloids Surf B Biointerfaces*, 2014, **1**(114): 324-33.

4. Amaral, PF, Coelho, MA., Marrucho, IM., Coutinho, JA. Biosurfactants from yeasts: characteristics, production and application. *Adv Exp Med Biol*, 2010, **672**: 236-49.

5. Arima, K., Kakinuma, A., Tamura, G. Surfactin, a crystalline peptidolipid surfactant produced by *Bacillus subtilis*: isolation, characterization and its inhibition of fibrin clot formation. *Biochem. Biophys. Res. Comm.*, 1968, **31**: 488-94.

6. Arutchelvi, J., Bhaduri, S., Uppara PV., Doble, M. Production and characterization of Biosurfactant from *Bacillus subtilis* YB7. *Journal of Applied Sciences*, 2009, **9**: 3151-3155.

7. Banat, IM. Biosurfactants production and possible uses in microbial enhanced oil recovery and oil pollution remediation: a review. *Bioresource Technology*, 1995, **51**: 1-12.

8. Barcelos, GS., Dias, LC., Fernandes, PL., Fernandes, RCR., Borges, AC., Kalks, KHM., Tótola, MR. Spray drying as a strategy for biosurfactant recovery, concentration and storage. *Springerplus*. 2014, **3**: 49.

9. Barkay, T., Navon-Venezia, S., Ron, EZ., Rosenberg E. Enhancement of Solubilization and Biodegradation of Polyaromatic Hydrocarbons by the Bioemulsifier Alasan. *Appl. Environ. Microbiol*, 1999, **65**(6): 2697-2702.

10. Bodour, AA., Drees, KP., Maier, RM. Distribution of biosurfactant-producing bacteria in undisturbed and contaminated arid Southwestern soils. *Appl Environ Microbiol*, 2003, **69**(6): 3280-7.

11. Burch, AY., Shimada, BK., Browne, PJ., Lindow, SE. Novel high-throughput detection method to assess bacterial surfactant production. *Applied and environmental microbiology*, 2010, **76**(16): 5363–5372.

12. Cameotra, SS., Makkar, RS., Kaur, J., Mehta, SK. Synthesis of biosurfactants and their advantages to microorganisms and mankind. *Adv Exp Med Biol*. 2010;**672**: 261-80.

13. Campos-Takaki, GM., Sarubbo, LA., Albuquerque, CD. Environmentally friendly biosurfactants produced by yeasts. *Adv Exp Med Biol*, 2010, **672**: 250-60.

14. Castillo, M., Alonso, MC., Riu, J., Barcelo, D. Identification of polar, ionic, and highly water soluble organic pollutants in untreated industrial wastewaters. *Environ. Sci. Technol*. 1999, **33**: 1300–1306.

15. Chamnrokh, P., Assadi, MM, Noohi, A., Yahyai, S. Emulsan analysis produced by locally isolated bacteria and *Acinetobacter calcoaceticus* Rag-1. *Iran. J. Environ. Health.Sci. Eng*, 2008, **5**(2): 101-108.

16. Chan, XY., Chang, CY., Hong, KW., Tee, KK., Yin, WF., Chan, KG. Insights of biosurfactant producing *Serratia marcescens* strain W2.3 isolated from diseased tilapia fish: a draft genome analysis. *Gut Pathog*, 2013, **5**: 29.

17. Chander, CRS., Lohitnath, TD., Kumar, JM., Kalaichelvan, PT. Production and characterization of biosurfactant from *Bacillus subtilis* MTCC441 and its

evaluation to use as bioemulsifier for food bio-preservative. *Advances in Applied Science Research*, 2012, **3** (3): 1827-1831.

18. Chandran, P., Das, N. Characterization of sophorolipid biosurfactant produced by yeast species grown on diesel oil. *I.J.S.N.*, 2011, **2**(1): 63-71.

19. Colla, LM., Rizzardi, J., Pinto, MH., Reinehr, CO., Bertolin, TE., Costa, JA. Simultaneous production of lipases and biosurfactants by submerged and solid-state bioprocesses. *Bioresource Technology*, 2010, **101**(21): 8308–8314.

20. Cooper, DG., Zajic, JE., Gerson, DE. Production of surface-active lipids by *Corynebacterium lepus*. *Appl Environ. Microbiol.*, 1979, **37**: 4-10.

21. de Lima CJB., Ribeiro, EJ., Sérvulo, EFC., Resende, MM., Cardoso, VL. Biosurfactant Production by *Pseudomonas aeruginosa* Grown in Residual Soybean Oil. *Applied Biochemistry and Biotechnology*, 2009, **152**(1): 156-168.

22. Desai, JD., Banat, IM. Microbial production of surfactants and their commercial potential. *Microbiol Mol Biol Rev*, 1997, **61**(1): 47-64.

23. Deziel, E., Paquette, G., Villemur, R., Lepine, F., Bisaillon, J. Biosurfactant production by a soil pseudomonas strain growing on polycyclic aromatic hydrocarbons. *J. Appl Environ Microbiol*, 1996, **62**(6): 1908-12.

24. Fiechter, A. Biosurfactants: moving towards industrial application. *Trends Biotechnol*, 1992, **10**(6): 208-17.

25. Folmsbee, M., Duncan, K., Han, SO., Nagle, D., Jennings, E., McInerney, M. Re-identification of the halotolerant, biosurfactant-producing *Bacillus licheniformis* strain JF-2 as *Bacillus mojavensis* strain JF-2. *Syst Appl Microbiol*, 2006, **29**(8): 645-9.

26. Fonseca, RR., Silva, AJR., De França, FP., Cardoso, VL., Sérvulo, EFC. Optimizing carbon/nitrogen ratio for biosurfactant production by a *Bacillus subtilis* strain. *Applied Biochemistry and Biotechnology*, 2007, **137** (1-12): 471.

27. Fontes GC, Filomena, P., Amaral, F., Nele, M., Alice, M., Coelho, Z. Factorial Design to Optimize Biosurfactant Production by Yarrowia lipolytica. *Journal of Biomedicine and Biotechnology*, 2010, 8.

28. Gobbert, U., Lang, S., Wagner, E. Sophorose lipid formation by resting cells of *Torulopsis bombicola*. *Biotech. Lett.*, 1984, **6**: 225-30.

29. Haghighat, S., Akhavan sepahy, A. Production of biosurfactant by indigenous *Bacillus licheniformis* and *Bacillus subtilis*, Application in oil industry. *SIM Annual Meeting and Exhibition*, 2009, 83.

30. Hitzman, DO. Review of microbial enhanced oil recovery field tests. In *Proc. Syrup. on Applying Microorganisms to Petroleum Technology*, ed. T. S. Burchfield and R. S. Bryant. NTIS, Springfield, USA, 1988, pp. VI-1-VI-41.

31. Hommel, RK. Formation and physiological role of biosurfactants produced by hydrocarbon-utilizing microorganisms. Biosurfactants in hydrocarbon utilization. *Biodegradation*, 1990, **1**(2-3): 107-19.

32. Horowitz, S., Gilbert, JN., Griffin, WM. Isolation and characterization of a surfactant produced by *Bacillus licheniformis* 86. *J. Ind. Microbiol.*, 1990, **6**: 243-8.

33. Hua, Z., Chen, J., Lun, S., Wang, X. Influence of biosurfactants produced by *Candida antarctica* on surface properties of microorganism and biodegradation of n-alkanes. *Water Research*, 2003, **37**(17): 4143–4150.

34. Itoh, S., Suzuki, T. Fructose-lipids of *Arthrobacter, Corynebacteria, Nocardia* and *Mycobacteria* grown on fructose. *Agr. Biol. Chem.*, 1974, **38**: 1443-9.

35. Javaheri, M., Jenneman, GE.,. Mcinerney, M J., Knapp, RM. Anaerobic Production of a Biosurfactant by *Bacillus licheniformis* JF-2. *Applied and Environmental Microbiology*, 1985,**50**(3): 698-700.

36. Joshi, S., Bharucha, C., Desai, AJ. Production of biosurfactant and antifungal compound by fermented food isolate *Bacillus subtilis* 20B. *Bioresource Technology*, 2008, **99**(11): 4603–4608.

37. Joshi, SJ., Desai, AJ. Biosurfactant's role in bioremediation of NAPL and fermentative production. *Adv Exp Med Biol*, 2010, **672**: 222-35.

38. Juwarkar, AA, Nair, A., Dubey, KV., Singh, SK., Devotta, S. Biosurfactant technology for remediation of cadmium and lead contaminated soils. *Chemosphere.* 2007 **68**(10): 1996-2002.

39. Kapadia SG., Yagnik, BN. Current Trend and Potential for Microbial Biosurfactants. *Asian J. Exp. Biol. Sci.* 2013, **4**(1): 1 – 8

40. Kappeli, O. and Finnerty, W. R., Partition of alkane by an extracellular vesicle derived from hexadecane grown *Acinetobacter. J. Bacteriol.*, 1979, **140**, 707–712.

41. Karanth, NGK., Deo, PG., Veenanadig, NK. Microbial Production of Biosurfactants and Their Importance. *Current Science*, 1999, **77**: 116 – 126.

42. Kim, S., Lim, E., Lee, S., Lee, J., Lee, T. Purification and characterization of biosurfactants from *Nocardia* sp. L-417, *Biotechnol. Appl. Biochem*, 2000, **31**: 249–253.

43. Kiran, GS., Hema, TA., Gandhimathi, R., Selvin, J, Thomas, TA, Rajeetha, RT, Natarajaseenivasan, K. Optimization and production of a biosurfactant from the sponge-associated marine fungus *Aspergillus ustus* MSF3. *Colloids and Surfaces B Biointerfaces*, 2009, **73** (2): 250–256.

44. Kiran, GS., Sabarathnam, B., Thajuddin, N., Selvin, J. Production of Glycolipid Biosurfactant from Sponge-Associated Marine Actinobacterium *Brachybacterium paraconglomeratum* MSA21. *Journal of Surfactants and Detergents*, 2014, **17** (3): 531-542.

45. Kiran, GS., Thomas, TA., Selvin, J., Sabarathnam, B., Lipton, AP. Optimization and characterization of a new lipopeptide biosurfactant produced by marine *Brevibacterium aureum* MSA13 in solid state culture. *Bioresource Technology*, 2010, **101** (7): 2389–2396.

46. Kiran, GS., Thomas, TA., Selvin, J. Production of a new glycolipid biosurfactant from marine *Nocardiopsis lucentensis* MSA04 in solid-state cultivation. *Colloids and Surfaces B: Biointerfaces*, 2010, **78**(1): 8–16.

47. Kitamoto, D., Yanaglshita, H., Shinbo, T., Nakane, T., Kamisava, C. and Nakahara, T. (). Surface-active properties and antimicrobial activities of mannosylerythritol lipids as biosurfactants produced by *Candida antarctica*. *J. Biotech.*, 1993, **29**: 91-6.

48. Krieger, N., Camilios, ND., Mitchell, DA. Production of microbial biosurfactants by solid-state cultivation. *Adv Exp Med Biol*, 2010, **672**: 203-10.

49. Kuyukina, MS., Ivshina, IB., Makarov, SO., Litvinenko, LV., Cunningham, CJ., Philp, JC. Effect of biosurfactants on crude oil desorption and mobilization in a soil system. *Environ Int*, 2005, **31**(2): 155-61.

50. Liu, JH., Chen, YT., Li, H., Jia, YP., Xu, RD., Wang, J. Optimization of fermentation conditions for biosurfactant production by *Bacillus subtilis* strains CCTCC M201162 from oilfield wastewater. *Environmental Progress and Sustainable Energy*, 2014, DOI: 10.1002/ep.1 2013.

51. Maier, RM., Chavez, GS. *Pseudomonas aeruginosa* rhamnolipids: Biosynthesis and potential applications. *Appl. Microbiol. Biotechnol.*, 2000, **54**: 625.

52. Makkar, RS., and Singh CS. Biosurfactant production by a thermophilic *Bacillus subtilis* strain. *Journal of Industrial Microbiology and Biotechnology*, 1997, **18** (1): 37-42.

53. Makkar, RS., Cameotra, SS., Banat, IM. Advances in utilization of renewable substrates for biosurfactant production. *AMB Express*, 2011, **1**: 5.

54. Marrakchi, S., Maibach, HI. Sodium lauryl sulfate-induced irritation in the human face: regional and age-related differences. *Skin Pharmacol Physiol*, 2006, **19**(3): 177-80.

55. Martins, VG., Kalil, SJ., Bertolin, TE., Costa JA. Solid state biosurfactant production in a fixed-bed column bioreactor. *Z Naturforsch C.* 2006, **61**(9-10): 721-6.

56. Menezes Bento, F., de Oliveira Camargo, FA., Okeke, BC., Frankenberger, WT Jr. Diversity of biosurfactant producing microorganisms isolated from soils contaminated with diesel oil. *Microbiol Res*, 2005, **160**(3): 249-55.

57. Miller, RM. Biosurfactant-facilitated remediation of metal-contaminated soils. *Environ Health Perspect*, 1995, **103**(1) : 59-62.

58. Mnif, I., Ellouze-Chaabouni, S., Ghribi, D. Optimization of Inocula Conditions for Enhanced Biosurfactant Production by *Bacillus subtilis* SPB1, in Submerged Culture, Using Box–Behnken Design. *Probiotics and Antimicrobial Proteins*, 2013, **5**(2): 92-98.

59. Morikawa, M., Daido, H., Takao, T., Marato, S., Shimonishi, Y., Imanaka, T. A new lipopeptide biosurfactant produced by *Arthrobacter* sp. Strain MIS 38. *J Bacteriol*, 1993, **175**: 6459–6466.

60. Moussa, TAA., Mohamed, MS., Samak, N. Production and characterization of Di-Rhamnolipid produced by *Pseudomonas aeruginosa* TMN. *Brazilian Journal of Chemical Engineering*, 2014, **31**(04): 867 – 880.

61. Mukerjee, P., Mysels, KJ. Critical micelle concentrations of aqueous surfactant systems. National Bureau of Standards of NSRDS-NBS 36, Washington, DC 20234, 1971, 227.

62. Mukherjee, S., Das, P., Sen, R. Towards commercial production of microbial surfactants Review. *Trends in Biotechnology*, 2006, **24**(11): 509–515.

63. Mulligan, CN., Chow, TY-K., Gibbs, BF. Enhanced biosurfactant production by a mutant *Bacillus subtilis* strain. *Applied Microbiology and Biotechnology*, 1989, **31**: 486–489.

64. Najafi, AR., Rahimpour, MR., Jahanmiri, AH., Roostaazad, R., Arabian, D., Soleimani, M., Jamshidnejad, Z. Interactive optimization of biosurfactant production by *Paenibacillus alvei* ARN63 isolated from an Iranian oil well. *Colloids and Surfaces B: Biointerfaces*, 2011, **82**(1): 33–39.

65. Neu, TR., Harmer, T., Poralla, K. Surface active properties of viscosin: a peptidolipid antibiotic. *Appl. Microbiol. Biotech*, 1990, **32**: 518-20.

66. Nitschke, M., Costa, SG., Contiero, J. Rhamnolipid surfactants: an update on the general aspects of these remarkable biomolecules. *Biotechnol Prog*, 2005, **21**(6): 1593-600.

67. Nitschke, M., Costa, SGVAO. Biosurfactants in food industry. *Trends in Food Science and Technology*, 2007, **18**(5): 252–259.

68. Nitschke, M., Pastore, GM. Biosurfactant production by *Bacillus subtilis* using cassava-processing effluent. *Appl Biochem Biotechnol*, 2004, **112**(3): 163-72.

69. Perfumo, A., Rancich, I., Banat, IM. Possibilities and challenges for biosurfactants use in petroleum industry. *Adv Exp Med Biol*, 2010, **672**: 135-45.

70. Peypoux, F., Bonmatin, J., Mand WJ. Recent trends in the biochemistry of surfactin. *Appl Microbiol Biotechnol*, 1999, **51**: 553–563.

71. Ramnani, PS., Kumar, S., Gupta, R. Concomitant production and downstream processing of alkaline protease and biosurfactant from *Bacillus licheniformis* RG1: Bioformulation as detergent additive. *Process Biochemistry*, 2005, **40**(10): 3352–3359.

72. Renner, R. European bans on surfactant trigger transatlantic debate. *Environ. Sci. Technol.*, 1997, **31**: A316–A320.

73. Robert, M., Mercade, M. E., Bosch, M. P., Parra, J. L., Espuny, MJ., Manresa, M.A., Guinea, J. Effect of the carbon source on biosurfactant production by *Pseudomonas aeruginosa* 44T 1. *Biotech. Lett.*, 1989, **11**: 871-4.

74. Rodrigues, L., Moldesb, Teixeiraa, JA., Oliveiraa, R. Kinetic study of fermentative biosurfactant production by *Lactobacillus* strains. *Biochemical Engineering Journal*, 2006, **28**(2): 109–116.

75. Rodrigues, LR., Teixeira, JA., Oliveira, R. Low-cost fermentative medium for biosurfactant production by probiotic bacteria. *Biochemical Engineering Journal*, 2006, **32**(3): 135–142.

76. Ron, EZ, Rosenberg, E. Biosurfactants and oil bioremediation. *Curr Opin Biotechnol*, 2002, **13**(3): 249-52.

77. Ron, EZ, Rosenberg, E. Natural roles of biosurfactants. *Environ Microbiol*. 2001, **3**(4): 229-36.

78. Rosenberg, E., Ron, EZ. High and low-molecular-mass microbial surfactants. *Appl Microbiol Biotechnology*, 1999, **52**(2): 154-62.

79. Rufino, RD., Moura de Luna, J., Takaki, GMC., Sarubbo, LA. Characterization and properties of the biosurfactant produced by *Candida lipolytica* UCP 0988. *Electronic Journal of Biotechnology*, 2014, **17**(1): 34–38.

80. Sachdev, DP., Cameotra, SS. Biosurfactants in agriculture. *Appl Microbiol Biotechnol*, 2013, **97**(3): 1005-16.

81. Sari, M., Kusharyoto, W., Artika, IM Screening for Biosurfactant-producing Yeast: Confirmation of Biosurfactant Production. *Biotechnology*, 2014, **13** (3): 106-111.

82. Scott, MJ., Jones, MN. The biodegradation of surfactants in the environment. *Biochimica et Biophysica Acta (BBA) Biomembranes*, 2000, **1508** (1–2): 235–251.

83. Selvin, J., Thangavelu, T., Kiran, GS., Gandhimathi, R., Shanmughapriya, S. Culturable heterotrophic bacteria from the marine sponge *Dendrilla nigra*: isolation and phylogenetic diversity of *Actinobacteria*. *Helgol Mar Res*, 2009, **63**: 239–247.

84. Shulga, AN., Karpenko, EV., Eliseev, SA., Turovsky, AA. (). The method for determination of anioiaogenic bacterial surface-active peptidolipids. *Microbiol. J.*, 1993, **55**: 85-8.

85. Shulga, AN., Karpenko, EV., Eliseev, SA., Turovsky, AA., Koronelli, TV. Extracellular lipids and surface-active properties of the bacterium *Rhodococcus erythropolis* depending on the source of carbon nutrition. *Mikrobiologya*, 1990, **59**: 443-7.

86. Simpson, DR., Natraj, NR., McInerney, MJ., Duncan, KE. Biosurfactant-producing *Bacillus* are present in produced brines from Oklahoma oil reservoirs with a wide range of salinities. *Appl Microbiol Biotechnol*, 2011, **91**(4): 1083-93.

87. Singh, P., Cameotra, SS. Potential applications of microbial surfactants in biomedical sciences. *Trends in Biotechnology*, 2004, **22**(3): 142–146.

88. Suwansukho, P., Rukachisirikul, V., Kawai, F., H-Kittikun, A. Production and applications of biosurfactant from *Bacillus subtilis* MUV4. *Songklanakarin J. Sci. Technol.*, 2008, **30** (1): 87-93.

89. Suzuki, T., Tanaka, H., Itoh, I. Sucrose lipids of *Arthrobacteria, Corynebacteria* and *Nocardia* grown on sucrose. *Agr. Biol. Chem.*, 1974, **33**: 190-5.

90. Symmank, H., Franke, P., Saenger, W., Bernhard F. Modification of biologically active peptides: production of a novel lipohexapeptide after engineering of *Bacillus subtilis* surfactin synthetase. *Protein Engineering*, 2002, **15**: 913–921.

91. Van der Vegt, W., Vander Mei, HC., Noordmans, J., Busscher, HJ. Assessment of bacterial biosurfactant production through axisymmetric drop shape analysis by profile. *Appl. Microbiol. Biotech.*, 1991, **35**: 766-70.

92. Wittgens, A., Tiso, T., Torsten, T., Wenk, AP., Hemmerich, J., Müller, C., Wichmann, R., Küpper, B., Zwick, M., Wilhelm, S., Hausmann, R., Syldatk, C., Rosenau, F., Blank, LM. Growth independent rhamnolipid production from glucose using the non-pathogenic *Pseudomonas putida* KT2440. *Microbial Cell Factories*, 2011, **10**: 80.

93. Xiao, Y., Gerth, K., Müller, R., Wall, D. Myxobacterium-produced antibiotic TA (myxovirescin) inhibits type II signal peptidase. *Antimicrob Agents Chemother*, 2012, **56**(4): 2014-21.

94. Youssef, N., Simpson, DR., Duncan, KE., McInerney, MJ., Folmsbee, M., Fincher, T., Knapp, RM. In situ biosurfactant production by *Bacillus* strains injected into a limestone petroleum reservoir. *Appl Environ Microbiol*, 2007, **73**(4): 1239-47.

95. Youssef, NH., Duncan, KE., Nagle, DP., Savage, KN., Knapp, RM., McInerney MJ. Comparison of methods to detect biosurfactant production by diverse microorganisms. *Journal of Microbiological Methods*, 2004, **56**(3): 339–347.

96. Zoller, U. Non-ionic surfactants in reused water: Are activated sludge/soil aquifer treatments sufficient? *Water Res.* 1994, **28**: 1625–1629.

Index

www.ingramcontent.com/pod-product-compliance
Lightning Source LLC
Chambersburg PA
CBHW050520190326
41458CB00005B/1612